MODERN ALGEBRA

MODERN ALGEBRA

SECOND EDITION

B S Vatsa
MSc., PhD
Ex-Professor & Head
Department of Mathematics
St. John's College
Agra

Suchi Vatsa
MSc.
Lecturer
Deptartment of Computer Science
APEEJAY Institute of Technology
Greater Noida (UP)

Publishing Globally

NEW AGE INTERNATIONAL (P) LIMITED, PUBLISHERS
LONDON • NEW DELHI • NAIROBI
Bangalore • Chennai • Cochin • Guwahati • Hyderabad • Kolkata • Lucknow • Mumbai
Visit us at **www.newagepublishers.com**

Published by New Age International (P) Ltd., Publishers
First Edition: 1999
Second Edition: 2010
Reprint: 2018

GLOBAL OFFICES

- **New Delhi** **NEW AGE INTERNATIONAL (P) LIMITED, PUBLISHERS**
  7/30 A, Daryaganj, New Delhi-110002, (INDIA)
  Tel.: (011) 23253771, 23253472, **Telefax:** 23267437, 43551305
  E-mail: contactus@newagepublishers.com • Visit us at www.newagepublishers.com
- **London** **NEW AGE INTERNATIONAL (UK) LTD.**
  27 Old Gloucester Street, London, WC1N 3AX, UK
  E-mail: info@newacademicscience.co.uk • Visit us at www.newacademicscience.co.uk
- **Nairobi** **NEW AGE GOLDEN (EAST AFRICA) LTD.**
  Ground Floor, Westlands Arcade, Chiromo Road (Next to Naivas Supermarket)
  Westlands, Nairobi, KENYA, **Tel.:** 00-254-713848772, 00-254-725700286
  E-mail: kenya@newagepublishers.com

BRANCHES

- **Bangalore** 37/10, 8th Cross (Near Hanuman Temple), Azad Nagar, Chamarajpet, Bangalore- 560 018
  Tel.: (080) 26756823, **Telefax:** 26756820, **E-mail: bangalore@newagepublishers.com**
- **Chennai** 26, Damodaran Street, T. Nagar, Chennai-600 017, **Tel.:** (044) 24353401, **Telefax:** 24351463
  E-mail: chennai@newagepublishers.com
- **Cochin** CC-39/1016, Carrier Station Road, Ernakulam South, Cochin-682 016, **Tel.:** (0484) 2377303, **Telefax:** 4051304
  E-mail: cochin@newagepublishers.com
- **Guwahati** Hemsen Complex, Mohd. Shah Road, Paltan Bazar, Near Starline Hotel, Guwahati-781 008,
  Tel.: (0361) 2513881, **Telefax:** 2543669, **E-mail: guwahati@newagepublishers.com**
- **Hyderabad** 105, 1st Floor, Madhiray Kaveri Tower, 3-2-19, Azam Jahi Road, Near Kumar Theater, Nimboliadda Kachiguda, Hyderabad-500 027, **Tel.:** (040) 24652456, **Telefax:** 24652457
  E-mail: hyderabad@newagepublishers.com
- **Kolkata** RDB Chambers (Formerly Lotus Cinema) 106A, 1st Floor, S N Banerjee Road, Kolkata-700 014
  Tel.: (033) 22273773, **Telefax:** 22275247, **E-mail: kolkata@newagepublishers.com**
- **Lucknow** 16-A, Jopling Road, Lucknow-226 001, **Tel.:** (0522) 2209578, 4045297, **Telefax:** 2204098
  E-mail: lucknow@newagepublishers.com
- **Mumbai** 142C, Victor House, Ground Floor, N.M. Joshi Marg, Lower Parel, Mumbai-400 013
  Tel.: (022) 24927869, **Telefax:** 24915415, **E-mail: mumbai@newagepublishers.com**
- **New Delhi** 22, Golden House, Daryaganj, New Delhi-110 002, **Tel.:** (011) 23262368, 23262370, **Telefax:** 43551305
  E-mail: sales@newagepublishers.com

ISBN: 978-81-224-2580-2

C-17-11-10840

Printed in India at Glorious Printers, Delhi.
Typeset at Kalyani Computer Services, New Delhi.

NEW AGE INTERNATIONAL (P) LIMITED, PUBLISHERS
7/30 A, Daryaganj, New Delhi-110002
Visit us at **www.newagepublishers.com**
(CIN: U74899DL1966PTC004618)

To

My Wife

&

Children

PREFACE TO THE SECOND EDITION

The second edition retains almost all the material in the first edition and includes everal new sections on abelian groups, cyclic groups, permutation groups and worked out xamples are added to homomorphisms and isomorphisms. A new section on finite abelian roups and on solvable groups, nilpotent groups and on perfect group have been included. section of special class of rings is added to the ring theory in which ordered fields are also iscussed. An important section of division ring and simple ring is included to the section ideals and quotient rings in which maximal ideal and prime ideals and theorem on these re discussed in detail. A new section on the field of quotients of an integral domain and its nportant results have been included and important concept of quotient field are treated in ngth.

Two new sections on inner product space and modules have been included to the hapter on vector space in which important inequalities Cauchy Schwarz inequlity Bassel's nequality and Gram-Schmidt process to form an orthonormal bases from a given basis of given vector space.

To a chapter on linear transformation, a section on matrices and linear transformation ave been included and properly discussed in detail. In the rest of chapters many solved xamples have been added at proper places which make the book more useful to the users. t the end of each section a graded exercise has been given. Major changes take place in ne chapters.

Above I have described what I have added. I did not enlarge the material on modules ubstantially. A mere addition of this new material is an adjunct with no application and no igher goal.

Many readers have written me about the first edition pointing out typographical mistakes making suggestion on how to improve the book. I thank them with my heart for their elp and kindness.

B.S. Vatsa
Suchi Vatsa

PREFACE TO THE FIRST EDITION

The purpose of this book is to provide a text for the graduate honours and Postgraduate modern (Abstract) Algebra course at the different institutes of technology and at different Indian universities. This text also suits all competitive examinations such as IAS, NET, etc. I have tried to make the book as self-contained as possible.

I believe a mathematics course should not give science, engineering or social science students a hodgepodge of techniques, but should provide them with an understanding of basic mathematical concept. On the other hand, I have been keenly aware of the wide range of backgrounds which the students may possess, in particular, of the fact that the students have very little experience with abstract mathematical reasoning. For this reason, I have introduced abstract ideas throughout the book very carefully through the numerical examples.

There is always a danger of introducing the abstract concepts too suddenly and without a sufficient base of examples. In order to mitigate it I have motivated the concepts beforehand. Since Abstract Algebra is a branch of mathematics in which we study algebraic systems with one or more than one binary operations defined on a non-empty set.

So sets and functions are main ingredients of the algebraic systems. Keeping this in view the sets, relation and functions have been given in chapter 1, though these topics are taught at junior level. Some useful results of number theory are also given here.

Chapter 2 deals with system of one binary operation, viz., groups, subgroups, normal subgroups, permutation group, cyclic groups and factor groups. It provides the students with some technique necessary to understand examples of the more abstract ideas occurring in the later chapters.

Chapter 3 treats homomorphism, isomorphism, automorphisms, endomorphisms and inner automorphisms.

Chapter 4 deals with structure of groups, conjugate elements and conjugate classes, Sylow's theorems and in Chapter 5 Jordan Holder theorem and solvable groups are discussed.

Chapter 6 deals with the system of two binary operations, viz., ring, integral domain, field, subring, ideals, quotientring, ring homomorphisms and isomorphisms, quotient field and direct product of rings.

Chapter 7 defines the algebra of polynomials over a ring field, ideals in that algebra and the prime factorization of a polynomial. It also deals with roots and greatest common division and least common multiple of two polynomials.

Chapter 8 treats factorization in integral domains, PID, euclidean domain, unique factorization domain.

Chapter 10 treats linear transformations, their algebra, isomorphisms, linear functionals and dual spaces. Transpose of a linear mapping and matrix representation of a linear transformation.

Chapter 11 deals with extension fields, algebraic extensions, splitting fields, separable and inseparable extensions.

Chapter 12 is devoted to Galois theory which provides us Galois groups, normal extension and its application in solving the polynomials, an application to geometry, Galois fields, radical extension and solvability by radicals.

Throughout the book have included a great variety of examples of the important concepts. The study of such examples is of fundamental importance and tends to minimize the number of students who can repeat definition, theorem, proof in logical order without grasping the meaning of the abstract concepts. The book also contains a wide variety of graded exercises ranging from routine applications to one which will extend the very best students.

I feel that the text has sufficient, proper, lucid, and coherent matter for all courses for which it is needed.

I am particularly indebted the Professor D.D. Joshi, Late Dr. P.B. Bhattacharya for making valuable suggestions in preparing this book. In addition, I would like to thank Mr. A. Mishra for getting it completed. Lastly, special thanks are due to Mrs. Savitri Vatsa for her constant encouragement.

B.S. Vatsa

CONTENTS

CHAPTER 1

SET THEORY

1.1 SETS

The main object of this book is to introduce the basic algebraic systems (mathematical systems)—groups, ring, integral domains, fields, and vector spaces. By an algebraic system we shall mean a non-empty set together with one or more than one binary operations defined on the set. The systems subsequently to be considered shall be divided according to the properties which they satisfy. Such a division leads to the branch of mathematics known as *Abstract Algebra.* Our object is to study a logical development of algebraic systems beginning with those with relatively little algebraic structure and progressing to systems rich in structure.

Thus the sets and binary operations (functions) are the fundamental and main ingredients of the algebraic systems. To develop an algebraic system we assume certain statements true without proof known as *axioms* or *postulates.* Thus this approach to the systems is known as axiomatic development of the systems.

Here it will be useful to discuss the ingredients of the systems - sets and operations on the sets. We shall not discuss the theory of sets through an axiomatic approach, but we shall adopt an informal approach to the subject. We shall also not list the undefined terms of the set theory such as sets, objects etc. and axiomatic relations.

By a set we mean intuitively a collection of object of any type whatsoever. The objects of the set are called its *elements* or *members.*

Sets will generally be designated by capital letters such as A, B, X and their elements by small case letters such as a, b, x.

The symbols N, Z, Q, R and C will stand for the set of natural numbers, set of integers, set of rational numbers, set of real numbers, and set of complex numbers, respectively.

If x is an element of the set A, then we write $x \in A$, where the symbol $\in$ (Greek letter epsilon) will mean "belongs to" or "is an element of. On the other hand, if x is not an element of the set A, *then we write* $x \notin A$, where $\notin$ will mean "does not belong to" or "is not an element of: For example, if we consider the set N of natural numbers, then $5 \in N$ but $-5 \notin N$ and $1/5 \notin N$.

There are generally two methods of designating a particular set. First, we may denote a set by listing all elements of the set within the braces such as $\{a, b, c\}$ and if such a listing is not practical (for example, the listing of all triangles in a plane is not practical), then we represent the set in terms of a common statement $P(x)$ which is true for all elements x of the set. Thus the set of all elements x for which the statement $P(x)$ is true is written as $\{\mid x \mid P(x)\}$. For example, $A = \{ x \mid x$ is a triangle in the plane$\}$. Certainly, certain sets may be written in both ways such as

$$A = \{1, 2, 3, 4, 5\}$$
$$= \{x \mid x \text{ is a natural number less than or equal to } 5\}$$
$$= \{x \mid x \in \text{N}, x \le 5\}.$$

Note 1. In listing method the elements are written in any order *i.e.*, $\{x, y, z\}$

$$= (z, x, y\}$$

2. Elements are not repeated *i.e.*, $\{1, 2, 2, 3\} = \{1, 2, 3\}$.

Definition 1.1.1: A set which has no element is called a *null set* or an *empty set* and is represented by the symbol ϕ.

Example 1.1.1: $\varphi = [x \mid x \in R; x^2 < 0\}$

or $\varphi = \{x \mid x$ is a real number whose square is less than zero$\}$.

Definition 1.1.2: A set which has only one element is called a *singleton* or a *unit set* and denoted by $\{x\}$.

Example 1.1.2: The set of planets on which we live is a singleton *i.e.*, this set contains only one elements, namely earth.

Note 1: We have $\{0\} \neq \phi$ since $\{0\}$ is not an empty set.

Definition 1.1.3: A particular set which contains all objects which are to be considered during some specific discussion, is called *universal set* for that particular discussion.

Definition 1.1.4: The set A is a *subset* of the set B, denoted by $A \subseteq B$, if every element of A is also an element of B. If there exists at least one element in B which is not in A, then A is a *proper subset* of B, denoted by $\text{A} \subset \text{B}$.

If A *is a* subset of B, it is also expressed by saying that A is *contained in* B or B is a *super set* of A. If A is not a subset of B, then we write $A \not\subseteq B$.

Example 1.1.3: Let $A = \{2, 3, 4, 5\}$

and $B = \{1, 2, 3, 4, 5, 6\}$

Then $A \subseteq B$ since every element of A is also element of B. In this case there exists 1 and 6 of B which are not in A. So $A \subset B$ (A is a proper subset of B).

Definition 1.1.5: Two sets A and B are said to be equal, written as $A = B$, if, and only if, every element of A is an element of B and every element of B is an element of A. That is, $A = B$ if A and B have exactly the same elements.

It is easy to see that $A \subseteq A$, $\phi \subseteq A$, and if $A \subseteq B$, $B \subseteq C$, then $A \subseteq C$.

1.2 SET OPERATIONS

Definition 1.2.1: Let A and B be two subsets of U. Then the union of A and B is

$$A \cup B = \{x \mid x \in A \text{ or } \in B\};$$

and the intersection of A and B is

$$A \cap B = \{x \mid x \in A \text{ and } x \in B\}.$$

If $A \cap B = \phi$, A and B are said to be disjoint sets.

Theorem 1.2.1: *Let A, B, C be three subsets of U, then*

(a) $A \cap A = A$; $A \cup A = A$; *(Idempotent law).*

(b) $A \cap B = B \cap A$; $A \cup B = B \cup A$; *(Commutative law)*

(c) $(A \cap B) \cap C = A \cap (B \cap C)$; $(A \cup B) \cup C = A \cup (B \cup C)$; *(Associative law).*

(d) $A \cap (B \cup C) = (A \cap B) \cup (A \cap C)$,

$A \cup (B \cap C) = (A \cup B) \cap (A \cup C)$, *(Distributive Laws)*

Definition 1.2.2: Let A be a subset of the universal set U. Then the complement of A is

$$A' = \{x \mid x \in U \text{ but } x \notin A\}.$$

The difference set (relative complement of B in A) is

$$A - B = [x \mid x \in A \text{ but } x \notin B\}$$

Definition 1.2.3: Let A and B be two subsets of the universal set U. Then the symmetric difference is

$$A \Delta B = \{x \mid \in A \cup B \text{ but } x \notin A \cap B\}. \text{ That is,}$$

$$A \Delta B = (A \cup B) - (A \cap B).$$

Theorem 1.2.2: *Let A, B and C be three subsets of the universal set U.*

(a) $(A \cup B)' = A' \cap B'$; $(A \cap B)' = A' \cup B'$; *(DeMorgan's Laws)*

(b) $(A \Delta B) = B \Delta A$;

(c) $(A \Delta B) \Delta C = A \Delta (B \Delta C)$;

(d) $(A - B) \cup (B - A) = (A \cup B) - (A \cap B)$.

Definition 1.2.4: The power set $P(A)$ of an arbitrary set A is the set of all subsets of A.

It can be seen than if A is a finite set with n elements, then $P(A)$ is a finite set with 2^n elements.

Definition 1.2.5: The ordered pairs of two elements a and b is defined by

$$(a, b) = \{\{a, b\}, \{a\}\}.$$

Theorem 1.2.3: *The ordered pairs (a, b) and (c, d) are equal if, and only if, a = c and b = d.*

Proof. If $a = c$ *and* $b = d$, *then*

$$\{a\} = \{c\} \text{ and } \{a, b\} = \{c, d\}, \text{ and consequently}$$

$$\{\{a, b\}, \{a\}\} = \{\{c, d\}\}. \{c\} \} = (a, b) = (c, d).$$

For converse, Let $\{ \{a, b\}, \{a\} \} = \{ \{c, d\}, \{c\} \}$ ....(1)

Case I: Let $a = b$, then

$$(a, b) = (a, a) \Rightarrow \{\{a, b\}, \{a\} \} = \{ \{a, a\}, \{a\} \} = \{ \{a\}, \{a\} \}$$
$$\Rightarrow \{\{a, b\}, \{a\}\} = \{\{a\}\}$$
$$\Rightarrow \{\{c, d\}, \{c\}\} = \{\{a\}\} \text{ from (1)}$$
$$\Rightarrow \{c, d\} = \{c\}\} = \{a\}$$
$$\Rightarrow a = c = d$$
$$\Rightarrow a = b = c = d, \text{ as } a = b.$$

Case II: Let $a \neq b$. Here $\{a\} \neq \{a, b\}$ and

$$\{c\} \neq \{a, b\}.$$

If $\{c\} = \{a, b\}$ then $a = c = b$

Hence contradiction to that $a \neq b$ from (1) we have

$$\{c\} \in \{\{c, d\}, \{c\}\} \Rightarrow \{c\} \in \{\{a, b\}, \{a\}\}$$
$$\Rightarrow \{c\} = \{a, b\} \text{ or } \{c\} = \{a\}.$$

Since $\{c\} \neq \{a, b\}$, then $\{c\} = \{a\} \Rightarrow a = c$. ...(2)

Again from (1), we have

$$\{a, b\} \in \{\{c, d\}, \{c\}\} \text{ with } \{a, b\} \neq \{c\}.$$

So $\{a, b\} = \{c, d\}$ and therefore $b \in \{c, d\} \Rightarrow b = c$ or d. If $b = c$, then from (2) $a = b = c$ which is a contradiction to that $a \neq b$, consequently $b = d$ ...(3)

From (2) and (3), we have

$$a = c \text{ and } b = d.$$

Hence the theorem.

Definition 1.2.6: The cartesian product of two nonempty sets A and B, denoted by A × B, is the set defined by

$$A \times B = \{a, b) \mid a \in A, b \in B\}.$$

Example 1.2.1: Let $A = \{1, 2, 3\}$ and $B = \{a, b\}$, then

$$A \times B = \{(1, a), (2, a), (3, a), (1, b), (2, b), (3, b)\}$$

and

$$B \times A = \{(a, 1), (a, 2), (a, 3), (b, 1), (b, 2), (b, 3)\}.$$

We observe that A contains 3 elements and B contains 2 elements. Then A × B and B × A contain 2 × 3 = 6 elements. In general, if A and B have m and n elements, respectively, then $A \times B$ and $B \times A$ will have mn elements.

Since, $(1, a) \neq (a, 1)$, $A \times B \neq B \times A$. In general $A \times B \neq B \times A$.

PROBLEMS

In the following exercises A, B, and C are subsets of some universal set U.

1. Suppose $A \subseteq B$, show that

(*a*) $A \cap C \subseteq B \cap C$.

(*b*) $A \cup C \subseteq B \cup C$.

2. Prove that $A \cap B \subseteq A \cup B$.

3. Show that

(*a*) If $A \cap B = B$, then $A \cup B = A$ and $B \subseteq A$.

(*b*) If $A \supseteq B$, then $A \cap B = B$ and $A \cup B = A$.

4. Prove that if $A \subseteq B$, then $A \cup B = B$.

5. Prove that

(*a*) $A - B = A \cap B'$,

(*b*) $A - B = B' - A'$,

(*c*) $A - B = \phi$ if and only if $A \subset B$,

(*d*) $A - B = A$ if and only if $A \cap B = \phi$,

(*e*) $A - B \subset A$,

(*f*) $B - A \subset B$,

(*g*) $(A \text{ U } B) \cap B' = A$ if and only if $A \cap B = \phi$.

6. Prove that

(*a*) $(A - B) - C = A - (B \cup C)$,

(*b*) $A - (B - C) = (A - B) \cup (A \cap C)$,

(*c*) $A \cup (B - C) = (A \cup B) - (C \cup A)$,

(*d*) $A \cap (B - C) = (A \cap B) - (A \cap C)$,

7. Prove that

(*a*) If $B \subseteq A$, then $B - C \subseteq A - C$,

(*b*) If $B \subseteq A$, then $A - (A - B) = B$,

(*c*) $A \cup B = A \cap B$ if, and only if, $A = B$,

(*d*) $A \cup B = (A - B) \cup B$,

(e) $(A - B) \cup B = A$ if, and only if , $A \supseteq B$,

(f) $A - B = A$ if, and only if, $A \cap B = \phi$.

(g) $(A \cup B) - C = (A - C) \cup (B - C)$,

(h) $A - (B \cap C) = (A - B) \cap (A - C)$.

8. Prove that

$A \cap (B \,\Delta\, C) = (A \cap B) \,\Delta\, (A \cap C)$.

9. Prove that

(a) $A \times (B \cup C) = (A \times B) \cup (A \times C)$,

(b) $A \times (B \cap C) = (A \times B) \cap (A \times C)$,

(c) If $A \subset B$, then $A \times C \subset B \times C$,

(d) $A \times (B - C) = (A \times B) - A \times C)$,

(e) $A \times B = \cup \{A \times \{b\} \mid b \in B\}$.

1.3 RELATIONS

Definition 1.3.1: A binary relation R from the set A to the set B is a subset of $A \times B$.

From this definition it is obvious that the empty set which is a subset of $A \times B$ $(\phi \subseteq A \times B)$ is a relation which is called a null *relation* in $A \times B$. Since $A \times B \subseteq A \times B$, $A \times B$ is called a *universal relation* from A to B.

If $A = B$, then any subset of $A \times A$ is a binary relation in A.

If $R \subseteq A \times B$, and $(a, b) \in R$, then we say that a has relation R to b, written as $a\,R\,b$. If $(a, b) \notin R$, *i.e.*, a has no relation R to b, then we write $a \not R\, b$.

Definition 1.3.2: The domain of a binary relation, denoted by $d(R)$, is the set of all first elements of the ordered pairs in the relation R. The range of a binary relation, denoted by $r(R)$, is the set of all second elements of the ordered pairs in the relation R.

Example 1.3.1: Let $S = T = \{1, 2, 3, 4\}$.

The elements $\{1, 1\}$, $\{2, 2\}$, $[3, 3]$, and $\{4, 4\}$ are diagonal elements in $S \times T = S \times S$

$R = \{(1, 1), (2, 2), (3, 3)\ (4, 4)\} \subseteq S \times S$ is a relation which can be described by a single formula.

$R = \{(x, y) \mid x = y\}$.

In this case $d(R) = r(R) = \{1, 2, 3, 4\}$.

Similarly, we can consider another relation

$R = \{(x, y) \mid x < y\}$ *in* $S \times S$, that is,

$<$ or $R = \{(1, 2), (1, 3), (1, 4), (2, 3), (2, 4), (3, 4)\}$.

Here $d(R) = \{1, 2, 3\}$ and $r\,(R) = \{2, 3, 4\}$.

Definition 1.3.3: *A* subset *R* of $A \times A$ is called an *equivalence relation* on *A* if and only if *R* satisfies the following three properties:

1. *Reflexive property for all* $a \in A$, $(a, a) \in R$.
2. *Symmetric property*–For $a, b \in A$, $(a, b) \in R \Rightarrow (b, a) \in R$,
3. *Transitive property*–For $a, b, c \in A$, $(a, b) \in R$ and $(b, c) \in R \Rightarrow (a, c) \in R$.

Wherever *R* is an equivalence relation and $x, y \in A$ such that $(x, y) \in R$, we say that *x* is equivalent to *y*.

Equivalence relations are generally denoted by the symbol ~ (pronounced "wiggle"). With this notation, the conditions of equivalence relation may be restated as follows:

1. $a \sim a$, for each $a \in A$,
2. $a \sim b \Rightarrow b \sim a$, for $a, b \in A$,
3. $a \sim b$, $b \sim c \Rightarrow a \sim c$, for $a, b, c \in A$.

Example 1.3.2: *Let* us consider the problem of defining an equivalence relation in the set A = {1, 3, 5, 7}. The relation R = {(1, 1), (3, 3), (5, 5), (7, 7)} is an equivalence relation in *A*, since $(a, a) \in R$, for all $a \in$ A; if $(a, b) \in R$, then $(b, a) \in R$; and no distinct pairs of the form (a, b) and (b, c) exist in *R*. We can restate this problem in the form:

Let *A* be a non-empty set, then $R = \{(x, y) \mid x = y, x, y \in A\}$ is an equivalence relation.

Example 1.3.3: Let *A* be a set of all straight lines in a plane. For $a, b \in A$, we define $a \sim b$ if, and only if, *a* is parallel to *b*. Then we observe that ~ is an equivalence relations.

Example 1.3.4: *Let A* be a set of all triangles in the plane. For $a, b \in A$, we define $a \sim b$ if, and only if, *a* and *b* are similar, (*i.e.*, *a* and *b* have corresponding angles equal). Then ~ is an equivalence relation.

Example 1.3.5: Let *Z* be the set of all integers and let $n > 1$ be a fixed integer. We define the relation $R = \{(a, b) \mid a - b = k,n$ for some integer $k\}$. This is an equivalence relation on *A*.

Note: If *n* divides $a - b$ $(n \mid a - b)$ without remainder, then we say that *a* is congruent to *b* (mod *n*) written as $a \equiv b \pmod{n}$. Thus ≡ is an equivalence relation which is known as the relation of *congruence modulo n*.

Definition 1.3.4: Let *A* be a given set and Let ~ be an equivalence relation defined on *A*. Let $a \in A$, then the set of all elements $x \in A$ which are equivalent to *a* $(x \sim a)$ is called the equivalence class, denoted by [*a*], that is,

$$[a] = \{x \in A \mid x \sim a\}.$$

The element $a \in A$ of equivalence class is called a representative for the set [a].

Example 1.3.6: Let Z be the set of all integers and let *R* be an equivalence relation defined by $(a, b) \in R$ if, and only *if*, $a - b = k \,.\, 3$, for some integer *k*.

We see that the relation *R* is an equivalence relation. Let us consider $0 \in Z$ and we form the subset [0] of elements of *Z* which are equivalent to 0, that is,

$$[0] = \{x \in Z \mid x \sim 0\}$$

or

$$[0] = \{x \in Z \mid x - 0 = k.\ 3,\ k \in Z\}$$

or

$$[0] = \{- - -, -9, -6, -3, 0, 3, 6, 9, \ldots.\}.$$

Similarly, let us form the set [1] of elements of Z which are equivalent to 1, that is,

$$[1] = \{x \in Z \mid x \sim 1\}$$

or $$[1] = \{x \in Z \mid x - 1 = k.\ 3,\ k \in Z\}$$

or $$[1] = \{x \in Z \mid x = 1 + k.3,\ k \in Z\}$$

or $$[1] = (\ldots.,\ -8,\ -5,\ -2,\ 1,\ 4,\ 7,\ 10\ldots\ldots.)$$

or

Similary, $$[2] = \{x \in Z \mid x \sim 2]$$

Similary, $$[2] = [x \in Z \mid x = 2 + 3k,\ k \in Z\}.$$

or $$[2] = \{\ldots,\ -7,\ -4,\ -1,\ 2,\ 5,\ 8,\ 11,\ldots\}$$

Again, if we form the equivalence class [3], then [3] = [0]. Similarly, [4] = [1], [5] = [2]. So there exist only three equivalence classes [0], [1], [2]. We observe the following points:

1. If $a \sim b$, $a, b \in$ Z then a, b must belong to the same class. If a is not equivalent to b, then a and b do not belong to the same class.
2. Every element of set Z is either in [0] or in [1] or in [2], that is, $[0] \cup [1] \cup [2] = Z$.
3. Any two equivalence sets are either identical sets or disjoint sets.
4. The integers of the form 3m, 3m + 1, 3m + 2 will give rise to the equivalence sets [0], [1] and [2] respectively.

Theorem 1.3.1: *'Let ~ be an equivalence relation defined on a set A.*

Let a, $b \in$ A, then

(1) $a \in [a]$.

(2) $a \sim b$ *if, and only if,* $[a] = [b]$.

(3) $a \nsim b$ *(a is not equivalent to b) if, and only if,* $[a] \cap [b] = \phi$

(4) $A = \cup\, a \in A\, [a]$.

Proof: **(1)** Since $a \sim a$, $a \in [a]$.

(2) Let $a \sim b$, then $b \in [b]$. Let $c \in [b]$.

$c \in [b] \Rightarrow b \sim c$. we have $a \sim b$, so

$a \sim b,\ b \sim c \Rightarrow a \sim c \Rightarrow c \in |a|$

Hence $c \in (b) \Rightarrow c \in [a] \Rightarrow [b] \subseteq [a]$,

Similarly, we can prove that

$$[a] \subseteq [b].$$

Thus $$[a] = [b].$$

Conversely, let $[a]$ = [b]. $c \in [a] = [b] \Rightarrow a \sim c$ and

$b \sim c \Rightarrow a \sim c$ and $c \sim b \Rightarrow a \sim b$.

(3) If a is not equivalent to b, then $b \notin [a]$. But $b \in [b]$.

So $$[a] \neq [b].$$

Conversely, if $[a] \cap [b] \neq \phi$ then if possible

$$x = [a] \cap [b] \Rightarrow x \in [a] \text{ and } x \in [b]$$
$$\Rightarrow x \sim a \text{ and } x \sim b$$
$$\Rightarrow a \sim x \text{ and } x \sim b$$
$$\Rightarrow a \sim b$$
$$\Rightarrow [a] = [b]$$

which is a contradiction to $[a] \neq [b]$ which implies

$$[a] \cap [b] = \phi$$

From this we conclude that either $[a] \cap [b] = \phi$ or $[a] = [b]$ (*i.e.*, any two equivalence classes are either disjoint or equal.)

(4) Let $B = \cup [a]$. Then it is clear that $B \subseteq A$ since every equivalence set $[a] \subseteq A$.

Conversely, since for each $a \in A$, there exists an equivalence set $[a] \subseteq B$. That is, if $a \in A$, then

$$a \in [a] \Rightarrow a \in B \Rightarrow A \subseteq B.$$

Hence A is the union of disjoint equivalence sets $[a]$.

Definition 1.3.5: The set of mutually disjoint equivalence sets of the set A with respect to the equivalence relation R defined on A is called the *Quotient set, denoted* by A/R, *of* A for the equivalence relation R.

Example 1.3.7: In example 1.3.6 the equivalence sets [0], [1], and [2] form a Quotient set. That is.

$$Z/R = \{[0], [1], [2]\}$$

Remark: In the theorem 1.3.1 we have seen that an equivalence relation R defined on a set A decomposes A into multually disjoint Equivalence sets. If we consider a set A and all disjoint subsets of A such that the union of all disjoint subsets of A is the set A itself, then the set P of all such disjoint subsets of A is called *the partition of A*. Thus, by the theorem 1.3.1 it is clear that an equivalence relation defined in A determines a partition of A which is *a* set of all equivalence sets with respect to the equivalence relation R. For examples, in example 1.3.6 the partition $P = \{[0], [1], [2]\}$.

Theorem 1.3.2: *Let P = $\{A_i\}$ be a partition of the set A. Then there exists an equivalence relation in A such that equivalence sets are precisely the sets A_i.*

Proof: For elements $a, b \in A$ we define $a \sim b$ if $a, b \in A_i$.

We have to verify that the relation $\sim$ is an equivalence relation. For $a, b, c \in A$ we have

1. *Reflexive property:* Since every element $a \in A$ is in some subset $A_i \in P$, there exists a subset A_i, such that $a \in A_i$. So $a \sim a$ by definition.

2. *Symmetric property:* If $a \sim b$, then there exist A_i, such that $a, b \in A_i$. $\Rightarrow b, a \in A_t$ $\Rightarrow b \sim a$.

3. *Transitive property:* If $a \sim b$ and $b \sim c$, then there exists A_i, and A_j with $i \neq j$ such that $a, b \in A_i$, and $b, c, \in A_j$.

$\Rightarrow b \in A_i \cap A_j \Rightarrow A_i = A_j \Rightarrow a, c \in A_j \Rightarrow a \sim c$.

For, if $i \neq j$, then $A_i \neq A_j$, by definition of partition,

and if $A_i \neq A_j$, then $A_i \cap A_j = \phi$,

and if $A_i \cap A_j \neq \phi$, then $A_i = A_j$,

which proves $\sim$ is an equivalence relation.

Now we have for $a \in A$, $[a] = [x \in A \mid a \sim x\}$.

Let us consider $a \in A_j$, then $b \in A_i$ if, and only if, $a \sim b$, that is, $a \sim b$ if and only if $b \in [a]$, which means $A_i = [a]$. So each member A_i *of* the partition is an equivalence set. This completes the proof of the theorem.

Definition 1.3.6: Let A be a non-empty set. The subset $R \subseteq A \times A$ is called partial order relation in the set A if R satisfies the following properties.

1. *Reflexive property: (a, a)* $\in R$, for all $a \in A$.
2. *Anti-symmetric property: (a, b)* $\in R$ and $(b, a) \in R \Rightarrow a = b$.
3. *Transitive property: (a, b)* $\in$. R and $(b, c) \in R \Rightarrow (a, c) \in R$.

A non-empty set A in which there is defined a partial order relation is called a *partially ordered set.*

Example 1.3.8: *Let N* be the set of positive integers. For $a, b \in N$, we define a relation "a divides b" denoted by $a \mid b$. Then for any $a, b, c \in N$, we have

1. *Reflexsive property:* Since $a = 1.a$, $a \mid a$, for all $a \in A$.
2. *Anti-Symmetric property:* If $a \mid b$ and $b \mid a$, then $a = b$
3. *Transitive property:* It $a \mid b$ and $b \mid c$, then $a \mid c$.

Hence the relation "a divides b" is a partial order relation. The set N is partially ordered.

Example 1.3.9: Let A be a non-empty set, then the relation "$\subseteq$" defined on *P(A),* the power set of A is a partial order relation. For $X, Y, Z \in P(A)$

1. $X \subseteq X$, for all $X \in P(A)$.
2. $X \subseteq Y$, and $Y \subseteq X \Rightarrow X = Y$.
3. $X \subseteq Y$, $Y \subseteq Z \Rightarrow X \subseteq Z$.

Hence the relation $\subseteq$ is a partial order relation.

1.4 MAPPINGS (FUNCTIONS)

The concept of function is widely used throughout mathematics. A particular type of functions (binary operations) are ingredient of the algebraic systems. So the function is a fundamental tool in dealing with algebraic systems. Here we shall define function in terms of ordered pairs.

Definition 1.4.1: Let A and B be two non-empty sets. A function f (or mapping) from the set A to the set B is a set of ordered pairs (a, b), where $a \in A$ and $b \in B$, such that no distint pairs have the same first elements, that is, if $(x, y_1) \in f$ and $(x, y_2) \in f$, then $y_1 = y_2$.

Thus *a* function f is a relation from A to B but *a* relation may not be a function. So sometimes the function is termed as a single valued relation.

For a given function f, when $(x, y) \in f$ we write $y = f(x,)$, $f(x)$ is called the image of x under the function f and x is called the pre-image of $f(x)$, $y = f(x)$ is a functional notation of the function f. As such $f(x)$ is also called the functional value of f at x.

From the definition it is clear that $f \subseteq A \times B$. The collection of all first components of ordered pairs belonging to f is called the domain of f and denoted by D_f. The collection of all second elements of ordered pairs of f is called the range of f and is denoted by R_f.

Definition 1.4.2: Two function f and g are said to equal if $D_f = D_g = A$ and $f(x) = g(x)$ for each $x \in A$.

Let $f: A \to C$ and $g : B \to C$. If $A \subset B$ and if $f(x) = g(x)$, $x \in A$, then we say g is an extension of f to B or that f is the restriction of g to A.

Example 1.4.1: *Let* $f = f(x, x^2) \mid 1 \leq x \leq 3\}$,

$$g = \{(x, x^2) \mid 1 \leq x \leq 4\}$$

Here g *is* an extension of f to $1 \leq x \leq 4$,

Definition 1.4.3: Let $f : A \to B$. If no two distint ordered pairs of f have the same second component, than f is called a *one to one mapping* (function); that is, f is one to one, if $(x_1, y) \in f$ and $(x_2, y) \in f$ implies $x_1 = x_2$.

In functional notation we can write that if $f(x_1) = f(x_2) \Rightarrow x_1 = x_2$, *for* $x_1, x_2 \in D_f$, then f is a one to one function. And if $f(x_1) = f(x_2)$ even if $x_1 \neq x_2$, then f is many to one mapping.

Remarks: A one to one function is also known by injection, an onto mapping by *surjection,* and one to one and onto mapping is called a bijection.

Definition 1.4.4: Let A be a non-empty set. The function $i_A : A \to A$ defined by $i_A(x) = x$, $x \in A$, is called identity function on A. That is, identity function is a set of ordered pairs $\{(x, x) \mid x \in A\}$.

Example 1.4.2: Let Q^+ and Z_0 be two sets of all positive rational numbers and all positive integers respectively. The function $f: Q^+ \to Z_0$ is defined by $f(m/n) = m + n$, where $m/n \in Q^+$, with $m, n \in Z_0$ and m, n have no common factor, f is onto since $\frac{1}{n} \in Q^+$ such that $f\left(\frac{1}{n}\right) = n + 1$, and f is not one to one since $f(2/3) = 2 + 3 = 3 + 2 = f(3/2)$ but $2/3 \neq 3/2$.

1.5 FINITE AND INFINITE SETS

Definition 1.5.1: A set A is said to be finite if it can be put into 1–1 correspondence with the set $\{1, 2, 3, \ldots, n\}$, where n is a positive integer. Empty set is also finite.

Definition 1.5.2: A set A is said to be infinite if there exists a proper subset B *of* A such that A can be put into 1–1 correspondence with the set B.

For example, Z is the set of all integers and the set E is the set of all even integers and we define a mapping $f: Z \to E$ *by* $f(n) = 2n$.

Now for $n_1, n_2 \in Z$ we assume that $f(n_1) = f(n_2)$. Then

$$f(n_1) = f(n_2) \Rightarrow 2n_1 = 2n_2$$
$$\Rightarrow n_1 = n_2$$

Therefore f is a one-to-one mapping. For an even integer $2n$ these exists an integer n such that

$$f(n) = 2n$$

which shows f is onto. Considering that f is both one to one and onto, Z and E can be put in one to one correspondence.

Hence Z is an infinite set.

1.6 INVERSE OF A FUNCTION

Definition 1.6.1: Let f be a one to one function from A onto (into) B. The inverse of f, denoted by f^{-1}, is

$$\{(y, x) \mid (x, y) \in f\}$$

If f is a one to one function from A onto B then f^{-1} is also a one to one function from B onto A, and if f is one to one function (from A into B, then f^{-1} is a one to one function) from a subset of B onto A. It is also clear that if $f: A \to B$ is one to one and onto, then $(f^{-1})^{-1} = f$.

Example 1.6.1: Let S = T = {1, 2, 3}. The function

$f = \{(1, 3), (3, 2), (2, 1)\}$ has its inverse

$f^{-1} = \{(3, 1), (2, 3), (1, 2)\}$ while the function

$g = \{(1, 2), (2, 1), (3, 1)\}$ does not have its inverse since it is not one to one.

Thus, we have seen that if f^{-1} exists of the function f, then $D_f = R_{f^{-1}}$ and $D_{f^{-1}} = R_f$.

Example 1.6.2: The function $f = \{(x, 5x - 2) \mid x \in R\}$ is one to one, for $5x_1 - 2 = 5x_2 - 2 \Rightarrow x_1 = x_2$, consequently f^{-1} exists, and $f^{-1} = \{(5x - 2, x) \mid x \in R\}$. In functional notation we can write, $f^{-}(x) = \frac{1}{5}(x + 2)$ for each $x \in R$ and $f(x) = 5x - 2$ for each $x \in R$.

1.7 COMPOSITION OF TWO FUNCTIONS

Definition 1.7.1: Let $f: A \to B$ and $g: B \to C$ be two functions, the composition of f and g, denoted by $g \circ f$, is a function

$g \circ f = \{(x, y) \mid \text{There exists } z \in B \text{ such that } (x, z) \in f \text{ and } (z, y) \in g\}$.

In terms of functional values we can write,

$(g \circ f)(x) = g(f(x))$, where $D_{g \circ f} = \{x \in D_f \mid f(x) \in D_g\}$. From this definition it is clear that $D_{g \circ f} \subseteq D_f$ and $R_{g \circ f} \subseteq R_g$.

Example 1.7.1: *Let $f: R \to R$* be a function defined by

$$f(x) = 1 + \sin x, \; x \in R,$$

and Let $g : R \to R$ be defined by

$$g(x) = x^2, \; x \in R.$$

Then

$$\begin{aligned}(g \circ f)(x) &= g(f(x) = g(1 + \sin x) \\ &= (1 + \sin x)^2 \\ &= 1 + \sin^2 x + 2 \sin x.\end{aligned}$$

fog is also meaningful

$$(f \circ g)(x) = f(g(x)) = f(x^2) = 1 + \sin x^2.$$

Thus *fog is* different from $g \circ f$.

Theorem 1.7.1: Let $f : S \to T$, $g : T \to U$, and $h : U \to V$ be functions, then

(a) $(hog) \circ f = h \circ (g \circ f)$,

(b) *if each f and g is one to one, then so is* $g \circ f$,

(c) *if each f and g is onto, then so is* $g \circ f$,

(d) *if each f and g is one to one and onto, then* $(g \circ f)^{-1} = f^{-1} \circ g^{-1}$.

Proof: ***(a)*** It is clear that $(h \circ g) \circ f$ and $h \circ (g \circ f)$ are functions from S to V. Now we have to prove that for any $x \in S$, the image of x under $((h \circ g) \circ f)$ is equal to the image of x under $(h \circ (g \circ f))$.

$$\begin{aligned}((h \circ g) \circ f)(x) &= (h \circ g) f(x) \\ &= h(g(f(x))) \\ &= h((g \circ f)(x)) \\ &= (h \circ (g \circ f))(x).\end{aligned}$$

which shows

$$(h \circ g) \circ f = h \circ (g \circ f).$$

(b) Let $x_1, x_2 \in S$, then

$$\begin{aligned}(g \circ f)(x_1) &= (g \circ f)(x_2) \\ &\Rightarrow g(f(x_1)) = g(f(x_2)) \\ &\Rightarrow f(x_1) = f(x_2) \text{ (since g is one to one)} \\ &\Rightarrow x_1 = x_2 \text{ (since } f \text{ is one to one)}\end{aligned}$$

Hence $g \circ f$ is one to one.

(c) by the definition of composition $gof : S \to U$ is a function. To prove that gof is onto we have to prove that every element $u \in U$ is an image element for some $x \in S$ under gof, since g is onto, there exists $t \in T$ such that $g(t) = u$. Again since f is onto from S to T, there exists $x \in S$ such that $f(x) = t$.

Now $(g \circ f)(x) = g(f(x)) = g(t) = u$.

which shows $go\,f$ is onto.

(d) We have $g \circ f : S \to U$ is a function. By (b) and (c) $g \circ f : S \to U$ *is* a one-to-one and onto. So its inverse $(g \circ f)^{-1}$ exists and $(g \circ f)^{-1} : U \to S$ is also a one to one and onto function.

Again, since f and g are one to one and onto functions, f^{-1} and g^{-1} exists and $f^{-1} : T \to S$ and $g^{-1} : U \to T$ are also one to one and onto. So $f^{-1} \circ g^{-1} : U \to S$ is also one to one and onto by (b) and (c).

Now to prove $(g \circ f)^{-1} = f^{-1} \circ g^{-1}$ we have to show that for every $u \in U$, the image of $u \in U$ under $(g \circ f)^{-1}$ and $f^{-1} \circ g^{-1}$ are equal.

Now, we have

$$g(t) = u \Rightarrow t = g^{-1}(u) \text{ since } g \text{ is one to one and onto}$$

$$f(x) = t \Rightarrow x = f^{-1}(t) \text{ since } f \text{ is one to one and onto}$$

$$(g \circ f)(x) = g(f(x)) = g(t) = u \Rightarrow x = (g \circ f)^{-1}(u) \quad \text{...(1)}$$

Since gof is one to one and onto.

Again

$$\begin{aligned}(f^{-1} \circ g^{-1})(u) &= f^{-1}(g^{-1}(u)) \\ &= f^{-1}(t) \\ &= x. \quad \text{...(2)}\end{aligned}$$

By (1) and (2), we have

$$(g \circ f)^{-1} = f^{-1} \circ g^{1}.$$

Corollary. The function $f : A \to B$ is one to one and onto if and only if there exists a function $g : B \to A$ such that $g \circ f: A \to A$ and $fog : B \to B$ are identity functions.

Definition 1.7.2: Let S be a non-empty set. The one to one function from S onto S itself is called a non-singular transformation or permutation of S.

The set of all permutations of S is denoted by $A(S)$.

Definition 1.7.3: If there exists a 1–1 mapping of A onto B, we say that A and B can be put in 1–1 correspondence, or briefly, that A and B are equivalent, and we write $A \sim B$.

Theorem 1.7.2: *(Schroder-Bernstein's Theorem) Let X and Y be two sets such that $X \sim A$ for some $A \subseteq Y$* and $Y \sim B$ for some $B \subseteq X$, then $X \sim Y$.

Proof: We assume that $f : X \to Y$ is a one to one mapping of X into Y, and $g: Y \to X$ is a one to one mapping of Y into X. Here we have to produce a mapping $F : X \to Y$ which is one to one and onto. Since f and g are one to one and into mappings, then the mappings

$f^{-1} : f(X) \to X$ and $g^{-1}: g(Y) \to Y$ are one to one and onto. We obtain the mapping F by decomposing both X and Y into disjoint subsets as follows:

If $x \in X$, then $g^{-1}(x) \in Y$ (if it exists) is called the first ancestor of x. The element x is called the zeroth ancestor of x. The element $f^{-1}(g^{-1}(x))$ (if it exists) is called the second ancestor of x, $g^{-1} f^{-1} (g^{-1}(x)) = (g^{-1} f^{-1} g^{-1})(x)$ (if it exists) is called third ancestor of x and so on.

Then there are three possibilities: (1) x has infinitely many ancestors. We denote by X_i the set of all $x \in X$ which are having infinitely many ancestors. (2) x has last ancestor in X, *i.e.*, X has an even number of ancesters and we denote by X_e the set of all element with even ancestors. (3) x has last ancestor in Y, *i.e.*, x has an odd number of ancestors and we denote by X_o the set of all elements with odd number of ancestors. The sets X_i, X_e, X_o are disjoint sets whose union is X. We decompose Y in just the way into three subsets Y_i, Y_e, Y_o. It is easy to see that f maps X_i onto Y_i and X_e on Y_o, and that g^{-1} maps X_o onto Y_e. Thus we define F by

$$F(x) = \begin{cases} f(x), \text{ if } x \in X_i \cup X_e \\ g^{-1}(x), \text{ if } x \in X_o \end{cases}$$

Hence $F : X \to Y$ is one to one mapping of X onto Y.

Hence $X \to Y$.

Definition 1.7.4: Let $f : X \to Y$ be a function. If $A \subseteq X$, the image of A, denoted by $f(A)$, is the subset of Y defined by $f(A) = \{f(x) \mid x \in A\}$.

On the other hand, if $B \subseteq Y$, then inverse image of B, denoted by f^{-1} (B), is the subset of X defined by

$$f^{-1}(B) = \{x \mid f(x) \in B\}.$$

Here f^{-1} does not mean the inverse of the function f. f^{-1} should be interpreted according to the context it is used.

Definition 1.7.5: Let Λ and A be two sets. Then a function $f : \Lambda \to A$ is called a family, each member of Λ is known as an index and the set Λ an index set. The range of f is known as an indexed set. If $i \in \Lambda$, than $f(i)$ or written as $f_i \in A$ is called the term of the family, and the indexed set is written as $\{f_i\}$, $i \in$ A. If f associates each $i \in \Lambda$ with a unique set A_i, then f is called the family of sets denoted by $\{A_i\}$ $i \in \Lambda$.

With the help of family of sets we can find out the union or intersection of more than two sets.

Let $\{A_i\}$ $i \in \Lambda$ be a family of sets. Than the union and intersection are defined by

$U \{A_i \mid i \in \Lambda\}$ and $\cap \{A_i \mid i \in \Lambda\}$ respectively.

Thus $U \{A_i \mid i \in \Lambda\} = \{x \mid x \in A_i$ for atleast one $A_i\}$

and $\cap \{A_i \mid i \in \Lambda\} = \{x \mid x \in A_i$ for all $A_i\}$.

PROBLEMS

1. Let S and T be two finite sets having m and n elements respectively. Then show that there exists 2^{mn} relations in $S \times T$.

2. If $S = \{1, 2, 3\}$ and $T = \{x, y\}$, list all relations in $S \times T$. And also list all functions in $S \times T$.

3. *Let* S and T be two finite sets with m and n elements respectively, than show that there are n^m functions of S into T.

4. Consider the following sets:

   (*a*) $\{(1, 2), (3, 2), (5, 5))$

   (*b*) $\{(1, 2), (2, 3), (5, 5)\}$

   (*c*) $\{(2, 1), (2, 3), (5, 5)$

   Are these sets the functions? Explain.

5. Let R be the set of real numbers and F that relation in $R \times R$ consisting of all pairs listed below. Determine which, if any, are functions. If f is not a function, exhibit two elements (x, y_1) and (x, y_2) in f, with $y_1 \neq y_2$.

   (*a*) $F = \{(x, x + 2)| \; x \in R\}$ (*b*) $F = [(x, 2x) | \; x \in R\}$

   (*c*) $F = \{(x, x^2) \mid x \in R\}$ (*d*) $F = ((x, a.x + b) \mid x \in R\}$

   (*e*) $F = \{(x, |x|) \; |x \in R\}$ (*f*) $F = \{(x, \sin x) \mid x \in R\}$

   (*g*) $F = \{(x, y) \mid x^2 + y^2 = 4, x, y \in R\}$ (*h*) $F = [(x, 0) \mid x \in R\}$

   (*i*) $F = ((x, \sqrt{x}) \mid x \in R\}$ (*j*) $F = \{(x, 1/x), \mid x \neq 0, x \in R\}$

   In case of functions determine which are one to one, many to one, into, and onto functions.

6. Let R be the set of real numbers. In the following $F \subset R \times R$, determine whether F has the inverse. If so, list the elements of f^{-1}. If no, exhibit that f is not one to one.

   (*a*) $F = \{(x, y) \mid x \leq y\}$ (*b*) $F = \{(x, y) \mid x < y\}$

   (*c*) $F = \{(x, y) \mid x > y\}$ (*d*) $F = \{(x, y) \mid x \geq y\}$

   (*e*) $F = \{x, x^2) \mid x \in R\}$ (*f*) $F = \{x, 2x) \mid x \in R\}$

7. Let X be a finite set and let $f : X \to Y$ be an onto function. Then show that the number of elements in Y cannot be greater than X.

8. Let $i_A : A \to A$ and $i_B : B \to B$ be two identity mappings and let $f : A \to B$ be a bijection. Then show that there exists a unique mappings $g : B \to A$, such that $g \circ f = i_A$ and $f \circ g = i_B$.

9. Let $f : X \to Y$ be a mapping. For any $x_1, x_2 \in X$ the relation $x_1 \sim x_2$ is defined by $f(x_1) = f(x_2)$. Show that $\sim$ is an equivalence relation on X.

10. Give an example of two functions f and g from R into R, the set of real numbers, with $f \neq g$ for which $f \circ g = g \circ f$.

11. Determine *fog* and *gof*, and their respective domains, given that

(*a*) $f = \{(x, x^2 + x) \mid x \in R\}$, $g = \{x, 3x + 4) \mid x \in R\}$

(*b*) $f = \left\{\left(x, \frac{x}{x^2+1}\right) \mid x \in R\right\}, g = \left\{\left(x, \frac{1}{x}\right) \mid x \in R\right\}$

12. Let $g : X \to Y$ and $f : Y \to Z$ be two functions. Show that the following statements are true:

(*a*) If *fog* is an onto function, then f is also.

(*b*) If *fog* is a one to one function, then g is also.

13. If the set A is finite, prove that:

(*a*) If $f : A \to A$ is onto function, then f is one to one.

(*b*) If $f : A \to A$ is one to one and into, then f is onto.

(*c*) Prove (*a*) and (*b*) false if A is infinite set by an example.

14. Prove that

(*a*) The set of rational number is infinite.

(*b*) The set of real numbers is infinite.

(*c*) If $A \subseteq X$ is infinite set, then X is infinite.

15. Let $f : X \to Y$ and let $A, B \subseteq X$. Then show that

(*a*) $f(A \cup B) = f(A) \cup f(B)$, (*b*) $f(A \cap B) \subseteq f(A) \cap f(B)$

(*c*) $f(A) - f(B) \subseteq f(A - B)$. (*d*) if $A \subseteq B$, then $f(A) \subseteq f(B)$

16. Let $f : X \to Y$ and $A, B \subseteq Y$ Then show that:

(*a*) $f^{-1}(A \cup B) = f^{-1}(A) \cup f^{-1}(B)$ (*b*) $f^{-1}(A \cap B) = f^{-1}(A) \cap f^{-1}(B)$

(*c*) $f^{-1}(A) - f^{-1}(B) = f^{-1}(A - B)$ (*d*) if $A \subseteq B$, then $f^{-1}(A) \subseteq f^{-1}(B)$

17. If $f : X \to Y$, prove that f is a one to one function if and only if $f(A \cap B) = f(A) \cap f(B)$, $A, B \subseteq X$.

18. A set A is said to be equipotent to a set B if there exists a bijection from A onto B. Symbolically, we can write $A \sim B$.

Prove:

(*a*) The relation of equipotent in the set of sets is an equivalence relation.

(*b*) A set can never be equipotent to its power set.

(*c*) Let A^B be the set of mappings of B onto A, then show that $A \sim A'$, $B \sim B \Rightarrow A^B \sim A'B'$.

19. Let X be a set. Let $\Delta = \{(x, x) \mid x \in X\}$ be the diagonal. If R is any relation on X, then $R^{-1} = \{(y, x) \mid (x, y) \in R\}$.

Prove:

(*a*) A relation R is relexive if and only if $\Delta \subseteq R$.

(*b*) A relation R is symmetric if and only if $R = R^{-1}$.

20. If $f : X \to Y$ is a function, then prove that $f^{-1} = \{(y, x) \mid (x, y) \in f\}$ need not be a function.

21. Let S be a finite set. Define the relation A ~ B if only if A and B have the same number of elements for all $A, B \in P(S)$. Show that ~ is an equivalence relation.

22. Prove that following relations ~ are equivalence relations in the cartesian product $R \times R$:

(*a*) $(a, b) \sim (c, d)$ if and only if $b - d = m(a - c)$, m a fixed real.

(*b*) $(a, b) \sim (c, d)$ if and only if $a + b = b + c$.

23. Let the following relations R to be defined in the set Z of integers. Classify whether they do or do not have the properties of being reflexive, symmetric, and transitive:

(*a*) $R = \{(a, b) \mid (a < b)\}$

(*b*) $R = \{(a, b) \mid a - b$ is an odd integer$\}$

(*c*) $R = \{(a, b) \mid (ab > 0)\}$,

(*d*) $R = \{(a, b) \mid a^2 = b^2\}$,

(*e*) $R = \{(a \,.\, b) \mid |a - b| < 1\}$.

24. Let X be a set of all functions of a set A into the set R of all real numbers. For any $f, g \in X$, define $f \leq g$ if and only if $f(a) \leq g(a)$ for all $a \in A$, then show that $\leq$ is a partial order relation on X.

25. Let A be a set of all straight lines. For $a, b \in A$, define $a \sim b$ if and only if a is perpendicular to b. Show that ~ is not an equivalence relation.

CHAPTER 2

GROUP THEORY

2.1 GROUPS

In this chapter, as also in the subsequent chapters, we shall deal with some important algebraic (mathematical systems). Here we shall confine our attention to one simple class of systems that are called groups which are defined by a set S together with one binary operation on S satisfying a set of postulates (axioms). We begin with the systems involving just one operation since they land themselves to the simplest formal description. Despite this simplicity the postulates permit the construction of a profuse and an elegant theory in which we will encounter many of the fundamental notions common to all algebraic systems.

Most sets dealt with in mathematics are sets which have an "algebraic structure" meaning that certain suitably restricted rules of combination or operation are defined on them. These rules enable us to combine the objects of the set in useful ways. Se we make use of a particular rule of combination called binary operation which combines any two elements $a, b \in S$ such that the combination $a * b$ *of* a and b also belongs to S. That is, we consider the binary operation as a function of the set $S \times S$ into S. Thus f: $S \times S \rightarrow$ S is a binary operation defined on S. Now we define mathematical systems.

Definition 2.1.1: A mathematical system consisting of a non-empty set S of arbitrary elements together with a single binary operation defined on S is called a groupoid and it is denoted by an ordered pair $(S, *)$.

A groupoid is determined not by its set of elements alone, but by the set together with the binary operation defined on it. Thus different groupoids may well have the same associated set of elements, and this is the reason for representing a groupoid as an ordered pair $(S, *)$ of set S and binary operation $*$defined on S.

Semi-Group 2.1.1: A semi-group is a groupoid (S, o) whose binary operation "o" satisfies the associative law, that is, a semi-group is a pair (S, o) consisting of a non-empty set and an associative binary operation o defined on S. Thus if $a, b, c \in S$, then (i) $a \; o \; b \in S$

(ii) $a \; o \; (b \; o \; c) = (a \; o \; b) \; o \; c$.

Monoid 2.1.1: *A* semi-group (S, o) is having an identity element with respect to the operation o is called a monoid. Thus, if $e \in S$, we have

$$a \; o \; e = e \; o \; a = a \text{ for all } a \in S.$$

Definition 2.1.2: *A* pair *(G, o)* is a group if, and only if, *(G, o)* is a semi group, with an identity, in which each element of *G* has an inverse in G.

While this definition is perfectly acceptable, we prefer to rephrase it in the following more detailed form merely as a matter of convenience.

Definition 2.1.3: Let *G* be a non-empty set and *'o' a* function with domain $G \times G$. Denote the image of a pair *(a, b)* in $G \times G$ by $a \; o \; b$. Then the pair *(G, o)* consisting of the non-empty set *G* and the function *o,* is a group if, and only if:

(*a*) (Closure Property) *G* is closed under the function *o*, that is, for all $(a, b) \in G \times G$ the image $aob \in G$.

(*b*) (Associative law) The function *o* is associative, that is, for all $a, b, c \in G$, $(aob) \; oc = ao \; (hoc)$.

(*c*) (Existence of Identity) *G* contains an identity element *e* for the function *o,* that is, there exists an element *e* in *G* such that $a \; o \; e = e \; o \; a = a$ for all $a \in G$.

(*d*) (Existence of Inverse)–For each element $a' \in G$, there exists an element $a' \in G$ such that $a \; o \; a' = a' \; o \; a = e$. That is, each element $a' \in G$ has an inverse $a' \in G$ with respect to the function *o.*

We will refer to *'o'* as the operation of the group, or as group multiplication. Although the definition requires only that *o* be a function from $G \times G$ into *G,* we note that it is always onto. For if, $x \in G$, an element $e \in G$ such that $xoe = x$. Hence (x, e) has *x* as its image under *o.*

Definition 2.1.4: If (G, *o*) is a group, then (G, *o*) is abelian (or commutative) if, and only if, the group operation *o* is commutative, that is, for all $a, b \in G$, $aob = boa$.

Finite and Infinite Group

Definition 2.1.5: *A* group (*G, o*) is called a finite or an infinite group according as the underlying set *G* is finite or infinite.

Order of a Group

Definition 2.1.6: The number of elements in a finite group is called the order of the group. The infinite set is said to be of infinite order. The order of a finite group is denoted by *o(G).*

Example 2.1.1: (*a*) Let *G* be a set of all integers and a binary operation* is defined by $a * b = a + b$, for all $a, b \in G$. Then the system $(G, *)$ is an infinite abelian group. The element 0 plays a role of identity element and $-a$ that of the inverse of $a \in G$.

(*b*) The pair *(Q, +), (R, +),* and *(C, +)*, where *Q, R,* and *C* are sets of rational, real, and complex numbers, respectively, and the operation in each pair is ordinary addition, are infinite abelian groups.

(*c*) The pair $(Q - \{0\},.)$, where $Q - \{0\}$ is the set of all non-zero rational numbers and the operation is ordinary multiplication, is an infinite abelian group. Similarly, the pairs $(R - \{0\},.)$ and $(C - \{0\},.)$ are infinite abelian groups.

(*d*) The set of all positive rational numbers together with the operation of multiplication is an infinite abelian group.

(*e*) The pair $(\{1,-1\},.)$ is a finite abelian group of order 2 under the operation of multiplication.

(*f*) The pair (G, o), where $G = \left\{1, \frac{-1+i\sqrt{3}}{2}, \frac{-1-i\sqrt{3}}{2}\right\}$ and o is the usual multiplication of complex numbers, is a finite group of order 3, The composition table is given below:

o	1	w	w^2
1	1	w	w^2
w	w	w^2	1
w^2	w^2	1	w

where $w = \frac{-1+i\sqrt{3}}{2}$,

$w^2 = \frac{-1-i\sqrt{3}}{2}$

(*g*) The pair (G, o) where $G = \{1, -1, i, -i\}$ and o is the usual multiplication of complex numbers, is a finite abelian group of order 4. The operation table is given below.

o	1	-1	i	$-i$
1	1	-1	i	$-i$
-1	-1	1	$-i$	i
i	i	$-i$	-1	1
$-i$	$-i$	i	1	-1

(*h*) Let G be the set of *nth* roots of unity. The system (G, o) is an abelian group, where o denotes the ordinary multiplication.

Example 2.1.2: Let six function be defined on the domain $R - \{0\}$ by

CHAPTER 2

$$f_1(x) = (x) \qquad f_2(x) = \frac{1}{x},$$

$$f_3(x) = 1 - x, \qquad f_4(x) = \frac{x-1}{x}, \quad x \in R - \{0\}$$

$$f_5(x) = \frac{x}{x-1}, \qquad f_6(x) = \frac{1}{1-x}$$

Let $G = \{f_1, f_2, f_3, f_4, f_5, f_6,\}.$

then (G, o) forms a group under functional composition.

Solution: We compute

$$f_1 \, o \, f_1(x) = f_1(f_1(x)) = f_1(x) = x$$

$$f_1 \, o \, f_2(x) = f_1 \,(f_2(x)) = f_1\left(\frac{1}{x}\right) = \frac{1}{x} = f_2(x)$$

$$f_1 \, o \, f_3(x) = f_1(f_3(x)) = f_1(1 - x) = 1 - x = f_3(x)$$

Similarly $f_1 \, o \, f_4 = f_4, f_1 \, o \, f_5 = f_5, f_1 o f_6 = f_6.$ (A)

$$f_2 \, o \, f_1(x) = f_2(f_1(x)) = f_2(x) = \frac{1}{x} = f_2(x)$$

Similarly $f_3 \, o \, f_1 = f_3, f_4 \, o \, f_1 = f_4, f_5 o f_1 = f_5, f_6 o f_1 = f_6$

$$f_2 \, o \, f_3(x) = f_2(f_3(x)) = f_2 \,(1 - x) = \frac{1}{1-x} = f_6(x)$$

$$f_2 \, o \, f_4(x) = f_2(f_4(x)) = f_2\left(\frac{x-1}{x}\right) = \frac{x}{x-1} = f_5(x)$$

and $f_3 \, o \, f_4(x) = f_3(f_2(x)) = f_3\left(\frac{1}{x}\right) = 1 - \frac{1}{x} = \frac{x-1}{x} = f_4(x)$

(1) Similarly we can compute the composition of two functions and see the resultant is a member of G, which shows the set G is closed.

(2) Since composition of functions is associative, the composition in the set G is also associative.

(3) The set G has an identity element f_1 for the operation, which is clear from the above computations in *(A)*.

(4) The inverse of every element also exists in G.

$$f_2^{-1} = f_2, f_3^{-1} = f_3, f_4^{-1} = f_6, f_5^{-1} = f_5, f_6^{-1} = f_4$$

(5) The composite of functions is not commutative since $f_2 \, o \, f_3 = f_6$ and $f_3 \, o \, f_2 = f_4$. Hence (G, o) is an non-commutative group. The composition table is given below:

o	f_1	f_2	f_3	f_4	f_5	f_6
f_1	f_1	f_2	f_3	f_4	f_5	f_6
f_2	f_2	f_1	f_6	f_5	f_4	f_3
f_3	f_3	f_4	f_1	f_2	f_6	f_5
f_4	f_4	f_3	f_5	f_6	f_2	f_1
f_5	f_5	f_6	f_4	f_3	f_1	f_2
f_6	f_6	f_5	f_2	f_1	f_3	f_4

Example 2.1.3: Consider the set G consisting of the lour functions f_1, f_2, f_3, f_4:

$$f_1(x) = x,\ f_2(x) = 1/x,\ f_3(x) = -x,\ f_4(x) - 1/x.$$

Prove that (G, o) is a commutative group, where o denotes functional composition.

Solution: We give below the operation table, and shall verify the group axioms.

Let $$G = \{f_1, f_2, f_3, f_4\}$$

o	f_1	f_2	f_3	f_4
f_1	f_1	f_2	f_3	f_4
f_2	f_2	f_1	f_4	f_3
f_3	f_3	f_4	f_1	f_2
f_4	f_4	f_3	f_2	f_1

We note the following things in the operation table:

That each element of the set appears once and only once in each row and in each column and

(1) Every entry in the table is an element of the set G. Hence the set G is closed for the operation.

(2) The table is symmetric about principal diagonal. Hence the operation is commutative.

(3) Since the composition of functions is associative, the operation is also associative in G.

(4) The first row and first column are identical with set G written horizontally and vertically. Therfore the element f_1 which appears at the intersection of the first row and first column is an identity element in the set G for the operation.

(5) At the entries (1, 1)th.. (2, 2)th, (3, 3)th, (4, 4)th the identify element f_1 is appearing

which implies $f_1^{-1} = f_1,\ f_2^{-1} = f_2,\ f_3^{-1} = f_3,\ f_4^{-1} = f_4$

Hence (G, o) is an abelian group of order 4.

Example 2.1.4: The pair (S_3, o), where underlying set is the set of all one to one mappings of the $S = \{x_1, x_2, x_3)$ onto itself and o denotes the composition of mappings, is a group.

Solution: The element of S_3 can be determined as follows:

f_1 :	$x_1 \to x_1$	$f_1 \, o \, f_1 = f_1$		
	$x_2 \to x_2$			
	$x_3 \to x_3$			
f_2 :	$x_1 \to x_2$	and $f_2 \, o \, f_2$:	$x_1 \to x_1$	
	$x_2 \to x_1$		$x_2 \to x_2$	
	$x_3 \to x_3$		$x_2 \to x_3$	Thus $f_2 \, o \, f_2 = f_1$
f_3 :	$x_1 \to x_2$	$f_3 \, o \, f_3$:	$x_1 \to x_3$	
	$x_2 \to x_3$		$x_2 \to x_1$	
	$x_3 \to x_1$		$x_3 \to x_2$	$f_3 \, o \, f_3 = f_6$
f_4 :	$x_1 \to x_3$	$f_4 \, o \, f_4$:	$x_1 \to x_1$	$f_4 \, o \, f_4 = f_1$
	$x_2 \to x_2$		$x_2 \to x_2$	
	$x_3 \to x_1$		$x_3 \to x_3$	
f_5 :	$x_1 \to x_1$	$f_5 \, o \, f_5$:	$x_1 \to x_1$	
	$x_2 \to x_3$		$x_2 \to x_2$	$f_5 \, o \, f_5 = f_1$
	$x_3 \to x_2$		$x_3 \to x_3$	
f_6 :	$x_1 \to x_3$	$f_6 \, o \, f_6$:	$x_1 \to x_2$	
	$x_2 \to x_1$		$x_2 \to x_3$	$f_6 \, o \, f_6 = f_3$
	$x_3 \to x_2$		$x_3 \to x_1$	

Similarly we can compute other compositions of functions.

(1) The set S_3 is closed for the composition of the functions, since every entry in the table is an element of S_3.

o	f_1	f_2	f_3	f_4	f_5	f_6
f_1	f_1	f_2	f_3	f_4	f_5	f_6
f_2	f_2	f_1	f_5	f_6	f_3	f_4
f_3	f_3	f_4	f_6	f_5	f_2	f_1
f_4	f_4	f_3	f_2	f_1	f_6	f_5
f_5	f_5	f_6	f_4	f_3	f_1	f_2
f_6	f_6	f_5	f_1	f_2	f_4	f_3

CHAPTER 2

(2) Since the composition of functions are associative the operation in S_3 is also associative.

(3) f_1 is the identity in S_3 for the operation *as* $f_1 \, o \, f = f \, o \, f_1 = f$ for any $f \in S_3$.

(4) Every element of S_3 is having inverse as it is shown that one to one function has its inverse function. In this case.

$f_1^{-1} = f_1,\ f_3^{-1} = f_6, f_2^{-1} = f_2, f_4^{-1} = f_4,$ and $f_5^{-1} = f_5$

(5) We compute $f_2 \, o \, f_3 = f_5$

and $f_3 \, o \, f_2 = f_4$

which shows $f_2 \, o \, f_3 \neq f_3 \, o \, f_2$.

Hence (S_3, o) is a non abelian group of order 6.

Example 2.1.5: To show that the power set $P(U)$ together with the operations of symmetric difference is a group.

Solution: We must show that the pair $P(U)$ together with the operations of symmetric difference is a group.

(1) Closure property. For any two sets A and B, $A \Delta B = (A - B) \, U \, (B - A) \in P(U)$.

Hence $P(U)$ is closed with respect to the operation of symmetric difference.

(2) Associative property. For any three sets A, B and C,

$$
\begin{aligned}
(A \Delta B) \Delta C &= [(A \cap B') \cup (B \cap A')] \Delta C \\
&= \{(A \cap B') \cup (B \cap A')] \cap C'\} \cup C \cap [(A \cap B') \cup (B \cap A')]' \\
&= (A \cap B' \cap C') \cup (B \cap A' \cap C') \cup (C \cap [(A' \cup B) \\
&\quad \cap (B' \cup A)] \\
&= (A \cap B' \cap C') \cup (B \cap A' \cap C') \cup (C \cap A' \cap B') \\
&\quad \cap (C \cap A \cap B) \qquad \ldots(1)
\end{aligned}
$$

which is the set of all the element which belong either to only one of the sets A, B, and C or to all of them.

$$\begin{aligned} A \,\Delta\, (B \,\Delta\, C) &= (B \,\Delta\, C) \,\Delta\, A \\ &= (B \cap C' \cap A') \cup (C \cap B' \cap A') \cup (A \cap B' \cap C') \\ &\quad \cap (A \cap B \cap C) \qquad ...(2) \end{aligned}$$

But (1) = (2) because operation of union is commutative.

(3) Existence of identity, the set P (U) has an identity element ϕ. For if $A \subseteq U$, then

$$A \,\Delta\, \phi = (A - \phi) \cup (\phi - A) = A \cup \phi = A$$

(4) Existence of inverse. Let $A \subseteq X$ and we observe that $A \,\Delta\, A = (A - A) \cup (A - A)$ $= \phi \cup \phi = \phi$.

This implies that each element $P(U)$ is its own inverse. Consequently, the mathematical system (P (U), Δ) is a commutative group.

PROBLEMS

1. Show that the system $(\{e\}, o)$ is a group, where o denotes the multiplication.
2. Prove that the set $R \times R$ together with the operation o defined by $(a, b)\ o\ (c, d)$ $= (a + c, b + d + 2bd)$ is a commutative semi-group with identity.
3. Show that (S, o), where $S = \{1, 2, 3, 4, 6\}$ and o binary operation is defined on S by.

   $aob = gcd\ (a, b)$, is a commutative semi-group.
4. *Is* (G, o) a commutative group if

   (*a*) (*i*) $G = R$, $a\ o\ b = a + b - ab$.

   (*ii*) $G = Z$, $a\ o\ b = a + b - 1$.

   (*b*) (*ii*) $G = Z$, $a\ o\ b = \min\ \{a, b\}$.

   (*ii*) $G = Z$, $a\ o\ b = a + b + 1$

   (*c*) $G = Z \times Z$, $(a, b)\ o\ (c, d) = (a + c, b + d)$.

   (*d*) $G = R \times R$, $(a, b)\ o\ (c, d) = (ac + bd, ad + bc)$.

   (*e*) $C = R \times R - \{o, o\}$, $(a, b)\ o\ (c, d) = (ac - bd, ad + bc)$.

   (*f*) $G = Q^+$, $a\ o\ b = \dfrac{a\,.\,b}{2}$

   (*g*) $G = K$ the set of all rational numbers of the form p/q, where q is odd integer and o denotes ordinary sum of two rational numbers.

   (*h*) $G = \{a \in R \mid -1 < a < 1\}$ $a\ o\ b = \dfrac{a + b}{1 + ab}$

   (*i*) $G = \{1, a, b, c\}$ and o binary operation is given by the following operation table

o	1	a	b	c
1	1	a	b	c
a	a	1	c	b
b	b	c	1	a
c	c	b	a	1

(This is known as K Lein's four group).

(j) $G = (a, b, c, d)$ and o binary operation is defined by the following table:

o	a	b	c	d
a	a	b	c	d
b	b	a	d	c
c	c	d	b	a
d	d	c	a	b

(This is known as K Lein's four group).

5. Show that the $(M_{m\times n}, o)$ where $M_{m\times n}$ is the set of all matrices of order $m \times n$ whose elements are integers (real or complex numbers) and o denotes the sum of two matrices of the same order, is a commutative group.

6. Show that (G, o), where

$$G = \left\{\begin{bmatrix}1 & 0\\0 & 1\end{bmatrix}, \begin{bmatrix}-1 & 0\\0 & 1\end{bmatrix}, \begin{bmatrix}1 & 0\\0 & -1\end{bmatrix}, \begin{bmatrix}-1 & 0\\0 & -1\end{bmatrix}\right\}$$

and o denotes matrix multiplications, is an abelian group.

7. Let $(S, *)$ be a semi-group without identity and e be any element not in S. Define the operation o on the set

$$S' = S \cup \{e\} \text{ by } a \ o \ b = a*b, \qquad \forall\ a, b \in S,$$
$$a \ o \ e = a = e \ o \ a, \qquad a \in S'.$$

Show that (S', o) is a semi-group with identity element e, (S', o) is said to be obtained by adjoining an identity.

8. Consider the set of matrices

$$G = \left\{A_\alpha = \begin{pmatrix}\cos\alpha - \sin\alpha\\ \sin\alpha \ \ \cos\alpha\end{pmatrix}, \alpha \in R\right\}$$

CHAPTER 2

Then prove that the set G under the operation of multiplication of two matrices is a group.

9. Show that the set G of all $n \times n$ non-singular matrices having their elements rational (Real, complex numbers) under the operation of multiplication of two matrices is a group.

2.2 ELEMENTARY PROPERTIES OF A GROUP

The definition of a group required that each group must contain an identity element e, but does not rule out the possibility that there may be two or more distinct identities. In the next theorem, we will show that this cannot happen. Similarly, axiom (4) requires that for each element in the group there is at least one inverse, but it does not 'specify that an element in a group never has more than one inverse. We give some theorems to clarify the situation.

Thorem 2.2.1:

(a) *The identity element of a group (G, o) is unique.*

(b) *Each element of a group (G, o) has a unique inverse.*

Proof: **(a)** Let a group (G, o) contain two identity elements e and e'. Then, since e is an identity element, $e \text{ o } e' = e'$. Similarly, e' is an identity element, $e \text{ o } e' = e$. Therefore, $e' = e \text{ o } e' = e$. That is, any two identity elements must be equal. Hence the identity element of a group is unique,

(b) Suppose that the element $a \in G$ has two inverses, a_1, and a_2.

According to the definition of inverse,

$$a \text{ o } a_1 = a_1 \text{ o } a = e, \text{ and}$$

$$a \text{ o } a_2 = a_2 \text{ o } a = e.$$

But the identity element is the same in both cases because identity element is unique in a group, so that

$$a \text{ o } a_1 = a \text{ o } a_2$$

Multiplying both sides of this equation on the left by a_1, (or by a_2) to get

$$a_1 \text{ o } (a \text{ o } a_1) = a_1 \text{ o } (a \text{ o } a_2)$$

$$\Rightarrow \quad (a_1 \text{ o } a) \text{ o } a_1 = (a_1 \text{ o } a) \text{ o } a_2, \text{ by associative law}$$

$$\Rightarrow \quad e \text{ o } a_1 = e \text{ o } a_2 \text{ since } a_1 \text{ o } a = e$$

$$\Rightarrow \quad a_1 = a_2 \text{ which means that } a \text{ has only one inverse.}$$

Corollary: *Each element of a monoid has at most one inverse.*

Theorem 2.2.2: *If $a \in G$ and (G, o) is a group, then $(a^{-1})^{-1} = a$*

Proof: Since a^{-1} is the inverse of a, *i.e.*, $a^{-1} \text{ o } a = a \text{ o } a^{-1} = e$. So a satisfies the definition of the inverse of (a^{-1}). Hence $(a^{-1})^{-1} = a$.

Theorem 2.2.3: If *(G, o) is a group and a, b* $\in$ *G, than* $(aob)^{-1} = b^{-1}\ o\ a^{-1}$.

Proof: According to the definition of inverse, we have to show that $(aob)\ o\ (b^{-1}\ oa^{-1}) = (b^{-1}\ o\ a^{-1})\ o\ (a\ o\ b) = e$,

where e is the identity of the group (G, o).

Now, we have

$$\begin{aligned} (a\ o\ b)\ o\ (b^{-1}\ o\ a^{-1}) &= a\ o\ ((b\ o\ b^{-1})\ o\ a^{-1}) \\ &= a\ o\ (e\ o\ a^{-1}) && \text{by definition of inverse} \\ &= a\ o\ a^{-1} && \text{by definition of } e \\ &= e && \text{by definition of inverse} \end{aligned}$$

Similarly,

$(b^{-1}\ o\ a^{-1})\ o\ (a\ o\ b) = b^{-1}\ o\ ((a^{-1}\ o\ a)\ ob) = b^{-1}\ o\ (e\ o\ b) = b^{-1}\ ob = e.$

This completes the proof of the theorem.

Theorem 2.2.4: *(Cancellation law)*

If a, b, c $\in$ G *and (G, o) is a group, then*

$$aoc = boc \Rightarrow a = b \text{ (Right cancellation law)}$$

$$coa = cob \Rightarrow a = b \text{ (Left cancellation law)}$$

Proof: Since $c \in G$, then $c^{-1} \in G$.

Multiplying the equation $aoc = boc$ on the right side by c^{-1}, we obtain

$$\begin{aligned} (a\ o\ c) = (b\ o\ c) \Rightarrow (a\ o\ c)\ oc^{-1} &= (b\ o\ c)\ o\ c^{-1} \\ \Rightarrow ao\ (c\ o\ c^{-1}) &= bo\ (c\ o\ c^{-1}) && \text{by associative law} \\ \Rightarrow a\ o\ e &= b\ o\ e && \text{by definition of inverse} \\ \Rightarrow a &= b && \text{by definition of } e \end{aligned}$$

Theorem 2.2.5: *In a group (G, o), the equations* $aox = b$ *and* $y\ o\ a = b$ *have unique solutions.*

Proof: Since $a \in G$, a^{-1} exists in G. Multiplying the equation $a\ o\ x = b$ on the left by a^{-1}, we obtain

$$\begin{aligned} & a^{-1}\ o\ (a\ o\ x) = a^{-1}\ ob \\ \Rightarrow\ & (a^{-1}\ o\ a)\ o\ x = a^{-1}\ ob \text{ by associative law} \\ \Rightarrow\ & e\ o\ x = a^{-1}\ ob \text{ by definition of inverse} \\ \Rightarrow\ & x = a^{-1}\ ob \text{ by definition of } e \end{aligned}$$

Since a^{-1}, $b \in G$, then $a^{-1}\ o\ b \in G$. Therefore $x \in G$.

Now we see that $a^{-1}\ o\ b$ is a solution of $a\ o\ x = b$. That is, $x = a^{-1}\ o\ b$ satisfies the group equation, since

$$a\ o\ (a^{-1}\ o\ b) = (a\ o\ a^{-1})\ o\ b = e\ o\ b = b$$

This shows that there is at least one solution. Now for uniquencess, suppose there exists another element $y \in G$ such that

$$a \; o \; y = b, \text{ then}$$

$$a \; o \; x = a \; o \; y \Rightarrow x = y \text{ by cancellation law}$$

Example 2.2.1: We see that in a semi-group the cancellation laws do not hold good. The semi-group (M_2, o), where M_2 is the set of all square matrices of order 2 and o denotes the multiplication of matrices, does not satisfy the cancellation laws.

Let

$$A = \begin{bmatrix} 1 & 0 \\ 0 & 0 \end{bmatrix}, B = \begin{bmatrix} 0 & 0 \\ 0 & 1 \end{bmatrix}, \text{and } C = \begin{bmatrix} 0 & 0 \\ 1 & 0 \end{bmatrix} \text{ belong to } M_2$$

Here we have

$$AB = AC \text{ but } B \neq C.$$

Example 2.2.2: We see that if in a semi group (G, o) the cancellation laws hold, (G, o) may not be a group. Let us consider the semi-group $(N, +)$ of natural numbers under addition. We have $a, b, c \in N$, $a + b = a + c \Rightarrow b = c$, but $(N, +)$ is not a group.

Theorem 2.2.6: *Let the both cancellation Laws hold in a finite semi-group (G, o), then (G, o) is a group.*

Proof: Let (G, o) be a finite semi-group, where

$$G = \{a_1, a_2, a_3, ..., a_n\}$$

in which both cancellation laws hold.

Let $a_r \in G$. Since (G, o) is a semi-group, the elements $a_1 \; o \; a_r, a_2 \; o \; a_r, \ldots, a_n \; o \; a_r$ are distinct elements of G. For any i and j with $i \neq j$ and $1 \leq i, j \leq n$.

$a_i \; o \; a_r = a_j \; o \; a_r \Rightarrow a_i = a_j$ by right cancellation law. Hence the set $\{a_1 \; o \; a_r, a_2 \; o \; a_r, \ldots, a_n \; o \; a_r]$ and the set $G = (a_1, a_2, \ldots a_n)$ are identical but the elements of the set $\{a, o \; a_1, a_2 \; o \; a_r, \ldots a_n \; o \; a_r\}$ may be in different order.

So for given $a_j \in G$, there exists $a_s, \in G$ such that

$$a_j = a_s \; o \; a_r$$

In particular for $a_r \in G$, there exists $a_k \in G$ such that

$$a_r = a_k \; o \; a_r$$

Then

$a_j \; o \; a_r = a_j \; o \; (a_k \; o \; a_r) = (a_j \; o \; a_k) \; o \; a_r \Rightarrow a_j = a_j \; o \; a_k, \forall j$ by right cancellation law.

Again we consider the elements

$$a_r \; o \; a_1, a_1 \; o \; a_2, \ldots, a_r \; o \; a_n.$$

Then for given $a_j \in G$, there exists $a_t \in \mathbf{G}$ such that

$$a_j = a_r \; o \; a_t.$$

In particular $a_r = a_r \ o\ a_s$ for some $a_s \in G$.

So

$$a_r \ o\ a_j = (a_r \ o\ a_s)\ o\ a_j = a_r \ o\ (a_s o a_j)$$

$$\Rightarrow \qquad a_j = a_s\ o a_j \text{ by left cancellation law.}$$

If $j = s$, then from $a_j = a_j\ o a_k$ we have

$$a_s = a_s \ o\ a_k$$

and if $j = k$, then from $a_j = a_s\ o a_j$ we have

$$a_k = a_s \ o\ a_k$$

Thus

$$a_s = a_s \ o\ a_k = a_k$$

So

$$a_j \ o\ a_k = a_j = a_k \ o\ a_j, \text{ that is, } a_k \text{ is an identity element in } G.$$

Let

$$a_k = e.$$

So for $e \in G$, there exists $a_n, a_m \in G$ such that

$$e = a_m \ o\ a_r \text{ and } e = a_r \ o\ a_n.$$

Then

$$a_n = e \ o\ a_n = (a_m \ o\ a_r)\ o\ a_n = a_m \ o\ (a_r\ o a_n)$$

$$= a_m o\ e = a_m$$

Hence

$$a_m \ o\ a_r = a_m \ o\ a_r = e \Rightarrow a_m = a_r^{-1} \in G.$$

This completes the proof of the theorem.

We also know by theorem 2.2.4 that if (G, o) is a group than both cancellation laws hold in G. Thus we have the following theorem.

Theorem 2.2.7: *A finite semi-group (G, o) is a group if and only if both cancellation laws hold in G.*

Theorem 2.2.8: *Let (G, o) be a semi-group, then the pair (G, o) is a group if and only if the equation aox = b and yoa = b have solution in G for given a, b ∈ G.*

Proof: Let (G, o) be a group, the equations $aox = b$ and $yoa = b$ have solutions in G by theorem 2.2.5.

Conversely, Let G be a non-empty set in which an associative binary operation o is defined and let the equations $aox = b$ and $yoa = b$ have solutions, then to prove (G, o) a group we have to show that *there exists* an identity element $e \in G$ and that for all $a \in G$, there exist $a^{-1} \in G$.

Existence of identity: Let $a \in G$. Since $aox = b$ has solution in G, $aox = a$ also has solution in G. Thus there exists an element $e \in G$ such that $aoe = a$. Now for $b \in G$, $b = yoa$, $y \in G$. Then

$$b \circ e = (y \circ a) \circ e = y \circ (a \circ e) = y \circ a = b$$

$$\Rightarrow \quad b \circ e = b, \text{ for all } b \in G. \quad ...(1)$$

Similarly the equation $yoa = a$ has solution for $a \in G$. Then there exists $e_i \in G$ such that $e_1 \circ a = a$. Now for $b \in G$, $b = aox$ when $x \in G$, then

$$e_1 \circ b = e_1 \circ (aox) = (e_1 oa)\ ox = aox = b$$

$$\Rightarrow \quad e_1 \circ b = b, \text{ for all } b \in G \quad ...(2)$$

putting $b = e_1$ in (1), we get $e_1 oe = e_1$ and $b = e$ in (2)

we get $e_1 \circ e = e$. So $e_1 = e_1 \circ e = e$.

Thus $\quad b \circ e = b = e \circ b$, for all $b \in G$. Hence e is the identity for G

Existence of inverse: Let $a \in G$. Since the equations $aox = e$ and $yoa = e$ have solutions in G, there exist $a', a'' \in G$ such that

$$a \circ a' = e \text{ and } a'' \circ a = e.$$

Then $\quad a' = e \circ a' = (a'' \circ a) \circ a' = a'' \circ (a \circ a') = a'' \circ e = a''$

Hence $\quad a \circ a' = e = a' \circ a$.

Thus $\quad a' = a^{-1} \in G$.

This proves the theorem.

On the basis of this theorem we can define the group in the following way.

Definition 2.2.1: A system (G, o), where G is a non-empty set and o is the binary operation defined on G, is called a group if and only if.

(*i*) $a \circ (b \circ c) = (a \circ b) \circ c$, for all $a, b, c, \in G$

(*ii*) The equation $a \circ x = b$ and $yoa = b$ have solutions in G.

Remark: Let (G, o) be a semi-group and let the equation $aox = b$ have solution in G for all $a, b \in G$. Then (G, o) may not be a group as is shown by an example.

Example 2.2.3: Let G be a non-empty set with at least two elements and let the binary operation o be defined by $aob = b$.

(*i*) Let $a, b, c \in G$, then $(a \circ b) \circ c = b \circ c = c$ and $ao\ (b \circ c) = a \circ c = c$. So $(a \circ b) \circ c = a \circ (b \circ c)$.

(*ii*) Let a and b be different elements of G. Since $a \circ b = b$, then $x = b$ is a solution of the equation $a \circ x = b$. So $aob = b$ which gives $a \circ b = b \circ b$ with $a \neq b$. So cancellation law does not hold. Here (G, o) is not a group.

Theorem 2.2.9: *If (G, o) is a group, then the unique solution of the group equation $xox = x$ is $x = e$.*

Proof: Since $x \in G$, then $x^{-1} \in G$

Multiplying the group equation $x \circ x = x$ by x^{-1} on the left hand side, we obtain

$$x^{-1} \circ (x \circ x) = (x^{-1} \circ x)$$

$$\Rightarrow \quad (x^{-1} \, ox) \, ox = x^{-1} \, ox$$

$$\Rightarrow \quad eox = e$$

$$\Rightarrow \quad x = e.$$

This completes the proof of the theorem.

Corollary: In an operation table for a group (G, o) each element appears exactly once in each row and in each column.

Proof: Let us suppose that the element b occured twice in the row headed by a. Then there exist two elements x_1 and x_2 with $x_1 \neq x_2$ such that

$$a \; o \; x_1 = b \text{ and } a \; o \; x_2 = b$$

$$\Rightarrow \quad a \; o \; x_1 = a \; o \; x_2$$

$$\Rightarrow \quad x_1 = x_2 \text{ by left cancellation law.}$$

which shows that there is a unique solution of the equation $aox = b$.

On parallel lines a proof for columns can be obtained.

Definition 2.2.2: An element $x \in G$ in a groupoid (G, o) is said to be idempotent if $xox = x$.

It shows that a group has one and only one idempoint element, namely, the group identity.

Definition 2.2.3: If there exists an element e_1 of the semi-group (G, o) such that $e_1 \, oa = a$ for every a of G, e_1 is called a left identity of (G, o).

Definition 2.2.4: If there exists an element e_2 of the semi-group (G, o) such that $a \; o \; e_2 = a$ for every a of G, e_2 is called a right identity of (G, o).

Exercise 2.2.1: If a semi-group (G, o) has both a left identity and a right identity then these two are equal. (This element which is both the left and right identity of (G, o) is called the identity element of the semi-group (G, o).

Definition 2.2.5: If for given $a \in G$, there exists an element $a^{-1} \in G$ of the semi-group (G, o) with an identity e such that $a^{-1} \; o \; a = e$, a^{-1} is called the left inverse of a.

Definition 2.2.6: If for given $a \in G$, there exists an element $a^{-1} \in G$ of the semi-group (G, o) with an identity e such that $a \; o \; a^{-1} = e$, a^{-1} is called the right inverse of a.

Exercise 2.2.2: If (G, o) is a semi-group with both a left inverse and a right inverse of an element a, then both these inverses are equal.

(This element which is both a left inverse and a right inverse of a is called the inverse element of a).

Theorem 2.2.10: *Let (G, o) be a semi-group which satisfies:*

(i) $e \in G$ such that $a \; o \; e = a$, for all $a \in G$,

(ii) for given $a \in G$, there exists $a' \in G$ such that $a \; o \; a' = e$.

Then (G, o) is a group.

CHAPTER 2

Proof: Let (G, o) be a semi-group satisfying the condition (i) and (ii). If $a \in G, \exists\, a'$ such that

$$a \; o \; a' = e \text{ by } (ii).$$

Again, for $a' \in G, \exists\, a'' \in G$. such that

$$a' \; o \; a'' = e \text{ by } (ii)$$

Then

$$\begin{aligned} a \; o \; a' &= (a' \; o \; a) \; o \; e = (a' \; o \; a) \; o \; (a' \; o \; a'') \\ &= (a' \; o \; (a \; o \; (a' \; o \; a''))) \\ &= (a' \; o \; (a \; o \; a') \; o \; a'') \\ &= a' \; o \; (e \; o \; a'') \\ &= (a' \; o \; e) \; o \; a'' \\ &= (a' \; o \; a'') \\ &= e \end{aligned}$$

Hence

$$a' \; o \; a = e = a \; o \; a'. \text{ Thus } a' = a^{-1}.$$

Further $\quad e \; o \; a = (a \; o \; a') \; o \; a = a \; o \; (a' \; o \; a) = a \; o \; e = a$

Hence $\quad e \; o \; a = a = a \; o \; e$ which shows that e is the identity in G.

Hence (G, o) is a group.

Remark: If (G, o) be a semi-group such that G contains left identity and every element of G may have right inverse, then (G, o) may not be a group. It is clear from example 2.2.3 where $eoa = a$, for all $a \in G$, that is, e is the left identity. For each $a \in G$, $a \; o \; e = e$, that is, e is the right inverse of a with respect to o, but (G, o) is not a group.

2.3 INTEGRAL EXPONENTS

For group $(G, .)$ written multiplicatively, we write the product aa as a^2, the product aaa as a^3, and so on, we write the inverse as a^{-1}, $a^{-1} a^{-1}$ as a^{-2}, and so on. Then it can be shown that usual laws of exponents hold, that is, if $a \in G$, a^n is defined for all integer and $a^n \in G$ since $(G, .)$ satisfies closure property under multiplication.

For groups $(G, +)$ written additively, we write $a + a$ as $2a$, the inverse of a as $-a$, and so on.

Definition 2.3.1: If a is an element of a group (G, o) then the integral power of a is defined by

(*a*) $a^0 = e$, where e is the identity of G.

(*b*) $a^k = a \; o \; a \; o \; a \; o \ldots \; o \; a$ (*k factors*),

(c) $a^1 = a$, and $a^{k+1} = a^k \; o \; a$ for every natural number k,

(d) $a^{-k} = (a^{-1})^k$ tor every natural number k.

Theorem 2.3.1: *Let (G, o) be a group, $a \in G$, and $m, n \in N$, then the powers of a satisfy the following laws of exponents:*

(a) $a^n \; o \; a^m = a^{n+m} = a^m o a^n$,

(b) $(a^n)^m = a^{nm} = (a^m)^n$,

(c) $a^{-n} = (a^n)^{-1}$,

(d) $e^n = e$.

The proofs of above results are left to the readers as exercises.

Example 2.3.1: Prove that if (G, o) is an abelian group, then for all $a, b \in G$ and integers n, $(aob)^n = a^n o b^n$.

Solution: We shall divide the proof of the statement in three parts according to $n > 0$, $n < 0$, and $n = 0$.

(i) $n > 0$: if $n = 1$, then $(a \; o \; b)^1 = a^1 \; o b^1$, which is abvious.

Let $n = k$, $(a \; o \; b)^k = a^k \; o \; b^k$ then

for $n = k + 1$, $(a \; o \; b)^{k+1} = (aob)^k \; o \; (aob)$

$$= (a^k \; o \; b^k) \; o \; (a \; o \; b)$$

$$= a^k \; o \; ((b^k \; o \; a) \; o \; b)$$

$$= a^k \; o \; ((a \; o \; b^k \; o \; b) \text{ since } G \text{ is abelain}$$

$$= (a^k \; o \; a) \; o \; (b^k \; o \; b)$$

$$= a^{k+1} \; o \; b^{k+1}.$$

The generalization for all $n > 0$ follows by induction,

(ii) $n < 0$, since $-$ n > 0. If follows from (1) that

$(aob)^{-n} = a^{-n} \; o \; b^{-n}$, and we also observe that

$$(a \; o \; b)^n \; o \; (a \; o \; b)^{-n} = (a^n \; o \; b^n) \; o \; (a^{-n} \; o \; b^{-n})$$

$$= (a^n \; o \; ((b^{-n} \; o \; a^{-n}) \; o \; b^{-n})$$

$$= (a^n \; o \; ((a^{-n} \; o \; b^n) \; o \; b^{-n})$$

$$= (a^n \; o \; a^{-n}) \; o \; (b^n \; o \; b^{-n})$$

$$= e \; o \; e = e$$

$$(a \; o \; b)^n \; o \; (a \; o \; b)^{-n} = e$$

Thus $(a \; o \; b)^n \; o \; (a \; o \; b)^{-n} = (a^n \; o \; b^n) \; o \; (a \; o \; b)^{-n}$

or $(a \; o \; b)^n = a^n \; o \; b^n$ by right cancellation law.

(iii) $n = 0$: $(a \; o \; b)^0 = e = ee = a^0 \; o \; b^0$

which completes the proof.

2.4 ORDER OF AN ELEMENT OF A GROUP

Definition 2.4.1: If a is an element of the group (G, o), the order of a is the least positive integer n, provided it exists, such that $a^n = e$. If no such integer exists; a is of infinite order. The order of a is denoted by $o(a)$.

According to the group operation a^n means $a \ o \ a \ o \ a \ o....o \ a$ (n factors). If addition operation is there, then $a + a + a \ ... + a$ (n times) $= na$.

Example 2.4.1: Find the order of the elements of the group $(z_4, +4)$.

Solution: Hence $Z_4 = \{[0], [1], [2], [3]\}$ and [0] is the identity in Z_4. Here [0] has order one.

$[1] +_4 [1] +_4 [1] +_4 [1] = [4] = [0]$. Thus adding [1] 4 times gives [0], the identity of Z_4.

$$[1] \neq [4]$$

But
$$[1] +_4 [1] = [2] \neq [4]$$
$$[1] +_4 [1] +_4 [1] = [3] \neq [4]$$

Hence [1] has order 4.

Similarly, $[2] +_4 [2] = [4] = [0]$, therefore [2] has order 2, and $[3] +_4 [3] +_4 [3] +_4 [3] = [12] = [4] = [0]$

Thus [3] has order 4.

Example 2.4.2: Find the order of the elements of the group $(\{1, w, w^2\},)$ where w is a cube root or 1.

Solution: Since $I^1 = 1$, 1 has order one.

$$w^1 \neq 1, \ w^1 w^1 = w^2 \neq 1, \ w^3 = 1, \ w \text{ has order } 3,$$
$$(w^2) \neq 1, \ (w^2)^3 = 1, \ w^2 \text{ has order } 3.$$

Example 2.4.3: In the group $(Z, +)$ only *0* has order and no element has finite order. Thus every element is of infinite order.

Theory 2.4.1: *The order of every element of a finite group (G, o) is finite.*

Proof: Let (G, o) be a finite group of order n. Let $a \in G$ consider powers of a, $a \ o \ a = a^2, a^3, ..., a^n$.

These are $n + 1$ elements. Therefore all elements cannot be distinct. Hence there are two elements a^r and a^s such that $a^r = a^s$. Since $a^r \in G$, then $a^{-r} \in G$.

$1 \leq r < s \leq n$ and by multiplying the equation $a^r = a^s$ *by* a^{-r} we get $a^r \ o \ a^{-r} = a^s \ o \ a^{-r}$

$$\Rightarrow \quad a^o = a^{s-r}$$

$$\Rightarrow \quad e = a^{s-r} \text{ Since } r \neq s$$

$$\Rightarrow \quad a^k = e, \ o < s - r < n \text{ and let } s - r = k$$

Hence $o(a)$ is finite if $o(G)$ is finite. The proof shows that the order of every element is less than or equal to the order of the finite group.

Theorem 2.4.1: *If $a \in (G, o)$, $o(a) = n$ and m is a non-negative integer, then $a^m = e$ if and only if m is a multiple of n, i.e., $m = nq$, $q \in N$.*

Proof: Since $o(a) = n$, there cannot exist another positive integer $r < n$ such that $a^r = e$. Thus if $a^m = e$, then $m \geq n$. Let us assume $a^m = e$.

If $m = n$, $a^m = a^n = e$

Now let $m > n$, there exist $q, r \in N$ such that

$$m = qn + r, \; o \leq r < n, \; 1 \leq q$$

Therefore $\quad e = a^m = a^{qn+r}$ since $a^m = e$ by assumption,

$$e = (a^n)^q \, oa^r$$

$\Rightarrow \quad e = a^r, \; o \leq r < n.$

It contradicts that n is the order of a unless $r = 0$

Hence we get $m = nq$, i.e., m is multiple of n.

Again, conversely, if $m = nq$, then

$$a^m = a^{nq}$$

$\Rightarrow \quad a^m = (a^n)q$

$\Rightarrow \quad a^m = e$ since $a^n = e$

$\Rightarrow \quad a^m = e$

The completes the proof of the theorem.

Theorem 2.4.2: $o(a^p) \leq o(a)$, *for all $a \in G$ and for all $p \in Z$, that is, the order of any power of any element of a group is less than equal to the order of the element.*

Proof: Let $o(a) = m$ and $o(a^p) = n$, $p \in Z$

Since $\quad o(a) = m$, we have

$a^m = e$, $(a^p)^m = a^{pm} = a^{mp} = (a^m)^p = e$ which shows

$$o(a^p) \leq m = o(a).$$

This completes the proof of the theorem.

Theorem 2.4.3: $o(a) = o(b^{-1} \, o \, a \, o \, b)$, *where $a, b \in G$*, and (G, o) is a group.

Proof: Let $\quad c = b^{-1} \, oaob$ and $a^n = e$.

If $\quad n = 2.$

$$\begin{aligned} & (b^{-1} \, o \, a \, o \, b)^2 \\ &= (b^{-1} \, o \, a \, o \, b) \, o \, (b^{-1} \, o \, a \, o \, b) \\ &= (b^{-1} \, o \, a \, o \, (bob^{-1}) \, o \, a \, o \, b) \\ &= b^{-1} \, o(a \, o \, e) \, o \, a \, o \, b) \\ &= b^{-1} \, o \, a^2 \, ob \end{aligned}$$

CHAPTER 2

In general we can prove by induction that

$$(b^{-1} \ o\ a\ o\ b)^k = b^{-1}\ o\ a^k\ o\ b.$$

$$(b^{-1}\ o\ a\ o\ b)^n = b^{-1}\ o\ a^n\ o\ b = b^{-1}\ o\ e\ o\ b, \text{ since } a^n = e.$$

$$= b^{-1}\ o\ b = e.$$

Hence the order of (b^{-1} *oaob*) cannot be greater than that of a, *i.e.*, $o(b^{-1}\ oaob) \leq o\ (a)$

Conversely, if $(b^{-1}\ o\ a\ o\ b)^n = e$, *the* $b^{-1}\ o\ a^n\ o\ b = e$

Multiplying $b^{-1}\ o\ a^n\ o\ b = e$ on the left by b we obtain

$$b\ o\ (b^{-1}\ o\ a^n\ o\ b) = boe$$

$$\Rightarrow \quad (b\ o\ b^{-1})\ o\ (a^n\ o\ b) = b$$

$$\Rightarrow \quad a^n\ o\ b = b$$

$$\Rightarrow \quad (a^n\ o\ b)\ o\ b^{-1} = b\ o\ b^{-1}$$

$$\Rightarrow \quad a^n\ o\ (b\ o\ b^{-1}) = b\ o\ b^{-1}$$

$$\Rightarrow \quad a^n = e$$

Hence $o\ (a) \leq (b^{-1}\ oaob)$

Thus $o(a) = o\ (b^{-1}\ oaob)$. Hence the theorem.

Theorem 2.4.4: *The order of a^{-1} is the same as that of $a \in$ G, i.e., $o(a^{-1}) = o(a)$.*

Proof: *Let $a \in G$ and let $o\ (a) = n$ and $o\ (a^{-1}) = m$*

We have

$$(a^{-1})^n = a^{-n} = (a^n)^{-1} = e^{-1} = e$$

Hence $o\ (a^{-1}) \leq o\ (a) \Rightarrow m \leq n$.

Similarly, $o\ ((a^{-1})^{-1}) \leq o\ (a^{-1})$

$\Rightarrow o(a) \leq o\ (a^{-1}) \Rightarrow n \leq m$

Hence from (1) and (2) $m = n$

Hence $o\ (a) = o\ (a^{-1})$.

Exercise 2.4.4: If in a group (G, o), $x\ o\ y^2 = y^3\ o\ x$ and $y\ o\ x^2 = x^3\ o\ y$,

Then show that $x = y = e$, where e is the identity in G.

Solution: We have

$$x\ o\ y^2 = y^3\ o\ x$$

$$\Rightarrow \quad x = y^3\ o\ x\ o\ y^{-2}$$

$$\Rightarrow \quad xox = x\ o\ y^3\ o\ x\ o\ y^{-2}$$

$$\Rightarrow \quad x^2 = x\ o\ y^2\ o\ y\ o\ x\ o\ y^{-2}$$

Since $\quad y^3\ o\ x = x\ o\ y^2$

$$\Rightarrow \quad x^2 = y^3\ o\ x\ o\ y\ o\ x\ o\ y^{-2}$$

$\Rightarrow \quad x^2 \text{ o } y = y^3 \text{ o } x \text{ o } y \text{ o } x \text{ o } y^{-1}$ (1)

Now $\quad y \text{ o } x^2 = x^3 \text{ o } y$

$\Rightarrow \quad y \text{ o } x^2 = x \text{ o } (x^2 \text{ o } y)$

$\Rightarrow \quad y \text{ o } x^2 = x \text{ o } (y^3 \text{ o } x \text{ o } y \text{ o } x \text{ o } y^{-1})$ by (1)

$\Rightarrow \quad x^2 = y^{-1} \text{ o } x \text{ o } y^3 \text{ o } x \text{ o } y \text{ o } x \text{ o } y^{-1}$

$\Rightarrow \quad x^2 \text{ o } y = y^{-1} \text{ o } x \text{ o } y^3 \text{ o } x \text{ o } y \text{ o } x$ (2)

From (1) and (2), we get

$$y^3 \text{ o } x \text{ o } y \text{ o } x \text{ o } y^{-1} = y^{-1} \text{ o } x \text{ o } y^3 \text{ o } x \text{ o } y \text{ o } x$$

$\Rightarrow \quad y^4 \text{ o } x \text{ o } y \text{ o } x = x \text{ o } y^3 \text{ o } x \text{ o } y \text{ o } x \text{ o } y$

$\Rightarrow \quad y^4 \text{ o } x \text{ o } y \text{ o } x = x \text{ o } y^2 \text{ o } y \text{ o } x \text{ o } y \text{ o } x \text{ o } y$

$\Rightarrow \quad y^4 \text{ o } x \text{ o } y \text{ o } x = y^3 \text{ o } x \text{ o } y \text{ o } x \text{ o } y \text{ o } x \text{ o } y$ Since $x \text{ o } y^2 = y^3 \text{ o } x$

$\Rightarrow \quad (y \text{ o } x) \text{ o } (y \text{ o } x) = (x \text{ o } y) \text{ o } (x \text{ o } y) \text{ o } (x \text{ o } y)$

$\Rightarrow \quad (y \text{ o } x)^2 = (x \text{ o } y)^3$ (3)

Interchange x and y in (3), we get

$$(x \text{ o } y)^2 = (y \text{ o } x)^3 \quad (4)$$

Now (3) and (4) imply

$$(x \text{ o } y)^2 = (y \text{ o } x)^3 = (y \text{ o } x)^2 \,.\, (y \text{ o } x) = (x \text{ o } y)^3 \text{ o } (y \text{ o } x)$$

$\Rightarrow \quad e = x \text{ o } y \text{ o } y \text{ o } x = x \text{ o } y^2 \text{ o } x$

$\Rightarrow \quad x^{-2} = y^2$

Further $\quad x \text{ o } y^2 = y^3 \text{ o } x \Rightarrow x \text{ o } x^{-2} = y \text{ o } x^{-2} \text{ o } x$

$\Rightarrow x^{-1} = y \text{ o } x^{-1}$

$\Rightarrow y = e$

Finally, $\quad y \text{ o } x^2 = x^3 \text{ o } y \Rightarrow e \text{ o } x^2 = x^3 \text{ o } e$

$\Rightarrow x = e.$

Hence $x = y = e$.

Exercise 2.4.5: Let (G, o) be a group and Let m and n be two co-prime integers such that $a^m\, ob^m = b^m\, oa^m$ and $a^n\, ob^n = b^n\, oa^n$ for all $a, b \in G$, then show that (G, o) in a abelian.

Solution: Since $(m, n) = 1$, there exist two integers x and y such that $mx + ny = 1$.

Now
$$(a^m \text{ o } b^n)^{mx} = a^m \text{ o } (b^n \text{ o } a^m)^{mx-1} \, ob^n$$
$$= a^m \text{ o } (b^n \text{ o } a^m)^{mx} \, o(b^n \text{ o } a^m)^{-1} \text{ o } b^n$$
$$= (b^n \, oa^m)^{mx} \text{ o } a^m \text{ o } a^{-m} \text{ o } b^{-n} \text{ o } b^n$$
$$= (b^n \text{ o } a^m)^{mx} \quad (1)$$

CHAPTER 2

Similarly, we can show that

$$(a^m \text{ o } b^n)^{ny} = (b^n \text{ o } a^m)^{ny} \qquad (2)$$

From (1) and (2), we get

$$a^m \text{ o } b^n = (a^m \text{ o } b^n)^{mx + ny} \qquad [\text{since } 1 = mx + ny]$$

$$= (b^n \text{ o } a^m)^{mx+ny}$$

$$= b^n \text{ o } a^m \qquad (3)$$

Now $\quad a \text{ o } b = a^{mx+ny} \text{ o } b^{mx+ny}$

$$= a^{mx} \text{ o}(a^{ny} \text{ o } b^{mx}) \text{ o } b^{ny}$$

$$= a^{mx} \text{ o } b^{mx} \text{ o } a^{ny} \text{ o } b^{ny} \quad \text{by (3)}$$

$$= b^{mx} \text{ o } a^{mx} \text{ o } b^{ny} \text{ o } a^{ny} \quad \text{by hypothesis}$$

$$= b^{mx+ny} \text{ o } a^{mx+ny} \quad \text{by (3)}$$

Since $mx + ny = 1$ $\quad = b \text{ o } a$

Hence (G, o) is abelian.

Example 2.4.6: (S, o) is a finite semi-group. Show that there exists an element $a \in S$ such that $a^2 = e$.

Solution: Let $x \in S$. Since S in finite, then $x, x^2, x^3,...$ all connot be distinct. So for some integer m and n $x^m = x^n$, where $m > n$. Let $m - n = k$. then

$$x^m = x^n$$

$\Rightarrow$ $\quad x^{n+k} = x^n$

Now $\quad x^{2n+k} = x^n \text{ o } x^{n+k} = x^n \text{ o } x^n = x^{2n}$

By principal of induction it can be proved that

$$x^{mn+k} = x^{mn} \textit{ for all } m \in N.$$

Also $\quad x^{mn+2k} = x^{mn+k} \text{ o } x^k$

$$= x^{mn} \text{ o } x^k = x^{mn+k} = x^{mn}.$$

Similarly, $\quad x^{mn+3k} = x^{mn+2k} \text{ o } x^k$

$$= x^{mn} \text{ o} x^k = x^{mn+k} = x^{mn} \text{ and so on.}$$

Again by principal of Induction we can prove that

$$x^{mn+pk} = x^{mn} \textit{ for any } p \in N.$$

In particular $\quad x^{nk+nk} = x^{nk}$

$\Rightarrow$ $\quad x^{2nk} = x^{nk}$

Put $\quad x^{nk} = a$, Then $a^2 = e$.

Exercise 2.4.7: Show that the equation $x^2 \text{ o } a \text{ o } x = a^{-1}$ is solvable for x in a group (G, o) if and only if a is the cube of some element in G.

Solution: Suppose the equation $x^2 \text{ o } a \text{ o } x = a^{-1}$ has solution for x in G, there exists an element $c \in G$ such that

$$c^2 \text{ o } a \text{ o } c = a^{-1}$$
$$\Rightarrow \quad c \text{ o } a \text{ o } c \text{ o } a = c^{-1}$$
$$\Rightarrow \quad (c \text{ o } a) \text{ o } (c \text{ o } a) \text{ o} c = e$$
$$\Rightarrow \quad (c \text{ o } a) \text{ o } (c \text{ o } a) \text{ o } (c \text{ o } a) = e \text{ o } a$$
$$\Rightarrow \quad (c \text{ o } a)^3 = a.$$

Conversely, let $a = b^3$ for some $b \in G$. Then $x = b^{-2}$ is a solution of $x^2 \text{ o } a \text{ o } x = a^{-1}$,

Since $\quad x^2 \text{ o } a \text{ o } x = b^{-4} \text{o } b^3 \text{ o } b^{-2} = b^{-3} = a^{-1}$, since $a^{-1} = b^{-3}$

CHAPTER 2

Exercise 2.4.8: Let (G, o) be a group and let $a \in G$. Define a binary operation $*$ on G such that $(G, *)$ is a group with a as its identity.

Solution: We define binary operation $*$ on G as follows for all $x, y \in G$.

$$x * y = (x \text{ o } a^{-1}) \text{ o } y.$$

We see that

(1) The operation $*$ is associative. For $x, y, z \in G$.

$$(x * y) * z = ((x \text{ o } a^{-1}) \text{ o } y) * z$$
$$= x \text{ o } a^{-1} \text{ o } y \text{ o } a^{-1} \text{ o } z \quad (1)$$
$$x * (y * z) = x * (yoa^{-1}oz)$$
$$= x \text{ o } a^{-1} \text{ o } y \text{ o } a^{-1} \text{ o } z \quad (2)$$

From (1) and (2), we have

$(x * y) * z = x * (y * z)$

(2) **Existence of Identity:** For any $x \in G$,

$x * a = x \text{ o } a^{-1} \text{ o } a = xoe = x$, where e in the identity in (G, o)

Similarly, $\quad a * x = a \text{ o } a^{-1} \text{ o } x = e \text{ o } x = x.$

This shows that a is the identity in $(G, *)$.

(3) **Existence of Inverse:** Let $x \in G$, then

$$x * (a \text{ o } x^{-1} \text{ o } a) = x \text{ o } a^{-1} \text{o } a \text{ o } x^{-1} \text{ o } a = x \text{ o } (a^{-1} \text{ o } a) \text{ o } x^{-1} \text{ o } a$$
$$= x \text{ o } e \text{ o } x^{-1} \text{ o } a$$
$$= e \text{ o } a = a.$$

and $\quad (a \text{ o } x^{-1} \text{ o } a) * x = a \text{ o } x^{-1} \text{ o } a \text{ o } a^{-1} \text{ o } x$

$$= a \text{ o } x^{-1} \text{ o } e \text{ o } x$$
$$= a \text{ o } e = a$$

Thus $\quad x^{-1} = a \text{ o } x^{-1} \text{ o } a.$

Hence $(G, *)$ is a group with a as its identity.

Definition 2.4.2: A semi-group (S, o) is called regular if for every $y \in S$ there exists $a \in S$ such that $y \text{ o } a \text{ o } y = y$.

Theorem 2.4.5: *Let (S, o) be a semi-group with at least three elements and $x \in S$ is such that $(S - \{x\}, o)$ is a group, then S is regular iff $x^2 = x$.*

Proof: Let $x^2 = x$. Then $x^3 = x^2 \, o \, x = x \, o \, x = x^2 = x$ *i.e.* $x \, o \, x \, o \, x = x$

Since $(S - \{x\}, o)$ is *a* group, for $y \in S - \{x\}$

$$y \, o \, y^{-1} \, o \, y = e \, o \, y = y, \text{ where } e \text{ in the identity of } S - \{x\}.$$

Hence, as $S = S - \{x\} \cup \{x\}$, S is regular.

Conversely, Let S be regular, we claim that for any $y \in S - \{x\}$, $y \, o \, x = x = x \, o \, y$.

Suppose $xoy \neq x$, then $xoy \in S - \{x\}$ and

$$(xoy)oy^{-1} = x \in S - (x) \text{ which is absurd.}$$

Hence $xoy = x$. Similarly, $y \, o \, x = x$.

Since S is regular, there exists an element $a \in S$ such that $xoaox = x$.

If $a \in S - \{x\}$, $ao \, x = x \Rightarrow x \, o \, x = x \Rightarrow x^2 = x$.

If $a \in S - \{x\}$, $a = x$, then $x \, o \, a \, o \, x = x \Rightarrow x^3 = x$.

Let $x^2 \neq x$, then $u = x^2 \in S - \{x\}$. Since $S - \{x\}$ has at least three elements, there is $v(\neq u)$ in $S - \{x\}$.

Now $zou = v$ is solvable in $S - \{x\}$. Because

$$z \in S - \{x\}, \; z \, o \, u = z \, o \, x^2 = (z \, o \, x) \, o \, x = x \, o \, x = x^2 = u.$$

i.e., $u = v$, which is absurd. Hence $x^2 = x$.

Theorem 2.4.6: *If a and b are non-commutative elements of a group (G, o), that is, $aob \neq boa$, then the elements of the set $\{e, a, b, a \, o \, b, b \, o \, a\}$ are all distinct.*

Proof: We shall examine the members of the set $\{e, a, b, aob, boa\}$ two at a time and show each of the ten possible equalities leads to a contradiction of the hypothesis $aob \neq boa$.

We observe the cases as follows:

(1) $a = e \Rightarrow a \, o \, b = e \, o \, b = b \Rightarrow a \, o \, b = b \, o \, e = b \, o \, a$

(2) $b = e \Rightarrow a \, o \, b = a \, o \, e = a \Rightarrow a \, o \, b = e \, o \, a = b \, o \, a$

(3) $a \, o \, b = e$, $a \, o \, e = e \, o \, a$

$$\Rightarrow \quad a \, o \, (a \, o \, b) = (a \, o \, b) \, o \, a$$

$$\Rightarrow \quad a \, o \, (a \, o \, b) = a \, o \, (b \, o \, a)$$

$$\Rightarrow \quad a \, o \, b = b \, o \, a.$$

(4) $b \, o \, a = e$, $e \, o \, a = a \, o \, e$

$$\Rightarrow \quad (b \, o \, a) \, o \, a = a \, o \, (b \, o \, a)$$

$$\Rightarrow \quad (b \, o \, a) \, o \, a = (a \, o \, b) \, o \, a$$

$$\Rightarrow \quad b \, o \, a = a \, o \, b.$$

(5) $a = b \Rightarrow a \text{ o } b = a \text{ o } a = b \text{ o } a$

(6) $a = a \text{ o } b \Rightarrow e = b$ reducing to case (2)

(7) $a = b \text{ o } a \Rightarrow e = b$ reducing to case (2)

(8) $b = a \text{ o } b \Rightarrow e = a$ reducing to case (1)

(9) $b = b \text{ o } a \Rightarrow e = a$ reducing to case (1)

(10) $a \text{ o } b = b \text{ o } a$ contradicts the hypothesis.

Theorem 2.4.7: *Any non-commutative group has at least six elements.*

Proof: If (G, o) is a non-commutative group, it must have a pair of non-commuting elements a and b. According to the previous, the set $\{e, a, b, a \text{ o } b, b \text{ o } a\}$ consists of distinct elements. We now establish that one of the group element aoa or $aoboa$ is distinct from these five; however it is not possible to specify abstractly whether it is aoa or $aoboa$.

Now we shall see that $a \text{ o } a$ is different from each element of $\{a, b, aob, boa\}$. We observe that:

(1) $a \text{ o } a = a \Rightarrow a = e$ reducing to case (1) of theorem 2.4.6.

(2) $a \text{ o } a = b \Rightarrow a \text{ o } b = ao\ (a \text{ o } a) = (a \text{ o } a) \text{ o } a = b \text{ o } a$

(3) $a \text{ o } a = a \text{ o } b \Rightarrow a = b$ reducing to case (5) of theorem 2.4.6.

(4) $a \text{ o } a = b \text{ o } a \Rightarrow a = b$ reducing to case (5) of theorem 2.4.6.

Thus either $aoa \neq e$, in which case $a \text{ o } a$ is the sixth distinct element of a or else $a \text{ o } a = e$.

Again we can show that $aoboa$ is distinct from each of e, a, b, aob, boa and consequently to be the sixth element. It is obvious that

$ao\ (a \text{ o } b \text{ o } a) = (a \text{ o } a) \text{ o } (b \text{ o } a) = e \text{ o } (b \text{ o } a) = b \text{ o } a$ as $a \text{ o } a = e$

(5) $a \text{ o } b \text{ o } a = e$, then $b \text{ o } a = a \text{ o } (a \text{ o } b \text{ o } a)$

$= b \text{ o } a = a \text{ o } e = a$ reducing to case (7) of theorem 2.4.6.

(6) $a \text{ o } b \text{ o } a = a \Rightarrow aob = e$, reducing to case (3) of theorem 2.4.6

(7) $a \text{ o } b \text{ o } a = b \Rightarrow a \text{ o } b = a \text{ o } (a \text{ o } b \text{ o } a) = b \text{ o } a$ if $a \text{ o } a = e$

(8) $a \text{ o } b \text{ o } a = a \text{ o } b \Rightarrow a = e$ reducing to case (1) of theorem 2.4.6.

(9) $a \text{ o } b \text{ o } a = b \text{ o } a \Rightarrow a = e$ reducing to case (1) of theorem 2.4.6.

Hence $aoboa$ is the sixth distinct element of $a \text{ o } a = e$.

PROBLEMS

1. Let a, b, c and d be the elements of the semi-group (G, o), prove that $((a \text{ o } b) \text{ o } c) \text{ o } d = a \text{ o } (b \text{ o } (c \text{ o } d))$.
2. If $(S, *)$ is a group with identity and G the set of all elements of S having inverses with respect to the operation*, show that $(G, *)$ is a group.

3. Prove that if $(G, *)$ is a group having more than two elements, then there exist $a, b \in G$, with $a \neq b$, $a \neq e$, $b \neq e$, such that $a * b = b * a$.

4. Let $(G, *)$ be a semi-group which satisfies in addition:

   (a) $\exists\ e \in G$ such that $e * a = a$, for all $a \in G$,

   (b) for given $a \in G$, $\exists\ b \in G$ such that

   $b * a = e.$

   Prove that $(G, *)$ must be group.

5. Let S be a set containing more than two elements and let binary operation $*$ be defined on S by

   $x * y = y$, for all $x, y \in S$. Prove that there exist several left identities, that is, $e * a = a$, for all $a \in S$, and every elements of S has a right inverse, that is, $a * a^r = e$, for all $a \in S$.

6. Prove that a finite semi-group $(G, *)$ with identity e is group if and only if G consists only one idempotent.

7. Prove that a semi-group $(G, *)$ with identity e is a group if for given $a \in G$, $\exists$ $a' \in G$ such that either $a * a' = e$ or $a' * a = e$.

8. If $(G, *)$ is a group in which $(a * b)^i = a^i * b^i$ for three consecutive integers i and for all $a, b \in G$. Show that $(G, *)$ is abelian. Show by an example that this result does not hold for semi-groups.

9. Show that the conclusion of problem 8 does not follow if we assume the relation

   $(a * b)^i = a^i * b^i$ for just two consecutive integers.

   Hint: Let $G = \{\pm 1, \pm i, \pm j, \pm k\}$ and binary operation $*$ be defined by

   $i^2 = j^2 = k^2 = -1$

   and $i * j = -j * i = k$,

   $j * k = -k * j = i$,

   and $k * i = -i * k = j$,

   Then $(G, *)$ is called Quaternion Group. Take $n = 4, 5$.

10. Let $(G, *)$ be a semi-group such that

    $x * y = y * z \Rightarrow x = z$, for all $x, y, z \in G$. then prove that $(G, *)$ is abelian.

11. In any group, prove that identity element is the only element whose order is 1.

12. If (G, o) is group of even order, prove that it has an element $a \neq e$, such that $a^2 = e$.

13. If every element of a group is its own inverse then show that the group is commutative.

14. Prove that a group is commulative if and only if

    $(a * b)^{-1} = a^{-1} * b^{-1}$

15. Show that the group $(G, *)$ is commulative if $(a * b)^2 = a^2 * b^2$ for all $a, b \in G$.

16. Show that the group is abelian if

(*i*) it has three elements

(*ii*) it has four elements

(*iii*) it has five elements

17. Prove that $o(a * b) = o(b * a)$ for any two elements $a, b \in G$, where $(G, *)$ is a group.

18. Prove that the order of any integral power of an element a of group G is less than or equal to the order of the element a, *i.e.*, $o(a^k) \leq o(a)$, for all $a \in G$ and $k \in N$.

19. If a is an element of a group (G, o) of order n, then $o(a^p) = n$, where p is a positive integer prime to n.

20. If $BA = A^m B^n$, prove that the elements

$A^m B^{n-2}$, $A^{m-2} B^n$, and AB^{-1} *have the same order.*

Hint: $A^m B^{-2} = BAB^{-2}$, So $A^m B^{n-2}$ has the same order as

$B^{-1}(BAB^{-2})B = AB^{-1}$

$A^{m-2} B^n = A^{-2}AB$, So $A^{m-2} B^n$ has the same order as

$A(A^{-2}BA)A^{-1} = A^{-1}B$ which is equal to the order of BA^{-1}.

21. Find the order of [3] in the multiplication group of integers modulo 5.

22. If the elements a, b and aob of a finite group (G, o) are each of order 2. Prove that $a\ o\ b = b\ o\ a$.

CHAPTER 2

2.5 GROUP OF INTEGERS OF MODULO

Definition 2.5.1: A binary operation $+_n$ may be defined on Z_n as follows: for each $[a], [b] \in Z_n$, let $[a] +_n [b] = [a + b]$. Now we see that the sum of two congruence classes is defined, in terms of representatives from these Classes. But we must show that the operation $+_n$ does not depend upon the two representatives chosen. That is, we must show that if $[a'] = [a]$ and $[b'] = [b]$, then $[a' + b'] = [a + b]$. Now $a' \in [a'] = [a]$ and $b' \in [b'] = [b]$ which implies

$$a' \equiv a \pmod n \text{ and } b' \equiv b \pmod n$$

$$a' + b' \equiv a + b \pmod n$$

$$\Rightarrow \quad a' + b' \in [a + b].$$

It follows that $[a' + b'] = [a + b]$

Example 2.5.1: We see below the case of modulo addition.

$$[3] +_7 [6] = [3 + 6] = [9] = [2]$$

$$[5] +_7 [2] = [15 + 2] = [7] = [0]$$

Theorem 2.5.1: *For each positive integer n, the mathematical system $(Z_n, +_n)$ forms a commutative group (known as the group of integers modulo n or the group of residue classes modulo n).*

Proof: Since Z_n is the set of residue classes mod n,

$Z_n = \{[0], [1], [2], \ldots [n - 1]\}$

Let $(Z_n, +_n)$. be the system. Now we have to check the group axioms on the system $(Z_n, +_n)$.

(1) Closure property, let $[a]$, $[b] \in Z_n$, then

$$[a] +_n [b] = [a + b] \in Z_n \text{ if } a + b < n.$$
$$= [r] \in Z_n, \text{ if } a + b \geq n$$

where r is the residue of $a + b$ modulo n.

(2) Associative law. If $[a]$, $[b]$, $[c] \in Z_n$, then

$$[a] +_n ([b] +_n [c]) = [a] +_n [b + c]$$
$$= [a + b] +_n [c] = [a + (b + c)]$$
$$= ([a] +_n [b]) +_n [c]$$

Hence $+_n$ is associative.

(3) Commutative law. If $[a]$, $[b] \in Z_n$, then

$$[a] +_n [b] = [a + b]$$
$$= [b + a]$$
$$= [b] +_n [a]$$

Thus $+_n$ is commutative.

(4) Existence of Identity. It is obvious that the system $(Z_n, +_n)$ has an identity [0], that is,

$$[a] +_n [0] = [a + 0] = [0 + a] = [a]$$

(5) Existence of Inverse, if $[a] \in Z_n$, then

$[n - a\} \in Z_n$ and

$$[a] +_n [n - a] = [a + (n - a)] = [n] = [0]$$

So that $-[a] = [n - a]$

Hence $(Z_n +_n)$ is a commutative group.

If we adopt the convention of designating each congruence class by its smallest non-negative representative then the operation table for, say $(Z_4 +_4)$, can be written as:

$+_4$	[0]	[1]	[2]	[3]
[0]	[0]	[1]	[2]	[3]
[1]	[1]	[2]	[3]	[0]
[2]	[2]	[3]	[0]	[1]
[3]	[3]	[0]	[1]	[2]

For simplicity it is convenient to remove the brackets in the representation of the congruence classes of Z_n, (that is, we often write:

$$Z_n = \{0, 1, 2, 3, \ldots\ldots n - 1\}$$

and the operation table assumes the form

$+_4$	0	1	2	3
0	0	1	2	3
1	1	2	3	0
2	2	3	0	1
3	3	0	1	2

Definition 2.5.2: A binary operation $\odot_n$ may be defined on Z_n as follows for each $[a], [b] \in Z_n$.

$$[a] \odot_n [b] = [ab] \in Z_n, \text{ if } 0 \le ab < n,$$
$$= [r] \in Z_n, \text{ if } 0 \le r < n.$$

where r is the residue of ab modulo n.

Example 2.5.2: We see below the case of modulo multiplication.

$$[2] \odot_6 [4] = [8] = [2],\ [2] \odot_6 [3] = [61 = [0].$$

Theorem 2.5.2: *The system $(Zp^+, \odot_p)$ is a commutative group, where the underlying set Z_p^+ is the set of all non-zero residue classes mod p, p is a positive prime number, and $\odot$ denotes the multiplication of residue classes.*

Proof: Let Z_p^+ be the set of all non-zero residue classes modulo a positive prime number p, that is,

Now we check the group axioms.

(1) Closure property. Let $[a], [b] \in Z^+_p$ then

$$[a] \odot_p [b] = [ab] \in Z_p^+ \text{ if } ab < p,$$
$$= [r] \in Z_p^+ \text{ if } ab \ge p$$

where r is the residue of ab modulo p.

The Z_p^+ is closed under the operation.

(2) Associative law. If $[a], [b], [c] \in Z_p^+$, then

$$([a] \odot_p [b]) \odot_p [c] = [ab] \odot_p [c]$$
$$= [(ab)c]$$
$$= [a\,(bc)]$$
$$= [a] \odot_p [bc]$$
$$= ([a] \odot_p ([b] \odot_p [c])$$

Hence $\odot_p$ is associative.

(3) Existence of identity. Z_p^+ has the identity element [1], that is, for any $[a] \in Z_p^+$ we have

$$[a] \odot_p [1] = [1] \odot_p [a] = [a]$$

(4) Existence of inverse. We also observe that every $[a] \in Z_p^+ \exists [a'] \in Z_p^+$ such that

$$[a] \odot_p [a'] = [aa'] = [1] \in Z_p^+$$

or $aa' \equiv 1 \pmod{p}$

or $ax = 1 \pmod{p}$ has a solution in Z_p^+.

Commutativity. It is obvious for any elements $[a], [b] \in Z_p^+$

$$[a] \odot_p [b] = [ab]$$
$$= [ba] = [b] \odot_p [a]$$

It completes the proof of the theorem.

Theorem 2.5.3: *The set of all non-zero residue classes modulo m(m is the composite number) together with the operation of multiplication of residue classes modulo m, is not a group.*

Proof: Let $Z_m^+ = \{[1], [2], [3], \ldots\ldots, [m-1]\}$.

Since m is composite number it has its factors, which are less than m. Therefore any two a, b factors of m such that $ab = m$, belong to the set $\{1, 2, 3, \ldots, (m-1)\}$, therefore,

$[a]\ [b] \in Z_m^+$. We observe here that

$[a] \odot_m [b] = [ab] = [m] = [0] \in Z_m^+$

Hence the set Z_p^+ is not closed.

Thus the system $(Z_m^+ \odot_m)$ is not a group.

PROBLEMS

1. Prove that the cancellation law for addition holds in Z_7.
2. Prove that, if $ab = 0$ module 7 ($a \odot_7 b = 0$) either $a = 0$ or $b = 0$.
3. Does the cancellation law for multiplication hold in Z_6^+ ?
4. Prove that the congruence modulo n is an equivalence relation in the set of integers, and this equivalence relation has n distinct equivalence classes.
5. Show that the residue classes [1], [3] [5] [7] modulo 8 together with the operation $\odot_8$ form a group.

2.6 SUB-GROUPS

In chapter I we have studied subsets of a set. In a similar fashion we shall now discuss subsets S of G, where (G, o) is a group, such that *(S, o)* is also group, where the group operation for S is inherited from G.

Definition 2.6.1: Let (G, o) be a group and $S \subseteq G$ be a non-empty subset of G. The ordered pair (S, o) is said to be sub-group of (G, o) if (S, o) is itself a group, using the same operation o of (G, o).

It is obvious that each group (G, o) has two sub-groups. For, if $e \in G$ is the identity element of the group (G, o), then both $(\{e\}, o)$ and (G, o) are sub-groups of (G, o). These two sub-groups are often referred to as the improper or trivial sub-groups of (G, o). All sub-groups between $(\{e\}, o)$ and (G, o) are called non-trival sub-groups. Any sub-group different from (G, o) is termed proper subgroup.

Example: 2.6.1: If E, S and Z denote the set of even integers, odd integers, and integers respectively, then $(E, +)$ is a sub-group of $(Z, +)$ while $(S, +)$ is not a sub-group of the group $(Z, +)$. It is obvious that under the operation of addition the sum of two add integers is an even integer which is not an element of the set S. Hence the set S does not satisfy the closure property under addition.

Example 2.6.2: Let $(Z, +)$ be a group of integers under addition. Let S be the set of positive integers. We see $S \subseteq Z$ but $(S, +)$ is not a sub-group of the group $(Z, +)$. The set is closed under addition but the inverse of positive integers do not belong to the set S, which shows $(S, +)$ is not a group.

Example 2.6.3: Let $(G, +)$ be a group of complex numbers under addition. Let $A = \{1, -1, i, -i\}$. Then $(A, .)$ is not a sub-group of the group $(G, +)$. Although A is non-empty subset of G and $(A, .)$ is a group under multiplication, the difficulty lies in the fact that the operation of the system $(A, .)$ is different from the operation of the system $(G, +)$.

Example 2.6.4: The set $A = \{1, -1, i, -i\}$ under multiplication which makes ihe system $(A, .)$ is a sub-group of the group $(G, .)$ of non-zero complex numbers under the operation of multiplication.

Example 2.6.5: We consider the group $(Z_6 +_6)$ of integers modulo 6. If $H = \{0, 2, 4\}$, then $(H, +_6)$, whose operation table is given below is a sub-group of $(Z_6 +_6)$.

Now we observe from the table that:

$+_6$	0	2	4
0	0	2	4
2	2	4	0
4	4	0	2

(1) Each entry of the table is an element of H.

(2) Each element of H is appearing once and only once in each row and in each column.

(3) It is obvious from the table that 0 is the identity for the operation $+_6$.

(4) It is also clear from the table that 0 is the inverse of 0, 2 is the inverse of 4, and 4 is the inverse of 2.

This shows that $(H, +_6)$ is a group under the operation inherited from the group $(Z_6, +_6)$.

Hence $(H, +_6)$ is a sub-group of the group $(Z_6, +_6)$.

***Theorem 2.6.1:** If (H, o) is a sub-group of the group (G, o), then identity in G is the identity of H.*

Proof: Let e be the identity of G and e' be the identity of H. Therefore $e' \in G$, since $H \subset G$.

$$e' \circ e' = e', \text{ as } e' \text{ is the identity of } (H, o) \quad ...(1)$$

and

$$e' \circ e = e', \text{ as } e \text{ is the identity of } (G, o) \quad ...(2)$$

From (1) and (2), we have

$$e' \circ e' = e' \circ e$$

which implies $e' = e$ by left cancellation law.

Now to establish that a given subset H of G, together with the operation of (G, o) called induced operation, is a sub-group, we must verify that all the conditions of the definition of sub-group are satisfied. We observe that the associativity of the operation o in H hold good, since $H \subset G$. It is necessary then to show only the following:

(1) $a, b \in H \Rightarrow a \circ b \in H$ (closure)

(2) $e \in H$, where e is the identity element of (G, o)

(3) $a \in H \Rightarrow a^{-1} \in H$.

***Theorem 2.6.2** Let (G, o) be a group and $\phi \neq H \subset G$, the (H, o) is a sub-group if G if and only if*

(1) $a, b \in H \Rightarrow a \circ b \in H$ (Closure)

(2) $a \in H \Rightarrow a^{-1} \in H$, for all $a \in H$, a^{-1} is the inverse of a.

Proof: This theorem has two parts—the first part of the theorem asserting that if (H, o) is a sub-group of (G, o), then conditions (1) and (2) are met.

(1) Since (H, o) *is* a group, then

$a, b \in H \Rightarrow a \circ b \in H$ *by* closure property.

Again, if $a \in H$, then $a^{-1} \in H$ by the existence of the inverse of an element of a group.

Conversely, H is a non-empty subset of G, and $a \circ b \in H$ whenever $a, b \in H$. We note that:

By (2) if $a \in H$, then $a^{-1} \in H$. By (1) we have if $a, a^{-1} \in H$, then $aoa^{-1} = e \in H$. Hence H has an identity.

(2) The operation o is associative in H since it is associative in G.

Thus (H, o) satisfies all group exioms and (H, o) is therefore a sub-group of (G, o).

A theorem which establishes a single convenient criterion for determining sub-groups is given below.

Theorem 2.6.3: *Let (G, o) be a group and $\phi \neq H \subset G$. The (H, o) is a sub-group of (G, o) if and only if a, b $\in$ H implies $aob^{-1} \in H$.*

Proof: The proof of the theorem constitutes of two parts. In first part we shall assume that (H, o) is a sub-group of (G, o) and prove that if $a, b \in H$, then $a\ o\ b^{-1} \in H$.

Let $a, b \in H$ and let (H, o) be a sub-group, that is, (H, o) is a group.

Therefore, if $b \in H$, $b^{-1} \in H$, by the inverse property. And if $a \in H$, $b^{-1} \in H$, $a\ o\ b^{-1} \in H$, by the closure property.

Conversely, suppose H is a non-empty subset of G, which contains the element aob^{-1} whenever $a, b \in H$, Now we observe the following:

(1) The operation o is associative in H since $H \subseteq G$ and the operation o is associative in G.

(2) For any $a \in H$, $aoa^{-1} = e \in H$, by applying the assumption to the pair $a, a \in H$. Hence e is the identity element in H.

(3) For every $b \in H$, we have

$e\ o\ b^{-1} = b^{-1} \in H$. Therefore every element b of H has unique inverse b^{-1}.

(4) Finally, if $a \in H$, $b^{-1} \in H$, then $a\ o\ (b^{-1})^{-1} \in H$ i.e., $a\ o\ b \in H$. Hence H is closed.

Hence all group axioms are satisfied, and (H, o) is therefore a sub-group of (G, o).

In the special case of a finite sub-group we have better condition that we can omit the second condition, for this we have the following theorem.

Theorem 2.6.4: *Let (G, o) be a finite group and $\phi \neq H \subseteq G$.*
Then (H, o) is a sub-group of (G, o) if H is closed, that is, if $a, b \in H$, $a\ o\ b \in H$.

Proof: Let (H, o) be a sub-group of (G, o). That is, (H, o) is a group.

Since (H, o) is a group, H must be closed under the operation. That is, if $a, b \in H$, $aob \in H$.

Conversely, H is a non-empty finite subset of G which contains $aob \in H$ whenever $a \in H$, $b \in H$. We note the following things:

(1) H is closed under the operation as it is given that if $a \in H$, $b \in H$, then $aob \in H$.

(2) The operation o is associative in H since it is associative in G.

(3) Suppose that $a \in H$, thus $a^2 = a\ o\ a \in H$, $a^3 = (a\ o\ a)\ oa = a^2\ o\ a \in H$,......, $a^m \in H$... since H is closed. In this manner we obtain that H contains $a, a^2, a^3,......a^m$, but H is finite set.

Therefore $a, a^2, a^3,....a^m$, cannot be all distinct elements. Thus there must be repetitions in the elements $a, a^2, a^3....a^m$, of the set H. That i.e., for some integers r, s with $r > s > 0$, $a^r = a^s$

$$\Rightarrow \qquad a^{r-s} = e$$

and if $r - s = k$, some positive integer, then $a^k = e$ and $a^k \in H$

Hence H contains an identity element.

(4) Since $k = r - s > 0$ and $k - 1 = r - s - 1 \geq 0$:

$$a^{k-1} \in H$$

and
$$a^k = e$$

$$\Rightarrow \qquad a \text{ o } a^{k-1} = e$$

$$\Rightarrow \qquad a^{-1} = a^{k-1} \in H$$

Thus every element $a \in H$ has inverse $a^{-1} \in H$. This completes the proof of the theorem.

2.7 UNION AND INTERSECTION OF SUB-GROUPS

In chapter one, intersection and union of sets were defined. If a group (G, o) is given, then a set G is also given. It might therefore be possible to extend the notion of intersection and union to groups. But in extending the notion of intersection and union we meet some difficulties. Let $(G, +)$ be the group of even integers under addition, and let $A = \{1, -1, i, -i\}$ and $(A, .)$ the group under multiplication. The set $G \cap A = \phi$. To overcome this difficulty, we shall always consider that the groups under discussion must be sub-groups of the same group.

Theorem 2.7.1: *If (G_1, o) and (G_2, o) are both sub-groups of the group (G, o) then $(G_1 \cap G_2, o)$ is also a sub-group.*

Proof: Since both G_1 and G_2 contain the identity e of (G, o), the intersection $G_1 \cap G_2$ of G_1 and G_2 is not empty and $e \in G_1 \cap G_2$.

Suppose $a, b \in G_1 \cap G_2 \Rightarrow a, b \in G_1$, and $a, b \in G_2$,

Since (G_1, o) and (G_2, o) are sub-groups,

then $aob^{-1} \in G_1$ and $aob^{-1} \in G_2$

Therefore $aob^{-1} \in G_1 \cap G_2$

Hence $(G_1 \cap G_2, o)$ is a sub-group of the group (G, o).

Remark: $(G_1, \cap G_2, o)$ is the largest sub-group which is contained in (G_1, o) and (G_2, o) and it contains every sub-group of (G_1, o) and (G_2, o).

Example 2.7.1: *Let* $(G, +)$ be the group of integers under addition. If $(S, +)$ and $(T, +)$ are sub-groups of $(G, +)$, where $S = \{6n \mid n$ is an integer$\}$ and $T = \{4n \mid n$ is an integer$\}$ then $S \cap T = \{12n \mid n$ is an integer$\}$ and Let $(S, +_{180})$ and $(T, +_{180})$ are sub-groups of $(G, +_{180})$, where $S = \{6n \mid n$ an integer$\}$, $T = \{10n \mid n$ an integer$\}$, then $S \cap T = \{30n \mid n$ an integer$\}$. The intersection of the sets S and T is $S \cap T = \{0, 30, 60, 90, 120, 150\}$. We see that $(S \cap T, +_{180})$ is a sub-group of $(G, +_{180})$.

Now the concept of union will be extended to sub-groups of a group. In order to motivate our choice of definition for union of sub-groups of fixed group let us consider some examples.

If $(Z, +)$ is the group of integers, S the set of integral multiples of 4, *i.e.*, $S = \{4n \mid n$ an integer$\}$ and $T = \{2n \mid n$ an integer$\}$, then $(S, +)$ and $(T, +)$ are sub-groups of $(Z, +)$.Further more $S \cup T = \{2n \mid n$ an integer$\}$, that is, the set of even integers, and $(S \cup T, +)$ is a sub-group $(Z, +)$. This might lead us to extend the concept of union to sub-groups in a natural way.

But from the above example we note that $S \subseteq T$ and $S \cup T = T$. The reader could have been misled if the example had been such that $T \subseteq S$ and $S \cup T = S$. It is only under one of these conditions, $S \subseteq T$ or $T \subseteq S$, that $(S \cup T, o)$ is a sub-group.

We will put on exhibit a group (G, o) with sub-groups (S, o) and (T, o) such that $(S \cup T, o)$ is not a sub-group of (G, o). Let Z be the set of integers, $S = \{2n \mid n$ an integer$\}$, and $T = \{5n \mid n$ an integer$\}$, then $(S, +)$ and $(T, +)$ are sub-groups of $(Z, +)$.

Now $2 \in S$ and $5 \in T$. If $(S \cup T, +)$ is a group, then closure axions must be satisfied. In particular, $2 \in S \cup T$, $S \in S \cup T$ but $2 + 5 = 7 \notin S \cup T$. It follows $(S \cup T, +)$ is not a group.

Another example we consider, Let $(G, +_6)$ be the group of integers modulo 6, $S = \{0, 2, 4\}$ and $T = \{0, 3\}$. Then $(S, +_6)$ and $(T, +_6)$ are sub-groups of $(G, +_6)$. We note here that $2 \in S$, $3 \in T$ and $2 +_6 3 = 5 \notin S \cup T$, therefore $(S \cup T, +_6)$ is not a group because the closure axiom is not satisfied.

Theorem 2.7.2: *Let (H_1, o) and (H_2, o) be sub-groups of group (G, o). The pair $(H_1 \cup H_2, o)$ is also a sub-group if and only if $H_1 \subseteq H_2$ or $H_2 \subseteq H_1$.*

Proof: First we show that if $H_1 \subseteq H_2$ or $H_2 \subseteq H_1$, then $(H_1 \cup H_2, o)$ is a sub-group.

Let $H_1 \subseteq H_2$, $H_1 \cup H_2 = H_2$

Since (H_2, o) is a sub-group so is $(H_1 \cup H_2, o)$.

Similarly, let $H_2 \subseteq H_1$, $H_1 \cup H_2 = H_1$

Again $(H_1 \cup H_2, o)$ is a sub-group since (H_1, o) is a sub-group. In either case $(H_1 \cup H_2, o)$ is a sub-group.

Conversely, it is enough to show that if $(H_1 \cup H_2, o)$ is a sub-group, then one of the sets H_1 or H_2 must be contained in the other, that is, either $H_1 \subseteq H_2$ or $H_2 \subseteq H_1$. We suppose that $H_1 \not\subset H_2$ and $H_2 \not\subset H_1$.

Then there exist elements $a \in H_1 - H_2$ and $b \in H_2 - H_1$.

Since $a \in H_1 - H_2$, then $a \in H_1$ and $a \notin H_2$

and $b \in H_2 - H_1$, then $b \in H_2$ and $b \notin H_1$.

But $a, b \in H_1 \cup H_2$. Since $(H_1 \cup H_2, o)$ is a sub-group, $aob \in H_1 \cup H_2 \Rightarrow a \text{ o } b \in H_1$ or $a \text{ o } b \in H_2$.

If $a \text{ o } b \in H_1$ and $a \in H_1$, then $a^{-1} \in H_1$ and $b = a^{-1} o\, (a \text{ o } b) \in H_1$ which is not true.

Again if $a \text{ o } b \in H_2$ and $b \in H_2$, $b^{-1} \in H_2$, $a = (a \text{ o } b) \text{ o } b^{-1} \in H_2$ which is also not true That is, the element $a, b \in H_1 \cup H_2$, but $a \text{ o } b \notin H_1 \cup H_2$. That is, the set $H_1 \cup H_2$ is not closed under the operation. Hence it contradicts the fact that $(H_1 \cup H_2, o)$ is a group. Therefore by assuming $H_1 \not\subset H_2$ and $H_2 \not\subset H_1$ we arrived at a contradiction which follows that the assumption is wrong. This completes the proof of the theorem.

Example 2.7.2: Let ((2*n*), +) and ((4*n*), +) be two sub-groups of the group (*Z*, +) of integers under addition, and (2*n*) contains the set (4*n*), that is, $(4n) \subset (2n)$. So $(4n) \cup (2n) = (2n)$, and ((2*n*), +) is a sub-group of (*Z*, +). Hence in this case $((4n) \cup (2n), +)$ is a sub-group of the group (*Z*, +) because $(4n) \subset (2n)$.

Example 2.7.3: Let $(\{0, 6\}+_{12})$ and $(\{0, 4, 8\}, +_{12})$ be two sub-groups $(Z_{12}+_{12})$ of integers modulo 12. The union $\{0, 6\} \cup \{0, 4, 8\} = \{0, 4, 6, 8\}$, together with operation $+_{12}$ does not form a group $(\{0, 4, 6, 8\}, +_{12})$ because the set {0, 4, 6, 8} is not closed for the operation as it is seen that, $4 \in \{0, 4, 6, 8\}$, $6 \in \{0, 4, 6, 8\}$ but $4 +_{12} 6 = 10 \notin (0, 4, 6, 8)$.

The theorem is verified that since $\{0, 6\} \not\subseteq \{0, 4, 8\}$ or $\{0, 4, 8\} \not\subseteq (0, 6)$, the union $(\{0, 4. 6, 8\}, +_{12})$ is not a group.

Theorem 2.7.3: *If (H_1, o) and (H_2, o) are sub-groups of (G, o), therefore there exists a smallest sub-group of (G, o) that contains both (H_1, o) and (H_2, o).*

Proof: There is one sub-group of (*G*, *o*) that contains both (H_1, o) and (H_2, o); namely *G* itself. (Every group is a sub-group of itself). Let *X* be the collection of all sub-groups of (*G*, *o*) which contains both (H_1, o) and (H_2, o). Since the intersection of all sub-groups is a sub-group of (*G*, *o*). Then let *M* be the intersection of all sub-groups belonging to *X*, then (*M.o*) is a sub-group.

We shall complete the proof of the theorem by proving that (*M*, *o*) is the smallest such sub-group.

Suppose (*L*, *o*) is a sub-group of (*G*, *o*) such that $H_1 \subseteq L$ and $H_2 \subseteq L$. But (*L*, *o*) is a member of *X* by the definition of *X*. Therefore $M \subseteq L$, since (*M*, *o*) is contained in every member of *X*. Thus (*M*, *o*) is the smallest such sub-group. Hence the theorem is proved.

Definition 2.7.1: A set generated by the union $H_1 \cup H_2$ of two sets H_1 and H_2 is designated by $\{H_1 \cup H_2]$, and it is equal to $H_1 + H_2$, *i.e.*, $\{H_1 \cup H_2\} = \{H_1 + H_2]$, if *o* is addition.

If (*G*, *o*) is a group and (H_1, o) and (H_2, o) are sub-groups of the group (*G*, *o*) then $(\{H_1 \cup H_2\},)$ is the smallest sub-group of (*G*, *o*) that contains both (H_1, o) and (H_2, o).

Example 2.7.4: For purpose of illustrating the above definiton let us consider the commutative gorup $(Z_{12} +_{12})$ of integers modulo 12 and the two sub-groups $(\{0, 6\}, +_{12})$ and $(\{0, 4, 8\}, +_{12})$. To obtain the smallest sub-group which contains {0, 6} and {0, 4, 8), it suffices merely to compute the sum of these subsets.

$\{0, 6\} +_{12} \{0, 4, 8\} = \{0 +_{12} 0, 0 +_{12} 4, 0 +_{12} 8, 6 +_{12} 0, 6 +_{12} 4, 6 +_{12} 8\}$

$= \{0, 4, 8, 6, 10, 2\}$

Hence the sub-group of $(Z_{12}, +_{12})$ generated by the union $\{0, 6\} \cup \{0, 4, 8\}$ is just $(\{0, 2, 4, 6, 8, 10,\}, +_{12}.)$

Now we consider subsets H and K of G if (G, o) is a group.

Definition 2.7.2: Let (G, o) be a group and H and K be non-emtpy subsets of G. The product of H and K, in that order, is the set

$$H \ o \ K = \{h \ o \ k \mid h \in H \text{ and } k \in k\}$$

Similarly, we can define $H \ o \ H$ which we shall denote by H^2. If one of the sets consists of a single element a, then we write $\{a\} \ o \ H$ or $a \ o \ H$.

Definition 2.7.3: Let (G, o) be a group and $H \subseteq G$, then $H^{-1} = \{h^{-1} \mid h \in H\}$. That is, the set H^{-1} is the set of all inverses of all elements of the set H. Since the group (G, o) has all inverses of its elements which shows that $H^{-1} \subseteq G$.

Theorem 2.7.4: *Let (G, o) be a sub-group and H ⊆ G. A necessary and sufficient condition for (H, o) to be a sub-group of (G, o) is that* $H \ o \ H^{-1} = H$.

The proof is left to readers.

CHAPTER 2

Theorem 2.7.5: *A necessary and sufficient condition for* $H \subseteq G$ *to be a sub-group (H, o) of the finite group (G, o) is that* $H \ o \ H = H$.

The proof is left to the readers.

At first sight, one might guess that whenever *(H, o)* and *(K, o)* are both sub-groups of *(G, o)*, the *(H o K, o)* will also be a sub-group. But in some cases $H \ o \ K$ is not even closed. Therefore we should impose conditions on $H \ o \ K$ to be a sub-group *(H o K, o)*.

Theorem 2.7.6: *If (H, o) and (K, o) are sub-groups of the group (G, o) such that* $H \ o \ K = K \ o \ H$, *then the pair (H o K, o) is also a sub-group.*

Proof: $H \ o \ K = K \ o \ H$ does not mean that each element of H is commutative with each element of K; it tells that whenever h and k are arbitrary elements of H and K, then there exist element $h' \in H$ and $k' \in K$ for which $h \ o \ k = k' \ oh'$.

It is obvious that $H \ o \ K$ is non-empty since H and K both contain the identity of *(G, o)*, *i.e.*, e and $e \ o \ e \in H \ o \ K$. Now let $a \in H \ o \ K$ *and* $b \in H \ o \ K$, then there exist $h, h_1 \in H$ and $k, k_1 \in K$ such that $a = h \ o \ k$ and $b = h_1 \ o \ k_1$.

Here our aim is to show that $a \ o \ b^{-1} \in H \ o \ K$ to prove *(H o K, o)* is a sub-group. We note that

$$\begin{aligned} a \ o \ b^{-1} &= (h \ o \ k) \ o \ (h_1 \ o \ k_1)^{-1} = (h \ o \ k) \ o \ (k_1^{-1} \ o \ h_1^{-1}) \\ &= h \ o \ (k \ o \ k_1^{-1}) \ o \ h_1^{-1}) \end{aligned}$$

Since (K, o) is a group, then $k \ o \ k_1^{-1} \in K$ and consequently

$(k \ o \ k_1^{-1}) \ o \ h_1^{-1} \in K \ o \ H$

But we are given $K \ o \ H = H \ o \ K$, then there exist $h_2 \in H$ and $k_2 \in K$ such that

$$(k \ o \ h_1^{-1}) \ o \ h_1^{-1} = h_2 \ o \ k_2$$

Thus we may conclude that

$$\begin{aligned} a \; o \; b^{-1} &= h \; o \; (k \; o \; k_1^{-1}) \; o \; h_1^{-1} \\ &= h \; o \; (h_2 \; o \; k_2) \\ &= (h \; o \; h_2) \; o \; k_2 \in H \; o \; K \end{aligned}$$

Since H is closed under o, then $h \; o \; h_2 \in H$. Hence $(H \; o \; K, o)$ is a sub-group.

Corollary: If (H, o) and (K, o) are sub-groups of the commutative group (G, o), then $(H \; o \; K, o)$ is again a sub-group.

Note: $\{H \cup K\} = H \; o \; K$

Here we note an interesting thing. Before dealing with it we should learn some important notations.

Remark: If (G, o) is an arbitrary group and $\phi \neq S \subseteq G$, the symbol (S) will represent the set

$$(S) = \cap \; \{H \,|\, S \subseteq H, \; (H, o) \text{ is a sub-group of } (G, o)\}$$

The set (S) clearly exists, for (G, o) is a sub-group of itself and $S \subseteq G$, *i.e.*, $G \in (S)$, for (G, o) is a trivial sub-group *of* (G, o) and $S \subseteq G$. In addition $S \subseteq (S)$, since S is contained in every sub-group whose set is a member of (S).

Theorem 2.7.7: *The pair ((S), o) is a sub-group of (G, o) (known as the sub-group generated by the set (S).*

Proof: By the definition of (S) it is obvious that $((S), o)$ is a sub-group, since the intersection $\cap H_i$ of the family of subsets $\{H_i\}$, where (H_i, o) is a sub-group of (G, o), is a sub-group $((\cap H_i), o)$ or the pair $((S), o)$ is the smallest sub-group containing a set S. If $(S) = G$, then (G, o) is said to be generated by the set S.

We shall give below the simpler defintion of (S).

$$(S) = \{a_1 \; o \; a_2 \; o \; a_3 ... o \; a_n \,|\, a_1, a_2, ... a_n \in S \cup S^{-1}; \; n \in \mathrm{N}\}$$

Proof of corollary

We observe that if H and K are sub-groups of group (G, o).

$$H = H \; o \; e \subseteq H \; o \; K \text{ and } K = e \; o \; K \subseteq H \; o \; K$$

From this follows $H \cup K \subseteq H \; o \; K$. Therefore we can say, *if* $(H \; o \; K, o)$ forms a sub-group of (G, o) it must be sub-group generated by $H \cup K$, or in symbols.

$$(H \; o \; K, o) = (\{H \cup K\}, o)$$

To see the validity of this result we know that sub-group generated by $H \cup K$ is the smallest sub-group contains this union. Thus if $(H \; o \; K, o)$ is a sub-group than it follows $H \cup K \subseteq H \; o \; K$.

Conversely, by the definition $\{H \cup K\}$ contains all elements of the term $h \; o \; k$ with $h \in H$, $k \in K$, which shows

$H \circ K \subseteq \{H \cup K\}$. Hence

$\{H \cup K\} = H \circ K$.

Definition 2.7.4: The centre of a group (G, o) denoted by $Z(G)$, is the set

$$Z(G) = \{c \in G \mid x \circ c = c \circ x, \text{ for all } x \in G\}$$

Thus, $Z(G)$ has those elements which are commutative with every element of G. Therefore the $Z(G)$ of the commutative group is the set G.

Theorem 2.7.8: *If (G, o) is a group, the pair (Z(G), o) is a sub-group of (G. o).*

Proof: We first observe that $Z(G)$ is non-empty since $e \in Z(G)$. Let $a, b \in Z(G)$. By the definition of $Z(G)$, we have

$$x \circ a = a \circ x, \text{ for all } x \in G \text{ and } x \circ b = b \circ x, \text{ for all } x \in G.$$

$$x \circ b = b \circ x \Rightarrow b^{-1} \circ (x \circ b) \circ b^{-1} = b^{-1} \circ (b \circ x) \circ b^{-1}$$

$\Rightarrow b^{-1} \circ x = x \circ b^{-1}$, for all $x \in G$

Thus, if $x \in G$, $(a \circ b^{-1}) \circ x = a \circ (b^{-1} \circ x)$

$$= a \circ (x \circ b^{-1}), \text{ since } x \circ b^{-1} = b^{-1} \circ x$$

$$= (a \circ x) \circ b^{-1}$$

$$= (x \circ a) \circ b^{-1}$$

$$= x \circ (a \circ b^{-1})$$

Hence $a \circ b^{-1} \in Z(G)$.

Thus, $(Z(G), o)$ is a sub-group of (G, o)

Definition 2.7.5: Let (G, o) be a group and $a \in G$. The centralizer or normalizer of a in G, denoted by $N(a)$, is the set

$$N(a) = \{x \in G \mid a \circ x = x \circ a\}$$

The $N(a)$ has those elements which commute with a, the fixed element $a \in G$.

Theorem 2.7.9: *Let (G, o) be a group and a ∈ G, then pair (N(a), o) is a sub-group of (G, o).*

Proof: We observe that $N(a)$ is a non-empty since $e \in N(a)$. Let $x, y \in N(a)$. By the definition of $N(a)$ we have

$$x \circ a = a \circ x, \text{ and } y \circ a = a \circ y \text{ or } a \circ y^{-1} = y^{-1} \circ a.$$

We know that $(x \circ y^{-1} \circ a) = x \circ (y^{-1} \circ a)$

$$= x \circ (a \circ y^{-1})$$

$$= (x \circ a) \circ y^{-1} = (a \circ x) \circ y^{-1}$$

$$= a \circ (x \circ y^{-1})$$

which implies $x \circ y^{-1} \in N(a)$. Hence the pair $(N(a), o)$ is a sub-group of each group (G, o).

PROBLEMS

1. In each of the following, determine whether *(S, o)* is a sub-group of (*G, o*). If it is not, give at least one reason why it is not.

(*a*) (*G, o*) is the group of complex numbers under addition, and *S,* is the set of rational numbers.

(*b*) (*G, o*) is the group of non-zero rational numbers under multiplication and *S* is the set of non-zero even integers.

(c) (*G, o*) is the group of non-zero real numbers under multiplication and *S* is the set of non-zero real numbers.

(*d*) (*G, o*) is the group of integers module 6 and $S = \{0, 1, 3, 5\}$.

2. Pove that the inverse of an element $a \in H$ is the same as the inverse of the same element *a* regarded as an element of the group (*G, o*).

3. If (*H, o*) is a sub-group of the group (*G, o*) and (*K, o*) is a sub-group of (*H, o*), then prove that (*K, o*) is a sub-group of (*G, o*).

4. Let (Z, +) be a group of integers under addition. Let S = ($sn \mid s$ is a fixed integer, n any integer), and $T = \{tn \mid t$ is a fixed integers, n any integer). Find $S \cap T$ and show that (S $\cap$ *T,* +) is a sub-group.

5. Let *P* be a collection of sub-groups of a group (*G, o*). Prove that the intersection of sub-groups in *P* is a sub-group of (*G, o*).

6. In each of the following (*G, o*) is a group, (*S, o*) and (*T, o*) are sub-groups of (G, *o*). Find ($S \cup T$, *o*) is a sub-group.

(*a*) (*G, o*) is the group of integers modulo 6, $S = \{0, 2, 4\}$ and $T = \{0, 3\}$,

(*b*) (*G, o*) is the group of integers modulo 16, $S = \{0, 8\}$ and $T = \{0, 3, 8, 12\}$.

7. Given a group (*G, o*) and $\phi \neq H \subseteq G$. Varify the following statements are equivalent.

(*a*) (*H, o*) is a sub-group of the group (*G, o*)

(*b*) $H \text{ o } H \subseteq H$ and $H^{-1} \subseteq H$,

(c) $H \text{ o } H^{-1} \subseteq H$.

8. Show that a proper sub-group of ($\{1, -1, i, -i\}$), is ($\{1, -1\}$,.).

Are there any others? Support your answer with arguments.

9. Show that in group (*G*, *), Z (*G*) $= \cap N(a),\ a \in G$.

10. In a commutative group (*G*, *), define the set *H* by

$H = \{a \in G \mid a^k = e \text{ for some } k \in N\}$

Detemine whether (*H*, *) is a sub-group of (G, *)

11. Let (*H*, *) be a sub-group of group (*G*, *) such that $H \neq$ G. Prove that the sub-group generated by the complement G – *H* is the group (*G*, *) itself.

12. If (*H* *) is a sub-group of the group (*G*, *) and $\phi \neq H \subseteq G$, prove that $H * K \subseteq H \Rightarrow K \subseteq H$.

13. Let $(H, *)$ and $(K, *)$ be sub-groups of the commutative group $(G, *)$ of order n and m respectively. Assuming $H \cap K = \{e\}$, verify that the order of the group $(H * K, *)$ is $n\ m$.

14. Let $(G, *)$ be a group of order n, where n is odd. Prove that each element of G is a square of some element of G. *That is, if* $x \in G$, $\exists\ y \in G$ *such that* $x = y^2$.

2.8 GROUPS OF PERMUTATIONS

We know that one to one mapping of a set S onto itself is called a permutation.

When the set S is a finite set, we speak of a permutation of n symnbols. It is not necessary to limit our discussion for finite sets. Hence we review carefully the notion and properties of permutations. The definition of equality of two permutations of infinite sets is the same as for finite sets.

Definition 2.8.1: Let f and g be two permutations of an arbitrary set A. f is said to be equal to g if for each $x \in A$, $f(x) = g(x)$.

Example 2.8.1: Let f and g be given by

$$f = \begin{pmatrix} 1 & 2 & 3 & 4 \\ 2 & 4 & 1 & 3 \end{pmatrix}, g = \begin{pmatrix} 2 & 1 & 3 & 4 \\ 4 & 2 & 1 & 3 \end{pmatrix}$$

We see that

$$f(1) = g(1) = 2$$
$$f(2) = g(2) = 4$$
$$f(3) = g(3) = 1$$
$$f(4) = g(4) = 3$$

This implies $f = g$

If the set S is finite, then we take the set of n symbols and more conveniently the set of n positive integers: $\{1, 2, 3, 4,..., n\}$ Let the $S = \{1, 2, 3,...,n\}$. Then the number of permutations of the set S which contains the finite numbers of elements, say, n is $n!$ The set of all permutations of the set S of n elements is called the symmetric set and denoted by S_n. Thus S_n contains $n!$ elements.

Let f be a permutation of S, that is, $f \in S_n$ that is, $f = \{(1, f(1)), (2, f(2)), (3, f(3))... (n, f(n))\}$, but it is written in a two line form

$$f = \begin{pmatrix} 1 & 2 & 3 & 4 & .... & n \\ f(1) & f(2) & f(3) & f(4) & .... & f(n) \end{pmatrix}$$

That is, in the first row the elements of the set are written and in the second row corresponding images appear below each integer. The order of elements in the first row of this symbol is immaterial but columns of this symbol should not be affected.

For instance $$f = \begin{pmatrix} 1 & 2 & 3 & 4 \\ 3 & 2 & 1 & 4 \end{pmatrix} = \begin{pmatrix} 2 & 4 & 3 & 1 \\ 2 & 4 & 1 & 3 \end{pmatrix}$$

Thus each permutation of S_n may be written in $n!$ different ways.

Let f and g be two arbitrary permutations of the integers 1, 2, 3,...n, then f can be written

$$f = \begin{pmatrix} g(1) & g(2) & g(3)... & g(n) \\ f(g(1)) & f(g(2)) & f(g(3))... & f(g(n)) \end{pmatrix}$$

follows from

$$f = \begin{pmatrix} 1 & 2 & 3 & ...n \\ f(1) & f(2) & f(3) & f(n) \end{pmatrix} \text{ and } g = \begin{pmatrix} 1 & 2 & 3 & n \\ g(1) & g(2) & g(3) & f(n) \end{pmatrix}$$

by replacing the elements of the first row of f by the corresponding images of g and then taking the images under f.

Example 2.8.2: $f = \begin{pmatrix} 1 & 2 & 3 \\ 2 & 3 & 1 \end{pmatrix}$ and $g = \begin{pmatrix} 1 & 2 & 3 \\ 3 & 1 & 2 \end{pmatrix}$

$f(1) = 2$, $f(2) = 3$, $f(3) = 1$, $g(1) = 3$, $g(2) = 1$, $g(3) = 2$.

We replace 1, 2, and 3 elements of the top row of f by $g(1)$, $g(2)$, and g(3) which are 3, 1, 2 respectively and then we take images of 3, 1 and 2 under f which are 1, 2 and 3 respectively.

Thus

$$f = \begin{pmatrix} g(1) & g(2) & g(3) \\ f(g(1)) & f(g(2)) & f(g(3)) \end{pmatrix} = \begin{pmatrix} 3 & 1 & 2 \\ 1 & 2 & 3 \end{pmatrix} = \begin{pmatrix} 1 & 2 & 3 \\ 2 & 3 & 1 \end{pmatrix}$$

Now with the help of this notation, we get the composite of two permutations. Let $f, g \in S_n$

$$f \circ g = \begin{pmatrix} 1 & 2 & 3 & ...n \\ f(1) & f(2) & f(3) & ...f(n) \end{pmatrix} 0 \begin{pmatrix} 1 & 2 & 3 & ...n \\ g(1) & g(2) & g(3) & ...g(n) \end{pmatrix}$$

$$= \begin{pmatrix} g(1) & g(2) & ...g(n) \\ g(g(1)) & f(g(2)) & ...f(g(n)) \end{pmatrix} 0 \begin{pmatrix} 1 & 2 & 3 & ...n \\ g(1) & g(2) & g(3) & ...g(n) \end{pmatrix}$$

$$= \begin{pmatrix} 1 & 2 & 3 & 4 & ...n \\ f(g(1)) & f(g(2)) & f(g(3)) & f(g(4)) & ...f(g(n)) \end{pmatrix}$$

Thus fog $\in S_n$.

Example 2.8.3: Let S = {1, 2, 3), then S_3 will contain 3! = 3. 2. 1 = 6 elements (Permutations of S).

$$f_1 = \begin{pmatrix} 1 & 2 & 3 \\ 1 & 2 & 2 \end{pmatrix}, f_2 = \begin{pmatrix} 1 & 2 & 3 \\ 2 & 3 & 1 \end{pmatrix}, f_3 = \begin{pmatrix} 1 & 2 & 3 \\ 3 & 1 & 2 \end{pmatrix}$$

$$f_4 = \begin{pmatrix} 1 & 2 & 3 \\ 1 & 3 & 2 \end{pmatrix}, f_5 = \begin{pmatrix} 1 & 2 & 3 \\ 2 & 1 & 3 \end{pmatrix}, f_6 = \begin{pmatrix} 1 & 2 & 3 \\ 3 & 2 & 1 \end{pmatrix}$$

We compute f_4 o f_6 as follows:

$$f_4 \; o \; f_6 = \begin{pmatrix} 1 & 2 & 3 \\ 1 & 3 & 2 \end{pmatrix} o \begin{pmatrix} 1 & 2 & 3 \\ 3 & 2 & 1 \end{pmatrix}$$

$$= \begin{pmatrix} 3 & 2 & 1 \\ 2 & 3 & 1 \end{pmatrix} o \begin{pmatrix} 1 & 2 & 3 \\ 3 & 2 & 1 \end{pmatrix}$$

$$= \begin{pmatrix} 1 & 2 & 3 \\ 2 & 3 & 1 \end{pmatrix} = f_2$$

$$f_6 \; o \; f_4 = \begin{pmatrix} 1 & 2 & 3 \\ 3 & 2 & 1 \end{pmatrix} o \begin{pmatrix} 1 & 2 & 3 \\ 1 & 3 & 2 \end{pmatrix}$$

$$= \begin{pmatrix} 1 & 3 & 2 \\ 3 & 1 & 2 \end{pmatrix} o \begin{pmatrix} 1 & 2 & 3 \\ 1 & 3 & 2 \end{pmatrix}$$

$$= \begin{pmatrix} 1 & 2 & 3 \\ 3 & 1 & 2 \end{pmatrix} = f_3$$

which shows $f_4 \; o \; f_6 \neq f_6 \; o \; f_4$ so that multiplication of permutations is not commulative.

Theorem 2.8.1: *The pair (s_n, o) forms a group, called as the symmetric group on n symbols, which is non-commutative for $n \geq 3$.*

Proof: Now we have to verify the group axioms for the pair (S_n, o), Let $f, g \in S_n$.

(1) Closure Property. Since f and g are one to one and onto functions from S onto itself; the composition fog is also one to one function of S onto itself. For, if $c \in S$, $\in \exists$ an element $b \in S$ such that $f(b) = c$. Similarly, again $\exists \; a \in S$ such that $g(a) = b$, thus $(f \; o \; g) \; a = f \; (g \; (a) = f \; (b) = c$,

which shows that $f \; o \; g$ is an onto function.

Again, let $a, b \in S$ with $a \neq b$, then

$g\,(a) \neq g(b)$ since g is one to one function.

Again $f(g(a) \neq f(g(b)$ since f is one to one.

Thus $f \circ g$ is one to one function. Hence $f \circ g \in S_n$.

Which shows S_n is closed.

(2) Associative Property. We know the composition of functions is associative.

(3) Existence of Identity. The set S_n contains an element e_s which is the identity function on S.

So that $f \circ e_s = e_s \circ f = f$ for all $f \in S_n$

(4) Existence of Inverse. For any $f \in S_n$, the inverse f^{-1} exists, and is a one to one function of S onto itself. Hence $f^{-1} \in S_n$.

(5) Commutative Property. We have seen that $f_i \circ f_j \neq f_j \circ f_i$ which shows the composition of function is not commutative. Hence the pair (S_n, o) is a non-abelian group.

We here recall again, that the identity element for the system (S_n, o) is the permutation.

$$\begin{pmatrix} 1 & 2 & 3 & ...n \\ 1 & 2 & 3 & ...n \end{pmatrix}$$

and if $f \in S_n$, then

$$f^{-1} = \begin{pmatrix} (f(1)) & f(2) & f(3) & ...f(n) \\ 1 & 2 & 3 & ...n \end{pmatrix}$$

Example 2.8.4: The pair (S_3, o) forms a group under the composition of functions.

Solution: The set S_3 contains $3\,! = 3.2.1 = 6$ elements which are:

$$f_1 = \begin{pmatrix} 1 & 2 & 3 \\ 1 & 2 & 3 \end{pmatrix}, f_2 = \begin{pmatrix} 1 & 2 & 3 \\ 2 & 3 & 1 \end{pmatrix}, f_3 = \begin{pmatrix} 1 & 2 & 3 \\ 3 & 2 & 1 \end{pmatrix},$$

$$f_4 = \begin{pmatrix} 1 & 2 & 3 \\ 1 & 3 & 2 \end{pmatrix}, f_5 = \begin{pmatrix} 1 & 2 & 3 \\ 2 & 1 & 3 \end{pmatrix}, f_6 = \begin{pmatrix} 1 & 2 & 3 \\ 3 & 2 & 1 \end{pmatrix},$$

We prepare the operation table. It is clear that all groups axioms are satisfied, which implies (S_3, o) is a non-abelian group.

0	f_1	f_2	f_3	f_4	f_5	f_6
f_1	f_1	f_2	f_3	f_4	f_5	f_6
f_2	f_2	f_3	f_1	f_5	f_6	f_4
f_3	f_3	f_1	f_2	f_6	f_4	f_5
f_4	f_4	f_6	f_5	f_1	f_3	f_2
f_5	f_5	f_4	f_6	f_2	f_4	f_3
f_6	f_6	f_5	f_4	f_3	f_2	f_1

Definition 2.8.2: A permutation f is said to be a transposition if f interchanges two numbers and leaves the other fixed, that is, there exist integers $i, j \in S$, $i \neq j$ such that $f(j) = i$ and $f(i) = j$ and $f(k) = k$ if $k \neq i$ and $k \neq j$.

for instance

$$f = \begin{pmatrix} 1 & 2 & 3 & 4 & 5 & 6 & 7 & 8 \\ 1 & 2 & 3 & 4 & 5 & 6 & 8 & 7 \end{pmatrix}$$

Here $f(k) = k$, for $k = 1, 2, 3, 4, 5, 6$, and $f(7) = 8$

$f(8) = 7$. Thus f is transposition.

Now we can see that

$$f^2 = f \circ f = \begin{pmatrix} 1 & 2 & 3 & 4 & 5 & 6 & 7 & 8 \\ 1 & 2 & 3 & 4 & 5 & 6 & 7 & 8 \end{pmatrix} = \text{I (identity)}$$

Theorem 2.8.2: *Every permutation may be expressed as the product of transposition.*

Proof: We shall prove our assertion by induction on n. For $n = 1$, there is nothing to prove. Let $n > 1$ and assume the assertion proved for $n - 1$. Let $f \in S_n$ and let $f(n) = k$. Let another permutation $g \in S_n$ such that $g(k) = n$, $g(n) = k$. Then $g \circ f$ is a permutation such that

$$(g \circ f)(n) = g(f(n)) = g(k) = n$$

In other words, $g \circ f$ leaves n fixed. We may therefore view $g \circ f$ as a permutation of $(n - 1)$ integers.

By induction we have the transposition $f_1, f_2, f_3 ... f_s$ of $(n - 1)$ integers, having n fixed, such that

$$g \circ f = f_1 \circ f_2 \circ f_2 \circ ... \circ f_s$$

or

$$f = g^{-1} \circ f_1 \circ f_2 \circ f_3 \circ ... \circ f^s,$$

which proves our assertion.

A direct check shows that

$$(1\ 2\ 3\ ...\ k) = (1k) \circ (1k - 1) \circ ... \circ (12).$$

Although as above the transpositions have an integer common, yet they do not, in general, commute and the decomposition is not unique.

Example 2.8.5: Let
$$f = \begin{pmatrix} 1 & 2 & 3 & 4 & 5 \\ 2 & 3 & 1 & 4 & 5 \end{pmatrix}$$

and let f_1 and f_2 be transpositions which are considered as follows:

$$f_1 = \begin{pmatrix} 1 & 2 & 3 & 4 & 5 \\ 2 & 1 & 3 & 4 & 5 \end{pmatrix},$$

$$f_2 = \begin{pmatrix} 1 & 2 & 3 & 4 & 5 \\ 3 & 2 & 1 & 4 & 5 \end{pmatrix},$$

We note that

$$f_2 \circ f_1 = \begin{pmatrix} 1 & 2 & 3 & 4 & 5 \\ 3 & 2 & 1 & 4 & 5 \end{pmatrix} \circ \begin{pmatrix} 1 & 2 & 3 & 4 & 5 \\ 2 & 1 & 3 & 4 & 5 \end{pmatrix}$$

$$= \begin{pmatrix} 2 & 1 & 3 & 4 & 5 \\ 2 & 3 & 1 & 4 & 5 \end{pmatrix} \circ \begin{pmatrix} 1 & 2 & 3 & 4 & 5 \\ 2 & 1 & 3 & 4 & 5 \end{pmatrix}$$

$$= \begin{pmatrix} 1 & 2 & 3 & 4 & 5 \\ 2 & 3 & 1 & 4 & 5 \end{pmatrix}$$

Similarly

and $$f_1 \circ f_2 = \begin{pmatrix} 1 & 2 & 3 & 4 & 5 \\ 3 & 1 & 2 & 4 & 5 \end{pmatrix}$$

Thus $$f_2 \circ f_1 \neq f_1 \circ f_2$$

Definition 2.8.3: Let $n_1, n_2, n_3 \ldots n_k$ be k distinct integers between 1 and n. If a permutation $f \in S_n$ is such that

$$f(n_1) = n_2$$
$$f(n_2) = n_3$$
$$f(n_3) = n_4$$
$$\ldots\ldots\ldots\ldots$$
$$f(n_k) = n_1$$

and $f(n) = n$ for $n \notin \{n_1, n_2, n_3, \ldots\ldots, n_k\}$,

then f is said to be k cycle, or a cycle of length k.

A cycle maps n_1, onto n_2, n_2 onto n_3,....and finally n_k onto n_1, and leaving all other elements fixed.

We write a cycle in condensed form $(n_1, n_2 \ldots n_k)$ indicating each element is replaced by its successor on the right and the last integers by the first, for example,

$$\begin{pmatrix} 1 & 2 & 3 & 4 & 5 \\ 1 & 5 & 2 & 4 & 3 \end{pmatrix} = (2\ \ 5\ \ 3).$$

In writing the permutation as a cycle we have omitted 1 and 4 which means 1 is mapped onto 1 and 4 is mapped onto 4. A given cycle can be written in more than one way.

$$(2\ \ 5\ \ 3) = (5\ \ 3\ \ 2) = (3\ \ 2\ \ 5).$$

Theorem 2.8.3: *Every permutation $f \in S_n$ can be expressed as commutaive product of disjoint cycles.*

Proof: First, we consider the set of images of 1 under the successive powers of f:

$f(1), f^2(1), f^3(1)...$, where $f^n = f \circ f \circ f \circ ... \circ f$, n times. Since the domain of f is finite. Then for some i and j, $i < j$ we have

$$f^i(1) = f^i(1), \text{ that is,}$$

$$f^{j-i}(1) = 1, \text{ where } 0 \le j - i = k < n$$

Thus $$f^{(k)}(1) = 1.$$

Thus we have first cycle $\{1, f(1), f^2(1),...f^{k-1}(1)\}$.

If the element 2 does not belong to this cycle, then we take the set of images of 2 under the successive powers of f: $\{2, f(2), f^2(2)...f^{j-1}(2)\}$, where j is the smallest positive integer such that $f^j(2) = 2$. In at most in such steps, this procedure must terminate. The order of multiplication of cycle is immaterial, for they have no elements in common.

Example 2.8.6: Let $f = \begin{pmatrix} 1 & 2 & 3 & 4 & 5 \\ 5 & 4 & 1 & 2 & 3 \end{pmatrix}$

Now we find out the images of 1 under the successive powers of f:

$$f(1) = 5$$

$$f^2(1) = fof(1) = f(5) = 3,$$

$$f^3(1) = fof^2(1) = f(3) = 1$$

Thus $= \{1, f(1), f^2(1)\} = \{1, 5, 3\}$

Since $2 \notin (1, 5, 3)$, then we compute

$$f(2) = 4$$

$$f^2(2) = f \circ f(2) = f(4) = 2$$

Hence the second cycle $= (2, f(2)) = (2\ 4)$.

Thus $$f = (1\ 5\ 3) \circ (2\ 4)$$

where $(1\ 5\ 3)$ and $(2\ 4)$ are disjoint cycles.

Remark: A cycle of length 2 is called a transposition.

We have already seen that a permutation can be written as the product of transpositions. So a cycle can also be written as the product of transpositions.

Example 2.8.7: *Let* $f = \begin{pmatrix} 1 & 2 & 3 & 4 & 5 & 6 & 7 & 8 \\ 1 & 4 & 3 & 6 & 5 & 8 & 7 & 2 \end{pmatrix}$

$$= (2\ 4\ 6\ 8)$$

$$= (6\ 8) \circ (4\ 8) \circ (2\ 8)$$

$$= (2\ 8) \circ (2\ 6) \circ (2\ 4)$$

$$= (4\ 8) \circ (2\ 4) \circ (6\ 8) \circ (2\ 8) \circ (2\ 6).$$

Theorem 2.8.4: *If a permutation of a finite set of n integers can be expresed as a product of u transposition and again as a product of v transpositions, then* $u \equiv v \bmod (2)$.

Proof: Let $f \in S_n$ and let

$$f = f_1 \circ f_2 \circ \ldots \circ f_u = g_1 \circ g_2 \circ \ldots \circ g_v,$$

where f_i and g_j are transpositions.

Let $P_n = \prod_{j<i}(i-j)$ be the product of all integers of the form $(i - j)$, where $1 \le i, j \le n$ and $j < i$. We define fP_n to be the product obtained from P_n by replacing the integers i and j in the product P_n by their respective images $f(i)$ and $f(j)$.

Thus $f P_n = \prod_{j<i}(f(i)) - f(j))$, for every $f \in S_n$.

In $P_n = \prod_{j<i}(i-j)$ there is a factor $(k - l)$, $l < k$ such that $f_u(k) - f_u(l) = l - k = -(k - l)$, and $f_u(i) = i$ and $f_u(j) = j$ where f_u is a transposition. So by one transposition we have $f_u P_n = - P_n$ and by another transposition f_v

$$(f_v \circ f_u) P_n = f_v(-P_n) = (-1)^2 P_n.$$

Therefore, if there are u transposition, then

$$fP_n = (f_1 \circ f_2 \circ \ldots \circ f_u) P_n = (-1)^u P_n \quad \ldots(1)$$

Similarly

$$f P_n = (g_1 \circ g_2 \circ \ldots \circ g_v) P_n = (-1)^v P_n \quad \ldots(2)$$

Hence from (1) and (2)

$$(-1)^u P_n = (-1)^v P_n$$

which implies that u and v both are either even or odd

Hence $u \equiv v \pmod 2$.

Definition 2.8.4: A permutation f is said to be even permutation if f is expressed as the product of even number of transpositions.

Definition 2.8.5: A permutation f is said to be odd permutation if f is expressed as the product of odd number of transpositions.

Remarks:

(1) From the definitions of odd and even permutations it follows that an even permutation cannot be equal to odd permutation.

(2) A transposition is always an odd permutation.

(3) An identity permutation is considered even permutation.

(4) The product of two even permutations is even.

(5) The product of two odd permutations is even

(6) The product of an even and an odd permutation is odd.

(7) The product of an odd and an even permutation is odd.

The verification of above statement is left to the readers.

Theorem 2.8.5: *Of the n! permutations on n symbols* $\frac{n!}{2}$ *are even permutations and* $\frac{n!}{2}$ *are odd permutations.*

Proof: Let $E_1, E_2, E_3, ..., E_s$ be even permutations and Let $0_1, 0_2, 0_3, ..., 0_t$ be odd permutations.

Since there are only $n!$ permutations of the set of n integers, then it is clear that

$$s + t = n!$$

Let us consider a transposition f and compute:

$f \circ E_1, f \circ E_2, ... f \circ E_s$. We see that all permutations are distinct. For, if $f \circ E_i = f \circ E_j$ $(f \circ f) \circ E_i = (f \circ f) \circ E_j$ by multiplying by f on the left,

$$I \circ E_i = I \circ E_j, \text{ since } f \circ f = f^2 = I$$

$$\Rightarrow \quad E_i = E_j$$

But we have $E_i \neq E_j$. Thus it contradicts the hypothesis. Hence $f \circ E_1, f \circ E_2, ..., f \circ E_s$ are distinct permutations, and these are all odd permutations since the product of odd and even permutations is an odd permutation. These are s in number but we have t odd permutations. Thus

$$s \leq t$$

Again, we compute:

$f \circ 0_1, f \circ 0_2, ..., f \circ 0_t$

These are distinct t even permutations. But we have s even permutations, so $t \leq s$ which shows $s = t$, Thus $s = t = n!/2$. This completes the proof.

Example 2.8.8: The pair (A_n, o) forms a group of even permutations under the composition of permutations. It is called Alternating group.

Solution: Let $E_1, E_2, E_3 .... E_s$ be the even permutations of n symbols. We observe that:

(1) Closure property. Since the product of two even permutations is an even, the set A_n is closed under the operation.

(2) Associative property. Since the composition of permutations is associative, the compositions of even permutations is associative.

(3) Existence of identity. The identity permutation is considered as even permutation which is also identity for even permutations.

(4) Existence of Inverse. Let $f \in A_n$, then f^{-1} exists. Since $f^{-1} \circ f = f \circ f^{-1} = I$ and f and I are even permutations, then f^{-1} will also be even permutation. Hence $f^{-1} \in A_n$.

(5) Since the composition is not commutative, the composition in A_n is not commutative.

(6) The set A_n contains $n!/2$ elements. Hence (A_n, o) is a non-abelian group of order $n!/2$.

CHAPTER 2

Example 2.8.9: Express the given permutation f as the product of disjoint cycles.

Where $f = \begin{pmatrix} 1 & 2 & 3 & 4 & 5 & 6 \\ 3 & 6 & 4 & 1 & 2 & 5 \end{pmatrix}$

Solution: Here f is the permutaitons of 6 positive numbers. First we determine the set of images of 1 under the successive powers of f.

$$f(1) = 3$$
$$f^2(1) = f \, o \, f(1) = f(3) = 4$$
$$f^3(1) = f \, o \, f^2(1) = f(4) = 1$$

Thus first cylce is (1 3 4).

Now, 2 does not belong to this cycle. Thus

$$f(2) = 6, f(6) = 5, f(5) = 2$$

Thus another cycle is (2 6 5).

But in the cycles (1 3 4) and (2 6 5) all elements are exhausted.

Therefore

$$f = \begin{pmatrix} 1 & 2 & 3 & 4 & 5 & 6 \\ 3 & 6 & 4 & 1 & 2 & 5 \end{pmatrix} (1\ 3\ 4)\ 0\ (2\ 6\ 5)$$

Example 2.8.10: Determine whether the following permutations are odd or even.

(*a*) $\begin{pmatrix} 1 & 2 & 3 & 4 & 5 \\ 3 & 2 & 4 & 1 & 5 \end{pmatrix}$ (*b*) (1 2 3) 0 (4 5) 0 (6 7 8)

Solution: (*a*) $\begin{pmatrix} 1 & 2 & 3 & 4 & 5 \\ 3 & 2 & 4 & 1 & 5 \end{pmatrix}$ can be written as the product of disjoint cycles.

Thus = $\begin{pmatrix} 1 & 2 & 3 & 4 & 5 \\ 3 & 2 & 4 & 1 & 5 \end{pmatrix}$ (1 3 4) 0 (2) 0 (5) = (1 3 4) = (14) 0 (13)

And every cycle can be written as the product of transposes. We have seen that it is the product of two transposes, therefore it is an even permutation.

(*b*) (1 2 3) 0 (4 5) 0 (6 7 8) can be written in terms of transposes.

Thus

(1 2 3) 0 (4 5) 0 (6 7 8) = (1 3) 0 (1 2) 0 (4 5) 0 (6 8) 0 (6 7) which are 5 in number. Hence it is an odd permutation.

PROBLEMS

1. Prove that the symmetric group on two symbols, (S_2, o), is commutative.
2. Demonstrate that the group (S_3, o) and hence any larger symmetric group, is non-commutative by considering the permutations:

$$\begin{pmatrix} 1 & 2 & 3 \\ 2 & 3 & 1 \end{pmatrix} \text{ and } \begin{pmatrix} 1 & 2 & 3 \\ 3 & 2 & 1 \end{pmatrix}$$

3. Express the following permutations as the product of transpositions:

$$\begin{pmatrix} 1 & 2 & 3 & 4 & 5 & 6 \\ 3 & 6 & 4 & 1 & 2 & 5 \end{pmatrix}, \begin{pmatrix} 1 & 2 & 3 & 4 & 5 & 6 & 7 \\ 5 & 7 & 6 & 2 & 1 & 3 & 4 \end{pmatrix}$$

4. Show that the cycle (1 2 3 4 5 6) may be written as a product of 3 cycles.
5. Consider the set G consisting of the four permutations:

$$\begin{pmatrix} 1 & 2 & 3 & 4 \\ 1 & 2 & 3 & 4 \end{pmatrix}, \begin{pmatrix} 1 & 2 & 3 & 4 \\ 2 & 1 & 4 & 3 \end{pmatrix}, \begin{pmatrix} 1 & 2 & 3 & 4 \\ 3 & 4 & 1 & 2 \end{pmatrix} \begin{pmatrix} 1 & 2 & 3 & 4 \\ 4 & 3 & 2 & 1 \end{pmatrix}$$

Show that (G, o) is a commutative group.

6. Form $G = \{f, f^2, f^3, f^4, f^5, f^6\}$, where f is the permutation

$$f = \begin{pmatrix} 1 & 2 & 3 & 4 & 5 & 6 \\ 2 & 3 & 4 & 5 & 6 & 1 \end{pmatrix}$$

and prove that (G, o) is a commutative group.

CHAPTER 2

2.9 CYCLIC GROUPS

In previous section we have defined the generated group by the set S. Now if we have a special case that S consists of a single element a. In this situation the sub-group generated by the single element a is called the cyclic sub-group $((a), o)$ generated by a or we can make a definition like this:

Definition 2.9.1: The group (G, o) is cyclic if it contains some one element a whose powers exhaust G; this element a is said to generate the group $((a), o) = (G, o)$.

The subset (a) is rather easy to describe; as all its product involve the element a or its inverse, the set *(a)* simply reduces to the integeral powers of a:

$$(a) = \{a^n \mid n \in Z\}$$

Theorem 2.9.1: *If $a \in G$ and (G, o) is a group, the set $H = (a) = [a^k \mid k$ is any integer$\}$, then (H, o) is a sub-group of (G, o).*

Proof: First H is not empty since $a = a^1 \in H$ and we have

(*i*) $a^0 = e$, for $0 \in Z$, $e = a^0 \in H$

(*ii*) Given any element $a^n \in H$, $n \in Z$, $a^{-n} \in H$, since $-n \in Z$.

Thus every element has its inverse.

(*iii*) Associative property exists as $H \subseteq G$ over which binary operation o is associative, which completes the proof of the theorem.

Example 2.9.1: Let $(Z, +)$ be the group of integers. In this case, since the group operation is addition, the product $a\ o\ a\ o\ a\ o ... oa$ is replaced by the ordinary sum $a + a + a + a + ... + a$. The subset generated by a

$(a) = \{na \mid n \in Z)$

Using this notation, we have

$\{0\} = \{0\}$, $Z = (I)$, $E = (2)$. We may thus conclude that both groups $(Z, +)$ and $(E, +)$ are cyclic, with the integers 1 and 2 as their respective generators. These are infinite cyclic groups.

Example 2.9.2: Let $(Z_{12} +_{12})$ be the group of integers modulo 12. We shall write $a +_{12} a +_{12} a +_{12} . . . +_{12} a$ for $a\ o\ a\ o\ a\ o....o\ a$. The subset generated by 3, $(3) = \{3n\ (m\ o\ d\ _{12}) \mid n \in Z.\} = \{0, 3, 6, 9\}$ and the cyclic group generated by 3 is $(\{0, 3, 6, 9), +_{12})$ As $(1) = Z_{12}$, then $(Z_{12}, +_{12})$ is itself a cylic group. These are finite cyclic groups.

Example 2.9.3: Let $(\{1, -1, i, -i),.)$ be a group. We see that $i^1 = i$, $i^2 = -1$, $i^3 = -i$, $i^4 = 1$, the identity of the group. Thus this group has order 4, and it is a finite cyclic group with i as generator. Similarly we can see that $-i$ is also a generator of this group.

Theorem 2.9.2: *If $((a), o)$ is a finite cyclic group of order n, then $(a) = \{e, a, a^2, ...a^{n-1}\}$.*

Proof: Since (a) is a finite set, all powers of the generator a are not distinct. Thus there must be repetition $a^i = a^j$ with $i \neq j$. On multiplying this equation by a^{-i}, we obtain

$$a^i\ oa^{-i} = a^j\ o\ a^{-i}$$

$$\Rightarrow \quad e = a^{j-i}, \text{ let } j - i = k$$

$$\Rightarrow \quad a^k = e$$

Thus the set of positive integers k for which $a^k = e$ is non-empty. Let m be the smallest positive integer with this property; that is, $a^m = e$, while $a^k = e$ for $0 < k < m$.

The set $S = \{e, a, a^2, . . . a^{m-1}\}$ consists of distinct element of (a). Suppose $a^r = a^s$ with $0 \leq r < s \leq m-1$ multiplying $a^r = a^s$ by a^{-r} on the right. We have

$$a^r\ o\ a^{-r} =- a^s\ o\ a^{-r} = a^{s-r}$$

$$\Rightarrow \quad a^{s-r} = e, \text{ when } 0 < s - r < m - 1,$$

which is contrary to the minimality of m. Hence no two elements of the set are equal.

Now each element a^k belongs to the set $S = \{a, a, a^2, . . . a^{m-1}\}$ If $k > m$ suppose there exist two numbers of q and r such that

$$k = qm + r,\ 0 \leq r < m$$

Hence
$$a^k = a^{qm+r} = (a^{mq})\ oa^r$$
$$= (a^m)^q\ oa^r$$
$$= a^r \in S, \text{ as } a^m = e$$

Hence $(a) = S = \{e, a, a^2, ..., a^{m-1}\}$,

and the subsequent equality $m = n$.

This completes the proof of the theorem.

If $a \in G$, where (G, o) is group, we define the order of a to be the order of the cyclic sub-group $((a), o)$ generated by a.

***Theorem 2.9.3:** If (G, o) be a finite group of order n with generator a. Then a^m is also a generator of G if and only if $m < n$ and $(m, n) = 1$ i.e..m is relatively prime to n.*

Proof: Let H be the set generated by a^m and

Let m be relatively prime n.

The H, C, F, of m, n is I. Therefore there exist two numbers p and q such that

$$pm + qn = 1$$

Now,
$$a^{pm+nq} = a^1$$
$$\Rightarrow \quad a^{pm} \; o \; a^{qn} = a$$
$$\Rightarrow \quad a^{pm} \; o \; (a^n)^q = a$$
$$\Rightarrow \quad a^{pm} \; o \; e^q = a, \text{ since } a^n = e$$
$$\Rightarrow \quad (a^m)^p = a$$
$$\Rightarrow \quad a^r = (a^m)^{pr}, \; 0 \le r \le n,$$

which implies every power of a^m belongs to the set G, thus every element $a \in G$ is some power of a^m. Thus $G \subseteq H$. We have $H \subseteq G$ which imples $H = G$.

Conversely, if a^m is a generator of G, then there exists some integer r such that

$$(a^m)^r = a$$
$$\Rightarrow \quad a^{mr} = a$$
$$\Rightarrow \quad a = a^{mr} \; o \; e = e^{mr} \; o \; (a^n)^s, \text{ since } a^n = e \text{ and } s \in Z.$$
$$\Rightarrow \quad a = a^{mr} \; o \; a^{ns}$$
$$\Rightarrow \quad a = a^{mr+ns}$$

which follows $I = mr + ns$, that is, m, n are prime to each other.

Note: If the *H.C.F.* of m, n is I, i.e., $(m, n) = 1$, then $\exists \; q, p \in Z$ such that $pm + qn = 1$. This result we have taken from number theory.

***Theorem 2.9.4:** If (G, o) is a finite cyclic group of order n and p is a divisor of n, then there exists one and only one sub-group of order p, which is also cyclic.*

Proof: Let a be the generator of the set G. Since p divides n, there exists $q \in Z$ such that $n = pq$.

Let a^q be the generator of set H. Then $H = (a^q)$ and (H, o) is a sub-group of order p. Therefore from this it is obvious that to get a sub-group of order p we chose an element of G of order p.

We also observe that there exists one and only one element of order p in G. For, if

$$o\,(a^m) = p \text{ and}$$

$$o(a^{m'}) = p$$

Then $$a^{mp} = a^{m'p} = a^n$$

which shows $$mp = m'p = n$$

or $$m = m'$$

This completes the proof of the theorem.

Example 2.9.4: Find all the sub-groups of a cyclic group of order 60.

Solution: We see that the integral divisiors of 60 are

1, 2, 3, 4, 5, 6, 10, 12, 15, 20, 30, 60.

Then there exists one and only one sub-group of each of these orders. Let a be the generator of the group and p is a divisor of 60.

Then there exists one and only one element in G whose order is p, that is, $a^{60/p}$. Thus all the elements of order, 1, 2, 3, 4, 5, 10, 12, 15, 20, 30, 60 will describe sub-groups. Thus

$(a^{60}) = (e)$, (a^{30}), (a^{20}) (a^{15}), (a^{12}), (a^{10}), (a^6), (a^5), (a^4), (a^3), (a^2), $(a^1) = G$.

Theorem 2.9.5: *If $a \in G$ is the generator of a cyclic group (G, o), then a^{-1} is also its generator.*

Let $a \in G$ be the generator of (G, o). Therefore all powers of a should belong to G.

If $a^n \in G$ for some $n \in Z$,

then $$a^{-n} = (a^n)^{-1} \in G$$

$$a^{-n} = (a^{-1})^n \text{ for some } n, -n \in Z$$

and $$a^n = (a^{-1})^{-n}$$

Hence a^{-1} is also a generator.

If $a = a^{-1} \Rightarrow a^2 = e$ which means a group of finite order 2. Thus $a \neq a^{-1}$.

Example 2.9.5: Show that $(U_n, .)$ is a cyclic group of nth roots of unity under multiplication.

Solution: Let U_n be the set of nth roots of unity, that is, roots of the equation $x^n = 1$

Thus $(1)^{1/n} = (\cos 0 + i \sin 0)^{1/m}$

$$= \cos\frac{2\pi r}{n} + i\sin\frac{2\pi r}{n}$$

$$= e\,\frac{2\pi i r}{n}$$

$$U_n = \left\{e\,\frac{2\pi i r}{n}\right\}, \text{ where } r = 0, 1, 2, 3, \ldots\ldots, n-1.$$

Here $e\,\dfrac{2\pi i}{n}$ is the generator of the group $(U_n, .)$ We have already seen that (U_n, o) is an abelian group. For the group (U_n, o) is cyclic we observe that for $o \leq r \leq n-1$

$$e^{\frac{2\pi i}{n}} \in U_n \text{ and } U_n = \left(e^{\frac{2\pi i}{n}}\right)^r$$

If $r > n$, then $\exists q$ and s such that $r = nq + s$ where $0 \leq s < n$ then

$$e^{\frac{2\pi ir}{n}} = e^{\frac{2\pi i(nq+s)}{n}}$$

$$= e^{2\pi iq} \cdot e^{\frac{2\pi is}{n}}.$$

$$= e^{\frac{2\pi is}{n}} \in U_n$$

since $e^{2\pi iq} = 1.$

Thus all powers of the generator $e^{\frac{2\pi i}{n}}$ belong to the set U_n. Hence (U_n, o) is a cyclic group and every cyclic group is abelian. Thus (U_n, o) is a finite abelian cyclic group.

PROBLEMS

1. Define a cyclic group and give an example.
2. Show that group $(\{a, a^2, a^3 = e\}, o)$ o is a cyclic group.
3. Show that the group $(\{0, 1, 2, 3, \ldots n - 1\}, t_n)$ is a cyclic group.
4. Show that the group $(\{[1], [3], [5], [7]\}, \odot_8)$ is not a cyclic group.
5. Find all sub-groups of a cyclic group of order 10.
6. Show that a finite group of order n containing an element of order n must be cyclic.
7. Does the group (G, o) have to be cyclic if all its proper sub-groups are cyclic? Explain.
8. If a group has no non-trivial sub-groups, prove that it must have prime order.
9. Prove that an infinite cyclic group has exactly two generators.
10. List all the sub-groups of a cyclic group of order 12.
11. Show that the four permutations I, $(a\ b\ c\ d)$, $(a\ c)\ o\ (b\ d)$; $(a\ b\ c\ d)$ form a cyclic group under the composite of permutations.
12. Show that every sub-group of a cyclic group is cyclic.
13. Show that every cyclic group is commutative.

2.10 COSETS OF A SUB-GROUP

We have seen in the chapter one that if R is an equivalence relation in the non-empty set G, the R induces a partition on the set. Here we shall investigate equivalence relation which arises due to an arbitrary sub-roup. The elements of the partition will be the coset which will be studied below.

Definition 2.10.1: Let (H, o) be a sub-group of the group (G, o); for $a, b \in G$ we say a is congruent to $b \; m \; o \; d \; H$, written as

$a \equiv b$ mod H if $a \; o \; b^{-1} \in H$.

Thus we can say that R is the relation in G × G given by $(a, b) \in R$ if, and only if $a \; o \; b^{-1} \in H$ for $a, b \in G$, or $R = \{(a, b) \mid a \equiv b \; m \; o \; d \; H$ or $a \; o \; b^{-1} \in H\}$.

Theorem 2.10.1: *The relation R defined by $a \equiv b$ mod H is an equivalence relation.*

Proof: It suffices to show that R is reflesive, symmetric and transitive, that is,

(1) $a \equiv a \; m \; o \; d \; H$, for all $a \in G$,

(2) $a \equiv b \; m \; o \; d \; H \Rightarrow b \equiv a \; m \; o \; d \; H$

(3) $a \equiv b \; m \; o \; d \; H$, $b \equiv c \; m \; o \; d \; H \Rightarrow a \equiv c \; mod \; H$.

(1) In showing reflexive property we have to prove that $a \; o \; a^{-1} \in H$, for all $a \in G$. Since (H, o) is a sub-group, then $e \in H$. Therefore, for all $a \in G$, $a \; o \; a^{-1} = e \in H \Rightarrow a \equiv a \; mod \; H$. Hence R is reflexive.

(2) Suppose $a \equiv b \; m \; o \; d \; H$, that is, $a \; o \; b^{-1} \in H$. Since (H, o) is a sub-group, then $(a \; o \; b^{-1})^{-1}$ must belong to H. That is, $(b^{-1})^{-1} \; oa^{-1} \in H$ or $b \; o \; a^{-1} \in H$ which means $b \equiv a \; m \; o \; d \; H$.

(3) Let $a \equiv b \; m \; o \; d \; H$ and $b \equiv c$ mod H which means $a \; o \; b^{-1} \in H$ and $b \; o \; c^{-1} \in H$. Since (H, o) is a sub-group, then if

$$a \; o \; b^{-1} \in H, \; b \; o \; c^{-1} \in H, \; (a \; o \; b^{-1}) \; o \; (b \; o \; c^{-1}) \in H$$

$\Rightarrow \quad a \; o \; (b^{-1} \; o \; b) \; o \; c^{-1} \in H$, since $b^{-1} \; o \; b = e$

$\Rightarrow \quad a \; o \; (e) \; o \; c^{-1} \in H$

$\Rightarrow \quad a \; o \; c^{-1} \in H$ which means $a \equiv c$ mod H.

Hence the relation $a \equiv b \; mod \; H$ is an equivalence relation.

Definition 2.10.2: Let (H, o) be a sub-group of the group (G, o), and let $a \in G$. The set

$$a \; o \; H = \{a \; o \; h \mid h \in H\}$$

is called a left coset of H in G. The element a is called the representative of $a \; o \; H$.

On parallel lines, we can define the right cosets $H \; o \; a$ of H. The right cosets of the same group in general are different from the left cosets. If the operation o is commutative in G, then $a \; o \; H = H \; o \; a$ for all $a \in G$.

We have established that the relation $a \equiv b \; m \; o \; d \; H$ is an equivalence relation. The equivalence class of a, denoted by $[a]$ is $[a] = \{x \in G \mid a \equiv x \; m \; o \; d \; H\}$. Now in the following theorem we see that $[a] = H \; o \; a$.

Theorem 2.10.2: *For all $a \in G$, $Hoa = [a]$.*

Proof: $[a] = \{x \in G \mid a \equiv x \; m \; o \; d \; H\}$. We have to show that $H \; o \; a = [a]$, For if $x \in H \; o \; a$, then there exists $h \in H$ such that $x = hoa$.

$$a \circ x^{-1} = a \circ (h \circ a)^{-1} = a \circ (a^{-1} \circ h^{-1}) = (a \circ a^{-1}) \circ h^{-1} = e \circ h^{-1} = h^{-1} \in H.$$

since (H, o) is a sub-group. That is, $aox^{-1} \in H$. Hence by the definition of congruence $x \in [a]$.

Thus, $H \circ a \subseteq [a]$ ...(1)

Again, let $x \in [a]$. Then $a \circ x^{-1} \in H$ *by* the definition of congruence. So $(a \circ x^{-1})^{-1} \in H$ since (H, o) is a sub-group.

That is, $x \circ a^{-1} \in H \Rightarrow (x \circ a^{-1}) \circ a \in H \circ a \Rightarrow x \in H \circ a$

Hence $[a] \subseteq H \circ a$ ...(2)

Thus from (1) and (2) we obtain $[a] = H \circ a$.

Now, it is clear that $[a]$, and thus *Hoa,* is the equivalence class of a in G, Therefore the equivalence classes yield a decomposition of G into disjoint classes. Thus, any two right cosets of H in G are either identical or disjoints.

In the following discussion, we will generally consider only right cosets of a sub-group. On parallel lines we may develop the theory of left cosets of a sub-group.

Before discussing other results of cosets we make several simple observations.

First, if e is the identity element of (G, o), then

$$H \circ e = \{h \circ e \mid h \in H\} = \{h \mid h \in H\} = H,$$

so that H itself is a right coset of H. Moreover, since $e \in H$ and (H, o) is a sub-group we have

$$a = e \circ a \in H \circ a$$

that is, the representative of the coset belongs to the coset of H.

We note further that there is one to one correspondence between the elements of H and those of any coset of H.

Let (G, o) be a group, $a \in G$ and $(H \circ a)$ a right coset of sub-group (H, o) in (G, o).

We define a mapping f: $H \rightarrow H \circ a$ by $f(h) = h \circ a$ for all $h \in H$.

We see that the right coset *Hoa* of H contains the elements of the form *hoa* for some $h \in H$. Which shows function f maps H onto *Hoa*.

For showing f one to one, let $h_1, h_2 \in H$.

We have $f(h_1) = f(h_2) \Rightarrow h_1 \circ a = h_2 \circ a \Rightarrow h_1 = h_2$ by right cancellation law. If (G, o) is the finite group, we may therefore conclude that any two right cosets of H have the same number of elements, namely, the number of elements in H.

Example 2.10.1: Consider $(Z_{15}, +_{15})$ the group of integers modulo 15. We take $\{0, 3, 6, 9, 12\}$ for the set H, then $(\{0, 3, 6, 9, 12\}, +_{15})$ is evenly a sub-group of $(Z_{15} +_{15})$. Find the right cosets of H.

Solution: The right cosets of H in Z_{15} are,

where $H = \{0, 3, 6, 9, 12\}$,

$$H +_{15} 0 = \{0, 3, 6, 9, 12\},$$

CHAPTER 2

$$H +_{15} 3 = \{3, 6, 9, 12, 0\},$$
$$H +_{15} 6 = \{6, 9, 12, 0, 3\},$$
$$H +_{15} 9 = \{9, 12, 0, 3, 6\},$$
$$H +_{15} 12 = \{12, 0, 3, 6, 9\},$$

Hence $H +_{15} 0 = \{0, 3, 6, 9, 12\} = H +_{15} 3 = H +_{15} 6 = H +_{15} 9 = H +_{15} 12.$

$H +_{15} 1 = (1, 4, 7, 10, 13\} = H +_{15} 4 = H +_{14} 7 = H +_{15} 10 = H +_{15} 13$

$H +_{15} 2 = (2, 5, 8, 11, 14) = H +_{15} 2 = H_{15} 5 = H +_{15} 8 = H +_{15} 14.$

We also observe that every coset of H and H have the same number of elements. If a, $b \in H \circ a$, then $H \circ a = H \circ b$ is also verified from the example.

Theorem 2.10.3: *If (H, o) is a sub-group of the group (G, o), then $Hoa = H$ if and only if $a \in H$.*

Proof: (*i*) Let $H \circ a = H$. Since (H, o) is a sub-group,

$e \in H, e \circ a \in H \circ a \Rightarrow a \in H \circ a$

Thus, $a \in H \circ a = H \Rightarrow a \in H$

(*ii*) Conversely, if $a \in H$, then $a^{-1} \in H$ since (H, o) is a sub-group. We note that each element $h \in H$ may be written as $h = h \circ (a^{-1} \circ a) = (h \circ a^{-1}) \circ a \in H \circ a$, since $h \circ a^{-1} \in H$ by the closure condition.

which follows $H \subseteq H \circ a$ .....(1)

Also, if $b \in H \circ a$, $b = h \circ a$ for some $h \in$ H

as $a \in H$, $b = h \circ a \in H$

Therefore, $H \circ a \subseteq H$ ....(2)

Thus, $H \circ a = H$ from (1) and (2).

Theorem 2.10.4: *If (H, o) is a sub-group of the group (G, o), then $H \circ a = H \circ b$, if and only if $b \circ a^{-1} \in H$.*

Proof: (*i*) Assume that $H \circ a = H \circ b$.

$a = e \circ a \in H \circ a = H \circ b$. Since (H, o) is a sub-group, therefore $\exists\ h \in H$ such that $a = hob$

So $$e = a \circ a^{-1} = (h \circ b) \circ a^{-1}$$
$$= h \circ (b \circ a^{-1})$$

which shows that $(b \circ a^{-1})$ is the inverse of $h \in H$ and hence $b \circ a^{-1} \in H$

(*ii*) If $b \circ a^{-1} \in H$,

Then $b = b \circ (a^{-1} \circ a) = (b \circ a^{-1}) \circ a \in H \circ a$.

But $b \in H \circ b$.

Thus $H \circ a$ and $H \circ b$ have a common element b. So $H \circ a = H \circ b$, since two cosets are either idenitcal or disjoint.

Definition 2.10.3: (*a*) The number of distinct right (left) cosets of any sub-group of a finite group is called the index of the same.

Theorem 2.10.5: *(Lagrange Theorem) The order of each sub-group of a finite group divides the order of the group.*

Proof: Suppose (G, o) is a finite group of order n, and (H, o) is a sub-group of (G, o) of order k. Then we can decompose the set G **into a union of a finite number of disjoint right cosets of** $\boldsymbol{H}$**:**

$$G = (H \ o \ a_1) \cup (H \ o \ a_2) \cup \ldots\ldots \cup (H \ o \ a_r)$$

thus there are r distinct cosets and each coset has the same number of element as the set H. Thus r cosets have $r.\ k$ elements which is the number of elements of G, that is, n.

From this discussion we conclude that

$$o\ (G) = n = r.k = r.\ o\ (H)$$

order of G = (index of H). (order of H)

which shows that $o\ (H)$ divides $o\ (G)$.

Example 2.10.2: Consider $(Z_{12}, +_{12})$ the group of integers modulo 12. If we take $\{0, 4, 8\}$ for the set H, then $(\{0, 4, 8\}, +_{12})$ is evidently a sub-group of $(Z_{12}, +_{12})$. Find left cosets of H in Z_{12}.

Solution:

$$0 +_{12} H = \{0, 4, 8\} = 4 +_{12} H = 8 +_{12} H,$$
$$1 +_{12} H = \{1, 5, 9\} = 5 +_{12} H = 9 +_{12} H,$$
$$2 +_{12} H = \{2, 6, 10\} = 6 +_{12} H = 10 +_{12} H,$$
$$3 +_{12} H = \{3, 7, 11\} = 7 +_{12} H = 11 +_{12} H.$$

In this way all the elements of Z_{12} are exhausted by the coset decomposition of H. Hence

$$Z_{12} = \{0, 4, 8\} \cup \{1, 5, 9\} \cup \{2, 6, 10\} \cup \{3, 7, 11\}$$

Hence Index of $H = 4$

order of H, $o\ (H) = 3$

order of Z_{12}, $o\ (Z_{12}) = 12$

Hence $o\ (Z_{12})$ = (Index of H). $o(H)$.

Example 2.10.3: If $(Z, +)$ is the group of integers together with addition and $(H, +)$ is the sub-group of $(Z, +)$ of even integers, then the cosets of H are

H and $H + 1$ such that $Z = H \cup (H + 1)$

$H + 1$ is the coset consisting of all odd integers.

Theorem 2.10.6: *If $(H, *)$ and $(K, *)$ are finite sub-groups of a group $(G, *)$ then*

$$o(H * K) = \frac{o(H).\ o(K)}{o(H \cap K)}$$

CHAPTER 2

Proof: Since $(H, *)$ and $(K, *)$ are sub-groups, $(H \cap K, *)$ is also a sub-group of $(G, *)$. Since $H \cap K \subseteq K$, $(H \cap K, *)$ is also a sub-group of $(K, *)$.

Let $D = H \cap K$ and let the index of D in K be r.

Then $K = D * k_1, \cup D * k_2 \cup ... \cup D * k_r = \bigcup_{i=1}^{r} D * k_i$, *That is,*

$$r = \frac{o(K)}{o(D)}$$

Further
$$\begin{aligned} H * K &= H * (D * k_1 \cup .... \cup D * k_r) \\ &= H * D * k_1 \cup .....\cup H * D * k_r \\ &= H * k_1 \cup .... \cup H * k_r, \text{ since } D \subseteq H \end{aligned}$$

Thus $H * k_1, H * k_2,....., H * k_r$ are right cosets of H.

Now we see that for some i and j,

$$\begin{aligned} H * k_i = H * k_j &\Rightarrow k_i * k_j^{-1} \in H \\ &\Rightarrow k_i * k_j^{-1} \in H \cap K = D, \text{ since } k_i * k_j^{-1} \in K \\ &\Rightarrow D * k_i = D * k_j \\ &\Rightarrow i = j \end{aligned}$$

Hence $H * k_1, H * k_{2,}......., H * k_r$ are pairwise disjoint right cosets $o\ f\ H$.

In each right coset of H there are $o\ (H)$ elements and there are r cosets.

$$H * K = H * k_1 \cup H * k_2 \cup ... \cup H * k_r$$

$$\Rightarrow \quad o\ (H * K) = r.o(H) = \frac{o(H).\ o(K)}{o(D)} = \frac{o(H).o(K)}{o(H \cap K)}, \text{ Since } r = \frac{o(K)}{o(D)}$$

Theorem 2.10.7: *(a) (G, o) is a group of order n, then the order of any element $a \in G$ is a factor of n. (b) $a^n = e$.*

***Proof:* (*a*)** To prove the result we shall make use of Langrange's theorem. We find a sub-group of (G, o) whose order is $o\ (a)$.

If $a \neq e$ and $a \in G$, then a generates the cyclic sub-group $((a), o)$, $H = (a) = \{a^k \mid k \in Z\}$. Thus (a) consists of $e, a, a^2....$ The order of (H, o) will be the order of a. For if $o\ (a) = m$, then $a^m = e$. If this sub-group contains fewer elements than this number of elements, then $a^i = a^j$ for some integers $0 \le i \le j < m$. Then $a^{j-i} = e$. Yet $0 < j - i < m$ which shows the order of a is $j - i$ which contradicts the definition of order of a. Thus cyclic sub-group generated by a has $o(a) = m$ elements. By Langrange's theorem $o(G)$ is divisible by $o(a)$.

(*b*) Just we have seen that the order $o\ (a)$ of a divides the order $o\ (G)$ of G. Thus $o\ (G) = k.\ o\ (a)$

or $n = k.m$ for some $k \in N$.

Hence $a^n = a^{km} = (a^m)^k = e^k = e$

Definition 2.10.4: The order of the group M of units of $(Z_n +_m)$ is the number of positive integers that are less than m and are relatively prime to m, $(a, m) = 1$. This number is denoted by $\phi(m)$ and the function of m thus determined is called Euler ϕ-function (totioint).

***Theorem 2.10.8:** (Euler-Fermat). If m is a positive integer and $(a, m) = 1$, that is, a is relatively prime to m, then $a^{\phi(m)} \equiv 1 \ mod \ m$.*

Proof: First, let us consider the group (M, o) of units $(Z_n +_m)$ which are less than m and relatively prime to m. Thus, the order of the group (M, o) of units is the number $\phi(m)$.

But we know, that if (G, o) is a finite group of order n, then $a^n = 1$ for every $a \in G$. Applying this result to the group (M, o) of units of order $\phi(m)$. If $a < m$, and $(a, m) = 1$ then $[a] \neq [m]$ and if a is a unit then $[a]$ is also a unit in $(Z_m, +_m)$, Hence $[a] \in M$, therefore

$$[a]^{\phi m} = [1], \text{ where } [1] \in M,$$

which implies $[a]^{\phi(m)} \equiv 1 \ m\ o\ d\ m$

That completes the proof of the theorem.

Example 2.10.4: We shall illustrate the Euler-Fermat's Theorem. *If* n is a positive integer and a is relatively to n, then $a^{\phi(n)} \equiv 1 \ mod \ n$, by a numerical example.

First, we should know $\phi(n)$, the Euler ϕ-function, which is defined for all integers n by: $\phi(1) = 1$, for $n > 0$, $\phi(n)$ = number of positive integers less than n and relatively prime to n. Thus for instance $\phi(16) = 8$, since 1, 3, 5, 7, 9, 11, 13, 15, are the numbers which are less than 16 and relatively prime to 16.

Since these numbers 1, 3, 5, 7, 9, 11, 13, 15 form a group under multiplication mod 16. This group has order $\phi(16) = 8$. Thus the set of positive integers which are less than n and relatively to n form a group of order $\phi(n)$ under multiplication *mod n*. Hence $a^{\phi(n)} = 1$ for $a \in G$.

or $a^{\phi(n)} \equiv 1 \ (m\ o\ d\ n)$

***Theorem 2.10.9:** (Fermat) If a is an integer, p a prime then $a^p = a \ (mod \ p)$.*

Proof: The group $\left(Z_p^+, \odot_p\right)$ is the multiplication group of integers modulo p (excluding zero). The set Z_p^+ has $(p - 1)$ elements 1, 2, 3,..., $(p - 1)$ which are relatively prime to p. The order of any element $a \in Z_p^+$ is a divisor of $(p - 1)$. So $a^{p-1} \equiv 1 \ mod \ p$. And $\phi(p) = p-1$. Then by the Euler theorem

$$a^{\phi(p)} \equiv 1 \ mod \ p$$

or

$$a^{p-1} \equiv 1 \ mod \ p$$

multiplying by $a \neq 0$, $a^p \equiv a \ mod \ p$.

This completes the proof of the theorem.

***Theorem 2.10.10:** If (G, o) is a finite group of composite order, then (G, o) has non-trival sub-groups.*

CHAPTER 2

Proof: If the group (G, o) is not cyclic, any element $a \in G$ with $a \neq e$ generates a non-trivial cyclic sub-group $((a), o)$. Thus we should consider (G, o) the cyclic group of composite order. To this end, suppose $G = (a)$, where the generator a has the order $mn (n, m \neq 1)$ then $(a^n)^m = a^{mn} = e$, while $(a^n)^{m'} \neq e$ for $0 < m' < m$. From this it is obvious that $((a^n), o)$ is a non-trivial sub-group of (G, o) with order m.

Theorem 2.10.11: *Every group (G, o) of prime order is cyclic.*

Proof: *If* $a \neq e$ and $a \in G$, then a generates cyclic sub-group (H, o), where $H = \{a^k \mid k \in Z\}$. The order of H divides p, a prime number, and hence the order of H must be equal to p. That *is*, $G = H$ and a is a generator of G.

Hence the theorem.

Theorem 2.10.12: *The only abstract groups of order 4 are the cyclic group of that order and the four groups.*

Proof: If (G, o) a group of order 4 contains an element of order 4, it is cyclic, otherwise, all elements of G except e must have order 2. Call them a, b, c. By cancellation law, aob cannot be either $a \; o \; e = a$, $e \; o \; b = b$ or $eoa = e$ Hence, $a \; o \; b = c$. By symmetry $a \; o \; c = c \; o \; a = b$, $b \; o \; c = c \; o \; b = a$, $b \; o \; a = c$. but these together with $a^2 = b^2 = c^2 = e$, and $e \; o \; x = x \; o \; e = x$ for all x, give the multiplication table of the four group.

Theorem 2.10.13: *Any non-commutative group has at least six elements.*

A group of prime order is a cycle group and is commutative. Thus, any group of order 2, 3 or 5 will be commutative. Suppose (G, o) is a group of order 4. By Lagrange's theorem every element of G except e has order 2 or 4. If one of them has order 4, (G, o) is a cyclic group of order 4 and therefore commutative. On the other hand, a group each of whose elements other than e has order 2 must be commutative as we have seen. This argument establishes that all groups of order less than 6 are commutative groups.

Exercise 2.10.5: If (H, o) and (K, o) are two sub-groups of a group (G, o), then for any $a, b \in G$ either $H \; o \; a \cap K \; o \; b = \phi$

or $$H \; o \; a \cap Kob = (H \cap K) \; o \; c, \text{ for some } c \in G.$$

Solution: Let $H \; o \; a \cap K \; o \; b \neq \phi$. Then there exists an element $c \in G$ such that $c \in H \; o \; a \cap K \; o \; b$. This implies that $c \in H \; o \; a$ and $c \in K \; o \; b$. But $c \in H \; o \; c$ and $c \in K \; o \; c$.

Thus $c \in H \; o \; a \cap H \; o \; c$ and $c \in H \; o \; b \cap K \; o \; c$.

This in turn implies that $H \; o \; a = H \; o \; c$, $H \; o \; b = K \; o \; c$.

Hence $$H \; o \; a \cap K \; o \; b = H \; o \; c \cap K \; o \; c$$

Now $$H \cap K \subseteq H \Rightarrow (H \cap K) \; o \; c \subseteq H \; o \; c$$

and $$H \cap K \subseteq K \Rightarrow (H \cap K) \; o \; c \subseteq K \; o \; c.$$

These togather implies

$$(H \cap K) \; o \; c \subseteq H \; o \; c \cap K \; o \; c. \qquad ...(1)$$

Again, $x \in H \circ c \cap K \circ c \Rightarrow x \in H \circ c$ and $x \in K \circ c$.

$$\Rightarrow x = h \circ c \text{ and } x = k \circ c, \text{ for some } h, \in H\ k \in K$$

$$\Rightarrow x = h \circ c = k \circ c$$

$$\Rightarrow h = x \circ c^{-1} = k \in H \cap K$$

$$\Rightarrow x \in (H \cap K) \circ c$$

Therefore $H \circ c \cap K \circ c \subseteq (H \cap K) \circ c$ ...(2)

From (1) and (2) we get

$$H \circ c \cap K \circ c = (H \cap K) \circ c.$$

Exercise 2.10.6: If (H, o) and (K, o) are two sub-groups of finite indices in (G, o) then show that $H \cap K$ is also of finite index in G.

Solution: Let $[G; H] = m$ and $[G; K] = n$, and let $H \circ a_1, H \circ a_2, \ldots, H \circ a_m$ and $K \circ b_1, \ldots, K \circ b_n$ be the distinct right cosets of H and K respectively. By above exercise, for any $1 \leq i \leq m$ and $1 \leq j \leq n$, either

$$H \circ a_i \cap K \circ b_j = \phi \text{ or } H \circ a_i \cap K \circ b_j = (H \cap K) \cdot c_{ij}$$

for some $c_{ij} \in G$.

Also $(H \cap K) \circ z = H \circ z \cap K \circ z$, so each right coset of $H \cap K$ is determined by the intersection of right coset of H and right coset of K. Hence the number of right cosets of $H \cap K$ is $m.n$. Hence $[G; H \cap K] = mm$.

Exercise 2.10.7: Let a be an element of order m in a group (G, o), the $o(a^k) = \dfrac{m}{(m, k)}$.

Solution: Let $(m, k) = d$, then $m = pd$ and $k = qd$ for some $p, q \in N$ such that $(p, q) = 1$.

Now $(a^k)^p = (a^{qd})^p = a^{pqd} = a^{mq} = (a^m)^q = e^q = e$;

where e in the identity of G.

Further if $(a^k)^l = e$, then $m \mid kl$ as $a^m = e$.

$$\Rightarrow pd \mid kl$$

$$\Rightarrow pd \mid qd\ l \Rightarrow p \mid ql.$$

But $(p, q) = 1$, then $p \mid l$. Hence $o(a^k) = p = \dfrac{m}{d}$.

Exercise 2.10.8: Let a be an element of order mn in any group $[G, o]$ and let $(m, n) = 1$. Then a has a unique representation $a = b \circ c = c \circ b$, where $b, c \in G$ and are of orders m and n respectively.

Solution: Since $(m, n) = 1$, then there exist two integer x and y such that $mx + ny = 1$.

Since $o(a) = mn$, $a^{mn} = e$, where e is the identity in G.

$$a^1 = a^{mx+ny}$$

$$\Rightarrow \quad a = a^{mx} \circ a^{ny}$$

CHAPTER 2

Put $b = a^{mx}$ and $c = a^{ny}$, then

$a = b \text{ o } c$ and $c^m = (a^{ny})^m = a^{mny} = e$

Let $c^k = e$, then $a^{nyk} = e$. Thus $mn \mid nyk$ as $a^{mn} = e$.

$\Rightarrow$ $m \mid yk$

Now $mx + ny = 1$

$\Rightarrow$ $mkx + nky = k$ and $m \mid mkx$, $m \mid n\ yk)$

$\Rightarrow$ $m \mid k$. Hence $o\ (c) = m$

Similarly, it can be prove that $o\ (b) = n$.

Clearly, $b \text{ o } c = c \text{ o } b$ as each is a power of a.

For uniqueness, let $a = x \text{ o } y = y \text{ o} x$, with $o\ (x) = m$,

$o\ (y) = n$. Then $x \text{ o } a = x \text{ o } (y \text{ o } x) = (x \text{ o } y) \text{ o } x = a \text{ o} x$

and $y \text{ o } a = y \text{ o } (x \text{ o } y) = (y \text{ o } x) \text{ o } y = a \text{ o } y$

Since c and b are powers of a, $x^{-1} \text{ o } b = b^{-1} \text{ o } x$ and $y \text{ o } c^{-1} = c^{-1} \text{ o } y$.

Now $a = b \text{ o } c = x \text{ o } y$

$\Rightarrow$ $x^{-1} \text{ o } b = y \text{ o } c^{-1}$.

Again $(x^{-1} \text{ o } b)^m = x^{-m} \text{ o } b^m = e$

and $(y \text{ o } c^{-1})^n = y^m \text{ o } c^{-n} = e$.

Thus we get $u^m = e$ and $u^n = e$, where $u = x^{-1} \text{ o } b = y \text{ o } c^{-1}$.

So $o(u) \mid m$ and $o(u) \mid n$. But $(m, n) = 1$, $o(u) = 1$

This in turn gives $x^{-1} \text{ o } b = y \text{ o } c^{-1} = e$. Hence $x = b$, $y = c$.

PROBLEMS

1. In each of the following find the left coset of H in G in each case:

(*a*) $(G, +_4)$ is the group of integers modulo 4 and $H = \{o\}$,

(*b*) $(G, +_{15})$ is the group of integers modulo 15 and $H = \{0, 3, 6, 9, 12\}$

(*c*) $(G, +_{15})$ is the group of integers modulo 15, $H = \{0, 5, 10\}$

(*d*) $(\{1, -1, i, -i\})$; and $H = \{1, -1\}$.

2. Let (H, o) be sub-group of (G, o). Prove that if (G, o) is commutative, then left coset aoH is equal to right coset Hoa.

3. Determine the coset decomposition of the sub-group

$H = \{(1), (12)\}, o)$ in S_3.

4. If (H, o) with order m is a sub-group of the group (G, o) with order n, then prove that $n = mj$, where j is the index of H in G.

5. Show that the order of a sub-group of finite group is a factor of the order of group.

6. If (H, o) is a sub-group of the group (G, o), then there is one to one correspondence between the set of the cosets of H in G and the right cosets of H in G.

7. Show that the relation $a \equiv b \bmod H$ if and only if $a \, o \, b^{-1} \in H$ *is* an equivalence relation.

8. Let (H, o) a sub-group of the group (G, o). Prove that there is one to one correspondence between any two right (left) cosets of H in G.

9. The order of every element of a finite group is a divisor of the order of the group.

10. If a is an integer and p a prime, then $a^p \equiv a \bmod (p)$.

11. Every group of composite order possesses a proper sub-group.

12. Every group of prime order is cyclic.

13. Let (G, o) be a finite group of prime order. Then prove that it has no proper sub-group.

14. Determine the coset decomposition of the sub-group $(\{f_1 \, f_2\}, o)$ of the following group functions:

$$f_1(x) = x, f_2(x) = \frac{1}{1-x}, f_3(x) = \frac{x-1}{x}, f_4(x) = \frac{1}{x}, f_5(x) = 1-x, f_6(x) = \frac{x}{x-1}$$

with respect to the operation defined by $(f_1 o f_j)\, x = f_i f_j(x))$.

2.11 NORMAL SUB-GROUPS

In this section we narrow the field and pay our attention to the restricted class of sub-groups which we shall refer to as normal sub-groups. Let us now proceed to develop this in detail. We have seen in the last section that each sub-group induces a decomposition of the elements of the parent group into disjoint subsets known as cosets.

Definition 2.11.1: Let (H, o) be a sub-group of the group (G, o). Then the sub-group (H, o) is said to be normal (invariant) in (G, o) if and only if left coset of H in G is also a right coset of H in G.

Thus, if (H, o) is normal and $a \, o \, H$ is left coset of H in G, then there exists some element $b \in G$ such that

$$a \, o \, H = H \, o \, b.$$

Since $a \in a \, o \, H$, then $a \in H \, o \, b$. If we have the right coset $H \, o \, a$ of H in G, then the right cosets $H \, o \, b$ and $H \, o \, a$ have common element a. We know if two right cosets have common element then they are identical, that is

$$H \, o \, a = H \, o \, b.$$

In this way we can conclude that if $a \, o \, H$ is a left coset of H in G, then it must be the right coset $H \, o \, a$. Thus, we reformulate the definition as follows:

Definition 2.11.2: A sub-group (H, o) is normal in the group (G, o) if and only if $a \, o \, H = H \, o \, a$ for every $a \in G$.

Thus, if (H, o) is a normal sub-group of (G, o) then the left and right cosets are not distinguished. The trivial sub-groups $(\{a\}, o)$ and (G, o) are always nromal sub-groups.

Definition 2.11.3: A group (G, o) is said to be simple if it has no normal sub-groups other than its trivial sub-groups.

In the definition of normal sub-group we know that $H \; o \; a = a \; o \; H$ does not mean that for any $h \in H$, $h \; o \; a = a \; o \; h$. Thus, for $h \in H$ there exists an element $h' \in H$ such that $h \; o \; a = a \; o \; h'$

If the (G, o) is commutative then every sub-group is a normal sub-group. But the converse of it is not true, that is, if every sub-group of the group (G, o) *is* normal, then the group (G, o) may not be an abelian.

Theorem 2.11.1: *The sub-group (H, o) is a normal sub-group of the group (G, o) if and only if for each element $a \in G$,*

$$a \; o \; H \; o \; a^{-1} \subseteq H.$$

Proof: First, we assume that $a \; o \; H \; o \; a^{-1} \subseteq H$ for every $a \in G$ and we shall prove that (H, o) is a normal sub-group, that is, $a \; o \; H = H \; o \; a$.

Let $aoh \in a \; o \; H$, for some $h \in H$.

Since $a \; o \; H \; o \; a^{-1} \subset H$, then

$a \; o \; h \; o \; a^{-1} \in H$

Therefore there exists some $h_1 \in H$ such that

$$a \; o \; h \; o \; a^{-1} = h_1$$

$$\Rightarrow \qquad a \; o \; h = h_1 \; o \; a$$

Since $h_1 \; o \; a \in H \; o \; a$, then $aoh \in H \; o \; a$.

Hence $a \; o \; H \subseteq H \; o \; a$.

To get the opposite inclusion we know that

$$a^{-1} \; o \; H \; o \; a = a^{-1} \; o \; H \; o \; (a^{-1})^{-1}$$

Since $a^{-1} \in G$, then $a^{-1} \; o \; H \; o \; a = a^{-1} \; o \; H \; o \; (a^{-1})^{-1} \subseteq H$

Thus, there exist h and h_1, such that

$$a^{-1} \; o \; h \; o \; (a^{-1})^{-1} = h_1$$

$$\Rightarrow \qquad a^{-1} \; o \; h \; o \; a = h_1$$

$$\Rightarrow \qquad h \; o \; a = a \; o \; h_1$$

Since $h \; o \; a \in H \; o \; a$, then $h \; o \; a \in a \; o \; H$.

Hence $H \; o \; a \subseteq aoH$

This proves that $H \; o \; a = a \; o \; H$.

Hence (H, o) is a normal sub-group.

Conversely, let $a \; o \; H = H \; o \; a$, for each $a \in G$.

Let $a \; o \; h_1 \; o \; a^{-1} \in a \; o \; H \; o \; a^{-1}$, for some $h_1 \in H$.

Then, since $a \; o \; H = H \; o \; a$, there exists an element $h_2 \in H$ such that $a \; o \; h_1 = h_2 \; o \; a$

$$\Rightarrow \quad a \; o \; h_1 \; o \; a^{-1} = (h_2 \; o \; a) \; o \; a^{-1}$$

$$\Rightarrow \quad a \; o \; h_1 \; o \; a^{-1} = h_2 \in H$$

which follows $a \; o \; h_1 \; o \; a^{-1} \in H$ and $a \; o \; H \; o \; a^{-1} \subseteq H$.

Example 2.11.1: We consider an example of the non-commutative group of order 6, given in the example 2.1.2. The elements of the set G are functions $f_1, f_2, f_3, f_4, f_5, f_6$. For convenience the operation table for the group (G, o) is given below:

0	f_1	f_2	f_3	f_4	f_5	f_6
f_1	f_1	f_2	f_3	f_4	f_5	f_6
f_2	f_2	f_1	f_6	f_5	f_4	f_3
f_3	f_3	f_4	f_1	f_2	f_6	f_5
f_4	f_4	f_3	f_5	f_6	f_2	f_1
f_5	f_5	f_6	f_4	f_3	f_1	f_2
f_6	f_6	f_5	f_2	f_1	f_3	f_4

We take the subset $H = \{f_1, f_4, f_6\}$ the pair (H, o) is a sub-group. Since the order of H and G are 3 and 6 respectively, the number of cosets of H in G will be 2. These are followings:

$$f_1 \; o \; H = \{f_1, f_4, f_6\} = H \; o \; f_1 = f_4 \; o \; H = H \; o \; f_4 = f_6 \; o \; H = H \; o \; f_6,$$

$$f_2 \; o \; H = \{f_2, f_3, f_5\} = H \; o \; f_2 = f_3 \; o \; H = H \; o \; f_3 = f_5 \; o \; H = H \; o \; f_5$$

which shows that (H, o) is a normal sub-group.

Now we take the sub-group $(\{f_1, f_2\} \; o)$ and we compute

$$f_4 \, f_2 \, o f_4^{-1} = (f_4, o f_2,) \, o f_4^{-1}$$

$$= f_3, o f_6, = f_5 \notin \{f_1, f_2\}$$

Hence $(\{f_1, f_2\}, o)$ is not normal.

Theorem 2.11.2: *If (H_1, o) and (H_2, o) are two normal sub-groups of (G, o) then $(H_1 \cap H_2, o)$ is also a normal sub-group.*

Proof: Since (H_1, o) and (H_2, o) are sub-goup of (G, o), then

$(H_1 \cap H_2, o)$ is also a sub-gourp of (G, o).

To prove that $(H_1 \cap H_2, o)$ is normal we must show that if $h \in H_1 \cap H_2$

$aohoa^{-1} \in H_1 \cap H_2$ for every $a \in G$.

Let $h \in H_1 \cap H_2$, then $h \in H_1$ and $h \in H_2$

Since $(H_1\ o)$ and (H_2, o) are normal sub-groups, then for every $a \in G$,

$a\ o\ h\ o\ a^{-1} \in H_1$, and $a\ o\ h\ o\ a^{-1} \cap \mathrm{H}_2$

Therefore, $a\ o\ h\ o\ a^{-1} \in H_1 \cap H_2$

Hence $(H_1 \cap H_2, o)$ is a normal sub-group.

***Theorem 2.11.3:** The intersection of any collection of normal sub-groups is itself a normal sub-group.*

Proof: Let $\{H_n\}$ be any collection of sets H_n, $n \in N$, where (H_n, o) is a normal sub-group for every $n \in N$.

Let $h \in \cap H_n$, $n \in N$, then

$h \in H_n$, $\forall\ n \in N$.

Since (H_n, o) is a normal sub-group,

$a\ o\ h\ o\ a^{-1} \in H_n\ \forall\ n \in N$.

Therefore $a\ o\ h\ o\ a^{-1}\ e \in H_n$.

Hence $(\cap\ Hn, o)$ is a normal sub-group.

Definition 2.11.4: If (H, o) is a normal sub-group of the group (G, o), then we shall denote the set of distinct right cosets *of H in G* by $G/H : G/H = \{a\ o\ H \mid a \in G\}$.

We have already defined the product of two sub-groups (H, o) and (K, o) of (G, o). Similarly we can define the product of two sets A and B of G, $A\ o\ B = \{a\ o\ b \mid a \in A,\ b \in B\}$.

If we take $A = B = H$, where (H, o) is a sub-gourp of (G, o), then $HoH = \{h_1oh_2 \mid h_2, h_2 \in H\}$. It is clear that $HoH \subseteq H$ since H is closed under the operation.

***Theorem 2.11.4:** The sub-group (H, o) of the group (G, o) is a normal sub-group if and only if the product of two right cosets of H in G is again a right coset of H in G.*

Proof: Let (H, o) be a normal sub-group of (G, o) and that $a, b \in G$.

We consider

$$\begin{aligned}(H\ o\ a)\ o\ (H\ o\ b) &= H\ o\ (a\ o\ H)\ o\ b\\ &= H\ o\ (H\ o\ a)\ o\ b, \text{ since } (H, o) \text{ is a normal, } H\ o\ a = a\ o\ H\\ &= (H\ o\ H)\ o\ (a\ o\ b)\\ &= H\ o\ (a\ o\ b), \text{ since } (H, o) \text{ is a sub-group, } H\ o\ H = H.\end{aligned}$$

Thus, the product of two right cosets is again a right coset when (H, o) is a normal sub-group.

Conversely, if $(H\ o\ a)\ o\ (H\ o\ b) = H\ o\ (a\ o\ b)$, then we shall prove that (H, o) is a normal sub-group.

We consider $h \in H$ and $a \in G$.

Since $a\ o\ h\ o\ a^{-1} = (e\ o\ a)\ o\ (h\ o\ a^{-1}) \in (H\ o\ a)\ o\ (H\ o\ a^{-1})$

$$= H\ o\ (a\ o\ a^{-1}) = H\ o\ e = H,$$

If $h \in H$ and $a \in G$, we have $a \text{ o } h \text{ o } a^{-1} \in H$

Hence (H, o) is a normal sub-group.

Theorem 2.11.5: *If (H, o) is a sub-group of (G, o) and (N, o) is a normal sub-group of (G, o), then $(H \cap N, o)$ is a normal sub-group of (H, o).*

Proof: We have already seen that the intersection of two sub-groups is again a sub-group. Now it remains to show that $(H \cap N, o)$ is a normal sub-group in H.

Since $H \cap N \subset H$, and then $(H \cap N, o)$ is a sub-group of (H, o)

Let $a \in H$ and let $h \in H \cap N$, then $h \in H, h \in N$

$a \text{ o } h \text{ o } a^{-1} \in N$, since (N, o) is normal.

Since $h \in H$, $a \in H$, then $a \text{ o } h \in H$.

Again $(a \text{ o } h) \in H$, $a^{-1} \in H$, the $(a \text{ o } h) \text{ o} a^{-1} \in H$

Since (H, o) is a sub-group.

Thus $a \text{ o } h \text{ o } a^{-1} \in N$

and $a \text{ o } h \text{ o } a^{-1} \in H$

which implies $aohoa^{-1} \in H \cap N$. Hence $(H \cap N, o)$ is a normal sub-group of (H, o).

Theorem 2.11.6: *If (H, o) is a normal sub-group of the group (G, o), then the pair $(G/H, o)$ forms a group (known as Quotient group of G by H).*

Proof: (1) First, let us observe the associativity of the operation o. Let $a \text{ o } H$, $b \text{ o } H$, $c \text{ o } H \in G/H$, $a, b, c \in G$.

$$\begin{aligned}\{(a \text{ o } H) \text{ o } (b \text{ o } H)\} \text{ o } (c \text{ o } H) &= \{(a \text{ o } b) \text{ o } H\} (c \text{ o } H)\\ &= ((a \text{ o } b) \text{ o } c) \text{ o } H\\ &= (a \text{ o } (b \text{ o } c)) \text{ o } H\\ &= (a \text{ o } H) \text{ o } \{(b \text{ o } c) \text{ o } H]\\ &= (a \text{ o } H) \text{ o } \{(b \text{ o } H) \text{ o } (c \text{ o } H)\}.\end{aligned}$$

(2) The coset $H = e \text{ o } H$ is the identity element for the operation.

$$\begin{aligned}(a \text{ o } H) \text{ o } (e \text{ o } H) &= (a \text{ o } e) \text{ o } H\\ &= a \text{ o } H\\ &= (e \text{ o } a) \text{ o } H\\ &= (e \text{ o } H) \text{ o } (a \text{ o } H).\end{aligned}$$

(3) It is a easily seen that the inverse of the coset aoH is $a^{-1} \text{ o } H$, where a^{-1} is the inverse of a in (G, o), which is clear from the computation.

$$\begin{aligned}(a \text{ o } H) \text{ o } (a^{-1} \text{ o } H) &= (a \text{ o } a^{-1}) \text{ o } H\\ &= e \text{ o } H\\ &= (a^{-1} \text{ o } a) \text{ o } H\\ &= (a^{-1} \text{ o } H) \text{ o } (a \text{ o } H).\end{aligned}$$

CHAPTER 2

Hence all the group exioms are fulfilled and the proof is complete.

In addition we can say that if (G, o) is finite, then the order of G/H is equal to the index of H in G, that is, $o\ (G/H) = \dfrac{o(G)}{o(H)}$.

Example 2.11.2: Let $(Z, +)$ be the add·tive group of integers and let $N = [3n \mid n$ is any integer}, then it is obvious $(N, +)$ is a normal sub-group of $(Z, +)$. We consider the cosets of normal sub-group $(N, +)$ in Z. These are N, $N + 1$, $N + 2$. We claim that these are all the cosets of N in Z. For, given $a \in Z$, $a = 3b + c$ where $b \in Z$ and $c = 0, 1, 2$ (c is the remainder of a on division by 3). Thus $N + a = N + 3b + c = (N + 3b) + c = N + c$ since $3b \in N$. This every coset is one of the cosets N, $N + 1$, $N + 2$, and $Z/N = [N, N + 1, N + 2\}$. Now the pair $(Z/N, +)$ is an abelian group, for any group containing 3 elements is abelian. If $N + 1 . N + 2 \in G/N$, $(N + 1) + (N + 2) = N + 3 = N \in Z/N$ *i.e.*, Z/N is closed. Addition is associative, N is the identify element in Z/N, N is the inverse of N, $N + 1$ is the inverse of $N + 2$, and the operation is commutative. Hence $(Z/N, +)$ is an abelian group.

Example 2.11.3: Let (H, o) be a sub-group of (G, o) and the set $N(H)$ be defined by $N(H) = \{a \in G \mid a\ o\ H\ o\ a^{-1} = H\}$.

Prove that (H, o) is normal if and only if $N(H) = G$.

Solution: Let (H, o) be a normal sub-group, then we show that $N(H) = G$. For, we have seen that $(N(H), o)$ is a sub-group of (G, o). That is, $N(H) \subseteq$ G.

Since (H, o) is normal, then for all $x \in$ G

$x\ o\ H = H\ o\ x$ which implies

$x \in N(H)$ by the definition of $N(H)$

which follows $G \subseteq N(H)$

Thus $N(H) = G$.

Conversely, let (H, o) be a sub-group and $N(H) = G$. Then every $x \in N(H)$ must belong to G, and $x \in$ G $\Rightarrow x \in N(H)$.

Since $x \in N(H)$, then $xoH = Hox$.

Hence every left coset of H in G is right coset.

Therefore (H, o) is a normal sub-group.

Example 2.11.4: Let (H, o) and (K, o) be two normal sub-groups of (G, o) and that $H \cap K = \{e\}$. Show that $h\ o\ k = k\ o\ h$ for any $h \in H$, $k \in K$.

Solution: Let (H, o) and (K, o) be normal sub-group such that $H \cap K = \{e\}$. Let $h \in H$, $k \in K$ and consider $h\ o\ k\ o\ h^{-1}\ o\ k^{-1} = h\ o\ (k\ o\ h^{-1}\ o\ k^{-1})$.

Since (H, o) is a normal, then for $h^{-1} \in H$ and $k \in G$,

$k\ o\ h^{-1}\ o\ k^{-1} \in H$.

Now, $h \in H$, $k\ o\ h^{-1}\ o\ h^{-1} \in H \Rightarrow h\ o\ (k\ o\ h^{-1}\ o\ h^{-1}) \in H$.

Thus, $h\ o\ k\ o\ h^{-1}\ o\ k^{-1} \in H$.

Again, $h\ o\ k\ o\ h^{-1} \in K$, since (K, o) is normal, and $h \circ k \circ h^{-1} \in K$, $k^{-1} \in K \Rightarrow h \circ (k \circ h^{-1} \circ k^{-1}) \in K$.

Therefore, $h \circ k \circ h^{-1} \circ k^{-1} \in H \cap K = \{e\}$

$\Rightarrow h \circ k \circ h^{-1} \circ k^{-1} = e$

$\Rightarrow h \circ k = k \circ h$, which proves the result.

We can see that the quotient group of an abelian group is abelain, but now we see whether a non-abelian group can have abelian quotient groups, and if so then under what conditions do they exists? So we give.

Definition 2.11.5: Let (G, o) be a group and let $a, b \in G$. Then product $a \circ b \circ a^{-1} \circ b^{-1}$ is called the commutator of a and b. The commutator of a and b is generally represented by $\{a, b\}$.

We see that $a \circ b = [a, b] \circ b \circ a$,

and $[a, b] = e$ if any only if $aob = boa$, the set of commutators is not a sub-group under the group operation o because the product of two commutators is not again a commutator. But the commutators do generate a sub-group which is called the *derived sub-group* or *commutator group* denoted by $([G, G]\ o)$.

Theorem 2.11.7: *The group $([G, G], o)$ is a normal sub-group of (G, o).*

Proof: We must show that for any $c \in [G, G]$, and $a \in G$, $aocoa^{-1}$ must belong to $[G, G]$.

Now we have

$$a \circ c \circ a^{-1} = (a \circ c \circ a^{-1} \circ c^{-1}) \circ c$$
$$= [a, c] \circ c.$$

the element $[a, c] \circ c$ is a finite product of commutators and so belongs to $[G, G]$. Hence the sub-group $([G, G].\ o)$ is normal.

Since $([G, G], o)$ is a normal sub-group of (G, o), $(G/[G, G], o)$ exists and it is called *commutator quotient group* or *abelianized group.*

Example 2.11.5: Prove that $(G/[G, G], o)$ is an abelian group.

Proof: Let $[G, G] = G'$. If $a \circ b \circ a^{-1} \circ b^{-1} \in G'$ for all $a, b \in G$,

$(a \circ b \circ a^{-1} \circ b^{-1}) \circ G' = G'$

$\Rightarrow (a \circ G') \circ (b \circ G') \circ (a^{-1} \circ G') \circ (b^{-1} \circ G') = G'$

$\Rightarrow (a \circ G') \circ (b \circ G') = (b \circ G') \circ (a \circ G')$

$\Rightarrow G/G'$ is abelian.

Theorem 2.11.8: *Let (H, o) be a normal sub-group of the group (G, o). The quotient group $(G/H, o)$ is commutative if and only if $[G, G] \subseteq H$.*

Proof: Let $a \circ H$ and $b \circ H$ be two elements of G/H we have $H = e \circ H$, the identity in the quotient group $(G/H, o)$. the group operation o is commutative if and only if $[a \circ H, b \circ H] = H$.

Thus $$H = [a \text{ o } H, b \text{ o } H] = (a \text{ o } H) \text{ o } (b \text{ o } H) \text{ o } (a \text{ o } H)^{-1} \text{ o } (b \text{ o } H)^{-1}$$
$$= (a \text{ o } b \text{ o } a^{-1} \text{ o } b^{-1}) \text{ o } H$$
$$= [a, b] \text{ o } H,$$

which implies $[a, b] = a \text{ o } b \text{ o } a^{-1} \text{ o } b^{-1} \in H$. That is, the quotient group (G/ H, o) is commutative if and only if all commutators belong to H. In other words, the quotient group $(G/H, o)$ is commutative if and only if $[G, G] \subseteq H$. Hence the theorem.

Exercise 2.11.6: If (H, o) is a sub-group of (G, o) such that $x^2 \in H$ for every $x \in G$, then prove that (H, o) is a normal sub-group of G.

Solution: For any $g \in G$, $h \in H$, $(g \text{ o } h)^2 \in H$ and $(g^{-1})^2 = g^{-2} \in H$. Since (H, o) is a sub-group, then

$h^{-1} \text{ o} g^{-2} \in H$ and so $(g \text{ o } h)^2 \text{ o} h^{-1} \text{ o } g^{-2} \in H$

$$\Rightarrow (g \text{ o } h) \text{ o } (g \text{ o } h) \text{ o } h^{-1} \text{ o } g^{-2} \in H$$
$$\Rightarrow g \text{ o } h \text{ o } g \text{ o } h \text{ o } h^{-1} \text{ o } g^{-2} \in H$$
$$\Rightarrow g \text{ o } h \text{ o } g \text{ o } g^{-2} \in H$$
$$\Rightarrow g \text{ o } h \text{ o } g^{-1} \in H.$$

Hence (H, o) is a normal sub-group of (G, o).

Exercise 2.11.7: If p is the smallest prime factor of the order of a finite group $[G, o)$, show that any sub-goup of index p is normal.

Solution: Let (H, o) be a sub-group of (G, o) of index p.

Here we shall claim that if $x \notin H$, then for all $i = 1, 2, \ldots, (p-1)$, $x^i \notin H$.

If our claim is not true then we can find k, $1 < k \leq (p-1)$ such that $x^k \notin H$. Let j be the least integer such that $x^j \in H$, *i.e.*, for $t = 1, 2, \ldots, (j-1)$, $x^t \notin H$.

Let $o(x) = m$ and $o(G) = n$. Then $m \mid n$ and $1 < j < (p-1)$, where p is the smallest factor of n, so we get that j cannot divide m. Then there are two integers q, r such that

$$m = q\,j + r, \; o < r < j.$$

Since $o(x) = m$, $x^m = e$, where e is the identity in G.

$$\therefore \qquad x^m = e$$
$$\Rightarrow \qquad x^{qj+r} = e$$
$$\Rightarrow \qquad (x^j)^q \text{ o } x^r = e$$

$\Rightarrow x^r \in H$ since $x^j \in H$.

This is against the choice of j. Hence our claim is proved.

To prove the main result, let (H, o) be not a normal sub-group of (G, o). Then there exist some $x \in G$ and $h \in H$ such that $x^{-1} \text{ o } h \text{ o } x \notin H$. Evidently $x \notin H$. We have proved $x^i \notin H$ for all $i = 1, 2, \ldots, (p-1)$.

Also for $1 \leq i < j \leq (p-1)$, $H \text{ o } x^i = H \text{ o } x^j$

$$\Rightarrow \qquad x^i = h_1 \text{ o } x^j \text{ for some } h_1 \in H.$$

$$\Rightarrow \quad x^{j-i} = h_1^{-1} \in H \text{ which is absurd.}$$

Thus we get H, $H \text{ o } x$, $H \text{ o } x^2$,...., $H \text{ o } x^{p-1}$ are all distinct right cosets of H. Let $y = x^{-1}$ oho x, then $y \notin H$ and in the same way we get that H, $H \text{ o } y$, $H \text{ o } y^2$,...., $H \text{ o } y^{p-1}$ are all right distinct cosets of H in G.

Therefore two sets $\{H, Hox, ..., Hox^{p-1}\}$ and $\{H, Hoy, ..., Hoy^{p-1}\}$ are equal. So for some $r, 1 \leq r \leq (p-1)$, $Hox = Hoy^r$

$$\Rightarrow \quad x = h' \text{ o } y^r \text{ for some } h' \in H.$$

$$\Rightarrow \quad x = h' \text{ o } (x^{-1} \text{ o } h \text{ o } x)^r$$

$$\Rightarrow \quad x = h' \text{ o } x^{-1} \text{ o } h^r \text{ o } x$$

$$\Rightarrow \quad x \text{ o } (h^r \text{ o } x)^{-1} = h' \text{ o } x^{-1}$$

$$\Rightarrow \quad h^{-r} = h' \text{ o } x^{-1}$$

$$\Rightarrow \quad x = h^r \text{ o } h' \in H. \text{ Which is absurd.}$$

Hence (H, o) is a normal sub-group of (G, o).

Remark: As a special case of this exercise we get that every sub-group of index 2 is a normal sub-group. We can give its alternative proof which is simple.

Let (H, o) be a sub-group of a group (G, o) of index 2. That is, there are only two right and two left cosets of H in G, namely, H and $H \text{ o } a$ or H and $a \text{ o } H$.

$$\text{Since} \quad G = H \cup H \text{ o } a = H \cup a \text{ o } H$$

$$\Rightarrow \quad G - H = \{H \cup H \text{ o } a\} - H = \{H \cup a \text{ o } H) - H$$

$$\Rightarrow \quad H \text{ o } a = a \text{ o } H.$$

Hence (H, o) is a normal sub-group.

Example 2.11.8: The alternating group (A_n, o) of the symmetic group $(S_n \; o)$ on n symbols $\{1, 2, 3, ..., n\}$ is a normal sub-group.

Since A_n is the set of all even permutations of S_n, and $o(S_n) = n!$ and $o(A_n) = \dfrac{n!}{2}$.

The index of A_n in S_n is $[S_n : A_n] = \dfrac{o(S_n)}{o(A_n)} = 2.$

Hence (A_n, o) is a normal sub-group of (S_n, o).

Hence quotient set $S_n/A_n = \{A_n, A_n \; of\}$, where f is an odd permutation.

PROBLEMS

1. Find an example of a group (G, o) having a sub-group (H, o) for which the product of two left cosets of H in G need not be a left coset of H.
2. Show by an example that if (H, o) is a normal sub-group of (G, o) and (K, o) is a normal sub-group of (H, o), then (K, o) may not be normal sub-group of (G, o).
3. Let (H, o) be a sub-group of the group (G, o), Let, for $g \in G$, $g \text{ o } H \text{ o } g^{-1} = [g \text{ o } h \text{ o } g^{-1} | h \in H\}$. Prove that $(g \text{ o } H \text{ o } g^{-1}, o)$ is a sub-group of (G, o).

CHAPTER 2

4. If (H, o) is sub-group of (G, o), let $N(H) = (a \in G | aohoa^{-1} = H\}$ Prove that

(1) $(N(H), o)$ is a sub-group of (G, o)

(2) (H, o) is a normal sub-group in $(N(H), o)$.

(3) If (H, o) is a normal sub-group of group (K, o) in group (G, o), then $K \subseteq N(H)$, that is, $(N(H), o)$ *is* the largest sub-group of (G, o) in which (H, o) is normal.

5. If a cyclic sub-group (T, o) of (G, o) is normal in (G, o), then show that every sub-group of (T, o) is normal in (G, o).

6. If (N, o) is normal in (G, o) and $a \in G$ with order $o(a)$, prove that the order, m, of $N\ o\ a$ in G/N is a divisor of $o(a)$.

7. Let (H, o) and (K, o) be two normal sub-group in a finite group (G, o) such that $o(H)$ and $o(G)$ are co-prime. Show that for any $h \in H$, $k \in K$, $h\ o\ k = k\ o\ h$.

8. Show that if a group (G, o) has a finite sumber of proper sub-groups, then (G, o) is finite.

9. Describe the quotient group of

(*a*) $(Z, +)$ in $(Q, +)$;

(*b*) $(E, +)$ in $(Z, +)$;

(*c*) $(\{0, 2, 4, 6, 8), +_{10})$ in $(Z_{10}, +_{10})$;

(*d*) $(\{1, -1),.)$ in $(\{1, -1, -1, i, -i\},.)$.

10. Let (G, o) be a clyclic group with generator a and (H, o) be any sub-group of (G, o). Prove that the quotient group $(G/H, o)$ is also cyclic group with generator aoH.

11. Prove that $(Z(G), *)$ is a normal sub-group of a group $(G, *)$.

12. If $(N, *)$ is a normal sub-group of $(G, *)$ and $(H, *)$ is a sub-group of $(G, *)$, then show that $(N * H, *)$ is a sub-group of $(G, *)$.

13. If $(N, *)$ and $(H, *)$ are normal sub-group of $(G, *)$, there prove that $(N * H, *)$ is a normal sub-group.

14. For every complex (subset) K of G and every normal sub-group $(N, *)$ of a group $(G, *)$ prove that $K * N = N * K$.

15. Prove that every sub-group of an abelian group is normal.

16. Prove that every sub-group of index 2 is a normal sub-group.

17. Prove that every quotient group of an abelian group is abelian.

18. Prove that every quotient group of a cyclic group is cyclic.

19. Z is the centre of group (G, o). Prove that if $(G/Z, o)$ is cyclic then (G, o) must be abelian.

Hint: Since $(G/Z, o)$ is cyclic, for some $g \in G$, $G/Z = \{(g\ o\ Z)^n\}$

Let $a, b \in G$. Then $a\ o\ Z = (g\ o\ Z)^k$ and $h\ o\ Z = (g\ o\ z)^l$

Then $a\ o\ Z = g^k\ o\ Z$ and $b\ o\ Z = g^l\ oZ$. For some $k, l \in Z'$.

For some $n_1, n_2 \in Z$,

$a = g^k\ o\ n_1$, $b = g'\ on_2$, and $a\ o\ b = g^{k+l}\ o\ n_1\ o\ n_2$ and $b\ o\ a = g^{l+k}\ o\ n_1\ o\ n_2 \Rightarrow a.b = b.a$

CHAPTER 3 HOMOMORPHISMS

3.1 HOMOMORPHISMS

We have studied in chapter 1 the mappings from one set to another set. Here we shall consider the mappings from one group to another group which will preserve the group operations. That's why sometimes it is called operation preserving function.

Definition 3.1.1: Let $(G, *)$ and (G', o) be two groups and f be a function from G into G' f : $G \to G'$, then f is said to be a homomorphism from $(G, *)$ into (G', o) if and only if.

$$f(a * b) = f(a) \, o \, f(b)$$

for every pair of elements $a, b \in G$, or we can say that the image of the product of two elements under f is equal to the product of the images.

If the requirement of homomorphism $f(a * b) = f(a)$ of (b) is satisfied by f, then this diagram is said to be commutative. This asserts that if we start with elements, $a, b \in G$ and move them to G' by either of the two roots indicated by the arrows

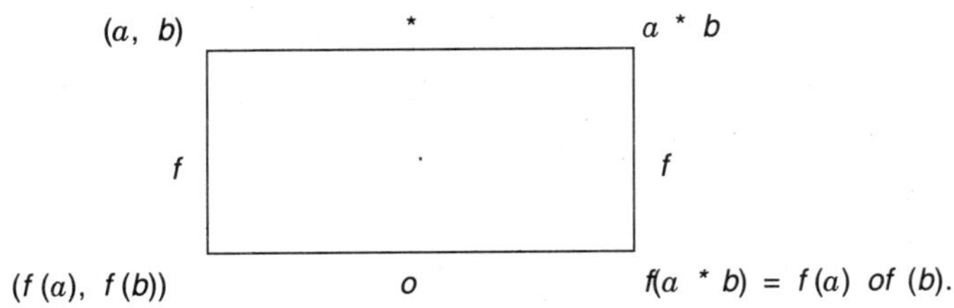

by first calculating the product $a * b$ and then applying f to it or first obtain the images $f(a)$ and $f(b)$ and then taking their product the result will be the same.

When f: $G \to G'$ exists, $f(G)$ is called homomorphic image of G.

Example 3.1.1: Let (G, +) (G′ +) be the groups of integers and even integers under additions. Define the function $f: G \to G'$ by $f(x) = 2x$ for all $x \in G$. Then f is a homomorphism that follows from $f(x + y) = 2(x + y) = 2x + 2y = f(x) + f(y)$.

Example 3.1.2: *Let* $(G, *)$ and (G', o) be two groups with identity elements e and e' respectively. The function $f: G \to G'$ given by $f(a) = e', \forall \, a \in G$ is a homomorphism:

$$f(a * b) = e' = e' \, o \, e' = f(a) \, of(b).$$

This is a constant function and is called trivial homomorphism.

Example 3.1.3: Consider the two groups $(R, +)$ and $(R - \{0\},.)$ for $a \in R$, define the function f by $f(a) = 2^a$. In order to verify that given function is a homomorphism we must check whether $f(a + b) = f(a) . f(b)$.

Thus, $f(a + b) = 2^{a+b} = 2^a. 2^b = f(a) . f(b)$. Hence f is a homomorphism of R into $R - \{0\}$.

Example 3.1.4: Let $(Z, +)$ be the group of integers under addition and $(Z_n, +_n)$ be the group of integers modulo n. Define $f : Z \to Z_n$ by $f(a) = [a]$; that is, f maps each integer into the congruence class containing it. The function is homomorphism that follows from the definition of modular addition.

$$f(a + b) = [a + b] = [a] +_n [b] = f(a) +_n f(b).$$

Theorem 3.1.1: *If f is a homomorphism from the group $(G, *)$ into the group (G', o), then*

(1) f maps the identity element e of $(G, *)$ onto the identity element e' of (G', o); $f(e) = e'$

(2) f maps the inverse of an element $a \in G$ onto the inverse of $f(a)$ in (G', o); $f(e^{-1}) = [f(a)]^{-1}$, for all $a \in G$.

Proof: **(1)** Let $f : G \to G'$ and

$$f(a * b) = f(a) \, o \, f(b).$$

Let $a \in G$ and $e' \in G'$, then

$$f(a) \, o \, e' = f(a) = f(a * e) = f(a) \, o \, f(e).$$

$$\Rightarrow \quad f(a) \, o \, e' = f(a) \, o \, f(e).$$

By left cancellation law, we have

$$e' = f(e).$$

(2) We observe that

$$f(a * a^{-1}) = f(a) \, o \, f(a^{-1})$$

and

$$f(a * a^{-1}) = f(e) = e' = f(a) \, o \, (f(a))^{-1}$$

Thus

$$f(a) \, o \, f(a^{-1}) = e' = f(a) \, o \, (f(a))^{-1}$$

By left cancellation we obtain

$$f(a^{-1}) = (f(a))^{-1}$$

Example 3.1.5: Let $(Z, +)$ and $(R - \{0\}.)$ be two groups and let f is defined by $f(n) = r^n$, $n \in Z$, $r \in R$. The function f is a homomorphism which follows from:

$$f(m+n) = r^{m+n} = r^m . r^n = f(m) . f(n)$$

We observe that $0 \in Z$, the identity of Z and $f(0) = r^0 = 1$, the identity of $R - \{o\}$ for multiplication, and $f(-n) = r^{-n}$, r^{-n} is the inverse of r^n in $R - \{o\}$ which clarifies the above theorem.

Theorem 3.1.2: *Let f be a homomorphism from the group $(G, *)$ into the group (G', o) then*

(1) For each sub-group $(H, *)$ of $(G, *)$, the pair $(f(H), o)$ is also a sub-group of (G , o).

(2) For each sub-group (H', o) of (G', o) the pair $(f^{-1}(H'), *)$ is a sub-group of $(G', *)$.

Proof: **(1)** We know the set $f(H) = (f(h) \mid h \in H]$ is the set of all images of the elements $h \in H$.

Now, let $f(h)$ *and* $f(k) \in f(H)$, then

$h, k \in H$ and $h * k^{-1} \in H$.

$f(h)\ o\ (f(k))^{-1} = f(h)\ of\ (k^{-1}) = f(h * k^{-1}) \in f(H)$.

It shows whenever $f(h), f(k) \in f(H)$, the $f(h)\ o\ (f(k))^{-1} \in f(H)$.

Hence $(f(H), o)$ is a sub-group of (G', o).

(2) In this case the set

$f^{-1}(H') = \{a \in G \mid f(a) \in H'\}$

Thus if $a, b \in f^{-1}(H')$, the images $f(a), f(b) \in H'$.

Since (H', o) is a sub-group,

$f(a), f(b) \in H' \Rightarrow f(a)\ o\ (f(b))^{-1} \in H'$,

it follows that

$f(a * b^{-1}) = f(a)\ of\ (b^{-1}) = f(a)\ o(f(b))^{-1} \in H'$

$\Rightarrow a * b^{-1} \in f^{-1}(H')$ which proves that $(f^{-1}(H'), *)$ is a sub-group of (G', *).

CHAPTER 3

Theorem 3.1.3: *Let f be a homomorphism from the group (G, *) into the group (G′, o), then*

(1) *If (H′, o) is a normal sub-group of (G′, o), the sub-group* $(f^{-1}(H'), *)$ *is a normal sub-group of (G, *).*

(2) *If f is onto mapping and (H, *) is a normal sub-group of (G, *), the sub-group (f(H), o) is a normal sub-group of (G′, o).*

Proof: **(1)** Here $f^{-1}(H') = \{a \in G \mid f(a) \in H'\}$.

Let $h \in f^{-1}(H')$, *then* $f(h) \in H'$ and let a be an arbitrary element of G, then

$f(a*h*a^{-1}) = f(a)\, of(h)\, o\, (f\, 1)\, (a)^{-1} \in H'$,

since (H', o) is a normal sub-group of (G', o).

Thus $a* h* a^{-1} \in f^{1}(H')$ which shows that $(f^{-1}(H'),*)$ is a normal sub-group of $(G, *)$.

(2) If $f(G) = G'$, then every element of G' is an image element $f(a)$ of some element $a \in G$. Since $(H, *)$ is normal, then $h \in H$ and $a \in G$ implies $a*h*a^{-1} \in H$ *and* $f(a*h*a^{-1}) = f(a)\ of\ (h)\ o(f(a))^{-1} \in f(H)$,

where $f(a) \in G', f(h) \in f(H)$.

Hence for all $f(a) \in G'$, $f(a)\ o\ f(h)\ o\ (f(a))^{-1} \in f(H)$.

Which is sufficient for $(f(\text{H}), o)$ to be a normal sub-group.

Definition 3.1.2: Let f be a homomorphism from the group $(G, *)$ into the group (G', o) and let e' be the identity element of (G', o). The Kernel of f, denoted by $Ker(f)$, is the set

$$Ker(f) = \{a \in G \mid f(a) = e'\}.$$

That is, $Ker(f)$ is the set of those elements of G which are mapped on the identity element e' of (G', o). It is obvious that $Ker(f)$ is not empty since $e \in Ker(f)$ which is mapped on e' under $f : f(e) = e'$.

Theorem 3.1.4: *If f is a homomorphism from the group (G, *) into the group (G′, o), then the pair (Ker (f), *) is a normal sub-group of (G, *).*

Proof: First, we shall prove that $(Ker(f), *)$ is a sub-group. For, $Ker(f)$ is non-empty since $e \in Ker(f)$ and $Ker(f) \subseteq G$.

Now, let $a, b \in ker(f)$, then

$f(a) = e', f(b) = e',$

and $f(b^{-1}) = (f(b)^{-1}) = e'^{-1} = e'$.

$f(a * b^{-1}) = f(a) \; o \; (f(b))^{-1} = e' \; o \; e' = e'$

Thus, if $a, b \in Ker(f)$, then $a*b^{-1} \in \text{Ker}(f)$.

Hence $(Ker(f), *)$ is a sub-group.

Further, for $a \in G$, and $k \in \text{Ker}(f)$ we observe that

$$\begin{aligned} f(a*k*a^{-1}) &= f(a) \; o \; f(k) \; o \; f(a^{-1}) \\ &= f(a) \; o \; f(k) \; o \; (f(a))^{-1} \\ &= f(a) \; o \; e' \; o \; (f(a))^{-1} = f(a) \; o \; (f(a))^{-1} = e'. \end{aligned}$$

That is, for all $a \in G$ and $k \in Ker\, f$,

$a*k*a^{-1} \in Ker(f)$ which proves that $(Ker(f), *)$ is a normal sub-group.

Example 3.1.6: Let f be a homomorphism from the group $(Z, +)$ into the group $(R - \{o\}, .)$ and defined by

$$\begin{aligned} f(n) &= 1 \text{ if } n \in E, \text{ the set of even integers,} \\ &= -1 \text{ if } n \in O, \text{ the set of odd integers.} \end{aligned}$$

In this 1 could serve the identity element in $R - \{o\}$ for multiplication.

Thus $Ker(f) = \{n \in Z \mid f(n) = 1\} = E$.

It is clear that $(E, +)$ is a normal sub-group of $(Z, +)$, and the direct image $f(Z) = \{1, -1\}$ which is a sub-group of $(R - \{o\}, .)$

Theorem 3.1.5: *Let f be a homomorphism from the group (G, *) into the pair (G′, o). The image (f (G), o) is a sub-group of (G′, o).*

Proof: We have to check up all group axioms on $f(G)$ under the group operation of (G', o).

(1) Let $x', y' \in f(G)$, then

$$\begin{aligned} x' &= f(x) \text{ for some } x \in G, \\ y' &= f(y) \text{ for some } y \in G. \end{aligned}$$

$$x' \text{ o } y' = f(x) \text{ o } f(y) = f(x*y) \in f(G) \text{ for } x*y \in G.$$

That is, $f(G)$ is closed under the operation.

(2) Let $x', y', z' \in f(G)$, then

$$x' = f(x)$$

$$y' = f(y),\ z' = f(z). \text{ For some } x, y, z \in G.$$

Now

$$\begin{aligned} x' \text{ o } (y' \text{ o } z') &= f(x) \text{ o } (f(y) \text{ o } f(z)) \\ &= f(x) \text{ o } (f(y*z)) \\ &= f(x*(y*z)) \\ &= f(x*y)*(z)) \\ &= f(x*y) \text{ o } f(z) \\ &= (f(x) \text{ o } f(y)) \text{ o } f(z) \\ &= (x' \text{ o } y') \text{ o } z'. \end{aligned}$$

Hence the operation of (G', o) is associative.

(3) Let e be the identity element of $(G, *)$ and let e' be the identity element of (G', o). Let $a' \in f(G)$, then $f(a) = a'$ for some $a \in G$

Now

$$\begin{aligned} f(a*e) &= f(a) \text{ o } f(e) \\ &= a' \text{ o } e' = a' \\ &= e' \text{ o } a' = f(e) \text{ o } f(a) = f(e*a) \end{aligned}$$

which shows that e' is the identity element in $f(G)$

(4) We have

$$a*a^{-1} = e$$

$$\Rightarrow \quad f(a*a^{-1}) = f(e) = e'$$

$$\Rightarrow \quad f(a) \text{ o}(f(a^{-1})) = e'$$

$$\Rightarrow \quad f(a) \text{ o } (f(a))^{-1} = e'$$

which implies $(f(a^{-1}) = (f(a))^{-1} \in f(G)$.

This completes the theorem.

Theorem 3.1.6: *Let $(H, *)$ be a normal sub-group of the group $(G, *)$, then the mapping $f: G \to G/H$ defined by $f(a) = a*H$ is a homomorphism from the group $(G, *)$ onto the quotient group $(G/H, *)$; the kernel of f is precisely the set H. (This homomorphism is known as natural homomorphism or canonical homomorphism.)*

Proof: The mapping is homomorphism which follows from the product defined in the Quotient group.

$$f(a*b) = (a*b) *H$$

$= (a*H) *(b*H)$

$= f(a) *f(b).$

Hence f is well defined homomorphism from the group $(G, *)$ onto the group $(G/H, *)$. To see that it is onto we have that every element of G/H is a coset aoH, where $a \in G$ and $f(a) = aoH$.

The group $(G/H, *)$ has an identity element H.

Now, the kernel of f is the set

$ker(f) = \{a \in G \mid f(a) = H\}$

$= \{a \in G \mid aoH = H\} = H$, since $aoH = H \Rightarrow a \in H$.

This completes the proof of the theorem.

Corollary: Let $(H, *)$ be a normal sub-group of $(G, *)$. Then there exists a group (G', o) and a homomorphism from $(G, *)$ onto (G', o) such that $ker(f) = H$.

Hint: Of course, we take (G', o) to be the Quotient group $(G/H, *)$ of G by H, and homomorphism f = natural mapping of G onto G/H.

Example 3.1.7: We consider $(Z, +)$ the group of integers under addition and we know that its normal sub-group is the cyclic group $((n), +)$, $n \in N$. Moreover the quotient group corresponding to any fixed $n \in N$ is simply $(Z_n, +_n)$, the group of integers modulo n. That is, $Z/(n) = Z_n$. The natural homomorphism $f : Z \to Z_n$ is defined by $f(a) = [a]$, that is, f sends every integer a to its congruence class modulo n.

3.2 ISOMORPHISMS

Definition 3.2.1: Two groups, $(G, *)$ and (G', o) are said to be isomorphic denoted by $(G, *) \cong (G, o)$, if there exists one to one homomorphism f of $(G, *)$ onto (G', o) that is, $f(G) = G'$. Such a homomorphism f is called an isomorphism, or isomorphic mapping of $(G, *)$ onto (G', o). That is, f is a one to one function from G onto G' such that

$$f(a * b) = f(a) \, o \, f(b),$$

for all elements $a, b \in G$.

Since f is one to one mapping from G onto G', then f^{-1} is also one to one mapping from G' onto G. Thus, we can say the group (G', o) is isomorphic to $(G, *) : (G', o) \cong (G, *)$.

Example 3.2.1: Consider the example of two groups $(Z_4, +_4)$ and $(G,.)$ where $G = \{1, -1, i - i\}$, and $i^2 = -1$.

The operation tables for these two systems are:

+4	0	1	2	3
0	0	1	2	3
1	1	2	3	0
2	2	3	0	1
3	3	0	1	2

⊙	1	–1	i	$-i$
1	1	–1	i	$-i$
–1	–1	1	$-i$	i
i	i	$-i$	–1	1
$-i$	$-i$	i	1	–1

We shall prove here that $(Z_4, +_4) \cong (G, o)$. To get this we must have a one to one homomorphism f from Z_4 onto G.

Since the preservation of identity elements is a general feature of any homomorphism, f must be such that $f(0) = 1$. Again let us suppose that $f(1) = -1$. The image of an inverse element must be equal to the inverse of the image. Thus since 3 is the inverse of 1, then

$$f(3) = f(1^{-1}) = (f(1))^{-1} = (-1)^{-1} = -1 = f(1)$$

$$\Rightarrow \qquad f(3) = f(1)$$

But we observe that the images of 3 and 1 under f are not distinct. Therefore f is not a one to one mapping.

Again if we take $f(1) = i$, then

$f(3) = f(1^{-1}) = (f(1))^{-1} = i^{-1} = -i$

and $f(2) = f(1 +_4 1) = f(1)\, of(1) = i.i. = -1$.

Thus the function f is defined by

$$f(0) = 1, f(1) = i, f(3) = -i, f(2) = -1.$$

Clearly this function is one to one function of Z_4 onto G. Here again we varify that function preserves the group operations, that is,

$$f(3) = f(1 +_4 2) = f(1)\, of\, (2) = i, (-1) = -i.$$

Here $$(Z_4 +_4) \cong (G,.)$$

We can say that two finite groups are isomorphic if it is possible to obtain each operation table from the others by merely renaming the elements. Thus we have the table for (G, o).

.	1	i	–1	$-i$
1	1	i	–1	$-i$
i	i	–1	–1	1
–1	–1	$-i$	1	i
$-i$	$-i$	1	i	–1

Apart from the particular symbols used, this group table is identical to that of $(Z_4, +_4)$ for corresponding elements appear at the same place in each table.

Example 3.2.2: Let $G = [e, a, b, c\}$ and the operation* be defined by the table

*	e	a	b	c
e	e	a	b	c
a	a	e	c	b
b	b	c	e	a
c	c	b	a	e

The two groups $(Z_4, +_4)$ and $(G, *)$ are not isomorphic.

We shall observe that every one to one function from Z_4 onto G fails to preserve the group operation. That is $(Z_4, +_4)$ and $(G, *)$ are two distinct algebraic structures.

CHAPTER 3

To see this point, if we consider the mapping

$$f(0) = e, f(1) = a, f(2) = b, f(3) = c,$$

then $$f(1 +_4 3) = f(1) * f(3) = a*c = b,$$

and $$f(1 +_4 3) = f(0) = e,$$

Hence $$f(1 +_4 3) = f(0) = e \neq b = a*c = f(1) * f(3).$$

Hence f is not a homomorphism. Yet f is a one to one function of Z_4 onto G.

Similarly we can test other one to one functions that do not preserve the group operation.

Example 3.2.3: The two groups $(Z,+)$ and $(Q - \{0\},.)$ are not isomorphic.

To see this suppose that there exists one to one function $f : Z \to Q - \{0\}$ with the property $f(a + b) = f(a).f(b), \forall a, b \in Z$.

Let $x \in Z$ such that $f(x) = -1$, then

$f(2x) = f(x + x) = f(x) . f(x) = (-1) . (-1) = 1.$

But we know that $0 \in Z$, the identity element in Z, is mapped on the identity element in $Q - \{0\}$. Thus,

$f(0) = f(2x) = 1 \Rightarrow 2x = 0 \Rightarrow x = 0. f(0 + 0) = f(0) . f(0) = (-1) . (-1) = 1$

Consequently $f(0) = 1$ and $f(0) = -1$, contradicting the fact that f is one to one. Thus $(Z, +)$ is not isomorphic to $(Q - \{0\},.)$.

Example 3.2.4: The function f defined by $f(x) = e^x$ is isomorphism of the additive group of integers onto the multiplicative group of positive real numbers.

We notice that for $x, y \in Z$

$$f(x + y) = e^{x+y} = e^x . e^y = f(x). (y).$$

It is the operation preserving function. It is also one to one and onto which follows from:

If $x, y \in Z$,

$$f(x) = e^x, f(y) = e^y$$

and $$f(x) = f(y) \Rightarrow e^x = e^y \Rightarrow x = y$$

Hence $(Z, +) \cong (R^+ .)$.

***Theorem 3.2.1:** The relation of isomorphism in the set of groups is an equivalence relation.*

Proof: To prove the theorem we have to show that the relation of isomorphism is reflexive, symmetric, and transitive. We see that:

(1) Reflexive: Let (G, o) be a group and let f be an identity mapping defined by $f(x) = x$, $\forall x \in G$. Therefore f is a one to one and onto mapping and let $x, y \in G$.

$$f(x \text{ o } y) = x \text{ o } y$$
$$= f(x) \text{ o } f(y),$$

Thus, f is an isomorphism from (G, o) onto itself, that is, $(G, o) \cong (G, o)$.

(2) Symmetric: Let f be an isomorphism from (G, o) onto $(G, *)$, that is, $(G, o) \cong (G', *)$.

Since f is one to one and onto, then f^{-1} is also one-to-one and onto, that is, f^{-1} is a one to one mapping of the set G' onto the set G.

Now, let $a, b \in G'$,

Such that $f(x) = a$, $f(y) = b$ for $x, y \in G$, then

$$x = f^{-1}(a),\ y = f^{-1}(b).$$

$$f(xoy) = f(x) * f(y) = a*b$$

$$\Rightarrow \quad xoy = f^{-1}(a*b).$$

Now,
$$f^{-1}(a*b) = xoy$$
$$= f^{-1}(a)\, o f^{-1}(b).$$

which shows f^{-1} is an isomorphism from $(G', *)$ onto (G, o). Hence $(G', *) \cong (G, o)$

(3) Transitive: Let f be an isomorphism from the group $(G, *)$ onto (G', o) and g be an isomorphism from the group (G', o) onto the group $(G'', \oplus)$. That is, $(G, *) \cong (G', o)$ and $(G'\ o) \cong (G'', \oplus)$.

Since f and g are one to one and onto mappings then their composite mapping $g o f$ is also one to one of the set G onto the set G''.

Now,
$$(g\ o\ f)(a*b) = g\,(f(a*b))$$
$$= g\,(f(a)\, o f(b))$$
$$= g(f(a)) \oplus g\,(fa)).$$

which shows that $g\ o\ f$ is an isomorphism from the group $(G, *)$ onto the group $(G'', \oplus)$.

This completes the proof of the theorem.

Theorem 3.2.2: *Let f be a homomorphism from the group $(G, *)$ into the group (G', o). Then f is one to one only if $Ker(f) = \{e\}$.*

Proof: First, let us suppose the function f is one to one. We must show that $Ker(f) = \{e\}$.

It is clear that $e \in Ker(f)$. Let $a \in Ker(f)$, $a \neq e$. Then

$$f(a) = e', \text{ and } f(e) = e'$$

Thus
$$f(a) = e' = f(e),$$

which contradicts that f is one to one. Hence $ker(f)$ contains only identity element e if f is one to one.

Conversely, $Ker(f) = \{e\}$. Let $a, b \in G$

and $f(a) = f(b) \Rightarrow f(a)\ o\ (f(b))^{-1} = e'$

$\Rightarrow f(a*b^{-1}) = e'$

Thus, $a*b^{-1} \in Ker(f) = \{e\}$

CHAPTER 3

$\Rightarrow a*b^{-1} = e$

$\Rightarrow a = b$ which shows f is one to one. Hence the theorem.

Theorem 3.2.3: *Every finite cyclic group of order n is isomorphic to* $(Z_n, +_n)$.

Proof: Let a cyclic group $((a), o)$ be generated by a of finite order n. Then the set (a) has n distinct elements which are the powers of a of the form a^k, $k = 0, 1, 2, \ldots, n - 1$, therefore the set

$$(a) = \{e, a^1, a^2, a^3, \ldots, a^{n-1}\}$$

Now we define a mapping f from (a) onto Z_n, $f : (a) \to Z_n$, by . $f(a^k) = [k]$, $0 \le k < n$. We see that the congruence class $[k] \in Z_n$ exists for some $a^k \in (a)$ which shows $f(a^k) = [k]$, Hence f is onto.

For, one to one, let

$$f(a^k) = f(a^j), \text{ where } 0 \le i < n,\ 0 \le j < n,$$

$$\Rightarrow \qquad [k] = [j].$$

$\Rightarrow k \equiv j \bmod n,$

$\Rightarrow a^k = a^j$ which establishes that f is one to one function.

Next we observe that it is operation preserving function: For, let $a^k, a^j \in (a)$.

$$f(a^k \, o \, a^j) = f(a^{k+j}) = [k + j] = [k] +_n [j] = f(a^k) +_n f(a^j).$$

which completes the proof of the isomorphism.

Hence $\qquad ((a), *) \cong (Z_n +_n).$

Theorem 3.2.4: *Every infinite cyclic group is isomorphic to* $(Z, +)$.

Proof: Let the cycilc group $((a), o)$ of infinite order be generated by a and we define a mapping f from Z onto (a) by $f(k) = a^k$. f is one to one since all powers of the generator are distinct.

Let $\qquad m, n \in Z$, then $f(m) = a^m$, and $f(n) = a^n$.

$$f(m) = f(n) \Rightarrow a^m = a^n$$

$$\Rightarrow \qquad m = n.$$

We also observe that each element a^k of (a) exists for some $k \in Z$ which reveals that f is onto.

Again

$$\begin{aligned} f(m + n) &= a^{m+n} \\ &= a^m \, o \, a^n \\ &= f(m) \, of(n) \end{aligned}$$

which means f is operation preserving function.

Hence $(Z, +) \cong ((a), o)$.

Theorem 3.2.5: *Any two cyclic groups of the same order are isomorphic.*

Proof: We know that any infinite cyclic group is isomorphic to $(Z, +)$. If (S, o) and $(T, *)$ are two infinite cyclic groups, then

$$(S, o) \cong (Z, +) \text{ and } (T, *) \cong (Z, +)$$

which implies $(S, o) \cong (T, *)$ by transitive property of isomorphism of groups.

If (S, o) and $(T, *)$ are two cyclic groups of the same order r, then set S and T are given by

$$S = \{e, a, a^2, \ldots\ldots a^{r-1}\} = (a)$$

and
$$T = \{e', b, b^2, \ldots\ldots b^{r-1}\} = (b).$$

where a and b are the generators of (S, o) and $(T, *)$ respectively and r is the smallest positive integer such that

$$a^r = e \text{ and } b^r = e'.$$

Now consider the map $f : (a) \to (b)$ by $f(a^n) = b^n$.

Let $m, n \in N$, then

and
$$f(a^n) = b^n$$

$$f(a^m) = f(a^n) \Rightarrow b^m = b^n$$

$$\Rightarrow \quad m = n$$

$$\Rightarrow \quad a^m = a^n.$$

Hence f is one to one mapping and we observe that it is also operation preserving mapping which follows from

$$\begin{aligned} f((a^m o a^n) &= f(a^{m+n}) = b^{m+n} \\ &= b^m * b^{*n} \\ &= f(a^m) * f(a^n). \end{aligned}$$

This completes the proof.

CHAPTER 3

Theorem 3.2.6: *(Cayley) if $(G, *)$ is an arbitrary group, then $(G, *) \cong (F_G, o)$, where for any fixed $a \in G$, define the left multiplication, $f_a : G \to G$ such that $f_a(x) = a * x$, $\forall x \in G$, and $F_G = \{f_a \mid a \in G\}$ and in the present context o indicates the operation of functional composition.*

Proof: First, we shall show that (F_G, o) is a group. Now for fixed $a \in G$, $f_a(x) = a*x$, $\forall x \in G$.

If $x \in G$, then

$x = a*(a^{-1} * x)$, $a^{-1} * x \in G$.

$\Rightarrow x = a*y = f_a(y)$, let $a^{-1} * x = y$,

so that f_a maps G onto itself. Moreover, f_a, is one to one, for if $x, y \in G$ with

$f_a(x) = f_a(y)$, then

$\Rightarrow a*x = a*y$

$\Rightarrow x = y$ left cancellation law.

This shows f_a, is one to one. Thus, F_G is the set of all transformations of G or (permutations of G). Now we note that

(1) Let $f_a, f_b \in F_G$,

$$\begin{aligned}(f_a \circ f_b)\, x &= f_a\,(f_b(x)) = fa(b^*x) = a^*\,(b^*x)\\ &= (a^*b)^*\, x\\ &= f_{\,a^*b}(x).\end{aligned}$$

This means that $f_a \circ f_b = f_{a^*b}$,

so that the set F_G is closed under the operation of functional composition.

(2) The composition of function is associative.

(3) Indeed, if e is the identity element of $(G, *)$, then f_e, is the identity element in F_G, since

$$f_a \circ f_e = f_{a^*e} = f_a = f_{e^*a} = f_e \circ f_a$$

(4) Moreover, $(f_a)^{-1} = f_{a-1}$, for we have

$$f_a \circ f_a^{-1} = f_{a^*a}^{\;-1} = f_e = f_a^{-1}{}_{*a} = f_a^{-1} \circ f_{\,a}.$$

Hence the pair (F_G, o) satisfies all group axioms. Hence (F_G, o) is a group.

To show $(G, *) \cong (F_G, o)$, we define the mapping $f : G \to F_G$ by $f(a) = f_a$, for all $a \in G$. Then the function is one to one which is obvious from

$$\begin{aligned}f(a) = f(b) &\Rightarrow f_a = f_b \Rightarrow f_a\,(x) = f_b\,(x)\\ &\Rightarrow a^*x = b^*x,\ \forall\ x \in G\\ &\Rightarrow a = b \text{ by right cancellation law.}\\ f_a\,(e) &= f_b\,(e) \Rightarrow\end{aligned}$$

In particular $a * e = b * e \Rightarrow a = b$ which shows that f is one to one. Again

$$f(a^*b) = f_{a^*b} = f_a \circ f_b = f(a) \circ f(b)$$

which proves that f is an operation preserving function.

Hence $(G, *) = (F_G, o)$.

Remark: We can state the theorem as follows:

Every finite group $(G, *)$ is isomorphic to sub-group of a permutation group.

Theorem 3.2.7: *(Factor Theorem). Let f be a homomorphism from the group (G, *) onto the group (G′ , o). Let (H, *) be a normal sub-group of the group (G, *) such that* $H \subseteq \ker(f)$. *Then there exists a unique homomorphism*

$\bar{f} : G/H \to G'$ *with the property.*

$f = \bar{f} \circ \phi$, *where* $\phi : G \to G/H$ *is a natural homomorphism given by* $\phi\,(a) = a^*H$.

Proof: First of all we define mapping $f : G/H \to G'$ by

$\bar{f}\,(a^*H) = f(a),\ a \in G.$

This mapping $\bar{f}$ is called *induced mapping.*

Now we shall see that $\bar{f}$ is well defined, that is $\bar{f}$ depends only on the cosets of H and independent of the representatives of the cosets of H. For, suppose a and b belong to the same coset of H. Then

$a^* H = b^*H \Rightarrow a^{-1} * b \in H \subseteq ker(f)$. It follows that $f(b) = f(a^* a^{-1} * b) = f(a) \circ f(a^{-1} * b) = f(a) \circ e' = f(a)$

$\Rightarrow \bar{f}(b * H) = \bar{f}(a * H)$

Hence $\bar{f}$ is well defined.

$\bar{f}$ is also operation preserving function. To see this we have

$$\begin{aligned}\bar{f}((a * H) * (b * H)) &= \bar{f}((a * b) * H)\\ &= f(a * b)\\ &= f(a) \circ f(b)\\ &= \bar{f}(a * H) \circ \bar{f}(b * H).\end{aligned}$$

It proves that $\bar{f}$ is a homomorphism.

Since f denotes the homomorphism from $(G, *)$ onto (G', o), for each element $a \in G$,

$$f(a) = \bar{f}(a * H) = \bar{f}(\phi(a)) = (\bar{f} \circ \phi)(a)$$

$$\Rightarrow \qquad f = \bar{f} \circ \phi.$$

For uniqueness, let $g : G/H \to G'$ be another function defined by $g(a * H) = f(a)$. Then

$$\bar{f}(a * H) = f(a) = g(a * H)$$

$\Rightarrow \bar{f} = g$. Hence this induced mapping $\bar{f}$ is unique homomorphism. This theorem tells that f can be factored through the quotient group $(G/H, *)$ or, f can be factored by the natural homomorphism ϕ.

This can be shown by the following diagram

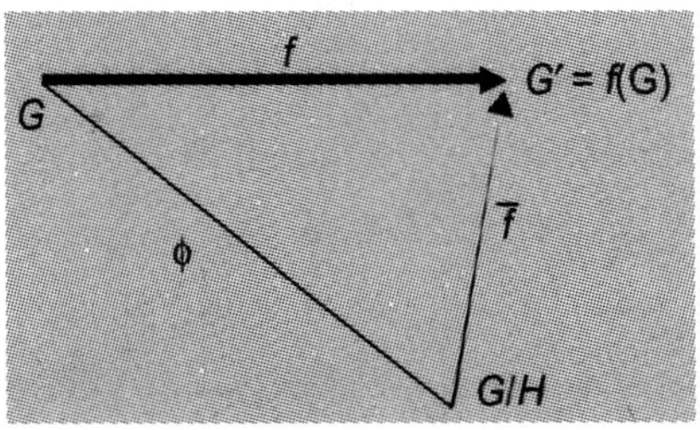

we say that the mapping of this diagram are commutative.

Corollary: The function $\bar{f}$ is one to one if and only if $Ker(f) \subseteq H$

Proof: We have
$$\begin{aligned}Ker(\bar{f}) &= [a^*H \mid f(a^*H) = e'\}\\ &= \{a^*H \mid f(a) = e'\}\\ &= \{a * H \mid a \in Ker(f)\}\end{aligned}$$

This set is the set of all image elements of $Ker(f)$ under the natural homomorphism ϕ, that is, $\phi(Ker(f))$. The function $\bar{f}$ is one to one if and only if $Ker(\bar{f}) = e * H = H$. Hence for all $a \in Ker(f)$, $[a^*H] \subset H \Rightarrow a \in Ker(f)$, $a^*H = H \Rightarrow a \in H$, which is equivalent to $Ker(f) \subseteq H$.

Theorem 3.2.8: *(Fundamental theorem of homomorphism): If f is a homomorphism from the group $(G, *)$ onto the group (G', o), then*

$(G/Ker(f)^*) \cong (G', o)$.

CHAPTER 3

Proof: To prove the theorem, we recall that the *ker (f)* is a normal sub-group of (*G*, *). Thus *(G/Ker (f)*, *) is a quotient group of *G* by *Ker (f)*. Now to show (*G/Ker(f)*), *) ≅ (*G′*, *o)* we must define an isomorphism from *(G/Ker (f)*. *) onto (*G′*, *o)*. Let a mapping ϕ be defined from *G/Ker (f)* onto *G′*, by ϕ $(K * x) = f(x)$, where *Ker (f) = K*, Since *f* is from *G* onto *G′*, then $f(x) \in G'$, and the elements of *G/K* are the cosets *K**x*, $\forall\ x \in G$. We can see that ϕ is well defined.

Let $K * x$, $K*y \in G/K$, then

$$\begin{aligned} \phi\ ((K*x) * (K*y)) &= \phi\ (K*(x*y)) \\ &= f\,(x*y) \\ &= f(x)\ o f(y) \\ &= \phi\ (K*x)\ o\ \phi\ (K * y). \end{aligned}$$

Hence the group operations are preserved by ϕ

Hence the mapping ϕ is a homomorphism.

To shows that ϕ maps *G/K* onto G′,

let $h \in G'$, then there exists some element $x \in G$ such that $f(x) = h$, since *f* is an onto mapping.

Now $K * x$ is an element of *G/K*, and by definition of ϕ $(K * x) = f(x) = h$.

Thus every element $h \in G'$ is an image element under ϕ. Hence ϕ is onto mapping.

Now we shall show that ϕ is one to one.

Let $(K * x)$, $(K *y) \in G/K$ such that

$$\begin{aligned} & \phi\ (K * x) = \phi\ (K * y), \text{ for some } x, y \in G. \\ \Rightarrow\quad & f(x) = f(y). \\ \Rightarrow\quad & f(x)\ o\ (f(y))^{-1} = e', \text{ since } f(x), f(y) \in G' \\ \Rightarrow\quad & f(x*y^{-1}) = e' \text{ by the property of homomorphism } f. \\ \Rightarrow\quad & x*y^{-1} \in \mathrm{K} \\ \Rightarrow\quad & K * x = K * y. \end{aligned}$$

That completes the proof of the theorem.

Example 3.2.5: Let (*Z*, +) be a group of integers under ordinary addition and let (*G*, *) be any arbitrary group. Define a mapping f: Z → *G* by $f(n) = a^n$, $n \in Z$ and *a* is a fixed element of *G*. Then ((*a*), *) ≅ (*Z*, +).

Solution: For $m, n \in Z$,

$$f(m + n) = a^{m+n} = a^m * a^n$$

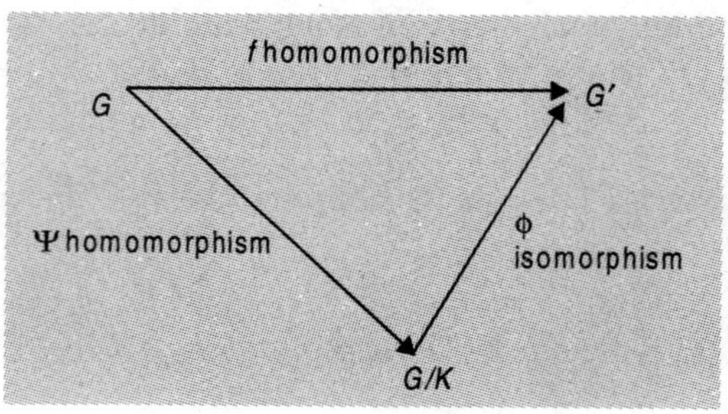

Further we see that for all $x \in G$

$$(\phi \text{ o } \Psi)(x) = \phi(\Psi(x)) = \phi(K * x) = f(x)$$

Hence $\phi \ o \ \Psi = f$.

It is shown in the diagram.

which shows f is a homomorphism from the additive group of integers $(Z, +)$ on the cyclic group $((a), *)$.

Here $Ker(f) = [n \in Z \mid a^n = e] = \{0\}$

If $Ker f = \{0\}$, then $a^n = e \Rightarrow n = 0$. The quotient group $(Z/Ker(f), +) \cong (Z, +)$.

If $Ker(f) \neq \{0\}$, then there exists some smallest positive integer n for which $a^n = e$. Then in the $Ker(f)$ we shall have the smallest positive integer n and all multiples of n. So $Ker(f) = (n)$.

Thus $(Z/Ker(f), +) \cong (Z/(n), +) \cong (Z_n, +_n)$.

Therefore *by* the theorem 3.2.8

$(Z/Ker(f), +) \cong ((a), *)$ which have two conclusions.

(1) If the generator a is of infinite order, then $((a),*) \cong (Z +)$, and .

(2) If the generator a is of finite order n, then $((a), *) \cong (Z_n, +_n)$.

CHAPTER 3

3.3 ENDOMORPHISM AND AUTOMORPHISM

Definition 3.3.1: A homomorphism f of a group (G, o) into itself is called an endomorphism.

If we study the set of endomorphisms f, f_1, f_2 ...... of Group (G, o) into itself, denoted *by* Hom (G,G), then we observe under the functional composition that

Theorem 3.3.1: *The pair Horn (G, G), o) is semi-group with identity under functional composition.*

Proof: **(1)**The set *Hom* (G, G) is closed under the functional composition, that is, if $x \in G$ and $f_1, f_2 \in$ *Hom* (G, G), then $(f_1 \ o \ f_2)(x) = f_1(f_2(x))$, $f_2(x) \in G$ let $f_2(x) = y \in G$.

$$= f_1(y) \in G,$$

(2) Since the composition of functions is associative, the functional composition is also associative in *Hom* (G, G).

(3) *Hom* (G, G) *has identity mapping* i_G.

But the functions which are not one to one do not have their inverse functions.

Thus *(Hom* $(G, G), o)$ is a semi-group with identity under functional composition.

Definition 3.3.2: An isomorphism of a group (G, o) onto itself is called an automorphism, and the set of all automorphisms is denoted by $A(G)$.

Theorem 3.3.2: *The pair (A (G), o) is a group, where A (G) is the set of all automorphisms of G onto itself and 'o' denotes the functional composition.*

Proof: The proof is left for the readers, as an exercise.

Remark: For given a fixed $a \in G$ define the mapping f_a by $f_a(x) = aoxoa^{-1}$, $\forall\ x \in G$, We study the properties of the function f_a, let $x_1, x_2 \in G$., then

$$\begin{aligned} f_a(x_1 \circ x_2) &= a \circ (x_1 \circ x_2) \circ a^{-1} \\ &= (a \circ x_1 \circ a^{-1}) \circ (a \circ x_2 \circ a^{-1}) \\ &= f_a(x_1) \circ f_a(x_2) \end{aligned}$$

Therefore f_a is a homomorphism of (G, o) into itself. Next we notice that f_a maps G onto itself.

For let $x \in G$, then

$$\begin{aligned} f_a(a^{-1} \circ x \circ a) &= a \circ (a^{-1} \circ x \circ a) \circ a^{-1} \\ &= x \end{aligned}$$

Further we mark that f_a is also one to one homomorphism. For, $x_1, x_2 \in G$ such that $f_a(x_1) = f_a(x_2)$

$$\Rightarrow a \circ x_1 \circ a^{-1} = a \circ x_2 \circ a^{-1} \Rightarrow x_1, = x_2,$$

by left and right cancellation laws.

Thus, f_a is an automorphism.

Definition 3.3.3: Functions of the form f_a with $a \in G$ are called *inner automorphisms* of the group (G, o). The set of all inner automorphisms is denoted by $I(G)$. Thus $I(G) = \{f_a \mid a \in G\}$. Since an inner automorphisms f_a is an isomorphism from the group $(G, *)$ onto itself, $f_a \in A(G)$, the set of all automorphison, which implies $I(G) \subseteq A(G)$.

Theorem 3.3.3: *The pair (I (G), o) is a group, known as the group of inner automorphism of (G, *). (I (G), o) is also a normal sub-group of the group (A(G), o) of all automorphisms.*

Proof: We see that the pair $(I(G), o)$ satisfies the following axioms:

(1) $\forall\ f_a, f_b \in I(G), f_a \circ f_b = f_{a*b} \in I(G)$.

(2) The functional composition o is associative in $I(G)$.

(3) $\forall\ f_a \in I(G), \exists\ f_e \in I(G)$ such that

$$f_a \circ f_e = f_{a*e} = f_a = f_{e*a} = f_e \circ f_a.$$

(4) For each $f_a \in I(G), \exists\ f_{a^{-1}} \in I(G)$ such that

$$f_a \circ f_{a^{-1}} = f_{a*a^{-1}} = f_e = f_{a^{-1}*a} = f_{a^{-1}} \circ f_a.$$

Since the functional composition is not commutative, the pair $(I(G), o)$ is non-commutative group. It also follows that $(I(G), o)$ is a sub-group of $(A(G), o)$.

We also have that for $f \in A(G)$, $f_a \in I(G)$,

$$(f \circ f_a \circ f^{-1})(x) = (f \circ f_a)(f^{-1}(x))$$

$$= f(f_a(f^{-1}(x))$$
$$= f(a * f^{-1}(x) * a^{-1})$$
$$= f(a) * f(f^{-1}(x)) * f(a^{-1})$$
$$= f(a) * x * f(a^{-1})$$
$$= f_{f(a)}(x)$$

This implies $f \; o \; f_a \; o \; f^{-1} = f_{f(a)} \in I(G)$.

This shows that the pair $(I(G), o)$ is a normal sub-group of $(A(G), o)$.

Theorem 3.3.4: *For each group $(G, *)$, $(G/Z(G), *) \cong (I(G), o)$*

Proof: We consider the function $f : G \to I(G)$ defined by

$f(a) = f_a$, $a \in G$, Now for all $a, b \in G$,

$f(a * b) = f_{a*b} = f_a \; o \; f_b$. This means that f is a homomorphism, Since for $f_a \in I(G)$ there exists $a \in G$ such that $f(a) = f_a$, f is onto mapping.

Now
$$Ker(f) = \{a \in G \mid f_a = f_e\}$$
$$= \{a \in G \mid f_a(x) = f_e(x), \forall \; x \in G\}$$
$$= \{a \in G \mid a * x * a^{-1} = x, \forall \; x \in G\}$$
$$= \{a \in G \mid a * x = x * a, \forall \; x \in G\}.$$

This means that $Ker(f)$ is the set of all elements $a \in G$ which commute with every element $x \in G$. That is, $Ker(f)$ is the centre $Z(G)$ of the group $(G, *)$. Since the pair $Z(G, *)$ is a normal sub-group of $(G, *)$, the quotient group $(G/Z(G, *)$ exists.

By Theorem 3.2.8 we have

$(G/Z(G, *) \cong (I(G), o)$. This proves the theorem.

Let f be a homomorphism from the group $(G, *)$ onto (G', o). If $(H, *)$ is sub-group of $(G, *)$, then $(f(H), o)$ is also a sub-group of (G', o). Now we see whether a one to one correspondence exists between the sub-groups of $(G, *)$ and the sub-groups of (G', o).

If $(H *)$ and $(K, *)$ are two sub-groups of $(G, *)$ with $H \subseteq K \subseteq H * Ker(f)$, then $(f(H), o) = (f(K), o)$.

$$H \subseteq K \Rightarrow f(H) \subseteq f(K),$$

and
$$K \subseteq H * Ker(f) \Rightarrow f(K) \subseteq f(H * Ker(f))$$
$$\Rightarrow f(K) \subseteq f(H) \; o \; f(Ker(f))$$
$$\Rightarrow f(K) \subseteq f(H) * e'$$
$$\Rightarrow f(K) \subseteq f(H)$$

This shows $f(H) = f(K)$. Hence $(f(H), o) = (f(K), o)$. From this we conclude that two distinct subsets of G have the same image set in G'. This can be over-come by taking $Ker(f) = \{e\}$ or considering the sub-groups $(H, *)$ with $Ker(f) \subseteq H$. So

$H \subseteq K \subseteq H * ker(f) \subseteq H \Rightarrow H = K.$

CHAPTER 3

If $Ker(f) = \{e\}$, then $(G, *) \cong (G', o)$ and if $Ker(f) \subseteq H$, then we have the correspondence theorem. Before it we prove.

Lemma 3.3.1: *If H is any subset of G such that $Ker(f) \subseteq H$, then $H = f^{-1}(f(H)$*

Proof: $a \in H \Rightarrow f(a) \in f(H) \Rightarrow a \in f^{-1}(f(H))$.

Therefore $H \subseteq f^{-1}(f(H))$

$b \in f^{-1}(f(H)) \Rightarrow f(b) \in f(H)$. Then for some

$$\begin{aligned} h \in H, f(b) = f(h) &\Rightarrow f(b) \; o \; (f(h))^{-1} = e' \\ &\Rightarrow f(b) \; of \; (h^{-1}) = e' \\ &\Rightarrow f(b * h^{-1}) = e' \\ &\Rightarrow b * h^{-1} \in Ker(f) \\ &\Rightarrow b * h^{-1} \in H \text{ (since } Ker(f) \subseteq H) \\ &\Rightarrow b \in H \text{ since } (H, *) \text{ is sub-group.} \end{aligned}$$

This shows $f^{-1}(f(H)) \subseteq H$ which proves

$$H = f^{-1}(f(H)).$$

Theorem 3.3.5: *(Correspondence Theorem): There is one to one correspondence between the set of the sub-groups (H, *) of the group (G, *) such that $Ker(f) \subseteq H$ and the set of all subgroups (H′, o) of the group (G′, o); specifically, $H' = f(H)$.*

Proof: First of all we see that if (H', o) is any sub-group of (G', o), we must produce some sub-group $(H, *)$ of $(G, *)$ with $Ker(f) \subseteq H$ such *that* $f(H) = H'$, For this we take $H = f^{-1}(H')$. Since (H', o) is sub-group, then by theorem 3.1.2 (2), the pair $(f^{-1}(H'), *)$ is a sub-group of $(G, *)$ and since $e' \in H'$.

$ker(f) = f^{-1}(e') \subseteq f^{-1}(H') = H$

Moreover, the function f is onto, so $f(f^{-1}(H')) = H'$.

Now we show that this correspondence is one-to-one. For this let $(H_1, *)$ and $(H_2\ *)$ be two sub-groups of $(G, *)$ with $ker(f) \subseteq H_1$ and $Ker(f) \subseteq H_2$.

Let us assume that

$f(H_1) = f(H_2) \Rightarrow f^{-1}(f(H_1) = f^{-1}(f(H_2) = H_1 = H_2$ by Lemma 3.3.1.

It follows that the correspondence $(H, *) \rightarrow (f(H), o)$ is one to one and onto. This completes the proof.

Theorem 3.3.6: *Let (H, *) be a normal sub-group of the group (G, *). There is one to one correspondence between those sub-groups (K, *) of (G, *) such that $H \subseteq K$ and the set of all sub-groups of the quotient group (G/H, *).*

Proof: Let Σ be the set of all sub-groups $(K, *)$ of $(G, *)$ such that $H \subseteq K$ and Let Σ' be the set of all sub-groups $(K/H, *)$ of the quotient group $(G/H, *)$.

Now we define the mapping $\phi : Z \rightarrow \Sigma'$ by

$$\phi (K) = K/H, K \in \Sigma.$$

First of all we see that $H \subseteq K$ and $(H, *)$ is normal in $(G, *)$, then $(H, *)$ is also normal in the sub-group $(K, *)$ of $(G, *)$. Thus the quotient group $(K/H, *)$ exists.

For $k_1, k_2 \in K, k_1 * H, k_2 * H \in K/H$, then

$(k_1 * H) \circ (k_2 * H)^{-1} = (k_1 * H) * (k_2^{-1} * H)$

$= (k_1 * k_2^{-1}) * H \in K/H$ since $k_1 * k_2^{-1} \in K$.

This proves that $(K/H, *)$ is a sub-group of $(G/H, *)$.

To prove ϕ is one to one, we suppose $(K_1, *)$ and $(K_2, *)$ are two sub-groups such that $H \subseteq K_1$ and $H \subseteq K_2$ then

$$\phi (K_1) = \phi (K_2) \Rightarrow K_1/H = K_2/H$$
$$\Rightarrow K_1 = K_2$$

Hence the correspondence $(K_1, *) \leftrightarrow (K_1/H, *)$ is one to one. We also observe that $(K, *)$ is normal if and only if $(K/H, *)$ is normal in $(G/H, *)$.

$x * k * x^{-1} \in K, \forall x \in G$ and $k \in K \leftrightarrow H * (x * k * x^{-1}) \in K/H$

$\forall H * x \in G/H, H * k \in K/H$

$\leftrightarrow (K/H, *)$ is normal sub-group.

Theorem 3.3.7: *Let f be a homomorphism from the group $(G, *)$ onto the group (G', o).*

If $(H, *)$ is any normal sub-group of the group $(G, *)$ such that $Ker (f) \subseteq H$, then $(G/H, *) \cong (G'/f(H), o)$.

Proof: By theorem 3.1.3 we know that if $(H, *)$ is normal sub-group of $(G, *)$, then $(f(H), o)$ is also normal sub-group of (G', o) which follows that the quotient group $(G'/f_{(H)}, o)$ exists.

Now we define the mapping $\bar{f} . G \rightarrow G'/f(H)$ by

$$\bar{f}(a) = f(a) \circ f(H).$$

If ϕ is a natural homomorphism from G' onto $G'/f(H)$, that is,

$$\phi (f(a)) = f(a) \circ f(H), \text{ then}$$
$$\bar{f}(a) = f(a) \circ f(H) = \phi (f(a)) = (\phi \circ f)(a)$$
$$\Rightarrow \quad \bar{f} = \phi \circ f.$$

Since ϕ and f are homomorphism, ϕ o f is *a* also homomorphism and consequently $\bar{f}$ is homomorphism. Now

$$ker(\bar{f}) = \{a \in G \mid \bar{f}(a) = f(H)\}$$
$$= \{a \in G \mid f(a) o f(H) = f(H)\}$$
$$= \{a \in G \mid f(a) \in f(H)\}$$

CHAPTER 3

$$= \{a \in G \mid a \in f^{-1}f(H)\}$$
$$= \{a \in G \mid a \in H\} \text{ by the lemma 3.3.1.}$$

Hence $ker(\bar{f}) = H$. Therefore by fundamental Theorem of homomorphism 3.2.8 we have

$$(G/H, *) \cong (G'/f(H), o).$$

Corollary: If (H', o) is any normal sub-group of the group (G', o), then $(G/f^{-1}(H'), *) \cong (G'/H', o)$.

Theorem 3.3.8: *Let (H, o) and (K, o) be two normal sub-groups of (G, o) such that $H \subset$ A. Then*

$$(G/K, o) \cong \left(\frac{G/H}{K/H}, o\right)$$

Proof: Since $H \subset K$, (H, o) is normal sub-group of (K, o) and since (H, o), (K, o) are normal sub-groups of (G, o), then the quotient groups $(G/K, o)$, $(G/H, o)$, and $(K/H, o)$ exist.

Now we define the mapping $\phi : G/H \to G/K$ by

$\phi(a \; o \; H) = a \; o \; K,\; a \in G$.

For $a \; o \; H,\; b \; o \; H \in G/H$,

$$\begin{aligned}\phi\,[c \; o \; H) \; o \; (b \; o \; H)\} &= \phi\,[(a \; o \; b) \; o \; H)] \\ &= (a \; o \; b) \; o \; K \\ &= (a \; o \; K) \; o \; (b \; o \; K) \\ &= \phi(a \; o \; H) \; o \; \phi(b \; o \; H)\end{aligned}$$

It shows ϕ is homomorphism and it is clearly onto. Next we have

$$\begin{aligned}ker(\phi) &= \{a \; o \; H \mid \phi(a \; o \; H) = K\} \\ &= \{a \; o \; H \mid a \; o \; K = K\} \\ &= \{a \; o \; H \mid a \in K\} \\ &= K/H.\end{aligned}$$

Therefore by fundamental theorem of homomorphism 3.2.8 we have

$$(G/K, o) \cong \left(\frac{G/H}{K/H}, o\right)$$

Theorem 3.3.9: *If $(H, *)$ and $(K, *)$ are sub-groups of the group $(G, *)$ with $(K, *)$ normal, then*

$$(H/H \cap K, *) \cong (H^* K/K, *).$$

Proof: By problem 12 on page 92 if $(H, *)$ and $(K, *)$ are sub-groups of $(G, *)$ with $(K, *)$ normal, then $(H * K, *)$ is a sub-group containing K and by theorem 2.11.5 $(H \cap K, *)$ is a normal sub-group contained in H. This follows that the quotient groups $(H/H \cap K, *)$ and $(H * K/K, *)$ exists.

Now we define the mapping $\phi: H \to H * K/K$ by

$$\phi(h) = h * K,\ h \in H.$$

For $h_1, h_2 \in H$,

$$\begin{aligned}\phi(h_1 * h_2) &= (h_1 * h_2) * K \\ &= (h_1 * K) * (h_2 * K) \\ &= \phi(h_1) * \phi(h_2)\end{aligned}$$

which shows ϕ is homomorphism and it is clearly onto. For the *ker* ϕ we have

$$\begin{aligned}ker(\phi) &= \{h \in H \mid \phi(h) = K\} \\ &= \{h \in H \mid h * K = K\} \\ &= \{h \in H \mid h \in K\} \\ &= \{h \in H \cap K\} = H \cap K.\end{aligned}$$

Therefore by fundamental theorem of homomorphism 3.2.8 we have

$$(H/H \cap K, *) \cong (H * K/K, *)$$

Theorem 3.3.10: *(Zassenhaus) Let $(H, *)$ and $(K, *)$ the two sub-groups of $(G, *)$ and let $(H_o, *)$ and $(K_o, *)$ are two normal sub-groups of $(H, *)$ and $(K, *)$ respectively, then*

$$\frac{Ho*(H \cap K)}{Ho*(H \cap Ko)} \cong \frac{Ko*(K \cap H)}{Ko*(K \cap Ho)}$$

Proof: Since $(H, *)$ and $(K, *)$ are sub-groups, $(H \cap K, *)$ is a sub-group and $H_o * (H \cap K)$ is a sub-group as $(H_o, *)$ is normal.

$H \cap K_o \subseteq H \cap K \Rightarrow H_o * (H \cap K_o) \subseteq H_o * (H \cap K)$. Since $(H \cap K_o, *)$ by theorem 2.11.5 and $(H_o\ *)$ are normal sub-groups, $(H_o * (H \cap K_o), *)$ is a normal sub-group of $(H_o * (H \cap K), *)$ by problem 13 on p. 92 Now we set $A = H_o^* (H \cap K_o)$ and $B = H \cap K$.

Then $A * B = H_o^* (H \cap K_o) * H \cap K = H_o * (H \cap K)$ since $H \cap K_o \subseteq H \cap K$.

Further, $\quad x \in A \cap B \Rightarrow x \in A$ and $x \in B$

$\Rightarrow x \in H_o * (H \cap K_o)$ and $x \in H \cap K$

$\Rightarrow x = h_o * h$ and $x = k$, for some $h_o \in H_o$,

$\Rightarrow h_o * h = k,\ h \in H \cap K_o,\ k \in H \cap K$

$\Rightarrow h_o = k * h^{-1} \in K$, since

$h^{-1} * H \cap K_o \subseteq H \cap K \subseteq \mathbf{K}$

$\Rightarrow h_o \in H_o \cap K = K \cap H_o.$

Therefore $x = h_o * h \in (K \cap H_o) * (H \cap K_o)$

Hence $A \cap B \subseteq (K \cap H_o) * (H \cap K_o)$

We see that $H \cap K_o \subseteq H \cap K,\ K \cap H_o \subseteq H \cap K$

$\Rightarrow (H \cap K_o) * (K \cap H_o) \subseteq H \cap K = B$ and $K \cap H_o \subseteq H_o$

and $(K \cap H_o) * (H \cap K_o) \subseteq H_o * (H \cap K_o) = A$

Therefore $(K \cap H_o) * (H \cap K_o) \subseteq A \cap B$

Hence $(K \cap H_o) * (H \cap K_o) = A \cap B$

Now from the theorem 3.3.9 we have

$\frac{A*B}{A} \cong \frac{B}{A \cap B}$ which gives

$$\frac{H_o*(H \cap K)}{H_o*(H \cap K_o)} \cong \frac{H \cap K}{(K \cap H_o)*(H \cap K_o)}$$

Interchange H and K, we get

$$\frac{K_o*(K \cap H)}{K_o*(K \cap H_o)} \cong \frac{K \cap H}{(H \cap K_o)*(K \cap H_o)}$$

Hence $\frac{H_o*(H \cap K)}{H_o*(K \cap H_o)} \cong \frac{K_o*(K \cap H)}{K_o*(K \cap H_o)}$

This completes the proof of the theorem.

Example 3.3.1: Let $(Z, +)$ be the additive group of integers. Consider the cyclic sub-groups $((3), +)$ and $((4), +)$ of $(Z, +)$. They are normal since $(Z, +)$ is a commutative group. In this case we have

$(3) \cap (4) = (12)$ and $(3) + (4) = Z$.

Then $((3) / (12), +) \cong (Z./(4), +)$ which is a simple illustration of the above theorem.

Theorem 3.3.11: *There are only two groups of order six, one is cycle and other is isomorphic to $(S_3\ o)$.*

Proof: Let $(G, *)$ be a group of order 6. There are five distinct elements except identity of the group. Since every element has its unique inverse, then $x \to x^{-1}$ is a one to one correspondence between the elements and their inverses, then there exists an element $a\ (\neq e) \in G$ such that $a^{-1} = a$ or $a^2 = e \Rightarrow o\ (a) = 2$. Let $(G, *)$ be abelian and Let (H, o) be the sub-group generated by a. So $H = (a)$ and $o\ (H) = 2$. The index $[G\ !$ $H]$ $\frac{o(G)}{o(H)} = 6/3 = 3$ *i.e.*, $o\ (G/H) = 3$ (prime) so $(G/H, *)$ is a cyclic group. Let $b * H \in G/H$ be a generator of G/H. Since $o\ (b * H) | o\ (b)$, then $o\ (b) = 3$ or $o\ (b) = 6$. If $o\ (b) = 6$ then $(G, *)$ is cyclic. If $o\ (b) = 3$, then $o\ (a * b) = 6$ as $a * b = b * a$ and $o\ (a)$ and $o\ (b)$ are co-prime. Hence $(G, *)$ is cycle group generated by $a * b$.

Suppose $(G, *)$ is not cyclic, $(G, *)$ cannot be abelian. Then all non-identity elements of G cannot be order 2. Thus there exists an element $c \in G$ such that $o\ (c) = 3$ and $aoc \neq coa$. Let $H = \{e, c, c^2\}$. As $[G\ !\ H\} = 2$, $(H, *)$ is a normal sub-group. Consequently $a^{-1} * c * a \in H$ so that $a^{-1} * c * a = c$ or $a^{-1} * c * a = c^2$. But as $a * c \neq c * a$ we get $a^{-1} * c * a = c^2 \Rightarrow c * a = a * c^2$, Now $a \notin H \Rightarrow G = H \cup a * H = \{e, c, c^2, a, a * c, a * c^2\}$.

We define mapping $f : G \to S_3$ by $f(e) = I$, $f(a) = (12)$, $f(c) = (123)$, $f(c^2) = (132)$, $f(a * c) = (23)$ and

$f(a * c^2) = (13)$ is an isomorphism.

Hence this proves the theorem.

With the help of this we shall prove that the converse of Lagrange's theorem is not true. Here we prove lemma.

Lemma 3.3.2: *(S_4, o) has no element of order 6.*

Proof: S_4 is the set of all permutations on the set {1, 2, 3, 4}.

Let $\phi\ (\neq I) \in S_4$, then $\phi = \phi_1\ o\ \phi_2\ o...o\ \phi_t$, where $\phi_1, \phi_2,..., \phi_t$ are pair wise disjoint cycles each of length > 1. Since the symbols which appear is ϕ_i are among {1, 2, 3, 4}, so no ϕ_i can be of length > 4 and no two $\phi_i\ \phi_j\ (i \neq j)$ have common symbol, we have $t \leq 2$. This ϕ is cyclic or ϕ can be expressed as a product of two disjoint cycles each of length 2. If ϕ is cyclic, then it is of length 2, 3, 4. Hence ϕ can be of order 2, 3 or 4. It ϕ is a product of two disjoint cycles, each of length 2, ϕ is of order 2. Hence S_4 has no element of order greater than 4.

CHAPTER 3

Theorem 3.3.12: *The alterrating group (A_4, o) has no sub-group of order 6.*

Proof: By above lemma A_4 has no element of order 6, and consequently A_4 has no cyclic sub-group of order 6. Suppose (A_4, o) has a sub-group of order 6, by theorem 3.3.1. It must be isomorphic to (S_3, o). Let (H, o) be a sub-group of (A_4, o) of order 6, then $(H, o) \cong (S_3, o)$. H contain 3 elements of order 2 each and 2 elements of order 3 each. The only elements of order 2 of A_4 are (12) *o* (3 4), (13) *o* (2 4) and (14) *o* (23). Let ϕ be an isomorphism between (H, o) and $S_3, o)$.

Now $\phi\ [(12)\ o\ (34)\ o\ (13)\ o\ (24)] = \phi\ [(14)\ o\ (23)]$

$(14)\ o\ (23)$ is of order 2 $\Rightarrow$ order of $\phi\ [(14)\ o\ (23)] = 2$

But in S_3, the product of two different permutations of order 2 is a permutation of order 3.

Thus order of $\phi\ [(14)\ o\ (23)] = 3$

Which is contradiction. So our supposition is wrong. Hence (A_4, o) has no sub-group of order 6.

Remark: We note that $o(S_4) = 4! = 24$ and $A_4 = 12$ and 6112, but (A_4, o) has no sub-group of order 6. Hence the converse of Lagrange's theorem is not true.

WORKED OUT EXERCISES

Exercise 3.3.1: Mapping $f : G \to G$ defined by $f(x) = x^{-1}$, for all $x \in G$ on a group $(G, *)$ is an auto morphism iff $(G, *)$ is abelian.

Solution: Let $f: G \to G$ be an auto morphism of a group $(G, *)$. Now for $x, y \in G$, since f is a homomorphism, then.

$$f(x * y) = (x * y)^{-1}$$

$$\Rightarrow \qquad f(x) * f(y) = y^{-1} * x^{-1}$$
$$\Rightarrow \qquad x^{-1} * y^{-1} = y^{-1} * x^{-1}$$
$$\Rightarrow \qquad x * y = y * x.$$

Hence $(G, *)$ is abelian.

Conversely, let $(G, *)$ be an abelian group, then

$$\begin{aligned} f(x * y) &= (x * y)^{-1} \\ &= y^{-1} * x^{-1} \\ &= x^{-1} * y^{-1} \text{ since } (G, *) \text{ is abelian} \\ &= f(x) * f(y) \end{aligned}$$

$\Rightarrow f$ is a homomorphism, moreover.

$$f(x) = f(y)$$
$$\Rightarrow \qquad x^{-1} = y^{-1}$$
$$\Rightarrow \qquad x = y$$

$\Rightarrow f$ in one to one

and each $x \in G$

$$f(x^{-1}) = (x^{-1}) = x \Rightarrow f \text{ is onto.}$$

Hence f is an automorphism.

Exercise 3.3.2: $(H, *)$ is a sub-group of a group $(G, *)$ and S is the set of all left cosets of H in G, then there is a homomorphism θ of G into $A(S)$ and the kernel of θ is the largest normal sub-group of $(G, *)$ which is contained in H.

Proof: Here $S = \{x * H \mid x * G\}$. Let us define a mapping

$\theta : G \to A(S)$ by

$\theta(g) = T_g$, where $T_g(x * H) = g * (x * H) = (g * x) * H$.

for all $x * H \in S$.

Firstly we show that $T_g \in A(S)$. Clearly T_g is mapping,

from S to S. Now for $x * H$, $y * H$,

$$T_g(x * H) = T_g(y * H)$$
$$\Rightarrow \qquad g * (x * H) = g * (y * H)$$
$$\Rightarrow \qquad (g * x) * H = (g * y) * H$$

$\Rightarrow (g * x)^{-1} * (g * y) \in H$

$\Rightarrow x^{-1} * g^{-1} * g * y \in H$

$\Rightarrow x^{-1} * y \in H$

$\Rightarrow x * H = y * H$

$\Rightarrow T_g$ is a one to one mapping.

Further, any coset $x * H$ can be written as $g * (g^{-1} * x * H)$,

so $$T_g (g^{-1} * x * H) = g * (g^{-1} * x * H) = x * H.$$

Which shows that T_g is onto and consequently $T_g \in A(S)$.

Again, we see that $\theta (g * h) = T_{g*h}$.

For $$T_{g*h} (x * H) = (g * h) * (x * H) = g * (h * (x * H))$$
$$= T_g (h * (x * H))$$
$$= T_g (T_h (x * H)) = T_g \, o \, T_h (x * H).$$

$$\Rightarrow \quad T_{g*h} = T_g \, o \, T_h$$

Hence $\theta (g*h) = T_{g*h} = T_g \, o \, T_h = \theta (g) \, o \, \theta (h)$

Thus θ is a homomorphism from G to $A(S)$.

Now, Let $g \in Ker\ \theta$, then $\theta (g) = I$, where I is the identity of $A(S)$. That is, $T_g (x * H) = I (x * H) = x * H$.

For all $x \in G$.

$$T_g (x * H) = x * H$$

$$\Rightarrow \quad g * (x * H) = x * H \text{ for all } x \in G \text{ and } g \in Ker\theta$$

On the other hand, if $b \in G$ such that

$$b * (x * H) = x * H, \text{ then } b \in Ker\ \theta.$$

Hence $$Ker\ \theta = \{b \in G \mid b * (x * H) = x * H \text{ for all } x \in G\}.$$

Now we claim that $Ker\theta$ is the largest normal sub-group contained in H. By largest normal sub-group we mean that $(N, *)$ is a normal sub-group of G which is contained in H, then $(N, *)$ must be contained in $Ker\theta$.

Since $Ker\theta$ is the kernel of homomorphism of G, $(Ker\ \theta, *)$ is a normal sub-group of $(G, *)$ we see that,

If $b \in Ker\theta$, $b * (x * H) = x * H$ for every $x \in G$, so in particular, $b * H = b * (e * H) = e * H = H \Rightarrow b \in H$

Hence $Ker\theta \subset H$.

Finally, if $(N, *)$ is normal sub-group of $(G, *)$ which is contained in H, if $n \in N$, $x \in G$, then

$$x^{-1} * n * x \in N \subset H \Rightarrow x^{-1} * n * x * H = H$$
$$\Rightarrow (n * x) * H = x * H \text{ for all } x \in G$$
$$\Rightarrow n \in K.$$

Hence $(Ker\theta, *)$ is the largest normal sub-group of (G, o). This proves the result.

Exercise 3.3.3: If $(G, *)$ is finite group, and $1 \neq G$ is a sub-group of $(G, *)$ such that $o(G) + i(H)$!, then H must contain a non trivial normal sub-group of $(G, *)$.

Proof: By above exercise $Ker\theta \subseteq H$. Since $H \neq G$, $Ker\theta \neq G$. Here $(H, *)$ is a sub-group whose index $i(H)$ satisfies $i(H)! < o(G)$. Further if $Ker\theta = \{e\}$, where e is the identity of G, then $\frac{G}{\text{Ker}\theta} \cong T$, where (T, o) is a sub-group of $A(S)$ gives that $o(G) = 0(T)$, must be a factor

CHAPTER 3

of o $(A(S))$. But $o[A(S)] = i(H)!$ This in turn implies that $o(G) \mid i(H)!$ which is contradiction to the given. Hence $Ker\theta \neq \{e\}$. Thus $(H, *)$ contains a non-trivial normal sub-group of G.

Exercise 3.3.4: (H, o) is a sub-group of (S_n, o) for some integer $n > 1$. If H contains an odd permutation, show that the set of all even permutations in H form a normal sub-group of H of index 2.

Solution: Let $X = [1, -1\}$, then the system $(X, .)$ is a sub-group of the group of real numbers under multiplication.

We define $\theta : H \to X$ by

$$\theta(f) = 1, \text{ if } f \text{ is even permutation}$$
$$= -1 \text{ if } f \text{ is odd permutation.}$$

For any permutations f and g of H, we have

$$\theta(fog) = 1 \text{ if } f \text{ and } g \text{ both are even}$$
$$= 1.1$$
$$= \theta(f)\,.\,\theta(g)$$

Again $$\theta(fog) = 1 \text{ if } f \text{ and } g \text{ both are odd}$$
$$= -1\,.\,-1$$
$$= \theta(f)\,.\,\theta(g).$$

Further, if $$\theta(fog) = -1 \text{ if one of them is even and other is odd}$$
$$= 1\,.\,(-1) \text{ or } (-1)\,.\,1$$
$$= \theta(f)\,.\,\theta(g) \text{ or } \theta(f)\,.\,\theta(g).$$

In all cases $\theta(fog) = \theta(f)\,.\,\theta(g)$.

Hence θ is a homomorphism and clearly onto.

$Ker\theta$ = set of even permutations of H.

As $(Ker\theta, o)$ is a normal sub-group of (H, o), and by fundamental theorem of homomorphism

$$\left(\frac{H}{Ker\theta}, o\right) \cong (X, .)$$

$$\Rightarrow o\left(\frac{o(H)}{o(Ker\theta)}\right) = o(X) = 2$$

$\Rightarrow Ker\theta$ is of index 2.

Exercise 3.3.5: $(G, *)$ is a group of order $2n$, where n is odd. Show that $(G, *)$ has a normal sub-group of order n.

Solution: G has at least one element of order 2, otherwise we can split G into union of disjoint pairs $\{a, a^{-1}\}$, $\{b, b^{-1}\}$, ..., etc. and singleton $\{e\}$, where e is the identity of G.

This implies that G has an odd number of elements.

Let $a \in G$ be of order 2. Then $a^2 = e$

Now consider mapping T_a ' $G \to G$ defined by $T_a\ (x) = a * x$, for all $x \in G$. we see that for any $x, y \in G$

$$T_a\ (x) = T_a\ (y)$$

$$\Rightarrow \qquad a * x = a * y$$

$$\Rightarrow \qquad x = y$$

$\therefore\ T_a$ is one to one mapping.

Again, for $a^{-1} * x \in G$, we have

$$T_a\ (a^{-1} * x) = a * (a^{-1} * x) = (a * a^{-1}) * x = e * x = x.$$

So T_a is onto and consequently T_a is a permutation.

Since
$$T_a^{\ 2}\ (x) = T_a \ o\ T_a\ (x)$$
$$= T_a\ (a * x) = a* (a * x)$$
$$= (a * a) * x = a^2 * x = e * x = x.$$
$$= T_e\ (x)$$

each orbit of T_a consists of two elements and the number of distinct orbit of $T_a = \frac{2n}{2} = n$. Since each orbit corresponds to a cycle, T_a has n cycles each of length 2.

Let $\sigma_1, \sigma_2, ..., \sigma_n$ be n cycles of length 2 (*i.e.*, transpositions) of T_a. So $T_a = \sigma_1\ o\ \sigma_2$ o..., oσ_n, since n is odd, T_a is an odd permutation in $A\,(G)$.

Let
$$K = \{T_g \mid g \in G\}.$$

For
$$g_1, g_2 \in G,$$

$$\left(T_{g_1}\ o\ T_{g_2}^{\ -1}\right) = T_{g_1}\left(T_{g_1}^{-1}\ (x)\right)$$

$$= T_{g_1}\left(T_{g_2^{-1}}\ (x)\right)$$

$$= T_{g_1}\left(g_2^{-1} * x\right)$$

$$= g_1 * \left(g_2^{-1} * x\right)$$

$$= \left(g_1 * g_2^{-1}\right) * x = T_{g^* g_2^{-1}}\ (x)$$

$$\Rightarrow \qquad T_{g_1}\ o\ T_{g_2}^{-1} = T_{g^* g_2^{-1}} \in K \text{ since } g_1 * g_2^{-1} \in G.$$

So (K, o) is a sub-group of $A\ (G)$

Since T_a is an odd permutation in K, K has a normal sub-group N of index 2. Thus G has a normal sub-group $H = f^{-1}\ (N)$ of index 2. But $o\ (H) = \frac{o(G)}{i(H)} = \frac{2n}{2} = n$. Hence $(G, *)$ has a normal sub-group of order n.

CHAPTER 3

Exercise 3.3.6: For $a, b \in R, a \neq 0$, define $f_{a,b}(x) : R \to R$ by $f_{a,b}(x) = ax + b$ for all $x \in R$. Let $G = \left\{f_{a,b} \mid a, b \in R, a \neq 0\right\}$ and $N = \left\{f_{1,b} \mid b \in R\right\}$. Prove that (N, o) is a normal sub-group of (G, o) and $(G/N, o) \cong (R - \{o\}, .)$

Solution: We see that for all $x \in R$

$f_{1,o}(x) = x$, that is, $f_{1,o} \in G$ is the identity in G.

Again, for $a, b, c, d \in R$

$$\begin{aligned}
(f_{a,b} \circ f_{c,d})(x) &= f_{ab}(f_{cd}(x)) \\
&= f_{ab}(cx + d) \\
&= a(cx + d) + b \\
&= acx + ad + b \\
&= f_{ac,\ ad+b}(x)
\end{aligned}$$

$$\Rightarrow \quad f_{ab} \circ f_{ed} = f_{ac,\ ad+b} \in G.$$

Suppose $\quad f_{ab}^{-1} = f_{cd}$, *then*

$$\begin{aligned}
f_{ab} \circ f_{cd} &= f_{10} \\
\Rightarrow \quad (f_{ab} \circ f_{cd})(x) &= f_{10}(x) \\
\Rightarrow \quad acx + ad + b &= x \\
\Rightarrow \quad ac &= 1 \text{ and } ad + b = 0 \\
\Rightarrow \quad c &= a^{-1} \text{ and } d = -a^{-1} b.
\end{aligned}$$

Hence $f_{a,b}^{-1} = f_{c,d} = f_{a^{-1},-a^{-1}b} \in G$.

Since composite of function is associative, therefore (G, o) is a group under the composition of functions.

Clearly $f_{10} \in N$ and $N \neq \phi$.

Again $f_{1,b}\ f_{1,d} \in N$, then $f_{1,b} \circ f_{1,d}^{-1} = f_{i,b} \circ f_{1,\ -d} = f_{1,\ -d+b} \in N$

Now for $f_{a,b} \in G$ and $f_{1,c} \in N$,

$$\begin{aligned}
f_{a,b} \circ f_{1,c} \circ f_{a,b}^{-1} &= f_{a,b} \circ f_{1,c} \circ f_{a^{-1},-a^{-1}b} \\
&= f_{a,b} \circ f_{a^{-1},-a^{-1}b+c} \\
&= f_{a,a^{-1},\ a(-a^{-1}b+c)+b.} \\
&= f_{1,-b+ac+b} \\
&= f_{1,\ ac} \in N.
\end{aligned}$$

Hence (H, o) is a normal sub-group of (G, o).

Finally, Let $(R -\{o\}$; be the group of non-zero real number under multiplication.

Define $\phi : G \rightarrow R - \{o\}$ by

$$\phi\,(f_{a,b}) = a, \text{ for all } f_{a,b} \in G.$$

Now for $f_{a,b}\ f_{c,d} \in G$

$$\begin{aligned}\phi\,(f_{a,b} \circ f_{c,d}) &= \phi\,(f_{ac,\ ad\ +\ b}) \\ &= ac \\ &= \phi\,(f_{ab},)\ .\ \phi\,(f_{c,d})\end{aligned}$$

This shows that ϕ is homomorphism.

Now we determine kernel of ϕ

$$\begin{aligned}ker\phi &= \{f_{a,b} \in G \mid \phi\,(f_{a,b}) = 1\} \\ &= \{f_{a,b}, \in G \mid a = 1\} \\ &= \{f_{1,b} \in G\} = N.\end{aligned}$$

Hence by fundamental theorem of homomorphism

$$(G/N,\ o) \cong (R - \{o\},\ .)$$

Exercise 3.3.7: If $(G, *)$ is a group such that $(G/Z(G), *)$ is cyclic, where $(Z/G), *)$ is the centre of $(G, *)$, show that $(G, *)$ is abelian.

Proof: Let $Z(G) = Z$. Since $(G/Z, *)$ is cyclic, then there exists a coset $g * Z$ which generate G/Z.

Let $a, b \in G$, $a * Z = (g * Z)^m$ and $b * Z = (g * Z)^n$

$= g^m * Z$ $\qquad\qquad$ $= g^n * Z$

for some integers m, n

Thus for some $z_1, z_2 \in Z$

$a = g^m * z_1$ and $b = g^n * z_2$

Now $a * b = (g^m * z_1) * (g^n * z_2) = g^{m+n} * z_1 * z_2$

and $b * a = (g^n * z_2) * (g^m * z_2) = g^{m+n} * z_1 * z_2$

Thus $a * b = b * a$.

Hence $(G, *)$ is abelian.

Exercise 3.3.8: $(R^*, .)$ is a group of non-zero real numbers under multiplication, then show that $(R^*, .)$ is not isomorphic to $(R, +)$.

Proof: Since for $-1 \in R^*$, $(-1)\ .\ (-1) = 1$ gives that order of (-1), $o\,(-1) = 2$. So If at all $(R^*, .) \cong (R, +)$, say under an isomorphism f, $f(-1)$ must be of order 2. But for every $a \in R$, $n.\ a = o \Rightarrow n = o$ or $a = o$.

So $(R, +)$ does not have any element of order 2.

Hence $(R^*, .)$ cannot be isomorphic to $(R, +)$.

Exercise 3.3.9: If (H, o) is a normal sub-group of (G, o), show that $W = \bigcap\limits_{g \in G} g \circ H \circ g^{-1}$ is a normal sub-group of (G, o).

Proof: Let for $g \in G$,

$x, y \in g \circ H \circ g^{-1}$, then for some h_1, h_2

CHAPTER 3

$$x = g \; o \; h_1 \; o \; g^{-1} \text{ and } y = g \; o \; h_2 \; o \; g^{-1}$$

and
$$x \; o \; y^{-1} = (g \; o \; h_1 \; o \; g^{-1}) \; o \; (g \; o \; h_2 \; o \; g^{-1})^{-1}$$
$$= (g \; o \; h_1 \; o \; g^{-1}) \; o \; (g^{-1})^{-1} \; o \; h_2^{-1} \; o \; g^{-1})$$
$$= g \; o \; h_1 \; o \; (g^{-1} \; o \; g) \; o \; h_2^{-1} \; o \; g^{-1}$$
$$= g \; o \; (h_1 \; o \; h_2^{-1}) \; o \; g^{-1} \in g \; o \; H \; o \; g^{-1}$$
Since $h_1 \; o \; h_2^{-1} \in H$.

Hence $(g \; o \; H \; o \; g^{-1}, o)$ is a sub-group.

Since each $(g \; o \; H \; o \; g^{-1}, o)$, $g \in G$ is a sub-group, (W, o) is a sub-group of (G, o),

$$w \in W \Rightarrow w \in go \; H \; o \; g^{-1}, \text{ for all } g \in G$$
$$\Rightarrow w = g \; o \; h \; o \; g^{-1}, \text{ for all } g \in G, \text{ for } h \in H.$$
$$\Rightarrow g^{-1} \; o \; w \; o \; g = h \in H, \text{ for all } g \in G.$$

Let $x \in G$, then $x \; o \; w \; o \; x^{-1} = x \; o \; (go \; h \; o \; g^{-1}) \; o \; x^{-1}$
$$= (x \; o \; g) \; o \; h \; o \; (x \; o \; g)^{-1} \in x \; o \; go \; H \; o \; (x \; o \; g)^{-1}$$

Now any element $y \in G$ can be written as $x \; o \; x^{-1} \; oy = xo \, g$ where $g = x^{-1} \; oy \in G$.

So $\qquad x \; o \; w \; ox^{-1} \in y \; o \; H \; o \; y^{-1}$ for all $y \in G$

$\Rightarrow \qquad x \; o \; w \; o \; x^{-1} \in W$.

Hence (W, o) is a normal sub-group of (G, o).

Exercise 3.3.10: Every element of A_n is a product of 2 cycles.

Proof: Let 1, α, β be three distinct elements of the set $\{1, 2, 3,\ldots\ldots, n\}$. Each $\phi \in A_n$ is expressible as a product of even number of transpositions and each transposition $(\alpha \; \beta)$ where $\alpha \neq 1$, $\beta \neq 1$ can be expressed as $(1 \; \alpha) \; (1 \; \beta) \; (1 \; \alpha)$. So ϕ it self can be written as $(1 \; \alpha_1)$ $(1 \; \alpha_2) \ldots (1 \; \alpha_r)$ where r is even and $\alpha_i \in \{2, 3,\ldots\ldots, n\}$. Further, $(1 \; \alpha) \; (1 \; \beta) = (1 \; \beta\alpha)$ and r is even

$\Rightarrow$ Every even permutation is a product as 2 cycles.

Remark: Here we require $n \geq 3$ (not $n \geq 5$).

Exercise 3.3.11: If a normal sub-group H of A_n contains a cycle of length 3, then $H = A_n$. $(n \geq 5)$.

Proof: Here we have to prove that every cycle of length 3 belongs to H.

Let (abc) be any cycle of length 3 and let $f = (xyz) \in H$.

That is, $f(x) = y$, $f(y) = z$, $f(z) = x$.

Consider $\phi \in S_n$ such that $\phi \; (x) = a$, $\phi \; (y) = b$ and $\phi \; (z) = c$.

Then $(\phi \; of \; o \; \phi^{-1}) \; (a) = (\phi \; of) \; (\phi^{-1} \; (a) = (\phi \; of) \; (x) = \phi \; (f(x)) = \phi \; (y) = b$

$(\phi \; of \; o \; \phi^{-1}) \; (b) = (\phi \; of \;) \; (\phi^{-1} \; (b)) = (\phi \; of) \; (y) = \phi \; (f(y)) = \phi \; (z) = c$

$(\phi \; of \; o\phi^{-1}) \; (c) = (\phi \; of) \; (\phi^{-1} \; (c)) = (\phi \; of) \; (z) = \phi \; (f(z)) = \phi \; (x) = a$

$\phi \; of \; o \; \phi^{-1} = \begin{pmatrix} a & b & c \\ b & c & a \end{pmatrix}$. Thus if $\phi \in A_n$, $(abc) \in H$ as H is a normal sub-group of A_n.

Let θ be an odd permutation, Then $\theta \notin A_n$, consider u, v in $\{1, 2, ..., n\}$ different from x, y, z (this is possible as $n \geq 5$).

Since θ is odd, $\theta \; o(u \; v) \in A_n$.

Now $[\theta \; o \; (uv)] \; o \; (x \; y \; z) \; o \; [\theta \; o \; (u, v]^{-1} \in H$

$\Rightarrow [\theta \; o \; (u \; v)] \; o \; (xyz) \; o \; [(u \; v) \; o \; \theta^{-1}) \in H$

$\Rightarrow \theta \; o \; (x \, o \, y \, o \, z) \, o \; \theta^{-1} \in H$ since (xyz) and (uv) and disjoint cycles and hence $(uv) \; o \; (xyz) \; o \; (uv) = (xyz) \; o \; (u \; v)^2 = (xyz)$.

$\Rightarrow (abc) \in H.$

This proves the lemma.

Exercise 3.3.12: If any normal sub-group H of (A_n, o) contains a product of two disjoint transpositions, then $H = A_n$ $(n \geq 5)$.

Proof: Let $(x \; y) \; o \; (u \; v) \in H$, where $(x \; y)$ and (uv) are disjoint. Let $w \in \{1, 2, 3, ... n\}$ different from x, y and u, v. This is possible if $n \geq 5$.

Put $\theta = (u \; v \; w)$. Clearly $\theta \in A_n$

so that $\theta \; o \; (x \; y) \; o \; (u \; v) \; o \; \theta^{-1} \in H$,

that is, $(u \; v \; w) \; o \; (x \; y) \; o \; (u \; v) \; o \; (w \; v \; u) \in H$

$\Rightarrow (xy) \; o \; (v \; w) \in H.$

However (H, o) is a subgroup, therefore

$(xy) \; o \; (u \; v) \; o \; (v \; w) \; o \; (x \; y) \in H$

$\Rightarrow (u \; v \; w) \in H$

Hence $H = A_n$, by above Exercise - 3.3.11.

Exercise 3.3.13: Prove that in S_n, the number of distinct cycles of length $r \leq n$ is $\frac{1}{r} \frac{n!}{(n-r)!}$

Solution: We know that the number of distinct permutations of r members taken out of n members at a lime is $\frac{n!}{(n-r)!}$

As each of the cycles $(a_1, a_2, ..., a_r)$, $(a_2, a_3, ..., a_r a_1)$,..... $(a_r \; a_1 \; a_2 ...., a_{r-1})$ is the same, so number of distinct r-cycles is equal to $\frac{1}{r} \frac{n!}{(n-r)!}$.

Exercise 3.3.14: A_4 has no subgroup of order 6.

Solution: Since 3-cycles of S_4 is an even premulation, all 3-cycles of S_4 belong to A_4. By exercise 3.3.13, the number of distinct 3-cycles in $A_n = \frac{1}{3} \frac{n!}{(n-\gamma)!} = \frac{4!}{3(4-3)!} = 8$.

Let (H, o) be a subgroup of A_4 of order 6. All 3-cycles of A_4 con not belong to H. Let ϕ be a 3-cycles of A_4 such that $\phi \notin H$. As $o(\phi) = 3$, the subgroup $K = \{I, \phi, \phi^2\}$ consists of three elements, where I is the identity of A_4.

Now $\phi \notin H, \phi^3 = I \Rightarrow \phi^2 = \phi^{-1} \notin H$

$$\Rightarrow H \cap K = \{I\}$$

$$\Rightarrow o(H\ o\ K) = \frac{o(H).o(K)}{o(H \cap K)} = \frac{6.3}{1} = 18$$

which is absurd since (HoK, o) is a subgroup of (A_4, o) and $o(A_4) = 12$. Hence our assumption is wrong and consequently (A_4, o) has no subgroup of order 6.

PROBLEMS

1. In the following situations, determine whether the indicated function f is a homomorphism from the first group into the second group.

   (a) $f(a) = |a|$, $(R - \{|o|\},.)$, (R_+) (b) $f(a) = a + 1$,

   $(Z, +)$, $(Z, +)$ (c) $f(a) = a^2$, $(R - \{o\},.)$, $(R +,.)$.

2. Let set $G = Z \times Z$ and the binary operation o on G is given by $(a, b)\ o\ (c, d) = (a + b, c + d)$. Varify that (G, o) is a group.

   (a) Show that the mapping $f : G \to Z$ given by $f(a, b) = a$ is a homomorphism from (G, o) onto $(Z, +)$.

   (b) Determine the kernel of this mapping.

   (c) If $H = \{(a, a)\}\ a \in Z\}$. Prove that (H, o) is a sub-group of (G, o), which is isomorphic to $(Z, +)$ under the function f.

3. Show that the two groups $(R, +)$ and $(R - \{o\},.)$ are not isomorphic.

4. Prove that all finite groups of order 2 are isomorphic.

5. Prove that if the group (G, o) is commutative (cyclic) and $(G, o) \cong (G', *)$, then the group $(G', *)$ is commutative (cyclic).

6. Consider the two groups $(Z, +)$ and $(\{1, -1, i, -i\},)$, where $i^2 = -1$. Shows that the mapping defined by $f(n) = i^n$ for $n \in Z$ is a homomorphism from $(Z, +)$ onto $(\{1, -1, i, -i),.)$, determine its kernel.

7. Suppose f is a homomorphism from the Group $(G, *)$ into the group (G', o): If e is the identity of $(G, *)$, show that the kernel of f may be described by $ker(f) = f^{-1}(f(e))$.

8. Let a be a fixed element of the group (G, o). Prove that $(G, o) \cong (G, o)$ under the mapping f defined by $f(x) = ao\ xoa^{-1}$, $x \in G$.

   What is the kernel of this function.

9. Let f and g be two homomorphisms from the group $(G, *)$ into the group (G', o). Define the function $\phi : G \to G'$ by $\phi(a) = f(a)\ og(a)$. Show that if the group (G', o) is commutative, then ϕ is also a homomorphism.

10. Prove the following generalizations of the factor theorem. Let f_1 and f_2 be homomorphisms from the group $(G, *)$ onto the groups (G_1, o_1) and (G_2, o_2) respectively. If $ker(f_1) \subseteq ker(f_2)$, then there exists a unique homomorphism $\overline{f} : G_1 \to G_2$ satisfying $f_2 = f_2\ o\ f_1$

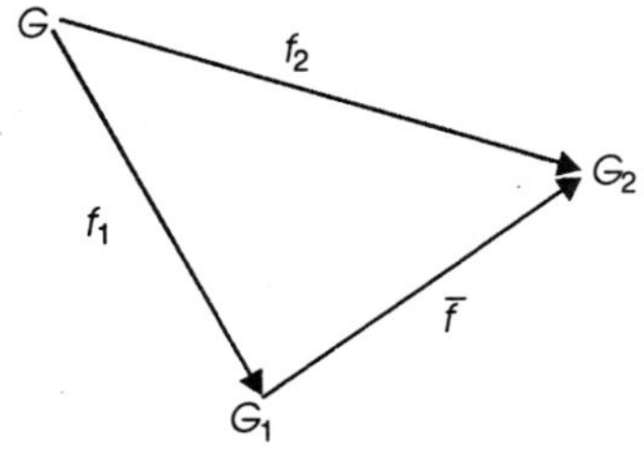

11. Given a group $(G, *)$, establish the following facts regarding the inner automorphisms of G.

(a) Whenever $(G, *)$, is commutative, $I(G) = [i_G]$.

(b) If $(H, *)$ is a sub-group of $(G, *)$, then the pair $(f_a(H), *)$ is also a sub-group for every $a \in G$.

(c) A sub-group $(H, *)$ is normal in $(G, *)$ if and only if the set H is mapped into itself, by each inner automorphism of G.

(d) If the element $x \in G$ is such that $f_a(x) = x$ for every $a \in G$, then the cyclic sub-group $((x), *)$ is normal in $(G, *)$.

12. Obtain the isomorphism $(S_3, o) \cong (I(S_3)\ o)$.

13. Let f be a homomorphism from the group $(G, *)$ into itself such that f commutes with inner automorphism, that *is*, $f_a\ of = fof_a$ for every $a \in G$. If H is defined by $H = \{x \in G \mid f(x) = x\}$, Prove that

(a) the pair $(H, *)$ is a normal sub-group of $(G, *)$,

(b) the quotient group $(G/H, *)$ is commutative.

14. Let $(H, *)$ be a normal sub-group in $(G, *)$ and Let $(K_1, *)$ and $(K_2, *)$ be two sub-groups of $(G, *)$ such that $H \subseteq K_1$, $H \subseteq K_2$. Prove that $K_1 \subseteq K_2$ if and only if $\phi(K_1) \subseteq \phi(K_2)$, where $\phi\ G \to G/H$ is a natural homomorphism.

In problems from 15 to 21 *f* denote a homomorphism from the group $(G, *)$ onto the group (G', o).

15. Show that the factor theorem implies that the function f can be expressed (non-trivially) as the composition of an onto function and a one to one function.

16. If $(H, *)$ is a sub-group of $(G, *)$ for which $H = f^{-1}(f(H))$; varify that $ker(f) \subseteq H$.

[Hint: $f^{-1}(f(H)) = H * ker(f)$]

17. If $(H, *)$ is any normal sub-group of the group $(G, *)$ such that $ker(f) \subseteq H$, then prove that $(G/H, *) \cong (G' \mid f(H), o)$ by considering the function $g : G/H \to G'/f(H)$ defined by $g(a * H) = f(a)\ of(H)$.

18. If $ker(f) \subseteq [G, G]$, prove that $(G/[G, G], *) \cong (G'/ [G', G'], o)$.

19. Suppose $(H, *)$ and $(K, *)$ are normal sub-groups of the group $(G, *)$ with $H \subseteq K$. Prove that

(a) $(H, *)$ is a normal sub-group of $(K, *)$,

(b) $(K/H, *)$ is a normal sub-group of $(G/H, *)$,

(c) $(G/H/K/H, *) \cong (G/K, *)$.

20. Let $(H, *)$ and $(K, *)$ be two sub-group of $(G, *)$ with $(K, *)$ normal in $(G, *)$ and $H \cap K = \{e\}$. Prove that $(H, *) \cong (H * K/K, *)$.

CHAPTER 3

21. In the symmetric group (S_4, o), let the set K consists of the four permutations.

$$\begin{pmatrix}1 & 2 & 3 & 4\\ 1 & 2 & 3 & 4\end{pmatrix}, \begin{pmatrix}1 & 2 & 3 & 4\\ 2 & 1 & 4 & 3\end{pmatrix}, \begin{pmatrix}1 & 2 & 3 & 4\\ 3 & 4 & 1 & 2\end{pmatrix}, \begin{pmatrix}1 & 2 & 3 & 4\\ 4 & 3 & 2 & 1\end{pmatrix}$$

(*a*) Show that the pair (K, o) is a normal sub-group of (S_4, o).

(*b*) Show that $(S_4/K, o) \cong (S_3, o)$.

22. Show that every homomorphic image of acyclic group is cyclic.

23. If $(G, *)$ is non-abelian group of order 6, prove that

$(G, *) = (S_3, o)$.

24. Let (G, o) be a finite group, T an automorphism of G with the property $T(x) = x$ if and only if $x = e$. Prove that

(*i*) every $g \in$ G can be expresed as $g = x^{-1}\,(T(x))$ for some $x \in G$.

(*ii*) If $T^2 = I$, prove that (G, o) must be abelian.

25. Let G be a finite group and suppose the automorphism T sends more than three quarters of the elements of G onto their inverses. Prove that $T(x) = x^{-1}$ for all $x \in G$ and that G is abelian.

26. Let (Z, o) be the centre of a group (G, o). If T is any automorphism of (G, o). Prove that $T(Z) \subset Z$.

27. Let (G, o) be a group and T an automorphism of (G, o). Prove that $N(T(a)) \subset: T(N(a))$, where,

for $a \in G$, $N(a) = \{x \in G \mid x \text{ o } a = a \text{ o } x)$.

28. Prove that every finite group having more than two elements has a non-trivial automorphism.

29. T is an automorphism of a group (G, o) and (N, o) is a normal sub-group of G such that $T(N) = N$. If for all $g \in G$, $T'(g \text{ o } N) = T(g) \text{ o } N$. Show that T' in an automorphism of G/N.

[**Hint:** For $x \text{ o } N$, $y \text{ o } N \in G/N$, $x, y \in$ G.

$$\begin{aligned} T'(x \text{ o } N) \text{ o } (y \text{ o } N) &= T'(x \text{ o } y \text{ o } N)\\ &= T(x \text{ o } y) \text{ o } N\\ &= T(x) \text{ o } T(y) \text{ o } N\\ &= (T(x) \text{ o } N) \text{ o } (T(y) \text{ o } N)\\ &= T'(x \text{ o } N) \text{ o } T'(y \text{ o } N).\end{aligned}$$

and

$$\begin{aligned} Ker\,T' &= \{x \text{ o } N \in G/N \mid T'(x \text{ o } N) = N\}\\ &= \{x \text{ o } N \in G/N \mid T(x) \text{ o } N = N)\\ &= \{x \text{ o } N \in G/N \mid T(x) \in N)\\ &= \{x \text{ o } N \in G/N \mid x \in N \text{ since } T(N) = N\\ &= N\end{aligned}$$

Hence T' is an automorphism]

CHAPTER 4 STRUCTURE THEORY OF GROUP

4.1 DIRECT PRODUCT OF GROUPS

In chapter 2 we have studied the sub-groups and quotient groups which have been constructed from a given group. Here we shall confine to what is called direct product of two groups. Let (G_1, o) and (G_2, o) be two groups. We shall see that the Cartesian product $G_1 \times G_2$ of the sets G_1 and G_2 under defined binary composition is a group. Thus we shall get a new class of groups which is different from sub-groups and quotient groups.

Theorem 4.1.1: *Let (G, o) and $(H, *)$ be two groups and let a binary operation be defined on the Cartesian product*
$G \times H = \{(g, h) \mid g \in G, h \in H\}$, as
follows for $(g_1, h_1), (g_2, h_2) \in G \times H$,
$(g_1, h_1). (g_2, h_2) = (g_1 \, o \, g_2, h_1 * h_2)$.
Then the pair $(G \times H,.)$ is a group, called the external direct product of (G, o) and $(H, *)$.

Proof: Let $a = (g_1, h_1)$, $b = (g_1, h_2)$, $c = (g_3, h_3) \in G \times H$.

1 . Closure: $a, b \in G \times H$, by definition $a.b = (g_1, h_1). (g_2, h_2) = (g_1 \, o \, g_2, h_1 * h_2) \in G \times H$, Since $g_1 \, o \, g_2 \in G$ and $h_1 * h_2 \in H$ as (G, o) and $(H, *)$ are groups.

This shows that $G \times H$ is closed under the operation.

2. Associativity: $\forall \; a, b, c \in G \times H$,

$$\begin{aligned}(a \,.\, b) \,.\, c &= [(g_1, h_1). (g_2, h_2)]. (g_3, h_3)\\ &= [g_1 \, o \, g_2, h_1 * h_2]. (g_3, h_3)\\ &= [g_1 \, o \, g_2) \, o \, g_3, (h_1 * h_2) * h_3]\\ &= [(g_1 \, o \, (g_2 \, o \, g_3), h_1 * (h_2 * h_3)]\\ &= (g_1, h_1) \,.\, [(g_2 \, o \, g_3), (h_2 * h_3)]\\ &= (g_1, h_1) \,.\, [(g_2, h_2) \,.\, (g_3, h_3)]\\ &= a \,.\, (b \,.\, c)\end{aligned}$$

Hence the operation is associative in $G \times H$.

3. Existence of identity: $\forall\ a \in G \times H$, $e = (e_1, e_2) \in G \times H$, where e_1 and e_2 are identity elements in the groups (G, o) and $(H, *)$ respectively, such that

$$\begin{aligned} a.\ e = (g_1, h_1)\ .\ (e_1, e_2) &= (g_1\ oe_1, h_1 * e_2) \\ &= (g_1, h_1) \\ &= (e_1\ og_1, e_2 * h_1) \\ &= (e_1, e_2)\ .\ (g_1, h_2) \\ &= e.a. \end{aligned}$$

Thus we get $a.\ e = a = e.a$

Therefore $e = (e_1, e_2)$ is an identity element in $G \times H$.

4. Existence of inverse: Since (G, o) and $(H, *)$ are groups, then if $g_1 \in G$, $g_1^{-1} \in G$ and if $h_1 \in H$, $h_1^{-1} \in H$, and $(g_1^{-1}, h_1^{-1}) \in G \times H$. Now $\forall a = (g_1, h_1) \in G \times H$,

$$\begin{aligned} d = \left(g_1^{-1}, h_1^{-1}\right) \in \text{G} \times \text{H such that } a.d &= (g_1, h_1)\ .\ \left(g_1^{-1}, h_1^{-1}\right) \\ &= \left(g_1\ o\ g_1^{-1}, h_1^{-1} * h_1^{-1}\right) \\ &= (e_1, e_2) \\ &= \left(g_1^{-1}\ o g_1, h_1^{-1} * h_1\right) \\ &= \left(g_1^{-1}, h_1^{-1}\right).\left(g_1 . h_1\right) \\ &= d\ .\ a. \end{aligned}$$

Thus, $a\ .\ d = e = d\ .\ a \Rightarrow a^{-1} = d$.

Hence every element $a \in G \times H$ has its inverse $a^{-1} \in G \times H$. This proves that $(G \times H,.)$ is a group.

Remark: (1) If $(G, *)$ and $(H, *)$ are both commutative groups, then $(G \times H,.)$ is also a commutative group for it is seen that

$a, b \in G \times H$

$$\begin{aligned} a\ .\ b &= (g_1, h_1)\ .\ (g_2, h_2) \\ &= (g_1\ o\ g_2, h_1 * h_2) \\ &= (g_2\ o\ g_1, h_2 * h_1) \\ &= (g_2, h_2)\ .\ (g_1, h_1) = b\ .\ a \in G \times H. \end{aligned}$$

(2) If (G, o) and $(H, *)$ are finite groups, then it is clear that $o\ (G \times H) = o\ (G)\ .\ o(H)$.

If $G \neq \{e_1\}$ and $H \neq (e_2)$ are finite groups, then the group $(G \times H, .)$ is neither isomorphic to (G, o) nor to $(H, *)$, because $o\ (G \times H) \neq o\ (G)$ and $o\ (G \times H) \neq o\ (H)$.

Corollary: Let $(G_1,.), (G_2,.), \ldots (G_n,.)$ be n groups, then Cartesian product $G = G_1 \times G_2 \times \ldots \times G_n = \{(a_1, a_2, \ldots a_n) | a_i \in G_i\}$ together with the binary operation defined by,

$\forall a = (a_1, a_2, \ldots a_n)$, $b = (b_1, b_2, \ldots b_n) \in G$

$a \cdot b = (a_1 \cdot b_1, a_2 \cdot b_2, \ldots a_n \cdot b_n)$, is a group, called external direct product of groups $(G_1,.)$, $(G_2,.)$ —, $(G_n, +)$.

If all group $(G_1, +)$, $(G_2, +)$,...., $(G_n, +)$ are abelian groups under addition composition, then the pair $(G, +)$, where $G = G_1 \times G_2 \times G_3, \times \ldots \times G_n$, is a group under component wise addition composition defined by $(a_1, a_2, \ldots, a_n) + (b_1, b_2, b_n) = (a_1 + b_1, a_2 + b_1 \ldots (a_n + b_n)$. This group $(G, +)$ is referred to as direct sum of $(G_1, +)$, $(G_2, +)$,..., $(G_n +)$ which is symbolically written as

$$G = G_1 \oplus G_2 \oplus \ldots \oplus G_n.$$

Example 4.1.1: Let $(Z, +)$ be a additive group of integers. Then $(Z \times Z, +)$ is a group, where the operation is defined by, (a, b), $(c, d) \in Z \times Z$, $(a, b) + (c, d) = (a + c, b + d)$. Here $(0, 0)$ is the identity and $(-a, -b)$ is the additive inverse of $(a, b) \in Z \times Z$.

Example 4.1.2: Let $(Z, +)$ be a additive group of integers and let $(C - \{0\},.)$ be a multiplicative group of non-zero complex numbers. Let $G = Z \times (C - \{0\})$. The operation 'o' in G is defined by

$$a \; o \; b = (n_1, z_1) \; o \; (n_2, z_2) = (n_1 + n_2, z_1.z_2),$$

$$a = (n_1, z_1), b = (n_2, z_2) \in Z \times (C - \{0\}).$$

The pair (G, o) is a group under the operation o. Here $(0, 1)$ is the identity for $(0, 1)$ $o \; (n, z) = (0 + n, 1.z) = (n, z)$

$$= (n + 0, z.\ 1) = (n, z) \; o \; (0, 1), \; \forall \; (n, z) \in G.$$

For each (n, z), $\exists \; (-n, z^{-1}) \in G$ which is the inverse of $(n, z) \in G$.

Example 4.1.3: Let (G, o) be a cyclic group of order 2 generated by g. That is, $G = [e, g\}$ where $g^2 = e$. Now $G \times G = \{(e, e), (e, g), (g, e), (g, g)\}$. The operation table for $G \times G$ is

	(e, e)	(e, g)	(g, e)	(g, g)
(e, e)	(e, e)	(e, g)	(g, e)	(g, g)
(e, g)	(e, g)	(e, e)	(g, g)	(g, e)
(g, e)	(g, e)	(g, g)	(e, e)	(e, g)
(g, g)	(g, g)	(g, e)	(e, g)	(e, e)

We note that the every element of $G \times G$ is of order 2. But we know that if any group $(G, *)$ of order 4 contains an element of order 4, it is cyclic:

Here $(G \times G,.)$ of order 4 does not contain any element of order 4. Therefore $G \times G$ is not cyclic. And it will not be isomorphic to any cyclic group of order 4. This group is called the *Klein four group,* or simply the four group.

Example 4.1.4: Let $(G_1 = (a))$, and $(G_2 = (b))$, be two cyclic groups of order 2 and 3 generated by a and b respectively $a^2 = e_1$, $b^3 = e_2$, that is, $G_1 = \{e_1, a\}$,

$G_2 = \{e_2, b, b^2\}$. Now $G_1 \times G_2 = \{(e_1, e_2, (e_1, b), (e_1, b^2), (a_1, e_2), (a, b), (a, b^2)\}$. For all $x = (a_1, b_1)$, $y = (a_2, b_2) \in G_1 \times G_2$ define $x.\ y = (a_1 \cdot a_2, b_1 \cdot b_2)$.

CHAPTER 4

Then $(G_1 \times G_2, .)$ is a group. Further if $u = (a, b)$, then

$u^2 = (a^2, b^2) = (e_1, b^2)$, $u^3 = (a^3, b^3) = (a, e_2)$, $u^4 = (a^4, b^4)$

$= (e_1, b)$, $u^5 = (a^5, b^5) = (a, b^2)$, $u^6 = (a^6, b^6) = (e_1, e_2)$. Thus we get

$G_1 \times G_2 = (u) = \{e, u, u^2, u^3, u^4, u^5\}$.

Hence $(G_1 \times G_2, .)$ is a cyclic group.

Theorem 4.1.2: *Let $(G \times H,.)$ be the direct product of two groups (G, o) and $(H, *)$, then*

(1) *The sets*

$G' = \{(g, e_2) \mid g \in G, e_2$ *is the identity of* $H\}$,

$H' = \{(e_1, h) \mid h \in H, e_1$ *is the identity of* $G\}$.

are normal sub-groups of $(G \times H,.)$ under the operation (.) and

$(G', .) \cong (G, o)$ and $(H', .) \cong (H, *)$.

(2) *If $a' = (g, e_2) \in G'$, $b' = (e_1\, h) \in H'$, then $a' . b' = b' . a'$.*

(3) *Each $a \in G \times H$ is uniquely expressible as $a' . b'$, where $a' \in G', b' \in H'$ in the sence that if*

$a' . b' = c' . d'$ with $a', c' \in G', b', d' \in H'$, then $a' = c'$ and *$b'. = d'$.*

(4) *$G \times H = G' . H'$ and $G' \cap H' = \{(e_1, e_2)\}$, and $(G \times H, .) \cong (H \times G, .)$.*

Proof: **(1)** The set G′ is elearly non-empty. If $(g_1, e_2); (g_2, e_2) \in G'$,

Then $(g_1, e_2) . (g_2, e_2)^{-1}$

$= (g_1, e_2).(g_2^{-1}, e_2^{-1}) = (g_1, e_2).(g_2^{-1}, e_2)$

$= (g_1 \; o \; g_2^{-1}, e_2 * e_2) = (g_1 \; o \; g_2^{-1}, e_2) \in G'$, Since $g_1 \; o \; g_2^{-1} \in G'$.

Hence (G', o) is a sub-group of $(G \times H,.)$

Similarly we can prove that $(H', *)$ is a sub-group of $(G \times H,.)$.

Again, $\forall$ $(g, h) \in G \times H$, and $(g_1, e_2) \in G'$

$$\begin{aligned}(g, h)^{-1} . (g_1, e_2). (g, h) &= (g^{-1}, h^{-1}) . (g_1, e_2) . (g_1, h)\\ &= (g^{-1} \; o \; g_1, h^{-1} * e_2) . (g, h)\\ &= (g^{-1} \; o \; g_1, h^{-1}) . (g, h)\\ &= (g^{-1} \; o \; g_1 \; o \; g, h^{-1} * h)\\ &= (g^{-1} \; o \; g_1 \; o \; g, e_2) \in G'. \text{ Since } g^{-1} \; o \; g_1 \; o \; g \in G\end{aligned}$$

Hence $(G' ,.)$ is a normal sub-group. Similarly we can show that $(H' ,.)$ is also normal sub-group of $(G \times H,.)$. Again, let us define $f : G \to G'$ by $f(g) = (g_1, e_2)$, $\forall$ $g \in G$

For $g_1, g_2 \in G$,

$f(g_1) = f(g_2) \Rightarrow (g_1, e_2) = (g_2 . e_2) \Rightarrow g_1 = g_2$

which shows f is one to one function.

For each $(g, e_2) \in G'$, $\exists\ g \in G$ such that $f(g) = (g, e_2)$.

Hence f is one-to-one function of the set G onto G'.

Now $\quad f(g_1 o g_2) = (g_1 o g_2, e_2) = (g_1, e_2) \cdot (g_2, e_2)$

$$= f(g_1) \cdot f(g_2),$$

which shows f preserves the group operation.

Hence $(G, o) \cong (G', .)$.

Similarly we can show that $(H, *) \cong (H', .)$.

(2) Let $a' = (g, e_2) \in G'$ and $b' = (e_1, h) \in H'$.

$$a' \cdot b' = (g, e_2) \cdot (e_1, h) = (goe_1, e_2 * h) = (e_1 o g, h * e_2)$$
$$= (e_1, h) \cdot (g, e_2)$$
$$= b'. a'.$$

(3) Let $a = (g, h) \in G \times H$. Then

$a = (g, h) = (g, e_2) \cdot (e_1, h) = a' \cdot b'$.

where $a' = (g, e_2) \in G'$, $b' = (e_1, h) \in H'$

If $a = c'. d'$, where $c' \in G'$, $d' \in H'$. Then

$c' = (g_1, e_2)$ and $d' = (e_1, h_1)$. So

$a = c' \cdot d' = (g_1, e_2) \cdot (e_1, h_1) = (g_1 \cdot h_1)$.

Thus we get

$a = (g, h) = (g_1, h_1) \Rightarrow g = g_1, h = h_1$

$\Rightarrow a' = c'$ and $b' = d'$.

(4) $(g, h) \in G' \cap H' \Rightarrow (g, h) \in G'$ and $(g, h) \in H'$

$\Rightarrow h = e_2$ and $g = e_1$

Hence $G' \cap H' = \{e_1, e_2)\}$.

Since G′ contain all elements of the form (g, e_2) and H' contains all elements of the form (e_1, h).

Again, if $(g, h) \in G \times H$, then

$(g, h) = (g, e_2)(e_1, h) \in G' \cdot H' \Rightarrow G \times H \subseteq G' \cdot H'$

If $(g, e_2) \in G'$ and $(e_1, h) \in H'$,

then $(g, e_2). (e_1, h) \in G' \cdot H'$.

and $(g, e_2). (e_1, h) = (g, h) \in G \times H \Rightarrow G' \cdot H' . \subseteq G \times H$.

Hence $G \times H = G' \cdot H'$.

Further we define $f : G \times H \Rightarrow H \times G$ by

$f(g, h) = (h, g), \forall\ (g, h) \in G \times H,$

CHAPTER 4

For (g_1, h_1), $(g_2, h_2) \in G \times H$,

$f(g_1, h_1) = f(g_2, h_2) \Rightarrow (h_1, g_1) = (h_2, g_2)$

$\Rightarrow h_1 = h_2$ and $g_1 = g_2$

$\Rightarrow (g_1, h_1) = (g_2, h_2)$.

Hence f is one to one mapping. It is clear that f is onto mapping. Again

$$\begin{aligned} f[(g_1, h_1) \,.\, (g_2, h_2)] &= f[g_1 \; o \; g_2 . \; h_1 * h_2)] \\ &= (h_1 * h_2, g_1 \; o \; g_2) \\ &= (h_1, g_1) \,.\, (h_2, g_2) \\ &= f(g_1, h_1) \,.\, f((g_2, h_2)) \end{aligned}$$

This completes the proof of $(G \times H, .) \cong (H \times G, .)$

We observe the following points from the above discussion:

(1) From two groups (G, o) and $(H, *)$ we form another group $(G \times H, .)$ called direct product of group.

(2) The group $(G \times H, .)$ has two sub-groups $(G', .)$ and $(H', .)$ such that $G' \cap H' = \{(e_1, e_2)\}$ the identity of $(G \times H, .)$,

(3) $\forall\, a' \in G', b' \in H', a' \,.\, b' = b' \,.\, a'$.

(4) Each element $a \in G \times H$ is uniquely expressible as $a = a' \,.\, b'$.

Now let (G , o) be any group having sub-groups.

(H_1, o) and (H_2, o) with the following properties:

(1) $\forall\, h_1 \in H, h_2 \in h_2, . \; h_1 \,.\, h_2 = h_2 \,.\, h_1$.

(2) every element $a \in G$ can be written uniquely as $a = h_1 \,.\, h_2$.

Generalisation: In similar fashion as ahove we can define the external direct product $G_1 \times G_2 \times .... \times G_n$ of any finite n groups $(G_1,.)$, $(G_2,..)$.... $(G_n,..)$ we also define the normal sub-groups

$(G_1',.)$, $(G_2',..,)$,... $(G_n',..)$ of $G_1 \times G_2 \times ....\times G_n$ such that

(1) $(G_i',..) \cong (G_i,..)$, $\forall\, i = 1, 2,... n$

(2) $\forall\, a_i \in G_i', a_j \in G_j', a_i \,.\, a_j = a_j \,.\, a_i$

(3) $G_j', \cap G_j' = \{e\}, i \neq j, e = (e_1, e_2..., e_n)$

(4) each element of $G_1 \times G_2 \times ....\times G_n$ is uniquely expressible as product of elements of $G_1',...G_n'$. That is,

$G_1 \times G_2 \times .....\times G_n = G_1' \,.\, G_2' \;....\; G_n'$.

Theorem 4.1.3: *Let (H, o) and (K, o) be two sub-groups of a group (G, o) such that $H \cap K = \{e\}$, the identity of G, and every element of H commute with those of K, and $H.K = G$, then $(G, o) \cong (H \times K,.)$*

Proof: First of all we shall show that every element $a \in G$ can be expressed uniquely as $a = h.k$, where $h \in H$ and $k \in K$.

Since $G = H.K$, $a = h \, . \, k$ for some $h \in H$ and $k \in K$. Now suppose $a = h_1 \, . \, k_1$ for some $h_1 \in H$ and $k_1 \in K$, then $h \; o \; k = h_1 \; o \; k_1 \Rightarrow h_1^{-1} \; o \; h = k_1 \; o \; k^{-1} \in H \cap K = \{e\} \Rightarrow h_1^{-1} \; o \; h = k \; o \; k^{-1} = e \Rightarrow h = h_1$ and $k = k_1$ since $h_1^{-1} \; o \; h \in H$ and $k_1 \; o \; k^{-1} \in K$.

Now we define $f : H \times K \to G$ by

$f(h, k) = h \; o \; k, \; \forall \; (h, k) \in H \times K.$

For $x = (h_1, k_1)$, $y = (h_2, k_2) \in H \times K$,

$f(x) = f(y) \Rightarrow f((h_1, k_1) = f(h_2, k_2)) \Rightarrow h_1 \; o \; k_1 = h_2 \; o \; k_2$

$\Rightarrow h_1 = h_2$ and $k_1 = k_2$ Since every element of G is uniquely expressible.

$\Rightarrow (h_1, k_1) = (h_2, k_2).$

$\Rightarrow x = y.$

Hence f is one-to-one function.

Since for $h \in H$ and $k \in K$, $h \; o \; k \in G$, $\exists \; (h, k) \in H \times K$

Such that $f((h, k)) = h \; o \; k$ which shows f is onto function.

Now for $x = (h_1, k_1)$, $y = (h_2, k_2) \in H \times K$,

$$\begin{aligned} f(x \, . \, y) = f(h_1, k_1)\,(h_2, k_2) &= f(h_1 \; o \; h_2, k_1 \; o \; k_2) \\ &= (h_1 \; o \; h_2) \; o \; (k_1 \; o \; k_2) \\ &= koh, \; \forall \; h \in H, \; k \in K. \\ &= h_1 \; o \; k_1 \; o \; h_2 \; o \; k_2, \text{ since } hok \\ &= f((h_1, k_1)) \; o \; f((h_2, k_2)) \\ &= f(x) \; o \; f(y). \end{aligned}$$

Hence f preserves the group operations which proves $(H \times K, .) \cong (G, o)$.

Definition 4.1.1: If (G, o) is a group with sub-groups (H, o) and (K, o) satisfying conditions:

(*i*) $h \; o \; k = k \; o \; h$, $\forall \; h, \in H$ and $k \in K$, and

(*ii*) every element $a \in G$ is a unique product of an element $h \in H$ and $h \in K$, (*i.e.*, if $a = h \; o \; k$, $h \in H$, $k \in K$;

and if $a = h_1 \; o \; k_1$, $h_1 \in H$, $k_1 \in H$, then $h = h_1$ and $k = k_1$.), then (G, o) is said to be the internal direct product of sub-group (H, o) and (K, o), and we write $G = H \otimes K$.

If $(G, +)$ is a group under the binary operation of addition, then we say that $(G, +)$ is the internal direct sum of (H, o) and (K, o) and we write $G = H \oplus K$.

Theorem 4.1.4: *If $G = H_1 \times H_2$, where (H_2, o), (H_2, o) are sub-groups of group (G, o), then (H_1, o) and (H_2, o) are normal sub-groups (G, o)* and *$(G/H_1, o) \cong (H_2, o)$ and $(G/H_2, o) \cong (H_1, o)$.*

CHAPTER 4

Proof: Now we define $f : G = H_1 \times H_2 \to H_1$ by

$f(g) = x_1, \forall g = (x_1, y_1) \in H_1 \times H_2$ such that $g = x_1 o y_1$.

Again if $g_1 \in G$, then

$f(g_1) = x_2$, where $g_1 = (x_2, y_2) \in H_1 \times H_2$ such that $g_1 = x_2 o y_2$

Now

$f((g o g_1))$

$= f(x_1 o y_1 o x_2 o y_2)$

$= f((x_1 o x_2) o (y_1 o y_2))$, since $xoy = yox, \forall x \in H_1, y \in H_2$.

$= x_1 ox_2 = f(g) of (g_1)$.

Hence f is homomorphism of (G, o) onto (H_1, o).

The *Ker* (f) is the set of those elements of $G = H_1 \times H_2$ which are mapped on the identity of (G, o). Then

$$ker(f) = \{g/f(g) = e, \forall g = (e, y) \in H_1 \times H_2 \text{ such that } g = eoy\}$$
$$= \{g / y \in H_2\}$$

or $Ker(f) = H_2$ and $(Ker\ f, o)$ is a normal sub-group of (G, o).

Hence by the theorem 3.2.8 of isomorphism $(G/H_2, o) \cong (H_1, o)$

Similarly we can show that $(G/H_1, o) \cong (H_2, o)$.

Theorem 4.1.5: *Let (H_1, o) and (H_2, o) be two sub-groups of (G, o).*
Then (G, o) is an internal direct product of (H_1, o) and (H_2, o) if and only if:
(1) *(H_1, o) and (H_2, o) are normal sub-groups,*
(2) $H_1 \cap H_2 = \{e\}$,
(3) $G = H_1 oH_2$.

Proof: **(1)** Let (G, o) be internal product of (H_1, o) and (H_2, o). Then $g \in G$ can be written as $g = h_1 o h_2$ and

$h_1 o h_2 = h_2 o h_1, \forall h_1 \in H_1, h_2 \in H_2$.

Let $h \in H_1, \forall g \in G$.

$g^{-1} o h og = (h_1 o h_2)^{-1} o h o (h_1 o h_2)$

$= (h_2^{-1} o h_1^{-1}) o h o (h_1 o h_2)$

$= h_2^{-1} o (h_1^{-1} o h o h_1) o h_2$

$= h_2^{-1} oh'oh_2$ since $h' = h_1^{-1} o h o h_1 \in H_1$, as $h', h_1^{-1}, h \in H_1$.

$= h' \, o \, (h_2^{-1} o h_2)$, since $h_1 \, o \, h_2 = h_2 \, o \, h_1, \forall \, h_1 \in H_1, h_2 \in H_2$

$= h' o \, e = h' \in H_1$

Hence (H_1, o) is a normal sub-group. This proves (1).

(2) Let $g \in H_1, \cap H_2$, then

$g \in H_1, g \in H_2 \Rightarrow g = hoe = eok$ for some $h \in H_1, k \in H_2$.

But every element is uniquely written as a product of an element of H_1 and element of H_2.

So this forces $h = e$ and $k = e$. Hence

$H_1 \cap H_2 = \{e\}$. This proves (2)

(3) Follows from the definition of internal direct product of sub-groups.

Conversely, Let (H_1, o) and (H_2, o) be two sub-groups of (G, o) satisfying all conditions (1) to (3).

Let $h_1 \in H_1, h_2 \in H_2$.

Then $\left(h_1^{-1}, o \, h_2^{-1}\right) o \, (h_1 \, o \, h_2) = h_1^{-1} \, o \left(h_2^{-1} \, o \, h_1 \, o \, h_2\right) \in H_1$ Since $h_2^{-1} \, o \, h_1 \, o \, h_2 \in H_1$ as (H_1, o) is normal.

Similarly, $\left(h_1^{-1} \, o \, h_2^{-1}\right) o \, (h_1 \, o \, h_2) = \left(h_1^{-1} \, o \, h_2^{-1} \, o \, h_1\right) o \, h_2 \in H_2$. Since $h_1^{-1} \, o \, h_2^{-1} \, o \, h_1 \in H_2$ as (H_2, o) is normal.

So $h_1^{-1} \, o \, h_2^{-1} \, o \, h_1 \, o \, h_2 \in H_1 \cap H_2 = \{e\} \Rightarrow h_1^{-1} \, o \, h_2^{-1} \, o \, h_1 \, o \, h_2 = e$

$\Rightarrow \left(h_1^{-1} \, o \, h_2^{-1}\right)^{-1} = h_1 \, o \, h_2$

$\Rightarrow h_2 \, o \, h_1 = h_1 \, o \, h_2$

Hence every element of H_1 commutes with those of H_2.

Since $G = H_1, oH_2$, each element of G is uniquely expressed as the product of an element in H_1 and an element in H_2. For the uniqueness we have that if possible

$h_1 \, o \, h_2 = k_1 \, o \, k_2, \; h_1, k_1 \in H_1; \; h_2, k_2 \in H_2$

Now $h_1 \, o \, h_2 = k_1 \, o \, k_2 \Rightarrow k_1^{-1} \, o \, h_1 = k_2 \, oh_2^{-1} \in H_1 \cap H_2 = \{e\}$

$\Rightarrow k_1^{-1} \, o \, h_1 = e$ and $k_2 \, oh_2^{-1} = e$

$\Rightarrow h_1 = k_1, \; h_2 = k_2$

This completes the proof of the theorem.

Generalisation: A group (G, o) is said to be internal direct product of sub-groups $(H_1, o), (H_2, o), \ldots, (H_1, o)$ of (G, o) if

CHAPTER 4

(1) each element of H_1 commutes with those of H_j, $i \neq j$.

(2) each element $g \in G$ is uniquely expressible as a product of

$h_1\ h_2, \ldots, h_n, h_i \in H_i, i = 1, 2, 3, \ldots n$

i.e., $g = h_1 \ o\ h_2 \ o\ h_3, o \ldots o\ h_n, h_i, \in H_i, i = 1, 2, 3, \ldots n$

Then we write

$G = H_1 \times H_2 \times \ldots \times H_n$

If $(G, +)$ is additive group we say that $(G, +)$ is the internal direct sum of its sub-groups $(H_1, +), (H_n, +) \ldots (H_n, +)$ and we write.

$G = H_1 \oplus H_2 \oplus \ldots \oplus H_n$

Theorem 4.1.6: *Let $(H_1, o), (H_2, o), \ldots (H_n, o)$ be n sub-groups of (G, o). Then (G, o) is an internal direct product if and only if*
(1) each $(H_i\ o)$, $1 \leq i \leq n$, is a normal sub-group of (G, o).
(2) $H_i \cap H_1 . H_2 \ldots H_{i-1} H_{i+1} \ldots H_n = \{e\}$, $1 \leq i \leq n$.
(3) $G = H_1, H_2 \ldots H_n$.

Proof: **(1)** Let $G = H_1 \times H_2 \times \ldots \times H_n$ and let $x \in G$.

Then $x = x_1 \ o\ x_2 \ o \ldots o\ x_n, x_i \in H_i, 1 \leq i \leq n$.

Let $a_k \in H_k$. Since $x_i \ o\ a_k = a_k \ o\ x_i, \forall\ i \neq k$,

Then $x^{-1} \ o\ a_k \ o\ x = (x_1 \ o\ x_2 \ o \ldots o\ x_n)^{-1} \ o\ a_k \ o\ (x_1 \ o x_2 \ o \ldots o\ x_n)$

$$= \left(x_n^{-1} \ o\ x_{n-1}^{-1} \ o \ldots x_1^{-1}\right) o\ a_k \ o \left(x_1 \ o\ x_2 \ o \ldots o\ x_n\right)$$

$$= x_n^{-1} \ o\ a_k \ o\ x_k \in H_k.$$

Hence (H_k, o) is a normal sub-group of (G, o). This eastablished (1)

(2) Now $x. \in H_i. \cap \prod_{j \neq i} H_j \Rightarrow x \in H_i$ and $x \in \prod_{j \neq i} H_j$

$\Rightarrow x = h_i$ and $x = h_1 \ o\ h_2 \ o \ldots h_{i-1} \ o\ h_{i+1} \ o \ldots o\ h_n$ for some $h_k \in H_k, \forall\ k = 1, 2 \ldots n$.

$\Rightarrow x = e \ o\ e \ o \ldots o\ h_i \ o\ e \ o \ldots o\ e$ and $x = h_1 \ o\ h_2 \ o \ldots o\ h_{i-1} \ o\ e \ o\ h_{i+1} \ o \ldots o\ h_n$

$\Rightarrow h_i = e, \forall\ i = 1, 2, \ldots n$.

by the uniqueness expression of x as products of $h_k \in H_k, \forall\ k = 1, 2, \ldots n$.

$\Rightarrow x = e \Rightarrow H_i \cap \prod_{j \neq i} H_j = \{e\}$.

This proves (2).

(3) Follows from the definition of internal direct product. Conversely, let $(H_1, o), (H_2, o), \ldots (H_n, o)$ be the sub-groups of group (G, o) satisfying given conditions (1) to (3).

Let $1 \leq k, l \leq n$ with $k \neq l$, by (2) we have

$H_k \cap \prod_{j \neq k} H_j = \{e\}$. Since $k \neq l$, $H_l \subseteq \prod_{j \neq k} H_j \Rightarrow H_k \cap H_l = \{e\}$, Let

$x_k \in H_k$, $x_l \in H_l$. Now consider the element x_k^{-1} o x_l^{-1} o x_k o x_l.

x_k^{-1} o x_l^{-1} o x_k o $x_l = x_k^{-1}$ o$\left(x_l^{-1}\text{ o } x_k \text{ o } x_l\right) \in H_k$, since

x_l^{-1} o x_k o $x_l \in H_k$ as (h_k, o) is normal sub-group.

Again

x_k^{-1} ox_l^{-1} o x_k o$x_l = \left(x_k^{-1}\text{ o } x_l^{-1}\text{ o } x_k\right)$ o $x_l \in H_l$. Since

x_k^{-1} o x_l^{-1} o $x_n \in H_l$ as (H_l, o) is normal sub-group.

Since $H_k \cap H_l = \{e\}$, then x_k^{-1} o x_l^{-1} o x_k o $x_l \in H_k \cap H_l = \{e\}$

This gives $x_k^{-1}\text{ o}x_l^{-1}\text{ o}x_k\text{ o}x_l = e \Rightarrow x_k\text{ o}x_l = \left(x_k^{-1}\text{ o}x_l^{-1}\right)^{-1} = x_1\text{ o}x_k$.

Hence every element of H_k is commutative with those of H_l, where $l \neq k$.

By (3) we have $G = H_1, H_2 \ldots H_n$. That is, if $x \in G$, then $x = x_1$ o x_2 o... ox_n, $x_i \in H_i$, $1 \leq i \leq n$.

Now we shall show that the representation of x in the above form is unique. For this, let

x_1 o x_2 o... o$x_n = y_1$ o y_2 o... oy_n, $y_i \in H_i$, $1 \leq i \leq n$.

Let l be a fixed integer with $l \neq k$, $\forall\ k$.

$$x_1 \text{ o} x_2 \text{ o } x_{k-1} \text{ o} x_k \text{ o} x_{k+1} \ldots\ldots\ldots x_n = y_1 \text{ o } y_2 \text{ o}\ldots\text{o } y_{k-1} \text{ o } y_k \text{ o } y_{k+1} \text{ o}\ldots y_n$$

$$\Rightarrow \left(x_1 \text{ o}\ldots\text{o } x_{k-1} \text{ o } x_k \text{ o } x_{k+1}\text{o}\ldots\text{o}x_n\right) \text{ o } \left(y_1 \text{ o } y_2 \text{ o}\ldots y_{k-1} \text{ o } y_k \text{ o } y_{k+1} \text{ o}\ldots y_n\right)^{-1} = e$$

$$\Rightarrow \left(x_1 \text{ o}\ldots\text{o } x_{k-1} \text{ o} x_k \text{ o } x_{k+1} \text{ o}\ldots\text{o } x_n\right) \text{ o } \left(y_n^{-1} \text{ o}\ldots y_{k+1}^{-1} \text{ o } y_k^{-1} \text{ o} y_{k-1} \text{ o}\ldots\text{o } y_1^{-1}\right) = e$$

$$\Rightarrow \left(x_1 \text{ o } y_1^{-1}\right) \text{o} \left(x_2 \text{ o } y_2^{-1}\right) \text{o}\ldots\text{.o} \left(x_{k-1} \text{ o } y_{k-1}^{-1}\right) \text{o} \left(k_{k+1} \text{ o } y_{k+1}^{-1}\right) \text{o}\ldots \text{ o} \left(x_n \text{ o } y_n^{-1}\right)$$

$$= \left(x_k \text{ o } y_k^{-1}\right)^{-1} = \left(y_k \text{ o } x_k^{-1}\right) \in H_k \cap \prod_{j \neq k} H_j = \{e\}$$

Since every element of H_l is commutative with every element of H_k for all $l \neq k$. $1 \leq l, k \leq n$.

$\Rightarrow y_k$ o $x_k^{-1} = e \Rightarrow x_k = y_k$, $\forall\ k$.

This completes the proof of the theorem.

Corollary: If (G, o) is the internal direct product of sub-groups (H_1, o).....(H_n, o) then $(G, o) \cong (H_1 \times H_2, \ldots \times H_n, o)$

Proof: It is left to the readers as an exercise.

CHAPTER 4

Remark: We know if (G, o) is the internal direct product of sub-groups (H, o),..... (H_n, o), then

$(G, o) \cong (H_1 \times H_2 \times .... \times H_n, o)$. Now if (G, o) is the external direct product of (G_1, o), (G_2, o),....(G_n, o), Then the group $(G, .)$ contains sub-group $(G_1', .)$, (G_2', o) ... (G_n', o) such that $(G_i', .) \cong (G_i', o)\ \forall i$. Since isomorphic groups are regarded equivalent, then we can say that (G, o) is internal direct product of $(G_1, o)...(G_n, o)$ Therefore we can treat internal direct product as external direct product and call it simple or direct product.

Example 4.1.5: Let (G, o) be a cyclic group generated by a of order 6 *i.e.*, $G = (a)$ and $o(G) = 6 = 3.2$

Let $H = (e, a^2, a^4)$ and $K = (e, .a^3)$. It is clear that (H, o) and (k, o) are sub-groups of (G, o) of order 3 and 2 respectively. We also have *(i)* $H \cap K = \{e\}$. *(ii)* Since (G, o) is abelian, (H, o), (K, o) are abelian and *(iii)* $H\ o\ K = (e\ o\ e, e\ o\ a^3, a^2\ o\ e, a^2\ o\ a^3, a^4\ o\ e, a^4\ o\ a^3)$

$= (e, a^3, a^2, a^5, a^4, a) = G$, Since $a^6 = e$

Hence (G, o) is the internal direct product of (H, o) and (K, o) and $(G, o) \cong (H \times K, .)$

Theorem 4.1.7: *If (G, o) is a finite cyclic group of order $n = P_1^{m_1}.P_2^{m_2}....P_n^{m_n}$, P_i prime, $P_i \neq P_j$ if $i \neq j$. Then (G, o) is the direct product of cyclic groups of order $P_i^{m_1}$, $i = 1, 2, ...n$.*

Proof: Let (H_i, o) be the sub-group of order $P_i^{m_1}$ of (G, o).

Since (G, o) is abelian, (H_i, o) is a normal sub-group. Then

$G' = H_1 o ....o\ H_i\ o\ ....\ o\ H_n$, and (G', o)

is a sub-group of (G, o) of order n', say, which means $n' | n$. Now every number of $P_1^{m_1}.P_2^{m_2}....P_n^{m_i}....P_n^{m_n}$ will divide the order of (G', o). That is,

$$P_i^{m_1} \Big| n',\ i = 1, 2...n \Rightarrow P_1^{m_1} P_2^{m_2} ... P_n^{m_n} \Big| .n'$$

$\Rightarrow n\ | n'$.

Thus, we conclude $n = n'$ from $n | n'$ and $n' | n$. Which follows $(G, o) \cong (G', o)$. From this we can also say that every element x of G is expressible as the products of elements $x_i \in H_i$, *i.e.*,

$x = x_1\ ox_2\ o\ ...\ ox_i\ o\ ...\ ox_n$.

Now consider the sub-group (L_i, o) of (G, o) of order $n/P_n^{m_1} = n_i$ since $(P_n^{m_1}, n_i) = 1$, then the order of sub-group $(L_i \cap H_i, o)$ is one which follows $L_i \cap H_i = (e)$. ...(1)

Since $L_i = H_1\ H_2\ ....\ H_{i-1}\ H_{i+1}\ ....H_n$, then

$$P_j^{m_j} \Big| n_i, \forall i \neq j \Rightarrow L_i \supseteq H_j .$$

Hence $L_i \supseteq H_1, H_2...H_{i-1}\ H_{i+1}...H_n$ ...(2)

From (1) and (2) we get

$H_i \cap H_1...H_2...H_{i-1}. H_{i+1} ....H_n = \{e\}$

Hence the conditions of the theorem are satisfied.

So we get

$G = H_1 \times ....\times H_{i-1} \times H_i \times H_{i+1} \times ....\times H_n.$

Hence the theorem.

Example 4.1.6: If (C_n, o) and $(C_m, *)$ are cyclic group of order n and m respectively and $(n, m) = 1$, then prove that

$(C_a \times C_m, .) \cong (C_{nm}, .)$ the cyclic group of order $n\ m$.

Solution: Let $C_n = (a)$ with $a^n = e$ and let $C_m = (b)$ with $b^m = e_1$. We consider the order of the element $(a, b) \in C_n \times C_m$.

Since $(a^n)^m = a^{nm} = e$ and $(b^m)^n = b^{mn} = e_1$, then $(a, b)^{mn} = (a^{mn}, b^{mn}) = (e, e_1)$, the identity of the group $(C_n \times C_m, .)$. This proves that order of (a, b) is mn. Since $o\ (C_n \times C_m) = mn$, then $(C_n \times C_m, .)$ is a cyclic group of order mn which is generated by (a, b), that is, $C_n \times C_m = ((a, b))$.

$(C_{mn}, .)$ is also cyclic group of order mn.

Therefore $(C_n \times C_m, .) \cong (C_{nm}, .)$, since two cyclic groups of the same order are isomorphic.

Example 4.1.7: Show that (C_{n^2}, o) is not isomorphic to $(C_n \times C_n, .)$, where (C_n, o) is the cyclic group of order n.

Solution: Since (C_{n2}, o) is the cyclic group of order n^2 and $n^2 = n.n$. So it has one and only one sub-group of order n. $(C_n \times C_n, .)$ is also cyclic group of order n^2 but it has two sub-groups of order n, namely $(((e, a)), .)$ and $(((a, e)), .)$. Where $(a) = C_n$. Since sub-groups of a given order are mapped onto sub-groups of the same order by any isomorphism, (C_{n2}, o) cannot be isomorphic to $(C_n \times C_n, .)$

Example 4.1.8: (M, o) and (N, o) are two normal sub-groups of a group (G, o). Show that $(GIM \cap N, o)$ is isomorphic to a group $(G/M \times G/N, .)$.

Proof: Let $\phi : G \to G/M \times G/N$ be defined by

$$\phi\ (x) = (M\ o\ x,\ N\ o\ x).$$

For $x, y \in G$,

$$\begin{aligned}\phi\ (xoy) &= (M\ o\ (x\ o\ y),\ N\ o\ (x\ o\ y)) \\ &= ((Max)\ o\ (Moy),\ (N\ o\ x)\ o\ (N\ o\ y)) \\ &= (M\ o\ x,\ N\ o\ x).\ (M\ o\ y,\ N\ o\ y) \\ &= \phi\ (x).\ \phi\ (y).\end{aligned}$$

And

$$\begin{aligned}ker\ \phi &= \{x \in G | \phi\ (x) = (M, N)\} \\ &= \{x \in G | (M\ o\ x,\ N\ o\ y) = (M, N)\} \\ &= [x \in G | M\ o\ x = M \text{ and } N\ o\ x = N\} \\ &= \{x \in G | x \in M \text{ and } x \in N\} \\ &= M \cap N\end{aligned}$$

Hence by fundamental theorem of homomorphism 3.2.8 $(G/M \cap N, o) \cong (G/M \cdot G/N.)$,

CHAPTER 4

PROBLEMS

1. If $(H, o) \cong (H', \odot)$ and $(K, *) \cong (K', \oplus)$, then

   Prove that $(H \times K, .) \cong (H' \times K', .)$.

2. Let $((a), o)$ and $((b), *)$ be two cyclic groups of order n and m respectively such that $(m, n) = 1$. Then show that $((a) \times (b).)$ is an abelian group of order mn which is not cyclic.

3. If (G_1, o), $(G_2, .)$, and $(G_{3,}\ \odot)$ are three groups, then show that

   (i) $o\ (G_1 \times G_2\ *) = o\ (G_1).\ o\ (G_2)$,

   (ii) $(G_1 \times G_2;) \cong (G_2 \times G_1, .)$,

   (iii) $G_1 \times (G_2 \times G_3) \cong G_1 \times G_2 \times G_3 \cong (G_1 \times G_2) \times G_3$.

4. If (G, o) is a finite group in which $a^2 = e, \ \forall\ a \in G$. Show that (G, o) is the internal direct product of a finite number of sub-groups each of order 2, and $o\ (G) = 2^n$ for $n \geq 0$.

5. Show that $(H \times K, .)$ is an abelian if and only if (H, o) and $(K, *)$ are both abelian groups.

6. Show that for any prime p there are exactly two non-isomorphic groups of order P^2.

7. Let (G, o) be an abelian group of order 4. Show that either (G, o) is cyclic or (G, o) is isomorphic to a direct product of two cyclic groups of order 2 each.

8. Express the cyclic group formed by tenth roots of unity of order 10 under multiplications as the direct product of two sub-groups of order 2 and 5.

9. Show that for every normal sub-group (H, o), there exists another normal sub-group (K, o) of (G, o) such that $G = H \times K$.

10. Consider the group $(C_2 \times K_4, .)$, $(C_4 \times C_2, .)$ $(C8, .)$, where $(C, .)$ is the cyclic group under multiplication composition and $(K_4, .)$ is four group, *i.e.*, the non-cyclic groups of order 4. Are any two of these groups isomorphic? Is any one non-abelian?

4.2 CONJUGATES AND CLASSES

Definition 4.2.1: Let $(G, *)$ be a group and $\phi \neq S \subseteq G$. Let x be a fixed element of G. Then the set of all elements of the form $x * s * x^{-1}$, $s \in S$, is called the transform set of S by x, denoted by S^x. That is $S^x = \{x * s * x^{-1} \mid s \in S \text{ and } x \text{ is fixed}\}$.

We can write it as $S^x = x * S * x^{-1}$

Lemma 4.2.1: *There exists a one-to-one correspondence between S and* S^x.

Proof: Let f be a function from S onto S^x defined by $f(s) = x * s * x^{-1}$, for $s \in S$, x is fixed.

Let $s_1, s_2 \in S$.

$f(s_1) = f(s_2) \Rightarrow x * s_1 * x^{-1} = x * s_2 * x^{-1}$

$\Rightarrow s_1 = s_2$, since left and right cancellation laws hold in G.

Hence f is one to one functions, every element $x * s * x^{-1}$ is an image element for some $s \in S$ such that $f(s) = x * s * x^{-1}$.

Hence f is onto function. This proves the lemma.

Definition 4.2.2: Let $(H, *)$ be a sub-group of $(G, *)$ and let S, S' be two subsets of G. If some $x \in H$ exists such that $S' = S^x = x * S * x^{-1}$, then we say that S and S' are conjugate under H.

Lemma 4.2.2: *The relation of conjugacy under H defined on the set G, where $(H, *)$ is a sub-group of $(G, *)$, is an equivalence relation.*

Proof: We see that the relation of conjugacy under H is reflexive, symmetric, and transitive.

(1) Since $e \in H$, $S = e * S * e^{-1}$, $\forall\ s \in S$. Hence S is conjugate to itself under H.

(2) For some $x \in H$, $S' = x * S * x^{-1} \Rightarrow x^{-1} * S' * x = S$

$$\Rightarrow x^{-1} * S' * (x^{-1})^{-1} = S$$

$$\Rightarrow y * S' * y^{-1} = S. \text{ Since } x \in H,\ x^{-1} = y \in H.$$

Hence the relation is symmetric.

(3) For some y, $x \in H$, $S' = x * S * x^{-1}$ and $S'' = y * S' * y^{-1}$

$$\Rightarrow S'' = y * (x * S * x^{-1}) * y^{-1}$$

$$\Rightarrow S'' = (y * x) * S * (x^{-1} * y^{-1})$$

$$\Rightarrow S'' = (y * x) * S * (y * x)^{-1}$$

Since x, $y \in H$, $y * x \in H$ and $(y * x)^{-1} \in H$.

Hence the relation is transitive.

Definition 4.2.3: The set of all S' conjugate to a given S is called a *class of conjugates.*

Thus the relation of conjugacy under H decomposes the family of all subsets of group $(G, *)$ into disjoint classes of conjugates.

Theorem 4.2.1:. *Let $(H, *)$ be a sub-group of group $(G, *)$ and $a \in G$. Then the pair $(a * H * a^{-1}, *)$ is also a sub-group of $(G, *)$ called the conjugate sub-group of $(H, *)$ determined by a.*

or

Any set conjugate to a sub-group is also a sub-group.

Proof: If x, $y \in a * H * a^{-1}$, $\exists\ h_1, h_2 \in H$ such that $x = a * h_1 * a^{-1}$ and $y = a * h_2 * a^{-1}$

CHAPTER 4

Then

$$\begin{aligned} x * y^{-1} &= (a * h_1 * a^{-1}) * (a * h_2 * a^{-1})^{-1} \\ &= (a * h_1 * a^{-1}) * (h_2 * a^{-1})^{-1} * a^{-1}) \\ &= (a * h_1 * a^{-1}) * (a * h_2^{-1} * a^{-1}) \\ &= a * h_1 * h_2^{-1} * a^{-1} \in a * H * a^{-1}, \text{ Since} \end{aligned}$$

$h_2, h_1 \in H, h_1 * h_2^{-1} \in H.$

Which shows $(a * H * a^{-1}, *)$ is a sub-group of $(G, *)$.

Definition 4.2.4: Let $(H, *)$ be a sub-group of $(G, *)$ and $\phi \neq S \subseteq G$. The set $N_H(S) = \{x \in \text{H} \mid x * S * x^{-1} = S\}$ is called the normalizer of S in H.

The set $C_H(S) = [x \in H \mid x * s * x^{-1} = s, \forall s \in S\}$ is called centralizer of S in H.

If $S = \{a\}$ the singleton, then the normalizer and centralizer of $\{a\}$ are identical.

Theorem 4.2.2: *Let $(H, *)$ be a sub-group of group $(G, *)$. Then for any sub-set $S \subseteq G$, $(N_H(S), *)$ is a sub-group of $(H, *)$.*

Proof: If $x, y \in N_H(S)$, $x * S * x^{-1} = S$ and

$$y * S * y^{-1} = S$$

Now we note that

$$\begin{aligned} y^{-1} * x * S * (y^{-1} * x)^{-1} &= y^{-1} * x * S * (x^{-1} * y) \\ &= y^{-1} * (x * S * x^{-1}) * y \\ &= y^{-1} * S * y = S. \end{aligned}$$

which shows $y^{-1} * x \in N_H(S)$. This proves the theorem.

Corollary: *$(C_H(S), *)$ is a sub-group of $(H, *)$.*

Proof: Let $x, y \in C_H(S)$, then $x * s * x^{-1} = s$ and

$$y * s * y^{-1} = s, \forall s \in S.$$

Now $y^{-1} * x * s * (y^{-1} * x)^{-1} = y^{-1} * x * s * x^{-1} * y$

$$\begin{aligned} &= y^{-1} * (x * s * x^{-1}) * y \\ &= y^{-1} * s * y = s, \forall s \in S. \end{aligned}$$

Hence $y^{-1} * x \in C_H(S)$. This proves the theorem.

Remark: $C_H(S)$ is always a subset of $N_H(S)$. For if $x \in C_H(S)$, then $x * s * x^{-1} = s, \forall s \in S$.

But $x * s * x^{-1} \in x * S * x^{-1} \Rightarrow x \in N_H(S)$.

Therefore $C_H(S) \subseteq N_H(S)$.

Conversely, if $y \in N_H(S)$, then $y * S * y^{-1} = S$. It shows that $\exists$ s_1 and s_2 such that $y * s_1 * y^{-1} = s_2 \Rightarrow y \notin C_H(S)$.

So $N_H(S) \notin C_H(S)$. Thus $C_H(S) \subset N_H(S)$.

Hence (C_H (S), *) is always sub-group of (N_H (S), *), where (H,*) is a sub-group of (G, *).

Now when $H = G$, then we write $N(S)$ for $N_G(S)$ and it is referred to the sub-group (N (S), *) simply as the normalizer of S. Similarly (C (S), *) is known as centralizer of S. By remark $C(S) \subseteq N(S)$ and (C (S), *) is a sub-group of (N (S), *)

if $S = \{a\}$, then $C(a) \subset N(a)$

Definition 4.2.5: $N(a)$ is the set of all elements in G which commute with a, that is, $N(a) = \{x \in | x * a = a * x\}$.

The centralizer of G in G, that is $C_G(G)$ or $(C(G)$ is called the centre of G, denoted by $Z(G)$, is the set of all elements in G which are commutative with every element of G. Thus

$$Z(G) = C(G) = \{x \in G | x * a * x^{-1} = a, \forall a \in G\}$$

$$= \{x \in G | x * a = a * x, \forall a \in G\}$$

Theorem 4.2.3: *For any $a \in G$, the pair ($N(a)$, *) is a sub-group of (G, *).*

Proof: Let $x, y \in N(a)$, then $x * a = a * x$ and $y * a = a * y$

$(x * y) * a = x * (y * a) = x * (a * y) = (x * a) * y = (a * x) * y = a * (x * y)$, which implies $x * y \in N(a)$

$x * a = a * x \Rightarrow a * x^{-1} = x^{-1} * a \Rightarrow x^{-1} \in N(a)$, This completes the proof.

Corollary: The pair ($Z(G)$, *) is a sub-group of group (G, *)

Proof: *Let $a, b \in Z(G)$, then* $a * x = x * a$ and $b * x = x * b$, $\forall x \in G$.

$(a * b) * x = a * (b * x) = a * (x * b) = (a * x) * b$

$= (x * a) * b = x * (a * b)$

$\Rightarrow a * b \in Z(G)$.

Further $a * x = x * a \Rightarrow x * a^{-1} = a^{-1} * x \Rightarrow a^{-1} \in Z(G)$.

Hence ($Z(G)$ *) is a sub-group of (G, *).

Lemma 4.2.3: *$a \in Z(G)$ if and only if $N(a) = G$. If G is finite, $a \in Z(G)$ if and only if $o(N(a)) = o(G)$.*

Proof: Let $a \in Z(G)$, then $a * x = x * a$, $\forall x \in G$.

$\forall x \in G$, $a * x = x * a \Rightarrow x \in N(a)$ when a is fixed.

Therefore $G \subseteq N(a)$.

Again $x \in N(a) \Rightarrow x * a = a * x \Rightarrow x \in G$ since $a \in Z(G)$

Therefore $N(a) \subseteq G$. Hence $N(a) = G$.

Conversely, if $N(a) = G$, then $\forall x \in G$, $a * x = x * a$.

That is, a commutes with every $x \in G$.

which implies $a \in Z(G)$.

If G is finite and $a \in Z$, then $N(a) = G \Rightarrow o(N(a)) = o(G)$.

Lemma 4.2.4: *If (G, *) is a group and if $a \in G$, then $Z(G) \subseteq N(a)$*

CHAPTER 4

Proof: If $x \in Z(G)$, then $x * b = b * x$, $\forall\, b \in G$. If we choose the particular element $a \in G$, *i.e.*, $b = a$, then $x * a = a * x \Rightarrow x \in N(a)$, therefore $Z(G) \subseteq N(a)$.

If $y \in N(a)$, then $a * y = y * a$, where a is fixed. That is, for other element $c \in G$, $c * y = y * c$ may not hold. Which means $y \notin Z(G)$. Therefore $N(a) \not\subset Z(G)$.

Hence $Z(G) \subset N(a)$.

Theorem 4.2.4: *The number of distinct conjugates of a sub-group (H, *) of (G, *) determined by the elements of a sub-group (K, *) is equal to* $\dfrac{o(K)}{o(N_k(H))}$ *the index of* $N_k(H)$ *in K.*

Proof: We shall establish the result by showing that there exists a one-to-one correspondence between the class T of conjugates of $(H, *)$ under K and the class F of left cosets of $N_k(H)$ in K.

Let $k_1, k_2 \in K$. Let $f : F \to T$ be defined by

$f(k_1 * N_k(H)) = k_1 * H * k_1^{-1}$, where

$k_1 * N_k(H) \in F$ and $k_1 * H * k_1^{-1} \in T$.

For this, $k_1 * N_k(H)$ and $k_2 * N_k(H) \in$ F, assume that

$f(k_1 * N_k(H)) = f(k_2 * N_k(H)) \Rightarrow k_1 * H * k_1^{-1} = k_2 * H * K_2^{-1}$

$$\Rightarrow H = \left(k_1^{-1} * k_2\right) * H * \left(k_2^{-1} * k_1\right)$$

$$\Rightarrow H = \left(k_1^{-1} * k_2\right) * H * \left(k_1^{-1} * k_2\right)^{-1}$$

$\Rightarrow k_1^{-1} * k_2 \in N_k(H)$ by the def. of $N_k(H)$.

$\Rightarrow k_1 * N_k(H) = k_2 * N_k(H)$.

Hence f is one to one function from F onto T. since, for $k *$ H $* k^{-1}$ there exists $k \in K$, consequently there exists $k * N_k(H)$ such that

$f(k * N_k(H)) = k * H * k^{-1}$

Hence the number of conjugates of $(H, *)$ determined by the elements of K is equal to the index of $N_k(H)$ in K.

Corollary: Let $(K, *)$ and $(K_1, *)$ be two sub-groups of the group $(G, *)$. If $(H, *)$ is invariant under n elements of K, that is, $x * H * x^{-1} = H$, $x \in K$ for n such $x'^s \in K$, then $(H, *)$ has $\dfrac{o(K)}{n}$ conjugate sub-groups by elements of K.

Proof: Corresponding to different n elements of K there are n different conjugates such that these n conjugates are equal to H. Hence $o(N_k(H)) = n$. By Lagrange's theorem $o(K)$ = (index of $N_k(H)$. $o(N_k(H))$.

Hence the index of N_k in $K = \dfrac{o(K)}{n}$.

By the above theorem index of H in K is equal to the number of conjugates of $(H, *)$ determined by the elements of K, which follows the corrollary.

4.3 CONJUGATE ELEMENTS

Definition 4.3.1: Let $(G, *)$ be a group and $a, b \in G$. Then the element a is said to be a conjugate of b in G if there exists some $x \in G$ such that $a = x * b * x^{-1}$. Symbolically, 'a is a conjugate to b' is written as $a \sim b$ which is referred to as the relation of conjugacy.

Comma 4.3.1: Conjugacy is an equivalence relation on the group $(G, *)$.

Proof: In order to prove the Lemma we shall prove that the relation of conjugacy is reflexive, symmetric, and transitive.

(1) Reflexive: Since $\forall\ a \in G$, $\exists$ an identity element $e \in G$ such that $a = e * a * e^{-1}$ which implies $a \sim a$.

(2) Symmetric: For $a, b \in G$, $a \sim b \Rightarrow a = x * b * x^{-1}$, for some $x \in G$

$$\Rightarrow b = x^{-1} * a * x = y * a * y^{-1}$$

where $y = x^{-1} \in G$

$$\Rightarrow b \sim a.$$

(3) Transitive: For $a, b, c \in G$, $a \sim b$ and $b \sim c \Rightarrow a = x * b * x^{-1}$

and $b = y * c * y^{-1}$

for some $x. y \in G$.

$$\Rightarrow a = x * (y * c * y^{-1}) * x^{-1}$$

$$\Rightarrow a = (x * y) * c * (y^{-1} * x^{-1})$$

$$\Rightarrow a = (x * y) * c * (x * y)^{-1} \Rightarrow a \sim c.$$

This completes the proof of the Lemma.

Definition 4.3.2: Let $(G, *)$ be a group and $a \in G$. Then the set $\{x \in G \mid a \sim x)$ is called the conjugate class of a in G, denoted by $C(a)$. That is, it is the set of all distinct elements of the form $y * a * y^{-1}$, $\forall\ y \in G$. Thus $C(a) = \{y * a * y^{-1} \mid y \in G)$, for some fixed $a \in G$. The element a is called the representative of the conjugate class $C(a)$.

From above discussion we have seen that the relation of conjugacy is an equivalence relation on G. So the relation of conjugacy will decompose the set G into disjoint equivalence classes—conjugate classes. Thus if $(G, *)$ is a group and there is defined a relation of conjugacy on G. then the set G can be written as the union of disjoint conjugate classes. Let $a_1, a_2, \ldots a_r \in G$ such that $C(a_1), \ldots C(a_r)$ are possible disjoint conjugate classes. Then

$$G = C(a_1) \cup C(a_2) \cup \ldots \cup C(a_r).$$

That is, the total number of elements in G is equal to the sum of the number of elements of all disjoint classes. If C_{a_i} denotes the number of elements in conjugate class $C(a_i)$, then $o(G) = \sum_{i=1}^{r} C_{ai}$

Now our immediate goal is to find out the number of element in $C(a)$, that is, C_a.

CHAPTER 4

Theorem 4.3.1: *Let (G, *) be a finite group, then* $C_a = \dfrac{o(G)}{o(N(a))}$*, That is, the number of elements in C(a) is equal to the index of the normalizer (centralizer) of a in G:*

Proof: Let $o(G) = n$. If $N(a)$ has r distinct left cosets : $x_1 * N(a), x_2 * N(a), ... x_r * N(a)$, then

$$r = \frac{o(G)}{o(N(a))}.$$

Note for $1 \le i, j \le r$,

$x_i * N(a) = x_j * N(a)$

$\Rightarrow x_i^{-1} * x_j \in N(a)$

$\Rightarrow x_i^{-1} * x_j = x$, for some $x \in N(a)$ for which $x * a = a * x$.

$\Rightarrow (x_i^{-1} * x_j) * a = a * (x_i^{-1} * x_j)$

$\Rightarrow x_j * a * x_j^{-1} = x_i * a * x_i^{-1} \in C(a)$

$\Rightarrow x_j = x_i \Rightarrow i = j$.

Since $C(a)$ is the set of distinct elements of the form $x * a * x^{-1}$.

Hence all r left cosets $x_1 * N(a), x_2 * N(a), ... x_r. * N(a)$ are distinct and consequently $x_1 * a * x_i^{-1}, x_2 * a * x_2^{-1} ... x_r * a * x_r^{-1}$ are all distinct conjugate elements of a.

If we show that these are only r conjugates of a, then the number of left cosets of $N(a)$ is equal to the number of conjugates of a. That is $C(a)$ will contain only r elements. Thus

$$C_a = r = \frac{o(G)}{o(N(a))}$$

Now consider $b \in G$ conjugate of a. Then there exists an element $x \in G$ such that $b = x * a * x^{-1}$.

Since $G = \bigcup_{i=1}^{r} x_i * N(a)$, $x = x_i * c$ for some $c \in N(a)$ and for some integer i, $1 \le i \le r$.

Therefore $b = x * a * x^{-1} = (x_i * c) * a * (x_i * c)^{-1}$

$= x_i * (c * a * c^{-1}) * x_i^{-1}$

$= x_i * a * x_i^{-1}$, since $c \in N(a)$,

then $c * a = a * c \Rightarrow a = c * a * c^{-1}$.

This proves that any conjugate b is equal to one of $x_i * a * x_i^{-1}$, conjugates of a. That is, $C(a)$ has only r elements of the form $x_i * a * x_i^{-1}$, $i = 1, 2, ... r$. Hence this completes the proof of the theorem.

Corollary 1: For any group $(G, *)$, $o(G) = \Sigma\, C_a$

$$= \Sigma \frac{o(G)}{o(N(a))},$$ where a runs over G. This is known as class equation.

Proof: Since $(G, *)$ is finite, there exists finite conjugate classes $C(a_1), C(a_2), C(a_3),\ldots C(a_n)$ of G. Since any two conjugate classes are disjoint, then $G = \bigcup_{i=1}^{n} C(a_i)$ which implies $o(G) = \Sigma\, C_{a_i}$. By the theorem 4.3.1,

$$C_{a_i} = \frac{o(G)}{o(N(a_i))}$$

Thus $o(G) = \Sigma C_{a_i} = \sum_{i=1}^{n} \frac{o(G)}{o(N(a_i))}$.

This completes the proof of the corollary.

Corollary 2: Let $(G, *)$ be a finite group and $Z(G)$ be its centre. Then $o(G) = o(Z(G)) + \sum_{a \notin Z} \frac{o(G)}{o(N(a))}$, where a runs over G.

Proof: Let $a \in Z(G)$, then $a * x = x * a \Rightarrow a = x * a * x^{-1}$, $\forall\, x \in G$. The conjugate ciass $C(a)$ of a is the set of all element of the form $x * a * x^{-1}$ which are all equal to a. Thus $C(a) = \{a\}$ is singleton.

If we add to $o(Z(G))$ the number of elements in the non-singleton conjugate classes, the elements of G will have been counted. Thus

$$o(G) = o(Z(G)) + \Sigma \frac{o(G)}{o(N(a_i))} \text{ where } a_i \notin Z(G)$$

and a_i runs over G.

This is also known as class equation.

Example 4.3.1: We consider the (S_3, o). Its elements are

$$e = \begin{pmatrix} 1 & 2 & 3 \\ 1 & 2 & 3 \end{pmatrix}, f_1 = \begin{pmatrix} 1 & 2 & 3 \\ 2 & 1 & 3 \end{pmatrix}, f_2 = \begin{pmatrix} 1 & 2 & 3 \\ 3 & 2 & 1 \end{pmatrix},$$

$$f_3 = \begin{pmatrix} 1 & 2 & 3 \\ 1 & 3 & 2 \end{pmatrix}, f_4 = \begin{pmatrix} 1 & 2 & 3 \\ 2 & 3 & 1 \end{pmatrix}, f_5 = \begin{pmatrix} 1 & 2 & 3 \\ 3 & 1 & 2 \end{pmatrix}$$

Now we compute the conjugate classes.

$$C(e) = \{f^{-1} \circ e \circ f \mid f \in S_3\} = \{e\},$$

$$C(f_1) = (f^{-1} \; o \; f_1 \; of | f \in S_3\}$$

$$= \left\{ f_1, \begin{pmatrix} 1 & 2 & 3 \\ 3 & 2 & 1 \end{pmatrix}^{-1} o \begin{pmatrix} 1 & 2 & 3 \\ 2 & 1 & 3 \end{pmatrix} o \begin{pmatrix} 1 & 2 & 3 \\ 3 & 2 & 1 \end{pmatrix}, \right.$$

$$\begin{pmatrix} 1 & 2 & 3 \\ 1 & 3 & 2 \end{pmatrix}^{-1} o \begin{pmatrix} 1 & 2 & 3 \\ 2 & 1 & 3 \end{pmatrix} o \begin{pmatrix} 1 & 2 & 3 \\ 1 & 3 & 2 \end{pmatrix},$$

$$\begin{pmatrix} 1 & 2 & 3 \\ 2 & 3 & 1 \end{pmatrix}^{-1} o \begin{pmatrix} 1 & 2 & 3 \\ 2 & 1 & 3 \end{pmatrix} o \begin{pmatrix} 1 & 2 & 3 \\ 2 & 3 & 1 \end{pmatrix},$$

$$\left. \begin{pmatrix} 1 & 2 & 3 \\ 3 & 1 & 2 \end{pmatrix}^{-1} o \begin{pmatrix} 1 & 2 & 3 \\ 2 & 1 & 3 \end{pmatrix} o \begin{pmatrix} 1 & 2 & 3 \\ 3 & 1 & 2 \end{pmatrix} \right\}$$

$= \{f_1, f_3, f_2\}$. Since

$$\begin{pmatrix} 1 & 2 & 3 \\ 3 & 2 & 1 \end{pmatrix}^{-1} o \begin{pmatrix} 1 & 2 & 3 \\ 2 & 1 & 3 \end{pmatrix} o \begin{pmatrix} 1 & 2 & 3 \\ 3 & 2 & 1 \end{pmatrix} = \begin{pmatrix} 1 & 2 & 3 \\ 1 & 3 & 2 \end{pmatrix} = f_3,$$

$$\begin{pmatrix} 1 & 2 & 3 \\ 1 & 3 & 2 \end{pmatrix}^{-1} o \begin{pmatrix} 1 & 2 & 3 \\ 2 & 1 & 3 \end{pmatrix} o \begin{pmatrix} 1 & 2 & 3 \\ 1 & 3 & 2 \end{pmatrix} = \begin{pmatrix} 1 & 2 & 3 \\ 3 & 2 & 1 \end{pmatrix} = f_2,$$

$$\begin{pmatrix} 1 & 2 & 3 \\ 2 & 3 & 1 \end{pmatrix}^{-1} o \begin{pmatrix} 1 & 2 & 3 \\ 2 & 1 & 3 \end{pmatrix} o \begin{pmatrix} 1 & 2 & 3 \\ 2 & 3 & 1 \end{pmatrix} = \begin{pmatrix} 1 & 2 & 3 \\ 3 & 2 & 1 \end{pmatrix} = f_2,$$

$$\begin{pmatrix} 1 & 2 & 3 \\ 3 & 1 & 2 \end{pmatrix}^{-1} o \begin{pmatrix} 1 & 2 & 3 \\ 2 & 1 & 3 \end{pmatrix} o \begin{pmatrix} 1 & 2 & 3 \\ 3 & 2 & 1 \end{pmatrix} = \begin{pmatrix} 1 & 2 & 3 \\ 1 & 3 & 2 \end{pmatrix} = f_3.$$

$$C(f_4) = \{f^{-1} \; o \; f_4 \; of | f \in S_3\} = \left\{ f_4, \begin{pmatrix} 1 & 2 & 3 \\ 2 & 1 & 3 \end{pmatrix}^{-1} o \begin{pmatrix} 1 & 2 & 3 \\ 2 & 3 & 1 \end{pmatrix} o \begin{pmatrix} 1 & 2 & 3 \\ 2 & 1 & 3 \end{pmatrix}, \right.$$

$$\begin{pmatrix} 1 & 2 & 3 \\ 3 & 2 & 1 \end{pmatrix}^{-1} o \begin{pmatrix} 1 & 2 & 3 \\ 2 & 3 & 1 \end{pmatrix} o \begin{pmatrix} 1 & 2 & 3 \\ 3 & 2 & 1 \end{pmatrix},$$

$$\begin{pmatrix} 1 & 2 & 3 \\ 1 & 3 & 2 \end{pmatrix}^{-1} o \begin{pmatrix} 1 & 2 & 3 \\ 2 & 3 & 1 \end{pmatrix} o \begin{pmatrix} 1 & 2 & 3 \\ 1 & 3 & 2 \end{pmatrix},$$

$$\begin{pmatrix}1 & 2 & 3\\ 3 & 1 & 2\end{pmatrix}^{-1} o \begin{pmatrix}1 & 2 & 3\\ 2 & 3 & 1\end{pmatrix} o \begin{pmatrix}1 & 2 & 3\\ 3 & 1 & 2\end{pmatrix}\Bigg\}$$

$$= \{f_4, f_5\}.$$

Thus, the set S_3 is divided into three disjoint conjugate classes. $C(e)$, $C(f_1)$, and $C(f_4)$. That is,

$\{e, f_1, f_2, f_3, f_4, f_5,\} = \mathrm{C}\ (e) \cup \mathrm{C}\ (f_1) \cup \mathrm{C}\ (f_4)$

Thus $o(S_3) = 6$, $C_e = 1$, $C_{f_1} = 3$, $C_{f_4} = 2$

Hence $o\ (S_3,) = 1 + 3 + 2 = \Sigma C_f = C_e + C_{f_1} + C_{f_4}$

Now again we compute the normalizers $N(e)$,

$N(f_1)$, and $N(f_4)$ of e, f_1, and f_4 in G respectively.

$N(e) = \{f \in S_3 \mid e \ o \ f = f \ o \ e\} = S_3$.

$N(f_1) = \{f \in S_3 \mid f \ o \ f_1 = f_1 \ o \ f\}$, to determine $N(f_1)$ we see which elements commute with f_1.

e commutes with f_1, so $e \in N(f_1)$

f_1 commutes with itself, so $f_1 \in N\ (f_1)$

we see

$$f_1 \ o \ f_2 = \begin{pmatrix}1 & 2 & 3\\ 2 & 1 & 3\end{pmatrix} o \begin{pmatrix}1 & 2 & 3\\ 3 & 2 & 1\end{pmatrix} = \begin{pmatrix}1 & 2 & 3\\ 3 & 1 & 2\end{pmatrix},$$

$$f_2 \ o \ f_1 = \begin{pmatrix}1 & 2 & 3\\ 3 & 2 & 1\end{pmatrix} o \begin{pmatrix}1 & 2 & 3\\ 2 & 1 & 3\end{pmatrix} = \begin{pmatrix}1 & 2 & 3\\ 2 & 3 & 1\end{pmatrix},$$

That is, $f_1 \ o \ f_2 \neq f_2 \ o \ f_1$... so $f_2 \notin N(f_1)$.

Similarly we can verify that $f_3 \ o \ f_1 \neq f_1 \ o \ f_3$, $f_4 \ o \ f_1 \neq f_1 \ o \ f_4$, and $f_1 \ o \ f_5 \neq f_5 \ o \ f_1$. So $f_3, f_4, f_5 \notin N\ (f_1)$.

Thus $N(f_1) = \{e, f_1\}$

$N(f_4) = \{f \in S_3 \mid f \ o \ f_4 = f_4 \ o \ f\}$.

e commutes with f_4 so $e \in N(f_4)$

f_4 commutes with f_4, so f_4

we see that

$$f_4 \ o \ f_5 = \begin{pmatrix}1 & 2 & 3\\ 2 & 3 & 1\end{pmatrix} o \begin{pmatrix}1 & 2 & 3\\ 3 & 1 & 2\end{pmatrix} = \begin{pmatrix}1 & 2 & 3\\ 1 & 2 & 3\end{pmatrix},$$

CHAPTER 4

$$f_5 \text{ o } f_4 = \begin{pmatrix} 1 & 2 & 3 \\ 3 & 1 & 2 \end{pmatrix} \text{o} \begin{pmatrix} 1 & 2 & 3 \\ 2 & 3 & 1 \end{pmatrix} = \begin{pmatrix} 1 & 2 & 3 \\ 1 & 2 & 3 \end{pmatrix}.$$ So $f_4 \text{ o } f_5 = f_5 \text{ o } f_4$.

Hence $f_5 \in N(f_4)$.

Similarly we can verify that $f_1 \text{ o } f_4 \neq f_4 \text{ o } f_1$, $f_2 \text{ o } f_4 \neq f_4 \text{ o } f_2$, $f_3 \text{ o} f_4 \neq f_4 \text{ o} f_3$. So $f_1, f_2, f_3 \notin N(f_4)$.

Thus, $N(f_4) = \{e, f_4, f_5\}$.

Now we verify the theorem that

$$C_e = \frac{o(G)}{o(N(e))} = \frac{6}{6} = 1,$$

$$C_{f_1} = \frac{o(S_3)}{o(N(f_1))} = \frac{6}{2} = 3,$$

and

$$C_{f_4} = \frac{o(S_3)}{o(N(f_4))} = \frac{6}{3} = 2$$

Hence the class equation $o(G) = \Sigma \dfrac{o(G)}{o(N(a))}$ is also verified that $6 = o(S_3) = C_e + C_{f_1} + C_{f_4} = 1 + 3 + 2 = 6$.

Note: In the decomposition *of* the set G into its conjugate classes, the conjugate classes are either identical or disjoint. But the normalizers are not disjoint for e belongs to every normalizer.

Theorem 4.3.2: *If $o(G) = p^n$, where p is a prime member, then $Z(G) \neq \{e\}$.*

Proof: Let $o(Z(G)) = z$. By the class equation we have $o(G) = o(Z(G)) + \Sigma \dfrac{o(G)}{o(N(a))}$, where sum runs over a, representative of the distinct conjugate classes which have more than one elements. That is, $a \notin Z(G)$. Because if $a \in Z(G)$, then $C(a) = \{a\}$.

Now for each $a \notin Z(G)$, $\exists\, N(a)$. Since $(N(a), *)$ is a sub-group, $o(N(a) \mid o(G) = p^n \Rightarrow o(N(a)) = p^{n_a}$, where $1 \leq n_a < n$. Since $p \mid p^n$ and

$$p \mid \frac{p^n}{p^{n_a}}, \text{ then } p \mid \left(p^n - \frac{p^n}{p^{n_a}}\right) \Rightarrow p \mid \left(p^n - \Sigma \frac{p^n}{p^{n_a}}\right)$$

$\Rightarrow p \mid z \Rightarrow Z(G) \neq \{e\}$.

Since $e \in Z(G)$, $z \neq 1$; Thus z is a positive integer divisible by the prime p. So $z > 1$.

Note: The theorem states that a group of prime power must always have a nontrivial centre.

Corollary: If $o(G) = p^2$, where p is prime, then $(G, *)$ is abelian.

Proof: We know that a group $(G, *)$ is abelian if and only if $Z(G) = G$. Now we show that $o(G) = p^2 \Rightarrow Z(G) = G$.

Let $o(G) = p^2$. Since $(Z(G), *)$ is a sub-group, then by Lagrange's theorem $o(Z(G)) \mid o(G) = p^2$. Therefore $o(Z(G)) = 1, p$ or p^2. By the theorem 4.3.2 $o(Z(G)) \neq 1$, so $o(Z(G)) = p$ or p^2. If $o(Z(G)) = p^2$, then $Z(G) = G$. Hence $(G, *)$ is abelian.

Now we suppose that $o(Z(G)) = p$.

Let $a \in G$, $a \notin Z(G)$ then by Lemma 4.2.4

$Z(G) \subset N(a) \Rightarrow p = o(Z(G)) < o(N(a))$. But by Lagrange's Theorem $o(N(a)) \mid o(G) = p^2 = o(N(a)) = p^2$ since $o(N(a)) > p$.

$o(N(a)) = p^2 \Rightarrow N(a) = G \Rightarrow a \in Z(G)$ by lemma 4.2.3. Thus we arrive at contradiction that $a \notin Z(G)$. Hence $o(Z(G)) \neq p$.

Therefore $o(Z(G)) = p^2$ and hence $Z(G) = G$.

This completes the proof of the theorem.

Partition of a Natural Number

Definition 4.3.3: For a given integer n the sequence of natural numbers $n_1, n_2, \ldots n_r$, $n_1 \leq n_2 \leq \ldots \leq n_r$ is called a partition of n if $n = n_1 + n_2 + \ldots + n_r$. The numbers of partitions of n will be denoted by $P(n)$.

Example 4.3.2: (*i*) Since $2 = 2$, $2 = 1 + 1$, then $P(2) = 2$

(*ii*) Since $3 = 3$, $3 = 1 + 2$, $3 = 1 + 1 + 1$, then $P(3) = 3$

(*iii*) Since $4 = 4$, $4 = 1 + 3$, $4 = 1 + 1 + 2$, $4 = 1 + 1 + 1 + 1$, $4 = 2 + 2$, then $P(4) = 5$

Similarly, we can determine $P(5) = 7$, $P(6) = 11$.

Theorem 4.3.3: *The number of conjugate classes in S_n is $P(n)$, the number of partitions of n.*

Proof: We shall see that corresponding to one partition of n there exists a conjugate class in S_n.

Let $\sigma \in S_n$. Since σ can be written as the product of disjoint cycles we obtain a partition of n. For if the cycles appearing have lengths $n_1, n_2 \ldots n_r$, respectively, $n_1 \leq n_2 \leq \ldots \leq n_r$, then $n = n_1 + n_2 + \ldots + n_r$.

Then the partition $\{n_1, n_2, \ldots n_r]$ is called cycle decomposition of $\sigma \in S_n$ if σ can be written as the product of disjoint cycles of lengths $n_1, n_2, \ldots, n_r$, $n_1 \leq n_2 \leq n_3 \leq \ldots \leq n_r$.

Now we prove that two permutations of S_n are conjugate if and only if they have the same cycle decomposition, that is, the corresponding to the one partition of n there exist some permutations having the same cycle decompositioin which form together the conjugate class in S_n.

Thus there will be $P(n)$ conjugate classes in S_n. For this, suppose $\sigma \in S_n$, $\theta \in S_n$ and that $\sigma(i) = j$, $\theta(i) = s$, $\theta(j) = t$, then $(\theta \circ \sigma \circ \theta^{-1})(s) = (\theta \circ \sigma)(\theta^{-1}(s)) = (\theta \circ \sigma)(i) = \theta(\sigma(i)) = \theta$

CHAPTER 4

$(j) = t$. In other words, to compute $\theta \circ \sigma \circ \theta^{-1}$ replace every symbol in σ by its image under θ. For example, to determine $\theta \circ \sigma \circ \theta^{-1}$ where $\theta = (1, 2, 3)(4,7)$ and $\sigma = (5, 6, 7)(3, 4, 2)$, then, since $\theta(5) = 5$, $\theta(6) = 6$, $\theta(7) = 4$, $\theta(3) = 1$, $\theta(4) = 7$, $\theta(2) = 3$, $\theta \circ \sigma \circ \theta^{-1}$ is obtained from σ by replacing in σ, 5 by 5, 6 by 6, 7 by 4, 3 by 1, 4 by 7, and 2 by 3, so that $\theta \circ \sigma \circ \theta^{-1} = (5, 6, 4)(1, 7, 3)$. We see that σ and its conjugate $\theta \circ \sigma \circ \theta^{-1}$ have the same cycle decomposition {1, 3, 3}.

It is clear that two permutations having the same cycle decomposition are conjugate. For if $\sigma = (a_1, a_2, ..., a_{n_1})(b_1, b_2, ..., b_{n_2}) ... (x_1, x_2, ..., x_{n_r})$ and

$$P = (\alpha_1, ...\alpha_2, ..., \alpha_{n_1})(\beta_1, ...\beta_2, ..., \beta_{n_2}) ... (\delta_1, ...\delta_2, ..., \delta_{n_r}), \text{ then}$$

$p = \theta \circ \sigma \circ \theta^{-1}$, where we can use

$$\theta = \begin{pmatrix} a_1 & a_2 .... a_{n_1} & b_1 ..... b_{n_2} ..... x_1 ..... x_{n_r} \\ \alpha_1 & \alpha_2 .... \alpha_{n_1} & \beta_1 ..... \beta_{n_2} ..... \delta_1 ..... \delta_{n_r} \end{pmatrix}$$

Thus there is 1–1 correspondence between the conjugate classes of S_n and partitions $P(n)$ of n.

Example 4.3.3: In S_9, $\sigma = \begin{pmatrix} 1 & 2 & 3 & 4 & 5 & 6 & 7 & 8 & 9 \\ 1 & 3 & 2 & 5 & 6 & 4 & 7 & 9 & 8 \end{pmatrix}$

$= (1)(2, 3)(4, 5, 6)(7)(8, 9)$

has the cycle decomposition {1, 1, 2, 2, 3}. Here $1 + 1 + 2 + 2 + 3 = 9$.

4.4 SYLOW THEOREM

By Lagrange's theorem we know that the order of a finite sub-group of a finite group divides the order of the group, but conversely, if $m \mid n$, we cannot be certain that a group of order n will possess at least one sub-group of order m. For example, the group (A_n, o) of order 12 does not possess any sub-group of order 6 in spite of $6 \mid 12$. On the other hand, in case of finite cyclic groups such sub-groups always do exist. In general, the problem of finding sub-groups of prescribed order of a finite group is difficult. So in this section we shall study that under what conditions a finite group will possess a sub-group of prescribed order. Before doing the theorems due to Norwegian mathematician Sylow we prove Theorem due to Cauchy.

Theorem 4.4.1: *(Cauchy's theorem for abelian groups): If $(G, *)$ is a finite abelian group whose order is divisible by a prime p, then there exists an element $a \neq e \in G$ such that $a^p = e$.*

Proof: We prove the theorem by mathematical induction over $o(G)$. That is, we prove that the result holds for $o(G) = 1$. Then we assume the result is true for all abelian groups of order less than $o(G)$. Assuming these two statements true we shall prove that the result is true for $(G, *)$.

Let $o(G) = 1$. Then there is nothing to prove.

If $o(G)$ is a prime number, then the group $(G, *)$ will not possess any proper sub-group except trivial sub-groups $(\{e\}, *)$ and $(G, *)$, and the group $(G,*)$ will be a cyclic group of prime order. Since $p \mid o(G)$ and $o(G)$ is prime, $o(G) = p$. So every element of G other than identity has order p, *i.e.*, $a \neq e \in G$ satisfies $a^p = e$.

Since $(G, *)$ is a finite group, the element $a \in G$ has a finite order n, *i.e.*, $a^n = e$. By Lagrange's theorem there exists a cyclic group $(N\ *)$ of $(G, *)$ generated by a whose order is equal to the order of a, that is, $o(N) = n$. Since $(N, *)$ is a sub-group of $(G, *)$, $o(N) < o(G)$ and $(N, *)$ is abelian.

Now either $p \mid n$ or p does not divide n In *case* $p \mid n = o(N)$, then there exists $b \neq e \in N$ such that $b^p = e$. Since $b \in N \subset G$, $b \neq e \in G$ satisfying $b^p = e$.

Suppose p does not divide n. Since $(G, *)$ is abelian and cyclic sub-group $(N, *)$ is abelian, $(G/N, *)$ is also an abelian group. By Lagrange's theorem, we have

$$o(G) = o\left(\frac{G}{N}\right).\, o(N)$$

Since $p \mid o(G)$, and $p \nmid o(N)$, $p \mid o(G/N)$ which implies $o\left(\frac{G}{N}\right) < o(G)$. Again by induction hypothesis, there exists an element $N * b \in G/N$ such that $(N * b)^p, = N$, where N is the identity of the group $(G/N, *)$. and $b \notin n$

$(N * b)^p = N * b^p = N \Rightarrow b^p \in N$, $b \in N$. Again using Lagrange's theorem we have $e = (b^p)^{o(N)} = (b^p)^n = (b^n)^p$.

Thus $b^n \in G$ such that $(b^n)^p = e$. We also show that $b^n \neq e$. For if $b^n = e$, then $(N * b)^n = N * b^n = N * e = N$ and we also have the $(N * b)^p = N$.

From $(N * b)^n = N$ and $(N * b)^p = N$. We have $p \mid n$ which is a contradiction to the assumption $p \nmid n$. Thus the element $b^n \in G$ satisfies $(b^n)^p = e$.

Corollary: Let $(G, *)$ be a finite abelian group and p a prime dividing $o(G)$. Then $(G, *)$ has, a sub-group of order p.

Theorem 4.4.2: *If $(G, *)$ is a finite abelian group whose order is divisible by a positive integer k, then there exists a sub-group of $(G, *)$ of order k.*

Proof: We prove the theorem by induction over $o(G)$. Let $o(G) = 1$. Since $k \mid o(G)$, then $k = 1$. Thus $(G, *)$ is a sub-group of itself. Now we assume that the result is true for abelian groups of order less than $o(G)$. Let $o(G) > 1$. Let us also suppose $k > 1$. For if $k = 1$, $(\{e\}, *)$ is the sub-group of order 1. Let p be a prime number such that $p| k$, $p| k$, $k|o(G) \Rightarrow p|o(G) \Rightarrow a \in G$ such that $a^p = e$ by Cauchy's theorem. Since $o(G)$ is finite, $\exists$ a cyclic sub-group $(H, *)$ generated by a, of order p *i.e.*, $H = (a)$. Since $(G, *)$ is abelian, then $(G/H, *)$ is abelian. By Lagrange's theorem we have $o\left(\frac{G}{H}\right).\ o(H) = o(G) \Rightarrow o\left(\frac{G}{H}\right) < o(G)$. Since

CHAPTER 4

$p \mid k$, $\exists\, k_1 \in Z$ such that $k = k_1 p$. $k \mid o\,(G) \Rightarrow k_1 p \mid o\,(G) \Rightarrow k_1 p \mid o \left(\frac{G}{H}\right) . \, p \Rightarrow k_1 \mid o \left(\frac{G}{H}\right)$ where $o(H) = p$.

So by induction there exists a sub-group $\left(\frac{K}{H}, *\right)$ of $\left(\frac{G}{H}, *\right)$ of order k_1 where $(K, *)$ is a sub-group of $(G, *)$ containing H. Then again by Lagrange's theorem we have $o(K) = o(H)$. $o\left(\frac{K}{H}\right) = p.\, k_1$. Hence $(K, *)$ is a required sub-group of $(G, *)$ of order k.

Hence the theorem is proved.

Theorem 4.4.3: *(Sylow's First Theorem): Let $(G, *)$ be a finite group of order $n = p^k q$ $(k \geq 1)$ where p is prime and $(p, q) = 1$. Then for each i $(1 \leq i \leq k)$, $(G, *)$ possesses at least one sub-group of order p^i.*

Proof: If $o\,(G) = p$, then $(G, *)$ is a cyclic group of prime order p.

Thus theorem is trivially true.

We shall prove the theorem by induction on $o\,(G)$. For $o\,(G) = 1$, the theorem is trivial to apply induction. We assume $n > 1$ and that the theorem is true for groups of order $< o\,(G)$.

Let $Z\,(G)$ be the centre of G. We have two possibilities.

(*i*) $p \mid o\,(Z(G))$ or (*ii*) $p \nmid o\,(Z\,(G))$.

(*i*) Let $p \mid o(Z\,(G))$. $(Z(G), *)$ is an abelian group, by Cauchy theorem, $a \in Z\,(G)$ such that $a^p = e$. So there exists a cyclic sub-group $(N. *)$ of $(Z\,(G), *)$ generated by a such that $o\,(N) = p$. It $k = 1$, $(N, *)$ is the desired sub-group of order p. Since $(N, *)$ is abelian, $(G/N, *)$ is abelian group. By Lagrange's theorem $o\,(G) = o(N).\, o\left(\frac{G}{N}\right) \Rightarrow p^k q = po\,(G/N) \Rightarrow o\left(\frac{G}{N}\right) = p^{k-1} q$.

Hence by induction hypothesis $\left(\frac{G}{N}, *\right)$ has a sub-group $(\overline{H}_i, *)$ of order p^i, $i = 1, 2, 3, \ldots k - 1$.

Now for each i, there exists a sub-group $(H_i, *)$ of $(G, *)$ containing N such that $\overline{H}_i = \frac{H_i}{N}$.

Again by Lagrange's theorem $o\,(H_i) = o\,(N).\, o\left(\frac{H_i}{N}\right) \Rightarrow o(H_i) = p.p^i = p^{i+1}$ $(i = 1, 2, \ldots k-1)$.

Thus in this case $(G, *)$ has a sub-group $(H_i, *)$ of order p^i, $\forall\, i = 1, 2 \ldots k$.

(*ii*) Let $p \nmid o\,(Z\,(G))$. We have the class equation for G.

$o\,(G) = o\,(Z\,(G)) + \sum_{N(a) \neq G} \frac{o(G)}{o(N(a))}$, where a runs over the representative elements of conjugate classes having more than one element.

Since $p \nmid o(Z(G))$, *then*, $p \mid \dfrac{o(G)}{o(N(a))}$ and $p \mid \Sigma \dfrac{(o(G))}{o(N(a))}$

$$\Rightarrow p \mid \left[o(G) - \Sigma \frac{(o(G))}{o(N(a))} \right]$$

$\Rightarrow p \mid o(Z(G))$ which is a contradiction to our assumption $p \nmid o(Z(G))$. Hence there exists an element $a \in G$ such that $N(a) \neq G$ and $p \nmid \dfrac{(o(G))}{o(N(a))}$

By Lagrange's theorem $o(G) = o\left(\dfrac{G}{N(a)}\right) \cdot o(N(a))$. $p \nmid \dfrac{(o(G))}{o(N(a))}$ and $p^k \mid o(G) \Rightarrow p^k \mid o(N(a))$. But $o(N(a)) < o(G)$. By induction assumption $(N(a), *)$ has sub-group $(H_i, *)$ of order p^i $(1 \leq i \leq k)$, which are also sub-groups of $(G, *)$. This completes the theorem.

Corollary 1: If $p^k \mid o(G)$, $p^{k+1} \nmid o(G)$, then $(G, *)$ has a sub-group of order p^k.

Corollary 2: (Cauchy's theorem of finite groups) If a prime p divides $o(G)$, then $\exists$ at least an element $a \in G$ such that $a^p = e$.

Proof: By Sylow's first theorem $(G, *)$ has a sub-group $(H, *)$ of order p. Since p is prime, the $(H, *)$ is a cyclic group generated by some $a \in H$, $H = (a)$ which shows $o(H) = p = o(a)$.

We shall continue our discussion by introducing new class of groups.

Definition 4.4.1: A group $(G, *)$ is said to be a p-group if the order of each element of G is a power of a fixed prime p, that is, p-group is a finite group whose order is a power of a prime p. (*i.e.*, p^n for some integer $n \geq 0$).

Let $(H, *)$ be a sub-group of finite group $(G, *)$. Then we call $(H, *)$ a p-group if $(H, *)$ is a p-group. That is, the order of every element of H is a power of a fixed prime p.

Example 4.4.1: Let $(G, *)$ be a cyclic group of order 8 generated by a. Since $o(G) = 8 = 2^3$, a power of prime 2, $(G, *)$ is a p-group.

Example 4.4.2: Let $(G, *)$ be a abelian group and the set H consists of those elements whose order are powers of a fixed prime p, (quite possibility $H = \{e\}$) then $(H, *)$ is a p-sub-group of group $(G, *)$.

Lemma.4.4.1 *$(G, *)$ be a finite p-group if and only if $o(G) = p^k$, for some $k > 0$.*

Proof: Suppose $(G, *)$ is a p-group, but $q \mid o(G)$ for some prime $q \neq p$. Then by Cauchy's theorem there exists an element of G of order q, contradicting the fact $(G, *)$ is a p-group (in *p-group* the order of each element is a power of prime p.). Thus, p is only the prime divisor of $o(G)$ which implies $o(G) = p^k$, $(k > 0)$.

Conversely, if $o(G) = p^k$, then from Lagrange's theorem that the order of every element of $a \in G$ will divide the order of $o(G)$, that is, $o(a) \mid o(G) = p^k$ which implies $o(a)$ is a power of p. Hence $(G, *)$ is a p-group.

CHAPTER 4

Definition 4.4.2: Let $(G, *)$ be a finite group and p a prime. If $p^k \mid (o(G)$ and $p^{k+1} \dagger o(G)$, $(k > 0)$, then a sub-group $(P, *)$ of group $(G, *)$ of order p^k is said to be a Sylow p-sub-group.

Example: Let (S_3, o) be a symmetric group, $o(S)_3 = 6 = 3.2$. Thus $3 \mid o(S_3)$ but $3^2 \mid o(S_3)$. By definition any Sylow p-sub-group (P, o) of (G, o) is of order 3. Now (P, o), where $P = \{I, (1\ 2\ 3), (1\ 3\ 2)\}$, is the only sub-group of order 3. Hence (P, o) is only Sylow p-sub-group of $(S_3\ o)$. Again $2 \mid o(S_3)$ and $2^2 \dagger o(S_3)$. Thus any Sylow p-sub-group of (S_3, o) is of order 2. Now the only sub-group of (S_3, o) of order 2 are (H_1, o), (H_2, o), and $(H_3\ o)$, where $H_1 = \{1, (12)\}$, $H_2 = \{1, (13)\}$, and $H_3 = \{1, (23)\}$. So these three sub-group are only Sylow 2-sub-groups of (S_3, o). Thus we can conclude that a Sylow p-sub-group of a given group need not be unique.

Example 4.4.3: Show that $(H, *)$ is a normal sub-group of $(N(H), *)$.

Solution: By the definition, $N(H) = \{x \in G \mid x * H * x^{-1} = H\}$. Therefore, if $a \in H$, $\forall$ $x \in N(H)$, $x * a * x^{-1} \in H$, which implies $(H, *)$ is a normal sub-group of $(N(H), *)$. $(N(H), *)$ is the largest sub-group of $(G, *)$ in which $(H, *)$ is normal.

Example 4.4.4: Show that any conjugate of a p-group is again a p-group.

Solution: Let $(x. * P * x^{-1})$ be a conjugate of p-sub-group $(P, *)$. We know that $o(a) = o(x * a * x^{-1})$, that is, the order of an element is equal to the order of its conjugate. Since in a p-group the order of every element is a power of p, p is prime, then the order of every elements of the form $x * a * x^{-1}$, where a is in p-group, is also a power of p. Since $x * a * x^{-1} \in x * P * x^{-1}$, $a \in P$, and $(P, *)$ is a p-group, then the conjugate $(x * P * x^{-1}, *)$ is also a p-group

Example 4.4.5: If $(S, *)$ is a Sylow p-sub-group of $(G, *)$ and $a \in G$, then show that conjugate sub-group $(a * S * a^{-1}, *)$ of $(S, *)$ is also a Sylow p-sub-group of $(G, *)$.

Solution: We have to show that $(a * S * a^{-1}, *)$ is Sylow p-sub-group if $(S, *)$ is a Sylow p-sub-group. Let $(S, *)$ be a Sylow p-sub-group of (G, o) of order p^m. Then $o(G) = p^m n$ with $(p, n) = 1$.

Since there exists a one-to-one correspondence between $(S. *)$ and $(a * S * a^{-1} *)$, $o(S) = o(a * S * a^{-1})$ Since the $o(S) = p^m$, $o(a * S * a^{-1}) = p^m$.

Hence $(a * S * a^{-1}, *)$ is a Sylow p-sub-group.

Example 4.4.6: If a group $(G, *)$ has a unique Sylow p-sub-group $(S, *)$, then $(S, *)$ is normal in G.

Solution: Since the conjugate sub-group $(a * S * a^{-1}, *)$ of a Sylow p-sub-group is a Sylow p-sub-group and it is unique, $a * S * a^{-1} = S$, $\forall$ $a \in G$ which shows $(S, *)$ is a normal sub-group in $(G, *)$.

Example 4.4.7: Let p be a prime number. Show that any group $(G, *)$, with $o(G) = 2p$, has a normal sub-group of order p.

Solution: Since $p \mid o(G) = 2p$, by Cauchy's theorem $\exists$ $a \in G$ such that $a^p = e$ which follows that there is a cyclic sub-group $(H, *)$ of $(G, *)$ generated by a of order p. *i.e.*, $H = (a)$ and $o(H) = p$, consequently it is abelian. Now by Lagrange's theorem index of $H =$

$$\frac{o(G)}{o(H)} = \frac{2p}{p} = 2.$$

Hence (*H*, *) is normal, since every sub-group of index 2 in *G* is normal in *G*.

We shall make use of the following Lemmas in proving the Sylow's theorem.

Lemma 4.4.2: *Let (H, *) be a normal sub-group of (G, *). If (H, *) and (G/H, *) are both p-groups, then (G, o) itself is a p-group.*

Proof: Since $\left(\frac{G}{H}, *\right)$ is *p*-group, $a * H \in G/H$, for any $a \in G$, of order of some power of *p*, say p^k.

Then $(a * H)^{p^k} = a^{p^k} * H = e * H \Rightarrow a^{p^k} \in H.$

Since (*H*. *) is p-group, the element a^{p^k} has also *p*-power order, say p^i. Thus $\left(a^{p^k}\right)^{p^j}$ $= a^{p^{k+j}} = e \Rightarrow o(a) = p^{k+j}$. But, since *p* is prime, it follows that *o* (*a*) is a power of *p*. This completes the Lemma.

Lemma 4.4.3: *Let (P, *) be a Sylow p-sub-group of (G. *) and* $a \in G$ *be any element whose order is a power of p. If* $a * P * a^{-1} = P$, *then* $a \in P$.

Proof: The condition $a * P * a^{-1} = P \Rightarrow a \in N(P)$. We have to prove that *N* (*P*)–*P* does not contain any element of *p*-power order. Suppose an element *a* of *p*-power order exists in *N* (*P*)–*P*.

By example 4.4.3, (*P*, *) is a normal sub-group in (*N*(*P*), *), So the quotient group (*N* (*P*)/ *P*, *) exists. Let $a * P \in N(P)/P$.

The order of $a * P$ divides the order of *a* (the representative of the coset *a***P*}. So *o* (*a***P*) is a power of *p* because *a* is of *p*-power order. This implies that the cyclic sub-group ((*a* * *P*), *) of (*N*(*P*)/*P*, *) has *p*-power order and consequently ((*a* * *P*). *) is a *p*-group. By correspondence theorem, (*N*(*P*), *) thus has a sub-group with $K \supseteq P$ and $K/P = (a * P)$. As $a \in P$, $P \subset K$, by the lemma 4.4.2, (*K*, *) must be a *p*-group, since both (*P* , *) and ((*a* * *P*), *) are *p*-groups. But this clearly contradicts the fact that (*P*, *) is a maximal *p*-sub-group. Accordingly, there can be no element of *N*(*P*)–*P* whose order is a power of *p*.

Theorem 4.4.4: *(Sylow-second theorem): Let (G, *) be a finite group and p a prime. Then the number of distinct Sylow p-sub-groups of (G, *) is congruent to 1 modulo p (i.e., 1 +mp) and is advisor of o(G) (i.e., (1 + mp) | o (G)).*

Proof: Let (*P*, *) be a fixed Sylow *p*-sub-group of (*G*, *) By example 4.4.4, the conjugate sub-group ($a * P * a^{-1}$, *) of (*P*, *) is also a sylow *p*-sub-group. So we have to count the number of distinct conjugates of (*P*, *). If (*P*, *) is a normal sub-group, *i.e.*, $a * P * a^{-1} = P$, then this number is 1. If (*P*, *) is not a normal sub-group, (*P*, *) has conjugate sub-group (P_1, *) with $P_1 \neq P$.

CHAPTER 4

Since any conjugate of $(P_1, *)$ is also the conjugate of $(P, *)$, we first consider conjugates of $(P_1, *)$ induced by elements from P. By the lemma 4.4.3

$$a * P_1 * a^{-1} = P_1 \Rightarrow a \in P_1,$$

and the elements $a \in P$ under which P_1 is invariant are of $P_1 \cap P$.

$$N_p\ (P_1) = \{a \in P \mid a * P_1 * a^{-1} = P_1\} = P_1 \cap P.$$

Since $(P_1 \cap P, *)$ is a p-sub-group of $(G, *)$, it follows that $o\,(P_1 \cap P)$ is a power *of p*. Now, from theorem 4.2.4, the number of distinct conjugates of $(P_1\ *)$ by elements of P is equal to

$$\frac{o(P)}{o(N_p(P_1))} = [P{:}N_p\ (P_1)] = o\,(P)/\ o\ (P_1 \cap P)$$

This shows that $o\ (p) |\ o\ (N_p\ (P_1))$ must be some power of p, say p^{k_1}. Furthermore $k_1 > 0$, for if $k_1 = 0$, then $o\,(P) = o\,(P_1 \cap P)$, so $P = P_1 \cap P \Rightarrow P \subseteq P_1$. But the fact that $(P, *)$ is a maximal p-sub-group would imply $P = P_1$, which is impossible. We also observe that $(P, *)$ will not be conjugate sub-group of $(P_1, *)$ by elements of P; If $P = a * P * a^{-1}$, for some $a \in P$, then $P_1 = a^{-1} * P * a \subseteq P$, again leading to a contradiction.

Now, if the conjugate sub-groups of $(P, *)$ are not exhausted by $(P, *)$ and p^{k_1} conjugates of $(P_1, *)$ induced by members of P, we choose another conjugate $(P_2, *)$ distinct from any yet considered. As above we can obtain the p^{k_2} , $k_2 > 0$ distinct conjugates of $(P_2, *)$ induced by P. No conjugate of $(P_2, *)$ by an element of P will also be an conjugate of $(P_1, *)$ by an element of P, for if

$$a * P_1 * a^{-1} = b * P_2 * b^{-1},\ a,\ b \in P,$$

then $(b^{-1} * a) * P_1 * (b^{-1} * a)^{-1} = P_2$, with $b^{-1} * a \in P$, contrary to the choice of P_2. If necessary, we repeat this argument once more. Since $(G, *)$ is a finite group, all the conjugates of $(P, *)$ are found after a finite number of steps, say n steps. The conjugate sub-groups of $(P, *)$ are these : $(P, *)$ itself, the p^{k_1} conjugates of $(P_1, *)$ by elements of P, the p^{k_2} conjugates of $(P_1, *)$ by elements P, etc. The total number of distinct conjugates is therefore

$$1 + p^{k_1} + p^{k_2} + ... + p^{k_n} = 1 + mp = 1(mod p), k_i > 0.$$

Finally, the number of distinct conjugates of the Sylow p-sub-group $(P, *)$ is, by theorem 4.2.4 $[G : N_a\ (P)]$ divides $o\,(G)$. This proves the theorem.

This theorem also proves.

Corollary 1: Any two Sylow p-sub-groups of $(G, *)$ corresponding to the same prime p are conjugate, hence isomorphic.

Corollary 2: The number of distinct Sylow p-sub-groups is $[G{:}\,N\,(P)]$, where $(P, *)$ is any particular Sylow p-sub-group of $(G, *)$.

PROBLEMS

1. Let $(H, *)$ and $(K, *)$ be two sub-groups of the group $(G, *)$. Show that $(H, *)$ is normal in $(K, *)$ if and only if $H \subseteq K \subseteq N(H)$.
2. For a finite group $(G, *)$, prove that
   (a) if $(G, *)$ has exactly two conjugacy classes, then $o(G) = 2$.
   (b) If $\exists\ a \in G$ with exactly two conjugates, $(G, *)$ contains a non-trivial normal sub-group.
3. Prove that a sub-group $(H, *)$ of $(G, *)$ is normal if and only if the set H is the union of conjugate classes of $(G, *)$.
4. Let $(G, *)$ be any group, a fixed element $a \in G$. Define the function f by $f(x * C(a)) = x * a * x^{-1}$, $\forall\ a \in G$. Prove that f is a one to one mapping from left cosets of $C(a)$ onto the set of distinct conjugates of a.
5. For a finite p-group $(G, *)$, prove the following:
   (a) Any homomorphic image of $(G, *)$ is again a p-group
   (b) $(G, *)$ has a normal sub-group of order p).

   [Hint: $Z(G)$ contains an element of order p.]
6. Prove that if $o(G) = p^n$, then the group $(G, *)$ has at least one normal sub-group of order p^k for all $0 \le k \le n$.
7. Let $(H, *)$ be a normal sub-group of a finite p-group $(G, *)$ and suppose that $o(H) = p$. Show that $H \subseteq Z(G)$.
8. Determine all Sylow p-sub-group of the group $(Z_{24}\ +_{24})$.
9. Let $(P, *)$ be a Sylow p-sub-group *of* $(G, *)$ of the finite order. Prove that if $(H, *)$ is any sub-group of $(G, *)$ with $P \subseteq N(P) \subseteq H$, then $N(H) = H$. In particular deduce that $N(N(P)) = N(P)$.
10. Prove that if $((H, *)$ is a p-sub-group of $(G, *)$ which is contained in exactly one Sylow p-sub-group $(P, *)$ of $(G, *)$, ihen $N(H) \subseteq N(P)$.
11. Let $(H, *)$ be a normal sub-group of the finite group $(G, *)$. Show that $(H, *)$ is a sub-group of each Sylow p-sub-group (for the same prime p) of $(G, *)$.
12. Prove that there are no simple groups of order 20, 30, or 56.
13. Assume $(G, *)$ is a group with $o(G) = p^n \cdot q$, where $(p, q) = 1$. Prove that a sub-group $(H, *)$ of $G, *)$ is a Sylow p-sub-group if and only if $o(H) = p^n$.
14. Let $(G, *)$ be a group of order $p\ q$, where $p < q$ are both primes.
    (a) Show that $(G, *)$ has precisely one normal sub-group of order q.
    (b) If $q \not\equiv 1 \pmod{p}$, prove that $(G, *)$ is cyclic.
15. Using Sylow's theorem or otherwise prove that every group of order 15 is cyclic.

    [Hint: Here $o(G) = 15 = 3.5$ Set $p = 3$, $q = 5 \not\equiv 1 \pmod{3}$. So by Q 14 (b) (G, o) is cyclic].

CHAPTER 4

16. $(G, *)$ is an abelian group of order p^n, p is a prime and n integer ≥ 1. If the number of elements of order p is strictly less then p, show that (G, o) must be cyclic.

17. Suppose (G, o) is a group of order p^2q, where p and q are primes, Prove that (G, o) is not simple.

4.5 FINITE ABELIAN GROUPS

Here we shall confine our study to the description of the structure of an arbitrary finite abelian group. The result which are shall obtain is a famous classical theorem, often referred to as the fundamental theorem on finite abelian groups. We now state this very fundament result. To prove it we use some results.

Lemma 4.5.1: *Let a, b be two elements of finite order, of a group $(G, *)$ such that $o(a)$ and $o(b)$ are co-prime and $a * b = b * a$, then $o(a * b) = o(a) \,.\, o(b)$.*

Proof: Let $o(a) = m$, $o(b) = n$, then $a^m = e$, $b^n = e$. Let

$H = (a * b)$ be the sub-group of generated by $a * b$.

Since $(a * b)^{mn} = a^{mn} * b^{mn} = (a^m)^n * (b^n)^m = e^n * e^m = e$.

We get $o(H) = o(a * b) \mid mn$.

Now $(a * b)^m = a^m * b^m = e * b^m = b^m \in H$

Since $a^m = e$. As $(m, n) = 1$, and $o(b) = n$, we have $o(b^m) = o(b) = n$.

We see that

$$b^m \in H \Rightarrow o(b^m) \mid o(H)$$
$$\Rightarrow n \mid o(H).$$

Similarly, we can show that $m \mid o(H)$.

Thus we have $mn \mid o(H)$. Hence $o(H) = mn = o(a)\, o(b)$.

This proves the lemma.

Definition 4.5.1: (Exponent of a group) Let $(G, *)$ be a finite group. If there exists an element $a \in G$ such that $o(a) = k$ and no element of G has order exceeding k, then k is called the exponent of G.

Lemma 4.5.2: *The order of any element of a finite abelian group $(G, *)$ divides the exponent of G.*

Proof: Let k be the exponent of a finite abelian group $(G, *)$, there exists an element $a \in G$ such that $o(a) = k$, and for every $b \in G$, $o(b) \leq o(a) = k$. Let b be an element such that $o(b) \dagger k$. This implies that there exists a prime number p such that for some positive integer t, $p^t \mid o(b)$ but $p^t \dagger o(a)$. This means that we can write $o(b) = p^\alpha q$ and $o(a) = p^\beta s$ such that $(p, q) = 1$, $(p, s) = 1$ and $\beta < \alpha$. Let $b_o = b^q$, $a_o = a^{p^\beta}$. Then $(b_o)^{p^\alpha} = (b^q)^{p^\alpha} = b^{p^\alpha . q} = e \Rightarrow$ $o(b_o) = p^\alpha$ and similarly, $o(a_o) = s$. By lemma 4.5.1 we have

$o(a_o\, b_o) = o(a_o) \,.\, o(b_o) = p^\alpha s > p^\beta s = o(a)$. This contradicts the assumption that $k = p^\beta s$ is the exponent of G. Hence $o(b) \mid k$. This proves the lemma.

Example 4.5.1: The symmetric group (S_3, o) is not commutative. The exponent of S_3 is 3. However (1.2) is an element of S_3 of order 2. *i.e.*, $\begin{pmatrix} 1\,2\,3 \\ 2\,1\,3 \end{pmatrix} \in S_3$ and

$$(12)^2 = \begin{pmatrix} 1 & 2 & 3 \\ 2 & 1 & 3 \end{pmatrix} o \begin{pmatrix} 1 & 2 & 3 \\ 2 & 1 & 3 \end{pmatrix} = \begin{pmatrix} 1 & 2 & 3 \\ 1 & 2 & 3 \end{pmatrix} = \text{I. But } o\ (12) \dagger 3.$$

This shows that above lemma does not hold good for non-abelian groups.

Theorem 4.5.1: *(Fundamental Theorem for Abelian Groups) Every finite Abelian group of order n is the direct product of cyclic groups i.e.,*

$G = G_1 \times G_1 \times G_2 \times \ldots \times G_t$,

where each G_i is a cyclic group of order n_i such that $n_{i+1} | n_i$ and the integers n_i are uniquely determined. Further $n = n_1 n_2 \ldots n_t$, and $o(G) > 1$ and $o(G_i) > 1$.

Proof: We prove the result by applying induction on $o(G)$. If $o(G) = 1$, the result holds trivially.

Suppose $o(G) = n > 1$, and the results holds for abelian groups whose order is less than n.

Let n_1 be the exponent of G and $g, \in G$ be an element of order n_1.

Let $G_1 = (g_1)$. If $G = G_1$, then G is itself is cyclic:

Let $G \neq G_1$. Then $1 < o(G/G_1) < n$. Then by induction

$$G/G_1 = \bar{H}_2 \times \bar{H}_3 \times \ldots\ldots \times \bar{H}_t,$$

where each $\bar{H}_i$ is a cyclic sub-group of $(G/G_1\ *)$, of order $n_i > 1$ such that $n_{i+1} | n_i$ for all

$$i = 2, 3, \ldots, t-1, \text{ and } \frac{n}{n_1} = o(\bar{G}) = n_2, n_3 \ldots n_t \qquad \ldots(1)$$

Now each $\bar{H}_i = H_i // G_1$ for some sub-group H_i of G containing G_1. Consider $h_i \in H_i$ such that $\bar{h}_i = h_i * G_1$ is a generator of $\bar{H}_i$. Then $h_i^{-n_i} = \bar{e} = \bar{G}_1 \Rightarrow h_i^{n_i} \in G_1 = (g_1)$

$\Rightarrow h_i^{n_i} = g_1^{m_i}$ for some m_i such that $1 \leq m_i \leq n_i$.

Let $\alpha_i = (m_i, n_i)$. Then $m_i = \alpha_i \beta_i$ for some $\beta_i \geq 1$,

$n_1 = \alpha_i \gamma_i$ for some $\gamma_i \geq 1$ and $(\beta_i, \gamma_i) = 1$.

Now $o(g_1) = n_i = \alpha_i \gamma_i \Rightarrow o\left(g_1^{\alpha_i}\right) = \gamma_i$

Since $(\gamma_i, \beta_i) = 1$, $o\left(g_1^{\alpha_i\beta_i}\right) = o\left(g_1^{\alpha_i}\right)$. However

$h_i^{n_i} = g_1^{\alpha_i\beta_i}$. So $o\left(h_i^{n_i}\right) = \gamma_i$. On the other hand

$$o(\bar{h}_i) \mid o(h_i) \Rightarrow n_i \mid o(h_i) \Rightarrow o(h_i^{n_i}) = \frac{o(h_i)}{n_i}$$

$$\Rightarrow o(h_i) = o(h_i^{n_i})\, n_i = \gamma_i\, n_i \qquad ...(2)$$

Since $o(h_i) \mid n_1$, exponent of G and $n_1 = \alpha_i\, \gamma_i$, we have $\gamma_i\, n_i | \alpha_i\, \gamma_i$ or $n_i | \alpha_i \Rightarrow \alpha_i = n_i\, \delta_i$ for some $\delta_i \geq 1$. Then $h_i^{n_i} = g_i^{n_i \delta_i \beta_i}$

Put $g_i = h_i (g_i)^{-\delta_i \beta_i}$ for all $i = 2, 3, ..., t-1$.

We see that $\bar{g}_i = g_i * G_i = \bar{h}_i$, $g_i^{n_i} = h_i^{n_i} (g_1)^{-n_i \delta_i \beta_i} = e$.

This yields $o(g_i) = n_i$

Define $H = (g_2)(g_3),...(g_t)$. $(H, *)$ is a sub-group of $(G, *)$ such that $o(H) \mid n_2\, n_3....n_t$.

Let $f: g \to G/G_1$ be the natural homomorphism.

Since $f(H) = (\bar{g}_2) \times (\bar{g}_3) \times .... \times (\bar{g}_t)$

$= \bar{H}_2 \times \bar{H}_3 \times ..... \times \bar{H}_t = G/G_t$

and also $f(H) = ((H * G_1)/G_1) \Rightarrow G = H * G_1 = G_1 * H$

$= (g_1)(g_2) .... (g_t)$.

This implies that $o(G) = n = n_1 n_2 ... n_{t_1}$

$= o(g_1)\, o(g_2) .... o(g_t)$.

This gives

$G = (g_1) \times (g_2) \times ..... \times (g_t)$.

Also $o(g_i) = o(g_i) = n_i$ such that $n_{i+1} \mid n_i$

Suppose $G = H_1 \times H_2 \times ..... \times H_i$ ...(3)

$= K_1 \times K_2 \times .... \times K_u$ ...(4)

be two decomposition of G into internal direct product of cyclic sub-groups, such that for $i = 1, 2, ..., t$,

$j = 1, 2, ..., u$, $o(H_i) = n_i$ and $o(K_j) = m_j$ and for $i \leq t - 1$, $n_{i+1} \mid n_i$ and for $j \leq u - 1$, $m_{j+1} \mid m_j$.

Consider $g \in G$, then (3) gives

$g = h_1 * h_2 * ..... * h_i$, $h_i \in H_i$.

Since $o(h_i) \mid o(H_i) = n_i$ and $n_i \mid n_1$, we get $h_i^{n_i} = e$.

Consequently $g^{n_1} = h_1^{n_1} * h_2^{n_1} * .... * h_i^{n_1} = e$

Thus $o(g) \le n_1$ for all $g \in G$. Further as $(H_1, *)$ is a cyclic group of order n_1, H_1 contains an element of order n_1. Hence n_1 is the exponent of G. Similarly, m_1 is the exponent of G. Thus $n_1 = m_1$. We have proved that $n_1 = m_1$, $n_2 = m_2, \ldots\ n_{i-1} = m_{i-1}$, for some i we shall prove that $n_i = m_i$.

Suppose contrary $n_i \ne m_i$ and to be definite let $n_i > m_i$. Define $K = \{x^{m_i} \mid x \in G\}$ For any $x, y \in G$,

$$x^{m_i} * y_i^{-m_i} = (x * y^{-1})^{m_i} \in K|$$

This shows that $(K, *)$ is a sub-group of $(G, *)$.

Suppose that for $k = 1, 2, \ldots, t$, $H_k = (a_k)$ and for each $j = 1, 2, \ldots, u$, $K_j = (b_j)$; Since for $j \ge i$,

$o(K_j) = m_j \mid m_i$, $b_j^{mj} = e$. Hence

$K = (b_1^{m_i}) \times (b_2^{m_i}) \times \ldots\ldots \times (b_{i-1}^{m_i})$.

Thus $o(K) = \dfrac{m_1}{m_i}\dfrac{m_2}{m_i} \ldots\ldots \dfrac{m_{i-1}}{m_i}$, since $o(b_j) = m_i$ for all i ...(5)

Also $G = (a_1) \times \ldots\ldots\ldots \times (a_t)$ from (3)

We have

$$K = (a_1^{m_i}) \times (a_2^{m_i}) \times \ldots\ldots \times (a_t^{m_i})$$

Now $$o(a_k^{m_i}) = \frac{n_k}{(m_i, n_k)} \text{ for all } k = 1, 2, \ldots\ t.$$

Consequently $$o(K) = \frac{n_1}{(m_i, n_1)} \cdot \frac{n_2}{(m_i, n_2)} \cdots \frac{n_i}{(m_i, n_i)} \cdot \frac{n_{i+1}}{(m_i, n_{i+1})} \cdots \frac{n_t}{(m_i, n_t)} \quad \ldots(6)$$

Now m_i/m_j and $m_j = n_j$ for all $j < i$ by our hypothesis.

Thus $$\frac{n_i}{(m_i, n_j)} = \frac{m_j}{(m_i, n_j)} = \frac{m_j}{n_i} \text{ for all } j < i$$

Hence from (5) and (6) we get

$$1 = \frac{n_i}{(m_i, n_i)} \cdots\cdots \frac{n_t}{(m_i, n_t)} \quad \ldots(7)$$

CHAPTER 4

However $m_i < n_i$ gives $\frac{n_i}{(m_i, n_i)} > 1$. As a consequence (>) cannot hold. This gives $m_i = n_i$. Hence by induction $m_i = n_i$, for all i. This completes the theorem.

Definition 4.5.2: If (G, o) is an abelian group of order p^n, p α prime, and $G = A_1 \times A_2 \times ...... \times A_k$, where each (A_i, o) is cyclic group of order p^{n_i} with $n_1 \geq n_2 \geq ..... \geq n_k > 0$, then the integers $n_1, n_2, ..., n_k$ are called invariants of G.

Before showing that invariants of G are unique and completely describe G, we note one thing about the invariants of G. If $G = A_1 \times A_2 \times .... \times A_k$, where A_i is cyclic of order p^{n_1}, $n_1 \geq n_2 \geq .... \geq n_k > 0$, then $o(G) = o(A_1).\ o(A_2) ....\ o(A_k)$, hence $p^n = p^{n_1} . p^{n_2} .... p^{n_k}$ which implies $n = n_1 + n_2 + ... + n_k$. In other words $n_1, n_2, ... n_k$ gives us the partition of n.

We shall make it absolutely clear that if $a_1, a_2, ..., a_k$ generate the cyclic sub-groups A_1, $A_2, ..., A_k$ respectively, are not unique. Let us see this in a very simple example. Let $G = \{e, a, a * b\}$ be a abelian groups of order 4, where $a^2 = b^2 = e$ and $a\ o\ b = b\ o\ a$. Then $G = A \times B$, where $A = (a) = \{e, a\}$ and $B = (b) = \{e, b\}$ and $A \times B = \{e, a\} \times \{e, b\} = \{e\ o\ e, e\ o\ b, aoe, aob\} = \{e, b, a, a\ o\ b\}$. A and B are cyclic groups of order 2.

But we have another decomposition of G as a direct product, namely, $G = C \times B$, where $C = (a, b)$ and $(B) = (b)$. $C \times B = \{(a, b), e\}\ o\ \{e, b\} = \{(a\ o\ b\ o\ e, (a\ o\ b)\ o\ b, e\ o\ e, e\ o\ b\}$.

$= \{a\ o\ b, a, e, b\}$ since $b^2 = e$.

Hence we see that the cyclic groups in the decomposition of the group are not unique but there orders are unique.

Definition 4.5.3: If (G, o) is abelian group and s is any integer, then $G(s) = \{x \in G | x^s = e\}$.

Theorem 4.5.2: *$(G(s), o)$ is a sub-group of (G, o) for any integer s.*

Proof: Hence $G(s) = \{x \in G | x^s = e\}$, for integer s.

Let $x, y \in G(s)$, then $x^s = e$, $y^s = e$

Now $(xoy^{-1})^s = x^s\ o\ y^{-s}$ since (G is abelian)

$= eoe = e$

Hence $xoy^{-1} \in G(S)$.

Hence $(G(s), o)$ is a sub-group.

Lemma 4.5.3: *If (G, o) and $(G', *)$ are isomorphic abelian groups, then for every integer s. $(G(s), o) \cong (G'(s), *)$*

Proof: Let $\phi : G \to G'$ be isomorphism. Here we shall prove that ϕ maps $(G(s))$ isomorphically onto $G'(s)$.

First we show that $\phi(G(s)) \subset G'(s)$

For if $x \in (G(s)$, then $\phi(x) \in \phi(G/s)$ and $x^s = e$

$\Rightarrow \qquad \phi(x^s) = \phi(e)$

$$\Rightarrow \qquad (\phi(x))^s = e'$$

$$\Rightarrow \qquad \phi(x) \in G'(s).$$

Therefore $\phi(G(s)) \subset G'(s)$

Conversely, Let $u' \in G'(s)$, then

$$(u')^s = e'.$$

But since ϕ is onto, there exists an element $y \in G$ such that $\phi(y) = u'$

Therefore $(u')^s = e'$

$$\Rightarrow \qquad (\phi(y))^s = e'$$

$$\Rightarrow \qquad (\phi(y^s)) = \phi(e)$$

$$\Rightarrow \qquad y^s = e \text{ since } \phi \text{ is one to one}$$

$$\Rightarrow \qquad y \in G|(s)$$

Thus ϕ maps $G(s)$ onto $G'(s)$.

Therefore since ϕ is one-to-one, onto, and a homorphism from $G(s)$ to $G'(s)$, we have

$$(G(s), o) \cong (G'(s), *)$$

Now we prove uniqueness of the invariants of an abelian groups of order p^n.

Theorem 4.5.3: *Two abelian groups of order p^n are isomorphic if and only if they have the same invariants.*

In other words, if (G, o) and $(G', *)$ are abelian groups of order p^n and $G = A_1 \times ...... \times A_k$, where each (A_i, o) is a cyclic sub-group of order p^{n_1}, $n_1 \geq .... \geq n_k > o$ and $G' = B' \times ...... \times B'_s$, where each B'_1 is a cyclic group of order p^{h_i}, $h_1 \geq .... \geq h_s > o$, then (G, o) and $(G', *)$ are isomorphic if and only if $k = s$ and for reach i, $n_i = h_i$.

CHAPTER 4

Proof: Let (G, o) and $(G', *)$ have the same invariants. Then we have to show that they are isomorphic. For then $G = A_1 \times .... \times A_k$, where each $A_i = (a_i)$ is a cyclic group of order p^{n_1} and $G' = B'_1 \times ...... \times B'_k$, where $B_i = (b'_1)$ is cyclic of order p^{n_1}. We define a mapping $\phi : G \to G'$ by

$$\phi\left(a_1^{\alpha_1} .... a_k^{\alpha_k}\right) = (b'_1)^{\alpha_1} ... (b'_k)^{\alpha_k}$$

For

$$x = a_1^{\alpha_1} .... a_k^{\alpha_k}, \; y = a_1^{\beta_1} .... a_k^{\beta_k}$$

$$\phi(x \,.\, y) = \phi\left(a_1^{\alpha_1} .... a_k^{\alpha_k}\right). \left(a_1^{\beta_1} .... a_k^{\beta_k}\right)$$

$$= \phi\left(a_1^{\alpha_1+\beta_1} .... a_k^{\alpha_k+\beta_k}\right)$$

$$= \left(b'_i\right)^{\alpha_1+\beta_1} .... \left(b'_k\right)^{\alpha_k+\beta_k}$$

$$= \left((b_i')^{\alpha_1} \ldots\ldots (b_k')^{\alpha_k}\right) \cdot \left((b_1')^{\beta_1} \ldots\ldots (b_k')^{\beta_k}\right)$$

$$= \phi(x) \cdot \phi(y).$$

Hence ϕ is a homomorphism

Again, Assume that

$\phi(x) = \phi(y)$

$$\Rightarrow \qquad \phi\left(a_1^{\alpha_1} \ldots\ldots a_k^{\alpha_k}\right) = \phi\left((a_1)^{\beta_1} \ldots\ldots a_k^{\beta_k}\right)$$

$$\Rightarrow \qquad (b_1')^{\alpha_1} \ldots (b_k')^{\alpha k} = (b_1')^{\beta_1} \ldots (b_k')^{\beta_k}$$

$$\Rightarrow \qquad \alpha_1 = \beta_1, \ldots\ldots, \alpha_k = \beta_k$$

$$\Rightarrow \qquad x = y.$$

$\therefore$ ϕ is one-to-one. It is clearly onto.

Hence ϕ is an isomorphism.

On the other hand, suppose that $G = A_1 \times \ldots \times A_k$, $G' = B_1' \times \ldots \times B_k'$, A_i, B_i' are cyclic groups orders p^{n_i}, p^{h_i} respectively, wheie $n_1 \geq \ldots \geq n_k > o$ and $h_1 \geq \ldots \geq h_s > o$. Here we shall show that if G and G' are isomorphic, then $k = s$ and each $n_i = h_i$.

If G and G' are isomorphic, then by,

Lemma 4.5.4: *$G(p^m)$ and $G'(p^m)$ are isomorphic for any integer $m \geq o$, hence they must have the same order. Specially, when $m = 1$, $o(G(p)) = o(G'(p))$. But by theorem 4.1.7, $o(G)^{(p)} = p^k$ and $o(G'(p)) = p^s$.*

Hence $p^k = p^s$ and so $k = s$. Hence the number of invariants for G and G' is the same.

If $n_i \neq h_i$ for some i, let t be the first i such that $n_t = h_t$; we may suppose that $n_t > h_t$.

Let $m = h_t$; we consider the sub-groups, $H = \{x^{p^m} | x \in G\}$ and $H' = \{(x')^{p^m}\} \; x' \in G'\}$ of G and G' respectively.

Since $(G, o) \cong (G', *)$ implies that

$(H, o) \cong (H', *)$, we now examine the invariants of H and H'.

Since $G = A_1 \times \ldots\ldots \times A_k$, where $A_i \; (a_i)$ is of order p^{n_1}, we get that

$$H = G_1 \times \ldots\ldots\ldots \times C_t \times \ldots\ldots \times C_r, \text{ where } C_i = \left(a_i^{p^m}\right)$$

is of order $p^{n_1 - m}$, and r is such that $n_r > m = h_r > n_{r-1}$.

Thus the invariants of H are $n_1 - m$, $n_2 -, \ldots, n_r - m$ and the number of invariants of H is $r \geq t$.

Again, since $G' = B_1' \times \ldots\ldots\ldots\ldots \times B_k'$ where $B_i = (b_i')$ is cyclic of order p^{h_i}, we get $H' = D_1' \times \ldots\ldots \times D_{t-1}'$, where $D_1 = ((b_i)^{p^m})$ is cyclic of order $p^{h_1 - m}$. Thus the invariants of H' are $h_1 - m, \ldots\ldots, h_{t-1} - m$ and so the number of invariants of H' is $t - 1$.

But H and H' are isomorphic which forces to have the same number of invariants. But assuming $n_i \neq h_i$ for some i led to a discrepancy in the number of there invariants. Hence each $n_i = h_i$ and the Theorem is proved.

Theorem 4.5.4: *The number of non isomprphie abelian groups of order p^n, p a prime, equals the number of partitions of n.*

Proof: If $n_1 \geq \ldots\ldots \geq n_k > o$, $n = n_1 + n_2 + \ldots + n_k$ is any partition of n, then we can easily construct on abelian group of order p^n whose invariants are $n_1 \geq \ldots \geq n_k > o$. To do this, A_i be a cyclic group of order p^{n_i} and let $G = A_1 \times \ldots \times A_k$ be the external direct product of $A_1, \ldots, A_k$. Then by definition, the invariant of G one $n_1 \geq \ldots \geq n_k > o$. So two different partitions of n given rise to non-isomorphic abelian groups of order p^n.

The above theorem 4.5.4 does not depend on the prime p. It only depends on the exponent n. For example the number of non isomorphic groups of order 2^4 equals that of order 3^4, or 5^4, etc. since there are five partitions of 4, namely: $4 = 4$, $4 = 3 + 1$, $4 = 2 + 2$, $4 = 2 + 1 + 1$, $4 = 1 + 1 + 1 + 1$, then there are five non-isomorphic abelian groups of order p^4 for any prime p.

Exercise 4.5.1: Let (G, o) be an abelian group and let a, b be elements of G of order m and n respectively. Show that there exists an element c of G such that $o(c) = k$, where k is L.C.M of m and n.

Proof: Let us first suppose that $(m, n) = 1$, then $k = mn$.

We put $c = a \,.\, b$. Then clearly $c^{mn} = (ab)^{mn}$

$$\therefore \quad c^{mn} = (a \text{ o } b)^{mn}$$

$$\Rightarrow \quad c^{mn} = a^{mn} \text{ o } b^{mn}$$

$$\Rightarrow \quad c^{mn} = (a^m)^n \text{ o } (b^n)^m \qquad \text{since } o(a) = m, \quad o(b) = n$$

$$\Rightarrow \quad c^{mn} = e \text{ o } e = e$$

Þ *o(c) is finite and o(c) £ mn* ···(1)

Let $o(c) = r$. Then

$$e = c^r = (a \text{ o } b)^r$$

$$\Rightarrow \quad e = a^r \text{ o } b^r$$

$$\Rightarrow \quad a^r = b^{-r}$$

$$\Rightarrow \quad (a^r)^m = (b^{-r})^m$$

$$\Rightarrow \quad a^{mr} = b^{-mr}$$

$$\Rightarrow \quad e = b^{-mr}$$

$$\Rightarrow \quad b^{mr} = e$$

$$\Rightarrow \quad n \mid mr \text{ since } b^n = e$$

$$\Rightarrow \quad n \mid r \text{ since } (m, n) = 1.$$

Similarly, it can be shown at $m \mid r$.

Again, since $(m, n) = 1$, $mn \mid r$, that is, $m \,.\, n \leq r$ ...(2)

From (1) and (2) we have $o\,(c) = m \,.\, n$.

Now suppose that $(m, n) > 1$, we can put $m = p_1^{\alpha_1} p_2^{\alpha_2} .....p_u^{\alpha_u}$ and $n = p_1^{\beta_1} p_2^{\beta_2} .....p_u^{\alpha_u}$ where $p_i^{'s}$ are distinct prime numbers and α_i, β_i are non-negature integers such that $\alpha_i \geq \beta_i$ for all $1 \leq i \leq t$ and $\alpha_j < \beta_j$. for all $t + 1 \leq j \leq$ u. Then

$$k = p_1^{\alpha_1}\, p_2^{\alpha_2} ....p_t^{\alpha_t}\; p_{t+1}^{\beta_{t+1}} .....p_u^{\beta_u}$$

Let
$$g = {}_a p_{t+1}^{\alpha_{t+1}} ...p_k^{\alpha_k}$$

and
$$h = b^{p_1^{\beta_1} p_2^{\beta_2} ......p_t^{\beta_t}}$$

Therefore
$$o\,(g) = p_1^{\alpha_1}\, p_2^{\alpha_2} ....p_t^{\alpha_t} \text{ and } o(h) = p_{t+1}^{\beta_{t+1}} ....p_u^{\beta_u}$$

It is clear that $o(g)$ and $o(h)$ are co-prime, hence by first case $c = gh$ has order equal to

$p_1^{\alpha_1}\, p_2^{\alpha_2} ....p_t^{\alpha_t} ....p_{t+1}^{\beta_{t+1}} ....p_u^{\beta_u}$, *i.e.*, $o\,(c) = k$.

Exercise 4.5.2: If n is the exponent of an abelian group G, then for every $a \in G$, $o(a) \mid n$.

Solution. Let $o\,(a) = m$. Then $a^m = e$. If $m \dagger n$, $[m, n]$ = L.C.M of m and n is strictly greater than n.

By exercise 4.5.1 there exists an element $c \in G$ such that $o\,(c) = [m, n] > n$. This is contradiction to the definition of exponent. Hence $o\,(a) \mid n$.

Exercise 4.5.3: Let (G, o) be a finite abelian group in which the number of solution of the equation $x^n = e$ is at most n in G for every positive integer n. Prove that G must be cyclic.

Solution. Let $o(G) = m$ Let n, be the exponent of G. Since for any $a \in G$, $o\,(a) \mid n$, $a^{n_1} = e$ for all $a \in G$. Hence $x^{n_1} = e$ has m solutions. Thus $m \leq n_1$ as $x^{n_1} = e$. has at most n_1 solutions, since n_1 is the exponent of G. There exists an element $b \in G$ such that $b^{n_1} = e$ or $o(b) = n_1$. But $o\,(b) \mid o(G) \Rightarrow n_1 \mid m \Rightarrow n_1 \leq m$. Hence $m = n_1$. *i.e.*, $o\,(b) = m = o\,(G)$. This results in that G is a cyclie group generated by b.

PROBLEMS

1. If (G, o) is an abelian group of order p^n, p a prime and $n_1 \geq n_2 \geq ...... \geq n_k > o$, are the invariants of G, show that maximal order of any element in G is p_n.
2. If (A_1, o),...... (A_k, o) are normal sub-groups of a group (G, o) such that $A_i \cap (A_1 \,o\, A_2\, o,..., o\, A_{L-1}) = (e)$ for all i, show that (G, o) is the direct product of (A_1, o),......, (A_k, o) if $G = A_1 \text{ o } A_2 \text{ o } .....o\, A_k$.
3. Prove that if a finite abelian group has sub-groups of orders m and n, then it has a sub-group whose order is the least common multiple of m and n.

4. Describe all finite abelian groups of order

(a) 2^6 (b) $2^4 . 3^4$. (c) 11^6 (d) $2^4 . 3^4$.

5. If G is an abelian group of order p^n with invariants $n_1 \geq n_2 \geq .... \geq n_k > o$ and $H \neq (e)$ is sub-group of G, show that if $h_1 \geq ..... \geq h_s > o$ are invariants of H, then $k \geq s$ and for each i, $h_i \leq n_i$, $i = 1, 2, ..., s$.

6. If (G, o) is a finite abelian group of exponent n,. then no proper sub-groups of G contains all elements of order n.

7. An abelian group has invariants p, p, where p is a prime. How many sub-groups of order it has? **Ans.** $(p + 1)$.

8. For $n = 20$, how many non-isomorphic abelian groups of order n exist? **Ans.** (Two)

9. An abelian group (G, o) of order p^3, p a prime, has invariants $p^2 . p$. Determine all its sub-groups of order p.

CHAPTER 5

JORDAN-HOLDER THEOREM AND SOLVABLE GROUPS

In this chapter, we shall establish the significant and historic Jordan-Holder Theorem. For the sake of simplicity we shall confine our attention to only finite groups. Before reaching the theorem we shall introduce some special terminology.

5.1 CHAINS (SERIES)

Definition 5.1.1: By a *chain* (Series) for a group $(G, *)$ we mean any finite sequence $(H_0, H_1, ..., H_n)$ of subsets $H_\alpha, H_1, ...H_n$ of G, with

$$G = H_0 \supset H_1 \supset H_2 \supset .... \supset H_n = \{e\}$$

descending from G to $\{e\}$ such that the all pairs $(H_i, *)$ are sub-groups of $(G, *)$. The integer n is called the length of the chain. If each group $(H_i, *)$ is a normal sub-group of $(H_{i-1}, *)$, then the chain $G = H_0 \supset H_1 \supset .... \supset H_n = \{e\}$ is called *normal chain* for $(G, *)$. There is always one normal chain for a finite group $(G, *)$ with two or more elements, viz., the trivial chain $G \supset \{e\}$.

Example 5.1.1: In the group $(Z_{12} +_{12})$ of integers modulo 12, the following chains are normal chains:

$Z_{12} \supset (6) \supset \{0\}$, $Z_{12} \supset (2) \supset (4) \supset \{0\}$,

$Z_{12} \supset (3) \supset (6) \supset \{0\}$, $Z_{12} \supset (2) \supset (6) \supset \{0\}$,

All sub-group $((2), +_{12})$, $((4), +_{12})$, $((3), +_{12})$, $((6), +_{12})$ are normal sub-groups, since $(Z_{12} +_{12})$ is abelian.

Definition 5.1.2: The chain $G = K_0 \supset K_1 \supset .... \supset K_{m-1} \supset K_m = \{e\}$ obtained by the insertion of admissible subsets in chain $G = H_0 \supset H_1 \supset ... \supset H_n = \{e\}$ is said to be a refinement of the chain $G = H_0 \supset H_1 \supset .... \supset H_n = \{e\}$.

It is clear that $n \le m$, since every chain is trivially a refinement of itself. A refinement is *termed* proper if the refinement contains a set not in the original chain, *i.e.*, when $n < m$.

Example 5.1.2: In the group $(Z_{12}, +_{12})$ of integers modulo 12 the normal chain $Z_{12} \supset (3) \supset (6) \supset \{0\}$ is the refinement of the normal chain $Z_{12} \supset (6) \supset \{0\}$, and the normal chain

$Z_{12} \supset (3) \supset (6) \supset \{0\}$ has no proper refinement.

Definition 5.1.3: In a group $(G, *)$, the descending sequence of sets $G = H_0 \supset H_1 \supset \ldots \supset H_n = \{e\}$ is called

a *Composition series if*

1. $(H_i, *)$ is a sub-group of $(G, *)$,
2. $(H_i, *)$ is a normal sub-group of $(H_{i-1}, *)$,
3. The normal chain $G = H_0 \supset H_1 \supset \ldots \supset H_n = \{e\}$ has no proper refinement.

Example 5.1.3: In the group $(Z_{24}, +_{24})$, the normal chain $Z_{24} \supset (2) \supset (12) \supset \{0\}$ is not a composition chain, since it has two refinements, viz.,

$Z_{24} \supset (2) \supset (4) \supset (12) \supset \{0\}$ and

$Z_{24} \supset (2) \supset (6) \supset (12) \supset \{0\}$.

On the other hand,

$Z_{24} \supset (2) \supset (4) \supset (8) \supset \{0\}$ and

$Z_{24} \supset (3) \supset (6) \supset (12) \supset \{0\}$ are both composition chains

for $(Z_{24}\ +_{24})$.

Example 5.1.4: *Let* (S_3, o) be a symmetric group on $S = \{1, 2, 3)$. Then the chain

$$S_3 \supset \left\{\begin{pmatrix}1 & 2 & 3\\1 & 2 & 3\end{pmatrix}, \begin{pmatrix}1 & 2 & 3\\2 & 3 & 1\end{pmatrix}, \begin{pmatrix}1 & 2 & 3\\3 & 1 & 2\end{pmatrix}\right\} \supset \left\{\begin{pmatrix}1 & 2 & 3\\1 & 2 & 3\end{pmatrix}\right\}$$

is a composition chain for (S_3, o).

Since $\left(\left\{\begin{pmatrix}1 & 2 & 3\\1 & 2 & 3\end{pmatrix}, \begin{pmatrix}1 & 2 & 3\\2 & 3 & 1\end{pmatrix}, \begin{pmatrix}1 & 2 & 3\\3 & 1 & 2\end{pmatrix}\right\}, o\right)$ is only normal sub-group in (S_3, o)

Example 5.1.5: We consider the group $(Z, +)$ of integers under addition. We have observed that the normal sub-groups of $(Z, +)$ are cyclic sub-group $((n), +)$.

Let $n \in N$, Since $(k\ n) \subseteq (n)$ holds for $k \in N$, there always exists a sub-group of any given group. Hence each chain for $(Z, +)$ may be refined indefinitely.

Hence every group need not possess a composition chain.

Definition 5.1.4: A. normal sub-group $(H, *)$ is said to be *maximal normal* sub-group of $(G, *)$ if $H \neq G$ and there exists no normal sub-group $(K, *)$ of $(G, *)$ such that $H \subset K \subset G$.

Here we can define the composition chain in terms of maximal normal sub-groups. That is, a chain

$G = H_0 \supset H_1 \supset \ldots \supset H_n = \{e\}$, is a composition chain for $(G, *)$ if each sub-group $(H_i, *)$ is a maximal normal sub-group of $(H_{i-1}. *)$.

Maximal normal sub-group $(H, *)$ does not mean that it is the sub-group of $(G, *)$ of largest order. A group may have many distinct maximal normal sub-groups. For example–

CHAPTER 5

in the group $(Z_{24}\ +_{24})$, the cyclic groups $((2), +_{24})$ and $((3), +_{24})$ are maximal normal sub-groups of $(Z_{24}, +_{24})$ of order 12 and 8 respectively.

Definition 5.1.5: A group $(G, *)$ is said to be simple if it has no proper normal sub-groups.

Theorem 5.1.1: *A normal sub-group $(H, *)$ of the group $(G, *)$ is maximal if and only if $(G/H, *)$ is simple.*

Proof: We know that to each normal sub-group $(K, *)$ of $(G, *)$ with $H \subseteq K$ there corresponds a normal sub-group $\left(\frac{K}{H}, *\right)$ of $\left(\frac{G}{H}, *\right)$ and this correspondence is one to one. Hence $(H, *)$ is a maximal normal in $(G, *)$ if and only if $\left(\frac{G}{H}, *\right)$ is simple.

We have seen in a composition chain

$G = H_0 \supset H_1 \supset .... \supset H_n = \{e\}$ for the group $(G, *)$, each sub-group $(H_i, *)$ is a maximal normal sub-group of $(H_{i-1}, *)$. That is, we can say that each factor group $\left(\frac{H_{i-1}}{H_i}, *\right)$ is a simple group in composition chain.

Example 5.1.6: In the group $(Z_{24}, +_{24})$ the following two chains:

$$Z_{24} \supset (2) \supset (4) \supset (8) \supset \{0\}, \quad ...(1)$$

$$Z_{24} \supset (3) \supset (6) \supset (12) \supset \{0\}, \quad ...(2)$$

are composition chains. Hence in (1) each factor group of groups

$\left(\frac{Z_{24}}{(2)}, *\right), \left(\frac{(2)}{(4)}, *\right), \left(\frac{(4)}{(8)}, *\right)$, and $\left(\frac{(8)}{(0)}, *\right)$ of orders 2, 2, 2, and 3 respectively, are simple, since the order of each factor group is prime.

The orders of factor groups in (2)

$\left(\frac{Z_{24}}{(3)}, *\right), \left(\frac{(3)}{(6)}, *\right), \left(\frac{(6)}{(12)}, *\right)$, and $\left(\frac{(12)}{(0)}, *\right)$ are 3, 2, 2, and 2 respectively.

Hence all factor groups are simple.

Example 5.1.7: In the group $(Z_{24},+_{24})$ there are two maximal normal sub-groups, viz $((3), +_{24})$ and $((2), +_{24}$ of orders 8 and 12 respectively. The factor groups $\left(\frac{Z_{24}}{(3)}, *\right)$, and $\left(\frac{Z_{24}}{(2)}, *\right)$ corresponding to maximal normal sub-groups

$((3), +_{24})$ and $((2), +_{24})$ are simple group of orders 3 and 2 respectively.

We know that every group of prime order is simple since it has no proper sub-groups. Now we shall see that there are infinitely many simple group of composite order. For this we shall prove that (A_n, o) Alternating groups, is simple for $n \geq 5$.

Lemma 5.1.1: *If $n \geq 3$ every element of A_n is a product of 3 cycles.*

Proof: Let 1, b, c be three distinct symbols of 1, 2, 3,..., n. Let $\alpha \in A_n$, then α can be written as a product of even number of transpositions (2-cycles) and each transposition $(b\ c) = (1\ b)\ o\ (1\ c)\ o\ (1\ b)$ if $b \neq 1$, $c \neq 1$.

Let $\alpha = (1\ a_1)\ o\ (1\ a_2)\ o\ ...\ o.\ (1\ a_r)$, where r is an even number and $a_1, a_2, ,a_r \in \{2, 3, ..., n)$.

Further $\beta = (1\ b)\ o\ (1\ c) = (1\ c\ b)$. That is, the product of two transposition can be expressed as one 3-cycle. Thus the product of r (r even) transposition can be expressed as a product of 3-cycles. This completes the proof.

Lemma 5.1.2: *If a normal sub-group (H, o) of (A_n, o) has a 3-cycle, then $H = A_n$.*

Proof: We have $H \subseteq A_n$. to prove $A_n \subseteq H$, it is sufficient to show that every 3-cycle belongs to H. by the lemma 5.1.1 every element of An is a product of 3-cycles.

Let $(a\ b\ c)$ be any 3-cycle and let $(x\ y\ z) \in H$. We choose $\alpha \in S_n$ such that $\alpha\ (x) = a$, $\alpha\ (y) = b$, and $\alpha\ (z) = c$.

Then

$\alpha\ o\ (xy\ z)\ o\ \alpha^{-1} = (a\ b\ c)$ which is obtained by replacing x by a, y by b, z by c. Since (H, o) is a normal sub-group, $\alpha \in A_n$, $(x\ y\ z) \in H$, then $\alpha\ o\ (x\ y\ z)\ o\ \alpha^{-1} = (a\ b\ c) \in H$. If $\alpha \notin A_n$, then p, q different from x, y, z can be taken (since $n \geq 5$) such that

$\alpha\ o\ (p\ q) \in A_n$.

Then $[\alpha\ o\ (pq)\}\ o\ (x\ y\ z)\ o\ [\alpha\ o\ (pq)\}^{-1} \in H$

$\Rightarrow\quad [\alpha\ o\ (pq)\}\ o\ (x\ y\ z)\ o\ [(pq)^{-1}\ o\ \alpha^{-1}] \in H$

$\Rightarrow\quad [\alpha\ o\ (pq)\}\ o\ (x\ y\ z)\ o\ [(pq)\ o\ \alpha^{-1}] \in H$ *as* $(pq)^{-1} = (pq)$

$\Rightarrow\quad \alpha\ o\ (xyz)\ o\ \alpha^{-1} \in H$ as $(pq)\ o\ (pq) = 1$

and (xyz) and (pq) are disjoint cycles. This proves the Lemma.

Theorem 5.1.2: *The alternating group (A_n, o) is simple if $n \geq 5$.*

Proof: In lemma 5.1.1 we have seen that every element $\alpha \in A_n$ is a product of 3-cycles. In lemman 5.1.2 we have also seen that if a normal sub-group (H, o) of the alternating group (A_n, o) contains one 3 cycle, then $H = A_n$.

Let (H, o) be a normal sub-group of (A_n, o). Let $H = \neq$ [I}. Then we have to show that H contains a 3-cycle, that is, $H = A_n$.

Let $\alpha \in H$ and let $\alpha \neq$ I and α leaves fixed as many elements as any other non-identity permutation $\beta \in H$. If α is not a 3-cycle, then there are two possibilities that either (i) α contains a cycle of length ≥ 3 and moves more than three elements.

or (ii) α is a product of at least two disjoint transpositions. Thus, according to (i) and (ii) posibilities we can choose that either

CHAPTER 5

$\alpha = (1, 2, 3.....)\ (\quad)....$

or $\alpha = (1\ 2)\ o\ (3\ 4)....$

Since α is an even permutation, then in the first case α moves at least two more elements, say, 4, 5. So

$\alpha = (1\ 2\ 3\ 4\ 5)$

Now let $\beta = (3\ 4\ 5)$, then

$$\alpha_1 = \beta\, o\, \alpha\, o\, \beta^{-1} = \begin{pmatrix} 1 & 2 & 3 & 4 & 5..... \\ 1 & 2 & 4 & 5 & 3..... \end{pmatrix} o \begin{pmatrix} 1 & 2 & 3 & 4 & 5..... \\ 2 & 3 & 4 & 5 & 1..... \end{pmatrix} o \begin{pmatrix} 1 & 2 & 3 & 4 & 5 \\ 1 & 2 & 5 & 3 & 4 \end{pmatrix}$$

$$= \begin{pmatrix} 1 & 2 & 3 & 4 & 5..... \\ 1 & 2 & 4 & 5 & 3..... \end{pmatrix} o \begin{pmatrix} 1 & 2 & 3 & 4 & 5..... \\ 2 & 3 & 1 & 4 & 5..... \end{pmatrix}$$

$$= \begin{pmatrix} 1 & 2 & 3 & 4 & 5..... \\ 2 & 4 & 1 & 5 & 3..... \end{pmatrix} = (1\ 2\ 4\ 5\ 3) \qquad ...(1)$$

If $\alpha = (1, 2)\ (3, 4)....$, then

$$\alpha_1 = \beta\, o\, \alpha\, o\, \beta^{-1} = \begin{pmatrix} 1 & 2 & 3 & 4 & 5..... \\ 1 & 2 & 4 & 5 & 3..... \end{pmatrix} o \begin{pmatrix} 1 & 2 & 3 & 4 & 5..... \\ 2 & 1 & 4 & 3 & 5..... \end{pmatrix} o \begin{pmatrix} 1 & 2 & 3 & 4 & 5.... \\ 1 & 2 & 5 & 3 & 4.... \end{pmatrix}$$

$$= \begin{pmatrix} 1 & 2 & 3 & 4 & 5 \\ 1 & 2 & 4 & 5 & 3 \end{pmatrix} o \begin{pmatrix} 1 & 2 & 3 & 4 & 5... \\ 2 & 1 & 5 & 4 & 3... \end{pmatrix}$$

$$= \begin{pmatrix} 1 & 2 & 3 & 4 & 5... \\ 2 & 1 & 3 & 5 & 4... \end{pmatrix} = (1\ 2)\ (4\ 5) \qquad ...(2)$$

Now it is clear that, if a number $i > 5$ is left fixed-by α, then it is also fixed by α_1 and it is left fixed by $\alpha_1\ o\ \alpha^{-1}$. But we see that if α is as in (1)

$$\alpha_1\ o\ a^{-1} = \begin{pmatrix} 1 & 2 & 4 & 5 & 3... \\ 2 & 4 & 5 & 3 & 1... \end{pmatrix} o \begin{pmatrix} 1 & 2 & 3 & 4 & 5... \\ 5 & 1 & 2 & 3 & 4... \end{pmatrix}$$

$$= \begin{pmatrix} 1 & 2 & 4 & 5 & 3... \\ 3 & 2 & 4 & 1 & 5... \end{pmatrix} = (1\ 3\ 4),$$

and if α is as in (2), we have

$$\alpha_1\ o\, \alpha^{-1} = \begin{pmatrix} 1 & 2 & 3 & 4 & 5... \\ 2 & 1 & 3 & 5 & 4... \end{pmatrix} o \begin{pmatrix} 1 & 2 & 3 & 4 & 5... \\ 5 & 1 & 2 & 3 & 4... \end{pmatrix}$$

$$= \begin{pmatrix} 1 & 2 & 3 & 4 & 5... \\ 4 & 2 & 1 & 3 & 5... \end{pmatrix} = (1\ 4\ 3).$$

Thus $\alpha_1 \, o \, \alpha^{-1}$ leaves more elements fixed than α. This contradicts that $\alpha_1 \, o \, \alpha^{-1}$ leaves fixed as many elements as α. Hence our choice of α is wrong. Therefore α is a 3-cycle. Thus H contains every three cycle. Hence $H = A_n$.

Remark: For $n = 3$, every permutation on three symbols 1, 2, 3 is a three-cycle. Hence (A_n, o) is simple. We can also see that (A_n, o) is simple for all natural number n except $n = 1$, 2 and 4.

Theorem 5.1.3: *Every finite group (G, *) with more than one element has a composition chain.*

Proof: If $(G, *)$ is simple, then the trivial chain $G \supset \{e\}$ is a composition chain. Let $(G, *)$ be not a simple group. Then there exist a normal sub-group $(H, *)$ of $(G, *)$. If $(H, *)$ is maximal in $(G, *)$, and $(\{e\}, *)$ is maximal in $(H, *)$, then

$$G \supset H \supset \{e\}$$

is a composition chain. If $(H, *)$ is not maximal, then there exists a larger normal sub-group $(K, *)$ with $H \subset K \subset G$. If $(H, *)$ is maximal in $(K, *)$ and $(K, *)$ is maximal in $(G, *)$, then

$$G \supset K \supset H \supset \{e\}$$

is a composition chain. Continuing this process we get a desired composition chain of finite length n, since $(G, *)$ is a finite group. This proves the result.

Definition 5.1.6: The composition chains for the group $(G, *)$,

$$G = H_0 \supset H_1 \supset \ldots\ldots \supset H_n = \{e\}$$

and
$$G = K_0 \supset K_1 \supset \ldots \supset K_n = \{e\}$$

are said to be equivalent or isomorphic if they have same length $(n = m)$ and if their associated factor groups are isomorphic in some order. That is,

$$\left(\frac{H_{i-1}}{H_i}, *\right) \cong \left(\frac{K_{j-1}}{K_j}, *\right), 1 \leq i, j, \leq n$$

Remark: If we replace composition chains by normal chains, then this definition becomes the definition of equivalent normal chains.

Example 5.1.8: In the group $(Z_{24}, +_{24})$ we have the following two composition chains:

$$Z_{24} \supset (2) \supset (6) \supset (12) \supset \{0\} \quad \ldots(1)$$

and
$$Z_{24} \supset (3) \supset (6) \supset (12) \supset (0) \quad \ldots(2)$$

The factor groups of (1) are

$$\left(\frac{Z_{24}}{(2)}, *\right), \left(\frac{(2)}{(6)}, *\right), \left(\frac{(6)}{(12)}, *\right) \text{ and } \left(\frac{(12)}{(0)}, *\right) \text{ of}$$

orders 2, 3, 2 and 2 respectively.

CHAPTER 5

The factor groups of (2) are

$$\left(\frac{Z_{24}}{(3)}, *\right), \left(\frac{(3)}{(6)}, *\right), \left(\frac{(6)}{(12)}, *\right) \text{ and } \left(\frac{(12)}{(0)}, *\right) \text{ of}$$

orders 3, 2, 2, 2 respectively.

The chains (1) and (2) have the same length, namely, 4. Since two cyclic groups of the same order are isomorphic, then

$$\left(\frac{(3)}{(6)}, *\right) \cong \left(\frac{Z_{24}}{(3)}, *\right), \left(\frac{(6)}{(12)}, *\right) \cong \left(\frac{(6)}{(12)}, *\right), \left(\frac{(12)}{(0)}, *\right) \cong \left(\frac{(12)}{(0)}, *\right)$$

and $$\left(\frac{Z_{24}}{(3)}, *\right) \cong \left(\frac{(2)}{(6)}, *\right)$$

Hence two composition chains (1) and (2) are equivalent or isomorphic.

Theorem 5.1.4: *(Schreier's refinement Theorem) any two normal chains of a group have equivalent refinement.*

Proof: Let $G = G_1 \supseteq G_2 \supseteq G_3 \supseteq .... \supseteq G_{s+1} = \{e\}$ ...(1)

$G = H_1 \supseteq H_2 \supseteq H_2 \supseteq ....H_{t+1} = \{e\}$ ...(2)

be two normal chains of a group $(G, *)$, we set

$G_{i.k} = G_{i+1} * (G_i \cap H_k),\ k = 1, 2, ...\ t + 1$ ...(3)

$H_{k.i} = H_{k+1} * (G_s \cap H_k),\ i = 1, 2, ...,\ s + 1$ ...(4)

Since $(G_i, *)$ and $(H_k, *)$ are normal sub-groups and $H_{k+1} \subseteq H_k$, $(G_{i.k}, *)$ and $H_{k\ i}, *)$ are normal sub-groups and $G_{i,\ k+1} \subseteq G_{i,k}$ and $H_{k.i\ +1} \subseteq H_{k.i}$.

Since $H_{t\ +1} = G_{s\ +1} = \{e\}$, then

$G_{s.t+1} = G_{s\ +1} * (G_s \cap H_{t\ +1}) = \{e\} * (G_i \cap \{e\}) = \{e\}$

and $H_{t,\ s+1} = H_{t\ +1} * (G_s \cap H_{t\ +1}) = \{e\} * (G_s \cap \{e\}) = \{e\}$.

Moreover

$G_{i.\ t+1} = G_{i+1} * (G_i \cap H_{t+1}) = G_{i\ +1} * (G_s \cap \{e\}) = G_{i+1}$ ...(5)

and $H_{k',\ s+1} = H_{k+1} * (G_{s+1} \cap H_k) = H_{s+1} * (\{e\} \cap H_k) = H_{k+1}$ ...(6)

for all i and k respectively.

Now we consider the two series

$G = G_{1,1} \supseteq G_{1,2} \supseteq ... \supseteq G_{1.t+1} = G_{2,1} \supseteq G_{2.2} \supseteq ... \supseteq G_{2,t+1} ...$

$= G_{s,\ 1} \supseteq G_{s,2} \supseteq ... \supseteq G_{s,t\ +1} = \{e\}$ ...(7)

$G = H_{1.1} \supseteq H_{1.2} \supseteq .... \supseteq H_{1,s+1} = H_{2.1} \supseteq H_{2.2} \supseteq .... \supseteq H_{2',s+1} ...$

$= H_{t.1} \supseteq H_{t.2} \supseteq ... \supseteq H_{t,\ s+1} = \{e\}$ ...(8)

We apply Zassenhaus theorem 3.3.10 to conclude that $(G_{i,\ k+1}, *)$ and $(H_{k.i+1}, *)$ are normal sub-groups of $(G_{i,\ k}, *)$ and $(H_{k.i}. *)$ respectively.

Then $\dfrac{(G_{i,k}, *)}{(G_{k+1.} *)} \cong \dfrac{(G_{i+1} * (G_i \cap H_k), *)}{(G_{i+1} * (G_i \cap H_{k+1}), *)} \cong \dfrac{(H_{k+1} * (G_i \cap K_k), *)}{(H_{k+1} * (G_{i+1} \cap H_k), *)} = \dfrac{(H_{kj,} *)}{(H_{kj+1}, *)}$ for all i, k.

Hence there exists 1–1 correspondence between the factor groups of (7) and (8) and hence (7) and (8) are quivalents.

Lemma 5.1.3: *If (H, *) and (K, *) are distinct maximal normal sub-group of the group (G, *), then*

(1) *The pair $(H \cap K, *)$ is a maximal normal sub-group of both (H, *) and (K, *).*

(2) $\left(\dfrac{G}{H}, *\right) \cong \left(\dfrac{K}{H \cap K,} *\right)$ *and* $\left(\dfrac{G}{K,} *\right) \cong \left(\dfrac{H}{H \cap K,} *\right)$

Proof: **(1)** Since (*H*, *) and (*K*, *) are both normal, the $(H \cap K, *)$ is also normal and it is the largest in (*H*, *) and (*K*, *).

(2) It is also clear that *(H * K, *)* is also a sub-group of (*G*, *). Then we have $H \subseteq H * K \subseteq G$. Since *(H, *)* is maximal normal, either $H = H * K$ or $H * K = G$.

Let $H * K = G$. If $H * K \neq G$, then

$H * K = H \Rightarrow K \subseteq H$. It is contradicting to the maximality of (*K*, *). Hence $H * K = G$. By isomorphism theorem 3.3.9, we have $\left(\dfrac{G}{K,} *\right) \cong \left(\dfrac{H}{H \cap K,} *\right)$. Since $\left(\dfrac{G}{K,} *\right)$ is simple, $\left(\dfrac{H}{H \cap K,} *\right)$ is also simple.

Which follows $(H \cap K, *)$ is maximal normal sub-group of *(H, *)*.

Theorem 5.1.5: *(Jordan-Holder). In a finite group (G, *) with more than one element, any two composition chains are equivalent. That is, if*

$$G = H_0 \supset H, \supset ..... \supset H_n = \{e\} \quad ...(1)$$

$$G = K_0 \supset K_1 \supset .... \supset K_m = \{e\} \quad ...(2)$$

Then $n = m$ and $\forall\ i = i, 2, ... n-1, n$

$$\left(\frac{H_{i-1}}{H_i}, *\right) \cong \left(\frac{K_{f(i)-1}}{K_{f(i)}}, *\right) \text{ where } f \text{ is any permutation on } S = (1, 2, ... n).$$

Proof: We shall prove the theorem by induction on *o* (*G*). If (*G*, *) is simple, then only possible composition chain is $G \supset \{e\}$. To apply induction let us assume that the theorem holds for all groups having order less then *o* (*G*). We have two cases:

Case I. If $H_1 = K_1$, then the composition chains (1) and (2) are

$$H_1 \supset H_2 \supset .... \supset H_n = \{e\}$$

CHAPTER 5

and $H_1 \supset K_2 \supset .... \supset K_m = \{e\}$

on deleting G from (1) and (2). Both represent the composition chains for the group $(H_1, *)$. Since theorem is true for $(H_1, *)$, whose order is less than $o(G)$ two composition chains are equivalent. But $\left(\frac{G}{H_1}, *\right) \cong \left(\frac{G}{K_1}, *\right)$ which follows (1) and (2) are equivalent.

Case II. $H_1 \neq K_1$. Then either $H_1 \cap K_1, = \{e\}$ or $(H_1 \cap K_1, *)$ is a finite group with more than one element. So there exists a composition chain for $(H_1 \cap K_1, *)$,

$H_1 \cap K_1 \supset L_{r-1} \supset ......L_{r-1} \supset L_r = \{e\}$. By the Lemma just established 5.1.3, $(H_1 \cap K_1, *)$ is a maximal normal sub-group of both $(H_1, *)$ and $(K_1, *)$. Then we have two composition chains

$$G \supset H_1 \supset H_1 \cap K_1 \supset L_1 \supset ..... \supset L_{r-1} \supset L_r = \{e\} \qquad ...(3)$$

and
$$G \supset K_1 \supset H_1 \cap K_1 \supset L_1 \supset .... \supset L_{r-1} \supset L_r = \{e\} \qquad ...(4)$$

By the Lemma 5.1.3, again we have

$$\left(\frac{G}{H_1}, *\right) \cong \left(\frac{H_1}{H_1 \cap K_1}, *\right)$$

and
$$\left(\frac{G}{K_1}, *\right) \cong \left(\frac{H_1}{H_1 \cap K_1}, *\right).$$

Hence the quotient groups obtained from (3) are isomorphic in pairs to the quotient groups obtained from (4). Hence (3) and (4) are equivalent. Next, we consider the following two composition chains for $(G, *)$,

$$G \supset H_1 \supset H_2 \supset ..... \supset H_n = \{e\}$$

and
$$G \supset H_1 \supset H_1 \cap K_1 \supset L_1 \supset .... \supset L_{r-1} \supset L_r = \{e\}.$$

By case I, these chains are equivalent, since first-two terms are common in these two chains. Similarly

$G \supset K_1 \supset K_2, \supset .... \supset H_n = \{e\}$

$G \supset K_1 \supset H_1 \cap K_1 \supset L_1 \supset .... \supset L_{r-1} \supset L_r = \{e\}$ are equivalent.

Hence (1), (2), (3), (4) are all equivalent, This proves the theorem.

Example 5.1.9: Let us consider the group $(Z_{60}, +_{60})$ of integers modulo 60 and the two composition chains

$$Z_{60} \supset (3) \supset (6) \supset (12) \supset \{0\}, \qquad ...(1)$$

$$Z_{60} \supset (2) \supset (6) \supset (30) \supset \{0\}, \qquad ...(2)$$

The factor groups of (1) are $\left(\frac{Z_{60}}{(3)} +_{60}\right), \left(\frac{(3)}{(6)}, +_{60}\right)$

$\left(\frac{(6)}{(12)}+_{60}\right), \left(\frac{(12)}{(0)},+_{60}\right)$ of orders 3, 2, 2 and 5 respectively.

The factor groups of (2) are

$(Z_{60}/(2), +_{60})$, $((2)/(6), +_{60})$, $((6)/(30), +_{60})$, $((30/(0), +'_{60})$ of orders 2, 3, 5 and 2 respectively.

Since two cyclic groups of the same order are isomorphic, then

$$\left(\frac{Z_{60}}{(3)}, +_{60}\right) \cong \left(\frac{(2)}{(6)}, +_{60}\right).$$

$$\left(\frac{(3)}{(6)}, +_{60}\right) \cong \left(\frac{Z_{60}}{(2)}, +_{60}\right),$$

$$\left(\frac{(6)}{(12)}, +_{60}\right) \cong \left(\frac{(30)}{(0)}, +_{60}\right).$$

$$\left(\frac{(12)}{(0)}, +_{60}\right) \cong \left(\frac{(6)}{(30)}, +_{60}\right).$$

All these quotient groups are simple, being cyclic groups of prime order: We note that the product of these orders is 60, *i.e.* 2, 2, 3, 5 = 60.

5.2 SOLVABLE GROUPS

Definition 5.2.1: A group $(G, *)$ is said to be solvable if it has a normal chain.

$G = H_0 \supset H_1 \supset \ldots\ldots \supset H_n = \{e\}$ in which every quotient group $\left(\frac{H_{i-1}}{H_i}, *\right)$ is abelian.

Such chain is called a solvable chain for $(G, *)$.

Since every finite group $(G, *)$ with more than one element has a composition chain, every finite abelian group is solvable.

Example 5.2.1: Show that (S_3, o) is solvable. In this example, we shall see that a non-abelian group is also solvable. Let (S_3, o) be a symmetric group for $n = 1, 2, 3$, It is non-abelian group. There exists a solvable chain for (S_3, o).

$$S_3 \supset \left\{\begin{pmatrix}1 & 2 & 3\\1 & 2 & 3\end{pmatrix}, \begin{pmatrix}1 & 2 & 3\\2 & 3 & 1\end{pmatrix}, \begin{pmatrix}1 & 2 & 3\\3 & 1 & 2\end{pmatrix}\right\} \supset \left\{\begin{pmatrix}1 & 2 & 3\\1 & 2 & 3\end{pmatrix}\right\}.$$

CHAPTER 5

The sub-group *(H, o)* of (S_3, o), where

$$H = \left\{\begin{pmatrix}1 & 2 & 3\\ 1 & 2 & 3\end{pmatrix}, \begin{pmatrix}1 & 2 & 3\\ 2 & 3 & 1\end{pmatrix}, \begin{pmatrix}1 & 2 & 3\\ 3 & 1 & 2\end{pmatrix}\right\}$$ is a normal sub-group, since index of *H* in *G* is 2 and any sub-group of index 2 is a normal sub-group of the group. The quotient groups $\left(\frac{S_3}{H}, o\right)$ and $\left(\frac{H}{(e)}, o\right)$ and are of prime orders 2 and 3 respectively. So both quotient groups are cyclic and consequently abelian. Hence (S_3, o) is solvable:

Example 5.2.2: Show that (S_4, o) is solvable.

Solution: The alternating group (A_4, o) is a sub-group of (S_4, o) of order 12. Then the index of A_4 in S_4 is 2. Which follows (A_4, o) is a normal sub-group of (S_4, o).

The group (A_4, o) has no sub-group of order 6. But (A_4, o) has a sylow 2-sub-group *(S, o)* of order 4. Since every group of order 4 is abelian, sylow 2-sub-group *(S, o)* is normal sub-group in (A_4, o), and the index of *S* in A_4 is 3. The group *(S, o)* has a normal sub-group *(K, o)* of order 2. The index of *K* in *S* is 2. Hence, we get a composition chain.

$$S_4 \supset A_4 \supset S \supset K \supset \{e\} \text{ for } (S_4, o).$$

As $\left(\frac{S_4}{A_4}, o\right), \left(\frac{A_4}{S}, o\right), \left(\frac{S}{K}, o\right)$ and $\left(\frac{K}{(e)}, o\right)$ are groups of prime orders 2, 3, 2, 2 respectively, (S_4, o) is solvable.

Theorem 5.2.1: *Every sub-group (H, o) of a solvable group (G, o) is solvable.*

Proof: Let $G = H_o \supset H_1 \supset .... \supset H_n = \{e\}$ be a fixed solvable chain for *(G, o)*. Let *(K, o)* be a sub-group of *(G, o)*, then we have to prove that there exists a solvable chain.

$K = K_o \supset K_1 \supset ..... \supset K_n = \{e\}$ *for (K, o).*

Let $K_i = K \cap H_i$ $(i = 0, 1, 2, ...n)$. First we shall show that (K_i, o) is normal in (K_{i-1}, o), For this

$$\begin{aligned} a \in K_{i-1} \text{ and } a\ o\ K_i\ o\ a^{-1} &= (a\ o\ K_i\ o\ a^{-1}) \cap K \\ &\subseteq (a\ o\ H_i\ o\ a^{-1}) \cap K \text{ since } K_i \subseteq H_i \\ &= H_i \cap K \text{ since } (H_i, o) \text{ is normal} \\ &= K_i \text{ as } H_i \cap K = K_i \end{aligned}$$

which shows (K_i, o) is normal in (K_{i-1}, o),

Again
$$\begin{aligned} K_i &= K \cap H_i \\ &= K \cap (H_{i-1} \cap H_i) \text{ Since } H_{i-1} \supset H_i \\ &= (K \cap H_{i-1}) \cap H_i \\ &= K_{i-1} \cap H_i, \end{aligned}$$

So that

$$\left(\frac{K_{i-1}}{K_i}, o\right) \cong \left(\frac{K_{i-1}}{K_{i-1} \cap H_i}, o\right)$$

By theorem 3.3.9:

$\left(\frac{K_{i-1}}{K_{i-1} \cap H_i}, o\right) \cong \left(\frac{K_{i-1} o H_i}{K_{i-1} \cap H_i}, o\right)$. But the quotient group $\left(\frac{K_{i-1} o H_i}{K_{i-1} \cap H_i}, o\right)$ is abelian, being sub-group of the abelian group $\left(\frac{K_{i-1}}{H_i}, o\right)$. This implies every quotient group $\left(\frac{K_{i-1}}{K_i}, o\right)$ is a abelian. Hence (K, o) is solvable.

Theorem 5.2.2: *Every homomorphic image of a solvable group (G, o) is solvable.*

Proof: Let f be a homomorphism from (G, o) onto $(G', *)$ and let $G = H_0 \supset H_1 \supset .... \supset H_n = \{e\}$

be a fixed solvable chain for (G, o). If we set $H'_i = f(H_i)$, $\forall\ i = 0, 1, 2, ..., n$. Then we get a chain for $(G', *)$.

$$G' = H'_0 \supset H'_1 \supset ...... \supset H'_n = \{e\}$$

Since (H_i, o) is normal in (H_{i-1}, o), then $(H'_i, *)$ is certainly normal in $(H'_{i-1}, *)$ by theorem 3.1.3 So the chain for $G', *)$ is a normal chain. To complete the proof we have to show that $\left(\frac{H'_{i-1}}{H'_i}, *\right)$ is abelian.

For this we define a mapping $f_i : \frac{H_{i-1}}{H_i} \to \frac{H'_{i-1}}{H'_i}$ by

$f_i\,(a\ o\ H_i) = f(a) * H'_i$, $a \in H_{i-1}$, $i = 1, 2, ...\ n$. The mapping f_i is well-defined, for if $a\ o\ H_i = boH_i$, then $a^{-1}\ o\ b \in H_i$. Hence $f(a^{-1}\ o\ b) = f(a)^{-1} * f(b) \in f(H_i) = H'_i$ which implies $f(a) * H'_i = f(b) * H'_i \Rightarrow f_i\,(a\ o\ H_i) = f_i\,(b\ o\ H_i)$. The mapping f_i is also a homomorphism. For this, if $a, b \in H_{i-1}$.

$$\begin{aligned} f_i\,((a\ o\ H_i)\ o\ (b\ o\ H_i)) &= f_i\,((a\ o\ b)\ o\ H_i) \\ &= f\,(a\ o\ b)) * H'_i \\ &= (f\,(a) * f(b)) * H'_i \\ &= (f\,(a) * H'_i) * (f(b) * H'_i) \\ &= f_i\,((a\ o\ H_i)) * f_i\,((b\ o\ H_i)). \end{aligned}$$

CHAPTER 5

Since $f(H_i) = H_i'$, the mapping f_i maps the set $\frac{H_{i-1}}{H_i}$ onto the set $\frac{H_{i-1}'}{H_i'}$, thus f_i is one to one and onto mapping.

Hence $\left(\frac{H_{i-1}}{H_i}, o\right) \cong \left(\frac{H_{i-1}'}{H_i'}, *\right)$.

Since $\left(\frac{H_{i-1}}{H_i}, o\right)$ is commutative, the quotient group $\left(\frac{H_{i-1}'}{H_i'}, *\right)$ is also abelian. Hence $(G', *)$ is solvable.

Corollary: *If (H, o) is a proper normal sub-group of a solvable group (G, o), then* $\left(\frac{G}{H}, o\right)$ *is solvable.*

Proof: It is left to the reader.

Theorem 5.2.3: *Let (H, o) be a normal sub-group of (G, o). If (H, o) and* $\left(\frac{G}{H}, o\right)$ *both are solvable groups, then (G, o) is itself solvable.*

Proof: Let $G/H = K_0' \supset K_1' \supset \ldots\ldots \supset K_n' = \{e \, o \, H\}$

be a solvable chain for the quotient group *(G/H, o)*. We know that if *(K, o)* is a sub-group of *(G, o)* containing *H*, then $\left(\frac{K}{H}, o\right)$ is a sub-group of $\left(\frac{G}{H}, o\right)$.

So by this correspondence we get a sequence of sets

$$G = K_0 \supset K_1 \supset \ldots\ldots\ldots \supset K_n = H.$$

By the theorem of natural homomorphism

$K_i = f^{-1}(K_i')$. Since (K_i', o) is normal in (K'_{i-1}, o), $(K_{i,} \,_o)$ is normal in (K_{i-1}, o).

Again $\left(\frac{K_{i-1}}{K_i}, o\right) \cong \left(\frac{K_{i-1}'}{K_i'}, o\right)$.

Since $\left(\frac{K_{i-1}'}{K_i'}, o\right)$ is commutative, $\left(\frac{K_{i-1}}{K_i}, o\right)$ is also commutative. Since *(H, o)* is solvable, it has a solvable chain from *H to* {*e*}.

$H = H_0 \supset H_1 \supset \ldots\ldots\ldots \supset H_m = \{e\}$. Thus we can get solvable chain

$G = K_0 \supset K_1 \supset \ldots\ldots\ldots \supset K_{n-1} \supset H \supset H_1 \supset \ldots \supset H_m = \{e\}$. for *(G, o)*. Therefore *(G, o)* is solvable.

Theorem 5.2.4: *Let (G, o) be a finite solvable group. Then the quotient groups of any composition chain for (G, *) are cyclic groups of prime order.*

Proof: We shall prove the result by induction on $o\ (G)$. If $o\ (G)$ is prime, there is trivial composition chain $G \supset \{e\}$ for (G, o), Let (H, o) be a proper normal sub-group of (G, o). The orders of $o\ (H)$ and $o\ (G/H)$ are less than the order of (G, o) *i.e.* $o\ (G)$. So by induction the theorem holds for the groups (H, o) and $(G/H, o)$. That is, (H, o) and $(G/H, o)$ are solvable and the quotient groups for their composition chains are of prime order: Let

$$H = H_0 \supset H_1 \supset \ldots\ldots \supset H_n = \{e\}$$

and $$G/H = K_0' \supset K_1' \supset \ldots\ldots \supset K_m' = \{e\ o\ H\}$$

be solvable chains for (H, o) and $(G/H, o)$ respectively.

By the theorem of natural homomorphism

$K_i = f^{-1}\ (K_i')\ (i = 1, 2, \ldots.n)$. So by correspondence theorem

$$G = K_o \supset K_i \supset \ldots \supset K_m = H,$$

where (K, o) is a normal sub-group of (K_{i-1}, o). And,

$$\left(\frac{K_{i-1}}{K_i}, o\right) \cong \left(\frac{K_{i-1}'}{K_i'}, o\right), \text{ where } \left(\frac{K_{i-1}'}{K_i'}, o\right)$$

is cyclic group of prime order. Combining these sequences together, we get a composition chain for (G, o), all of whose quotient groups are cyclic groups of prime order.

Theorem 5.2.5: *If $n \geq 5$, then (S_n, o) and $(A_n\ o)$ are not solvable groups.*

Proof: We know that symmetric group (S_n, o) has a normal sub-group (A_n, o) of even permutations of index 2. This sub-group is called alternating group on n symbols. If $n \geq 5$, (A_n, o) is a simple group, consequently its normal chain $A_n \supset \{e\}$. Thus we are led to conclude that for $n > 5$, $(A_n\ o)$ is solvable group if and only if (A_n, o) is commutative.

Since (A_n, o) is not a commutative group for $n \geq 4$, $(A_n\ o)$ is not solvable for $n \geq 5$. Consequently the symmetric group (S_n, o) is not solvable.

Theorem 5.2.6: *A group (G, o) is solvable of and only if $G^{(n)} = \{e\}$, for some $n \geq 1$.*

Proof: Let $G^{(n)} = (e)$ for some n. By theorem 2.11.7 $(G^{(i+1)}, o)$ is a normal sub-group of $(G^{(i)}, o)$ and $(G^{(i)}/G^{(i+1)}, o)$ is abelian for each i, $(o \leq i \leq i-1)$ by example 2.11.5.

Hence the series

$$G \supset G^{(1)} \supset G^{(2)} \supset \ldots. \supset G^{(i)} \supset G^{(i+1)} \ldots. \supset G^n = (e).$$

satisfies the conditions (i) and (ii) for solvability of G.

Conversely, assume that (G, o) is solvable. So there exists a descending chain

$$G = G_o \supset G_1 \supset \ldots. \supset G_i \supset G_{i+1} \supset \ldots.. \supset G_n = (e).$$

Such that (G_{i+1}, o) is normal in (G_i, o) and $(G_i/G_{i+1}, o)$ is abelian, for $o \leq i \leq n-1$.

We prove by induction on i that $G^{(i)} \subset G_i$ for all i. Since $(G/G_i, o)$ is abelian, by theorem 2.11.5, $G^{(1)} \subset G_1$. Assuming $G^{(k)} \subset G_{k+1}$, we have to show that $G^{(k+1)} \subset G_{k+1}$ since G_k/G_{k+1} is abelian.

CHAPTER 5

$G_{k+1} \supset G_k^{(k)}, \supset \left(G_k^{(k)}\right)^{(1)} = G^{k+1}$

In particular $G^{(n)} \subset G_n = (e)$.

Hence this completes the proof.

Note: $G(n)$ is the nth commutator of G.

Theorem 5.2.7: *A group of prime power order is solvable.*

Proof: Let (G, o) be a group of order p^n $(n \geq 1)$.

By theorem 4.3.2 if $o(G) = p^n$, p is a prime number, then $Z(G) \neq \{e\}$. If $Z(G) = G$, then G is abelian and hence G is solvable.

If $G \neq Z(G) = G_1$, say, then consider G/G_1. This is a p-group and hence has a non-trivial centre of the form G_2/G_1, where $(G_2\ o)$ is normal in (G, o). If $G_2 \neq G$, Let (G_3, o) be the sub-group of (G, o) Such that G_3/G_2 is the centre of G/G_2. Continue the process. Since G is finite, it stops after a finite number of steps and we have a sequence.

$(e) \subset G_1 \subset G_2 \ldots\ldots \subset G_n = G.$

that is, $G \supset G_{n-1} \supset \ldots\ldots \supset G_1 = (e).$

with $(G_i/G_{(-1)}, o)$ is abelian for all i. Hence G is solvable.

Theorem 5.2.8: *Any group of order pq., p and q being distinct primes, is solvable.*

Proof: If $p = q$, that is $o(G) = p^2$, then G is abelian and hence G is salvable. If $p \neq q$, assume $p < q$. By theorem 4.4.3 (G, o) has a unique Sylow q-sub-group Q which is normal in G. The normal series $G \supset Q \supset (e)$ has abelian quotient. Hence (G, o) is solvable.

Note: A very important and famous theorem of Burnside states that any group of order $p^n q^m$, p, q primes $(n \geq 1, m \geq 1)$ is solvable. The proof of this theorem is beyond the scope of the book.

W. Feit and J.G. Thompson have proved (Pacific Jr. Maths Vol. 13, pp, 775-1029) that every group of odd order is solvable. This is very important theorem on finite groups. Because it gives immediately many examples of solvable group. Again if (G, o) is a non-abelian group of even order, $G^{(1)}$ is its commutator sub-group, then $G = G^{(1)}$ and hence (G, o) is simple. Thus $G^{(n)} = G$, for all n, implies that (G, o) is not solvable and by Feit-Thompson's theorem G must be of even order.

We know study another class of groups which is more general then solvable groups.

Definition 5.2.2: Let (G, o) be a group. The upper central series of G is the sequence of sub-groups $\{C_n\}$ such that

(i) $\{e\} = C_0 \subset G_1 \subset \ldots \subset C_n \subset \ldots.$

$$(ii)\ \frac{C_i}{C_{i-1}} = Z\left(\frac{G}{C_{i-1}}\right)\ (i \geq 1).$$

Definition 5.2.3: A group (G, o) is said to be nilpotent if it has normal series

$\{e\} = C_0 \subset G_1 \subset \ldots. \subset C_n = G$

Such that

$$\frac{C_i}{C_{i-1}} = Z\left(\frac{G}{C_{i-1}}\right) (i \geq 1)$$

Remark: For some sub-group (H, o) and (K, o) of (G, o) we denote the sub-group generated by all the commutators $a \ o \ b \ o \ a^{-1} \ ob^{-1}$, $a \in H$, $b \in K$ by $[H, K]$.

Definition 5.2.4: The lower central series of G is the sequence of sub-groups $\{Z^n\}$ such that

(*i*) $G = Z^\circ \supset Z^1 \supset .... \supset Z^i \supset Z^{i+1} \supset .....$

(*ii*) $Z^{i+1} = [G, Z^i]$, $i \geq 0$.

***Theorem 5.2.9:** A group (G, o) is nilpotent if and only if $Z^n = \{e\}$ for some n.*

Proof: Let (G, o) be nilpotent. Then for some n there exists a sub-group $C_n = G$ in the upper central series. We claim that $Z^i \subset C_{n-i}$, for all $i \leq (o \leq i \leq n)$. We see that for $i = 0$, $Z^\circ = G \subset C_n = G$.

This shows that $Z^\circ \subset C_{n-o} = C_n$.

Assume that $Z_i \subset C_{n-i}$

Then $Z^{i+1} = [G, Z^i] \subset [G_i, C_{n-i}] \subset C_{n-i-1}$ as

$$\frac{C_{n-i}}{C_{n-i-1}} = Z\left(\frac{G}{C_{n-i-1}}\right).$$

In particular, for $i = n$, $Z^n \subset C_o = \{e\}$, that is, $Z^n = \{e\}$.

Conversely, assume that $Z^n = \{e\}$ for some n. Now we shall show that G is nilpotent. Here we claim that $Z^{n-i} \subset C_i$, for all i, $(o \leq i \leq n)$. This is clearly true for $i = 0$, because $Z^n = \{e\} \subset C_o = \{e\}$.

Assume that $Z^{n-i} \subset C_i$ and consider Z^{n-i-1}.

We see that

$[Z^{n-i-1}, G] = Z^{n-i} \subset C_i$ and since

$\frac{C_{i+1}}{C_i} = Z\left(\frac{G}{C_i}\right)$, we have

$Z^{n-i-1} \subset C_{i+1}$ In particular, for $i = n$,

$Z^\circ = G \subset C_n$, *i.e.* $G = C_n$. This implies that G is inpatient.

Example 5.2.3: (*i*) Every abelian group (G, o) is nilpotent since $Z^1 = [G, G] = \{e\}$.

(*ii*) The symmetric group (S_3, o) is not nilpotent because for $i \geq 1$ $Z^i = A_3$. But (S_3, o) is a solvable group.

(*iii*) We see that every nilpotent group is solvable, because $G^{(i)} \subset Z^i$ for all $i \geq o$ and

CHAPTER 5

$Z^n = \{e\}$ implies $G^{(n)} = \{e\}$. But by example (II) it is clear that a solvable group need not be nitpotent.

Theorem 5.2.10: *Any sub-group of a nilpotent group is nitpotent.*

Proof: Let (H, o) be a sub-group of a nilpotent group (G, o) clearly $Z^i (H) \subset Z^i (G)$. If $Z^n (G) = \{e\}$, then $Z^n (H) = \{e\}$ *i.e.*, (H, o) is nilpotent.

Theorem 5.2.11: *Any homomorphic image of nilpotent group is nilpotent.*

Proof: Let $f : G \to G'$ be a homomorphism of group $(G, *)$ onto a group (G', o). That is,

$$f(G) = G'.$$

Clearly, $f(Z^i (G)) = Z^i (f(G) = Z^i (G')$, for all $i \geq 1$.

If $Z^n (G) = \{e\}$, then $Z^n (G') = \{ f(e)\} = \{e'\}$.

Hence (G', o) is nilpotent.

Example 5.2.4: Let $G = S_n$ and $H = A_3$. Here H and G/H are both nilpotent but G is not nilpotent. This example shows the analogue of theorem that "If a sub-group (H, o) and quotient group $(G/H, o)$ are solvable, then (G, o) is solvable," is not true.

We may however prove the following:

Theorem 5.2.12: *Let (G, o) be a group and Let $H \neq \{e\}$ be a sub-group contained in the centre $Z(G)$ of (G, o) such that $(G/H, o)$ is nilpotent, then (G, o) is nilpotent.*

Proof. Since $(G/H, o)$ is nilpotent, $Z^n (G/H) = \{e\} = \{H\}$, for some n. It implies that $Z^n (G) \subset H$. Since $H \subset Z (G)$, $Z^{n+1} (G) = [Z^n (G), G] = \{e\}$. Hence (G, o) is nilpoltent.

This results in the following:

Theorem 5.2.13: *Any group of prime order is nilpotent.*

Proof: Let (G, o) be a group of order p^n, p-rime and $n \geq 1$. We prove it by induction applying on = 1. For $n = 1$, $o (G) = p$ and for $n = 2$, $o (G) = p^2$, (G, o) is abelian group and hence it nilpotent.

Let $n > 1$. Then $Z (G) \neq \{e\}$ and $(G/Z(G), o)$ has order less them p^n, By induction. It is nilpotent, and by above theorem (G, o) in nilpotent.

Definition 5.2.5: A sub-group (H, o) of a group (G, o) is called a characteristic sub-group of (G, o) if $\phi (H) \subset$ H for any automorphism ϕ of G.

Such a sub-group is clearly normal because $i_x (H) = x \; o \; H \; o \; x^{-1} \subset H$, for all $x \in G$.

Theorem 5.2.14: *If (H, o) is a characteristic sub-group of a normal sub-group (N, o) of a group (G, o), then (H, o) is normal is G.*

Proof: Since (N, o) is normal in G, then for any $x \in G$, $i_x (N) \subset H$. That is, i_x induces an automorphism of N. Since (H, o) is a characteristic in (N, o) $i_x (H) \subset H$. This shows that (H, o) is normal in (G, o).

***Theorem 5.2.15:** If a Sylow p-sub-group P of a finite group (G, o) is normal in G, then (P, o) is a characteristic sub-group of (G, o).*

Proof: Let $\phi: G \to G$ be an automorphism of G. Since (P, o) is a Sylow p-sub-group, then $(\phi(P), o)$ is also group and $\phi(P)$ has the same order as P, *i.e.*, $o(\phi(P) = o(P)$.

This means that $\phi(P)$ is a Sylow p-sub-group.

Since P is normal, the Sylow p-sub-group is unique.

Hence $\phi(P) = P$. Hence P is a characteristic sub-group of G.

***Theorem 5.2.16:** Any proper sub-group H of a nilpotent group is properly contained in its normalizer.*

Proof: Since G is nilpotent, then there exists a central series

$$C_o = \{e\} \subset G \subset \ldots \subset C_n = G$$

with $C_{c+1} = Z(G/C_i)$. Since $H \neq G$, there exists some j such that $C_j \subset H$ and $C_{j+1} \not\subset H$.

Then $[C_{j+1}, H] \subset [C_{j+1}, G] \subset C_j \subset H$. Hence, there is an element $Z \in C_{j+1}$, $Z \notin H$ with $[Z, H] \subset H$. This implies that $Z \in N(H)$. Thus $H \neq N(H)$.

Definition 5.2.6: A group (G, o) is said to be perfect if $G = G'$.

***Theorem 5.2.17:** If (G, o) is a perfect group, then centre of G/Z(G) is trivial.*

Proof: Hence we shall write $Z = Z(G)$. Suppose the centre of G/Z is non-trivial. Then there exists a coset $Zoa \in G/Z$ such that $a \notin Z$.

We define a mapping $f: G \to G$ by

$$f(x) = a \text{ o } x \text{ o } a^{-1} \text{ o } x^{-1}.$$

We claim that $a \text{ o } x \text{ o } a^{-1} \text{ o } x^{-1} \in Z$ for all $x \in G$.

Since $Z \text{ o } a \in G/Z$, $(Zoa) \text{ o } (Zox) = (Zox) \text{ o } (Zoa)$

$$\Rightarrow Zo\,(aox) = Zo\,(xoa)$$

$$\Rightarrow (aox) \text{ o } (xoa)^{-1} \in Z$$

$$\Rightarrow a \text{ o } x \text{ o } a^{-1} \text{ o } x^{-1} \in Z.$$

This shows that f is mapping from G onto Z. Again, we see that

$$\begin{aligned} f(xoy) &= a \text{ o } (xoy) \text{ o}a^{-1} \text{ o } (xoy)^{-1} \\ &= a \text{ o } x \text{ o } y \text{ o } a^{-1} \text{ o } y^{-1} \text{ o } x^{-1} \\ &= a \text{ o } x \text{ o } (x \text{ o } a)^{-1} \text{ o } (xoa) \text{ o } y \text{ o } a^{-1} \text{ o } y^{-1} \text{ o } x^{-1} \\ &= (a \text{ o } x \text{ o } a^{-1} \text{ o}x^{-1}) \text{ o } x \text{ o } (a \text{ o } y \text{ o } a^{-1} \text{ o } y^{-1}) \text{ o } x^{-1} \\ &= (a \text{ o } x \text{ o } a^{-1} \text{ o } x^{-1}) \text{ o } (a \text{ o } y \text{ o } a^{-1} \text{ o}y^{-1}) \text{ o } x \text{ o } x^{-1}, \\ &\quad \text{as } a \text{ o } y \text{ o } a^{-1} \text{ o } y^{-1} \in Z \\ &= f(x) \text{ o } f(y). \end{aligned}$$

CHAPTER 5

This shows that f is a homomorphism we know that Ker f is a normal sub-group of (G, o).

If Ker $f = G$, then for all $x \in G$,

$$f(x) = e$$

$$\Rightarrow \quad a \circ x \circ a^{-1} \circ x^{-1} = e$$

$$\Rightarrow \quad a \circ x = x \circ a$$

$\Rightarrow a \in Z$ which is a contradiction.

Hence $Ker\ f \subset G$.

Since $f: G \to Z$ is a homomorphism,

$G/Ker\ f \cong T$, where (T, o) is a sub-group of (Z, o).

As (Z, o) is abelian, $G/Ker f$ is abelian. Hence $G' \subseteq Ker f$.

But G is prefect, $G = G'$. This in turn implies that $G = Ker f$ which is absurd. Hence $a \circ Z = Z$, as a consequence, the centre of G/Z consists of Z only.

Theorem 5.2.18: *(Welson's theorem), if p is a prime number, then $(p-1)! \equiv -1 \pmod p$.*

Proof: Consider the symmetric group (Sp, o)

Since $p \mid p!$, $p^2 \dagger p!$, (S_p, o) has sylow p-sub-group of order p. The number of sylow p-sub-groups of S_p is $1 + kp$. Since each sylow p-sub-group is of order p and no two sylow p-sub-group have any element common except identity.

Then the total number of elements of order p is $(1 + kp)(p-1)$.

The elements of order p in S_p are precisely the cycles of length p. There number is $(p-1)!$

Hence
$$(p-1)! = (1 + kp)(p-1)$$
$$= (p + k p^2 - k p - 1$$
$$\Rightarrow (p-1)! \equiv -1 \pmod p$$

Example 5.2.6: Let f and g be two distinct cycles in S_n of order m and k respectively. Show that

$$o(fg) = d \text{ where } d = [m, k].$$

Solution: Since $o(f) = m$ and $o(g) = k$, then

$$f^m = I \text{ and } g^k = I.$$

Since f and g are distinct,

$$fg = gf.$$

$$\text{So } (fg)^d = f^d g^d = II = I \text{ as } m \mid d,\ k \mid d$$

Further, Let $(fg)^r = I$, then

$$f^r = g^{-r}.$$

If f^r is not identity permutation, there exists an element $a \in S$ such that

$$f^r(a) \neq a.$$

So $f^r(a)$ belongs to the cycle of f.

But $f^r(a) = g^{-r}(a)$

$\Rightarrow$ $g^{-r}(a)$ belongs to the cycle of f.

This is an absurd since f and g are distinct disjoint cycles. Hence our assumption is wrong. Thus

$$f^r = I = g^{-r} \Rightarrow f^r = I \text{ and } g^r = I$$
$$\Rightarrow m \mid r \text{ and } k \mid r.$$

Now $d = [m, k]$. Consequently $d \mid r$. Hence $o(f g) = d$.

Example 5.2.7: A group G of order 42 cannot be simple.

Proof: Since $42 = 2 \cdot 3 \cdot 7$, G has sylow 7-sub-groups.

The number of 7-sub-groups is congruent to 1 modulo 7. It also divides 6, the index of a 7-sylow group. Thus $1 + 7m$ divides 6. This implies $m = 0$. Hence G has a unique sylow 7-sub-groups which is normal by example 4.4.6. Hence G is not simple.

(*i*) A group of order 56 cannot be simple.

Proof: Since $56 = 2^3.7$, G has a sylow 2-sub-groups and sylow 7-sub-groups. If G has a unique sylow 7-sub-group, then it is normal in G and hence G is not simple. Otherwise, G will have 8 sylow 7-sub-groups P_i accounting for 49 elements of G as $P_i \cap P_j = \{e\}$, $i \neq j$. Since G has a sylow 2-sub-group of order 8, the remaining seven elements togather with e must be this sylow 2-sub-group. Hence G has a unique sylow 2-sub-group and G is not simple.

Remark: From above two examples, it is clear that there exist finite groups which are simple. The following theorem shows that there are infinitely many finite simple groups.

Theorem 5.2.19: *Show that (Q, +) has no maximal sub-group.*

Proof: Let $(H, +)$ be a maximal normal sub-group of $(Q, +)$. There exists a rational number $\frac{r}{s}$ $(r, s \in Z, s \neq 0)$ such that $\frac{r}{s} \notin H$. Since $(Z, +)$ is a sub-group of $(Q, +)$, and $Z \neq Q$, $H \neq \{0\}$. Let $\frac{m}{n} \in H$, $m, n \in Z$, $mn \neq 0$.

Now $\frac{m}{n} \in H \Rightarrow n \frac{m}{n} \in H \Rightarrow m \in$ H.

Maximality of H implies $H + \left(\frac{r}{s}\right) = Q$. As a consequence $\frac{r}{sms} = h + t\frac{r}{s}$ for some $h \in H$, $t \in Z$.

Thus $\frac{r}{s} = msh + (tr)\, m \in H$ as $m \in H$, sh, $t_r \in Z$.

This is a contradiction.

Hence $(Q, +)$ has no maximal sub-group.

Example 5.2.8: Prove that the centre of a group (G, o) is properly contained in every maximal sub-group of G having a composite index.

CHAPTER 5

Solution: Let (H, o) be a maximal sub-group (G, o) with composite index.

Let (Z, o) be the centre of the group (G, o) and let Z be not contained in H. Then there exists an element $a \in Z$ and $a \notin H$.

Consider (H, a), *i.e.*, a sub-group generated by H and a. Since (H, o) is maximal, $H \subset (H, a)$ and $a \notin H \Rightarrow (H, a) = G$. So any $g \in G$ can be written as $g = a^i h$ for some $h \in H$, $a \in Z$ and $i \in Z$.

Now for any $y \in H$, $g \in G$,

$$\begin{aligned} g \text{ o } y \text{ o } g^{-1} &= (a^i \text{ o}h) \text{ o } y \text{ o } (a^i \text{ o } h)^{-1} \\ &= a^i \text{ o } h \text{ o } y \text{ o } h^{-1} \text{ o } a^{i-1} \\ &= h \text{ o } y \text{ o } h^{-1} \text{ as } a \in Z. \end{aligned}$$

But $h, y, h^{-1} \in H \Rightarrow g \text{ o } y \text{ o } g^{-1} \in H$.

Hence (H, o) is a normal sub-group of (G, o).

As $(G/H, o)$ has a composite order, it has a non-trival sub-group $(K(H), 0)$.That is,

$$H \subset K/H \subset G/H$$

$$\Rightarrow \quad H \subset K \subset G$$

which is against the maximality of H. Hence $Z \subseteq H$.

PROBLEMS

1. Check that the following chains represent composition chains for the indicated group,

   (*a*) For $(Z_{36}, +_{36})$, the group of integers modulo 36.

   $Z_{36} \supset (3) \supset (9) \supset (18) \supset \{0\}$.

   (*b*) For $((a), o)$, a cyclic group of order 30.

   $(a) \supset (a^5) \supset (a^{10}) \supset \{e\}$.

2. Show that no infinite cyclic group has a composition chain.

3. Show that if (G, o) is a finite cyclic group of order 2^n, then there is exactly one composition chain for (G, o).

4. Show that if the quotient group $(G/H, o)$ is of prime order, then (H, o) is a maximal normal sub-group of (G, o).

5. Prove that the cyclic group $((n), *)$ is a maximal normal sub-group of $(Z, +)$ if and only if n is prime.

6. Establish that the following two composition chains for $(Z_{24} +_{24})$ are equivalent:

   $Z_{24} \supset (3) \supset (6) \supset (12) \supset \{0\}$,

   $Z_{24} \supset (2) \supset (4) \supset (12) \supset \{0\}$.

7. Find all composition chains for $(Z_{30}, +_{30})$ and varify the Jordan-Holder theorem. (There are four composition chains).

8. Suppose (H, o) is a proper normal sub-group of the group (G, o) and
$G \supset K_1 \supset ... \supset K_r \supset H \supset L_1 \supset ... \supset L_n = \{e\}$
is a composition chain for (G, o). Prove that
$G/H \supset K_1/H \supset ... \supset K_r/H \supset \{H\}$
represents a composition chain for the quotient group. $\left(\frac{G}{H}, o\right)$.

9. Prove that if $(G, *)$ and $G', o)$ are solvable groups, then their direct product $(G \times G', o)$ is also solvable group.

10. Let (H, o) and (K, o) be two solvable sub-group of the group (G, o), with (H, o) normal in (G, o) Verify that $(H, o\ K, o)$ is also solvable.

11. Prove that a simple group with more than one element is solvable if and only if it is of prime order.

12. Prove that if an Abelian group has a composition chain then it must be finite.
[Hint. Let $G = H_o \supset H_t \supset H_2 \supset ... \supset H_k = (e)$ be a composition chain of G. Since H_i/H_{i+1} is a simple commutative group, it is of prime order, say p_i.

$$\text{Then } o\ (G) = \frac{o(G)}{o(H_i)} \frac{o(H_1)}{o(H_2)} .... \frac{o(H_{k-1})}{o(H_k)}$$

$$= o\ (G/H_1)\ o\ (H_1/H_2)\ ....\ o\ (H_{k-1}/H_k)$$

$$= p_o\ .\ p_1 ... p_{k-1} \text{ is a positive integer.}$$

Hence (G, o) is a finite group].

13. Prove that a group (G, o) is solvable if and only if $G^{(k)} = (e)$ for some integer $k \geq 1$. $(G'^{(k)}$ is the kth derived or k^{th} commutator sub-group of G).

14. Let (G, o) and $(H, *)$ be two groups having composition series. Prove that $(G \times H)$ also has a composition series.

CHAPTER 5

5.3 SURVEY OF ABSTRACT GROUPS WITH ORDER NOT EXCEEDING 8

We shall see that given a positive integer n there exists only finite number of different abstract groups of order n. We have also seen that two finite groups of the same order n need not be isomorphic. For example, the group (S_3, o) of all permutations of $S = \{1, 2, 3\}$ under functional composition and the cyclic group $((a), *)$, where $a^6 = e$, are of the same order 6 but they are not isomorphic. Given a positive integer n how many non-isomorphic groups of order n exist? The problem of the determination of the actual number of different abstract groups of finite order yet awaits solution. Here we shall find out all possible abstract groups of order n, $n \leq 8$.

Let p be a prime number. We know that every group of prime order p is cyclic and unique. So there exists one and only one abstract cyclic group whose order is 1, 2, 3, 5, or 7.

Now we shall consider the groups of orders 4, 6, 8.

Lemma 5.3.1: *A group (G, o) is an abelian group if its each element has order 2 and if (G, o) is finite then it is expressible as a direct product of cyclic groups of order 2 and the order of the group is some power of 2.*

Proof: Let $a, b \in G$, then $a^2 = e$, $b^2 = e$, and $(a \text{ o } b)^2 = e$.

We have

$$(a \text{ o } b)^2 = (a \text{ o } b) \text{ o } (a \text{ o } b) = a \text{ o } (b \text{ o } (a \text{ o } b)) = a \text{ o } ((b \text{ o } a) \text{ o } b)$$

So

$$\begin{aligned} a \text{ o } ((b) \text{ o } a) \text{ o } b = e &\Rightarrow (aoa) \text{ o } (boa) \text{ o } (bob) = a \text{ o } e \text{ o } b \\ &\Rightarrow a^2 \text{ o } b \text{ o } a \text{ o } b^2 = a \text{ o } b \\ &\Rightarrow e \text{ o } b \text{ o } a \text{ o } e = a \text{ o } b \\ &\Rightarrow b \text{ o } a = a \text{ o } b. \end{aligned}$$

Hence (G, o) is an abelian group.

Let $a_1 \in G$. Then $a_1^2 = e$. Let $H_1 = \{e, a_1\}$.

(H_1, o) is a sub-group of (G, o) of order 2.

If $H_1 = G$, there is nothing to prove. Let $a_2 \in G$, $a_2 \neq a_1$ and $H_2 = \{e, a_2\}$. Then (H_2, o) is a sub-group of (G, o) of order 2. Since $(H_1\ o)$ and (H_2, o) are both normal sub-groups, then by theorem 4. 10 $(H_1 \text{ o } H_2, o)$ is a group of order 4. If $G = H_1 \text{ o } H_2$, we have nothing to prove. Let $a_3 \in G$ and $a_3 \notin H_1 \text{ o } H_2$. Let $H_3 = \{e, a_3\}$, then (H_3, o) is sub-group of (G, o) of order 2. Again $(H_1 \text{ o } H_2 \text{ o } H_3, o)$ will be a group of order 8.

Proceeding in this manner, we shall arrive at the required result in a finite number of steps.

Groups of order 4. We have read the following two groups of order 4:

(*i*) If $G_1 = \{1, -1, i, -i\}$, then $(G_1\ .)$ is a commutative group of order 4. This is a cyclic group.

$$(ii) \text{ If } G_2 = \left\{\begin{pmatrix}1&2&3&4\\1&2&3&4\end{pmatrix}, \begin{pmatrix}1&2&3&4\\2&1&3&4\end{pmatrix}, \begin{pmatrix}1&2&3&4\\1&2&4&3\end{pmatrix}, \begin{pmatrix}1&2&3&4\\2&1&3&4\end{pmatrix} \text{o} \begin{pmatrix}1&2&3&4\\1&2&4&3\end{pmatrix}\right\}$$

then (G_2, o) is a group of permutations under the operation of composite of permutations. This is a commutative group but non-cyclic.

It is obvious that $G_2 = G_2' \text{ o } G_2''$, where

$$G_2' = \left\{\begin{matrix}1&2&3&4\\1&2&3&4\end{matrix}\right\}, \begin{pmatrix}1&2&3&4\\2&1&3&4\end{pmatrix}, G_2'' = \left\{\begin{pmatrix}1&2&3&4\\1&2&3&4\end{pmatrix}, \begin{pmatrix}1&2&3&4\\1&2&4&3\end{pmatrix}\right\}$$

and (G_2', o) and (G_2'', o) are sub-groups of (G_2, o) of order 2. We shall see that there are only two abstract groups of order 4 which are isomorphic to the above two groups respectively.

Let $(G_3, *)$ be any group of order 4. Then each non-identity element is either of order 2 or 4. If $a \in G_3$, $a^4 = e$, then $G_3 = [a, a^2, a^3, a^4 = e]$ and $(G_3, *)$ will necessarily be cyclic.

If each non-identity element is of order 2 by Lemma 5.3.1 then $(G_4, *)$ is a direct product of its two sub-groups of order 2 each and it is abelian. Thus, we have a cyclic group of order 4, if the group has a non-identity clement of order 4, and an abelian group of order 4 but non-cyclic since the elements of G are of order 2. This abelian non-cyclic group of order 4 is known as the four group or the Klien group.

Here $(C_1, .) \cong (G_3, *)$ and $(G_2, o) \cong (G_4, *)$.

Groups of order 6. We have come across the following two groups of order 6.

(*i*) The set of the six sixth roots of unity together with ordinary multiplication is a commutative group of order 6. This is also cyclic. Here

$$G_1 = \left\{\pm 1, e^{\frac{\pi i}{3}}, e^{\frac{-\pi i}{3}}, e^{\frac{2\pi i}{3}}, e^{\frac{-2\pi i}{3}}\right\} \text{ and } (G_1, .) \text{ is a cyclie group of order 6:}$$

(*ii*) The permutation group (S_3, o) of three symbols 1, 2, 3 under composite of functions is a non-commutative group of order 6.

We shall see that there are only two abstract groups of order 6 which are isomorphic (to the above two groups respectively.

We know that the order of an element of a group divides the order of the group. So non-identity elements of a group of order 6 are of order 2, 3, or 6.

If $a \neq e$ is an element of a group $(G_2, *)$ and if $a^6 = e$, then $(G_2, *)$ is necessarily the cyclic group of order 6.

If the group $(G_3, *)$ has no element of order 6, then $(G_2, *)$ will not by cyclic and the order of its elements will be 2 or 3.

If each element of $(G_3, *)$ is of order 2, then by Lemma 5.3.1 the group $(G_3, *)$ will be of order of some power of 2. Thus we shall not have the required group $(G_3, *)$ of order 6. Therefore, there exists at least one element $a \in G_3$ such that $a^3 = e$. Then the set $H = \{a, a^2, a^3 = e)$ together with the operation $*$ is a sub-group $(H, *)$ of $(G_3, *)$.

Let $b \notin H$ and $b \in G$, then the right coset of H,

$H * b = \{b, a * b, a^2 * b\}$

Since $H \cup H * b = G_3$, the three elements of $H * b$ are different elements of G_3, Since $(H, *)$ has index 2, it is a normal sub-group of $(G_3, *)$. So its cosets do not have any element common. That is, elements of H are different from those of $H * b$.

Thus $G_3 = \{e, a, a^2, b, a * b, a^2 * b\}$.

Now $b \in G_3, a \in G_3 \Rightarrow b * a \in G_3$, Then $b * a$ must be equal to one of the six elements of G_3.

Since $b * a = e \Rightarrow b * a^3 = e * a^2 \Rightarrow b * e = a^2$ since $a^3 = e$

$$\Rightarrow b = a^2, \text{ but } b \neq a^2,$$

$$b * a = a \Rightarrow b = e, \textit{ but } b \neq e,$$

$$b * a = a^2 \Rightarrow b = a, \text{ but } b \neq a,$$

$$b * a = b \Rightarrow a = e, \textit{ but } a \neq e.$$

CHAPTER 5

Therefore

$b * a = a * b$ or $b * a = a^2 * b.$

Again $b \in G \Rightarrow b^2 \in G$ and b^2 must be equal to the one of six elements of G.

Since $b^2 = b \Rightarrow b = e$, $b^2 = a * b \Rightarrow b = a$, $b^2 = a^2 * b \Rightarrow b = a^2$ but $b \neq e$, $b \neq a$, $b \neq a^2$, then we see that

$$b^2 \neq b,\ b^2 \neq a * b,\ b^2 \neq a^2 * b.$$

Again, $b^2 = a \Rightarrow b^3 = a * b \neq e,$

$$b^4 = b^2 * b^2 \Rightarrow b^4 = a^2 \neq e,\ b^5 = (b^2)^2 * b = a^2 * b \neq e,$$

This shows that $b^6 = e$ and we have already excluded this case.

Similarly, we can shows that *if* $b^2 = a^2$, then $b^6 = e$. Thus we have $b^2 = e$. So we have the following two possibilities:

$$b * a = a * b,\ b^2 = e;\ b * a = a^2 * b,\ b^2 = e.$$

If $b * a = a * b$, $b^2 = e$; then $(b * a)^2 = (b * a) * (b * a)$

$$= (a * b) * (b * a)$$

$$= a * b^2 * a$$

$$= a^2$$

$$(b * a)^3 = b,\ (b * a)^4 = a,\ (b * a)^5 = b * a^2,\ (b * a)^5 = e.$$

This shows that $(b * a)^6 = e$. Again we shall exclude this case. Thus, we have remaining possibility.

$$b * a = a^2 * b,\ b^2 = e. \text{ We have } a^3 = e$$

Now we give the operation table

$*$	e	a	a^2	b	$a*b$	$a^2 * b$
e	e	a	a^2	b	$u* b$	$a^2 * b$
a	a	a^2	e	$a * b$	$a^2 * b$	b
a^2	a^2	e	a	$a^2 * b$	b	$a * b$
b	b	$a^2 * b$	$a * b$	e	a^2	a
$a * b$	$a * b$	b	$a^2 * b$	a	e	a^2
$a^2 * b$	$a^2 * b$	$a * b$	b	a^2	a	e

From this table $(G, *)$ is a group of order 6. This group is non-cyclic and unique. This group is generated by two elements a, b which are subjected to the defining conditions $a^3 = e$, $b^2 = e$, $b * a = a^2 * b$.

This group is isomorphic to (S_3, o), and this varification is left to the readers. The group of the six sixth roots of unity under multiplication is isomorphic to the cyclic group of order 6.

Theorem 5.3.1: *Any group of order p^2, p prime, is either cyclic or a direct product of two cyclic groups of order p each.*

Proof: We know that any group $(G, *)$ of order p^2, p is prime, is an abelian group. If $(G, *)$ has an element of order p^2 then it is a cyclic group of order p^2. Let $(G, *)$ be not cyclic. Then it has a non-identity element a such that $o(a) | o(G) = p^2$. This gives $o(a) = p$. The set $H = \{a, a^2, ..., a^p = e\}$ together with group operation $*$ is a cyclic group of order p.

Let $b \in G$ and $b \notin H$. Since $o(b) | o(G)$, then $o(b) = p$. The set $K = \{b, b^2, ... b^p = e\}$ is again a cyclic group of order p under the group operation*. Here $H \cap K = \{e\}$.

Consequently $o(H * K) = \dfrac{o(H)\, o(K)}{o(H \cap K)} = \dfrac{p.p}{1} = p^2 = o(G)$. Hence $G = H \times K$. Thus $(G, *)$ is a direct product of two cyclic groups $(H, *)$ and $(K, *)$ of order p each.

Groups of order 8: We are familiar with the following groups of order 8.

(*i*) The set of the eight eighths roots of unity together with multiplication forms a group. This group is cyclic.

(*ii*) The sub-group of the group (S_6, o) which is the direct product of its sub-groups of order 2.

If $H_1 = \{I, (12)\}$, $H_2 = [I, (1, 3, 4)\}$, $H_3 = \{I, (5, 6)\}$ then (H_1, o), (H_2, o) and (H_3, o) are sub-groups of order 2 and the required group is the direct product of these sub-groups.

(*iii*) The sub-groups of (S_6, o) which is the direct product of its two sub-groups of order 4 and 2.

If $H_1 = \{I, (1\ 2\ 3\ 4), (1\ 4\ 3\ 2), (1\ 3)\ o\ (2\ 4)\}$ and

$H_2 = [I, (56)\}$, then (H_1, o) and (H_2, o) are

sub-groups of order 4 and 2 respectively and the required group is the direct product of (H_1, o) and (H_2, o).

All these groups discussed above are abelian. First one is cyclic, and (*ii*) and (*iii*) are not cyclic.

Definition 5.3.1: For any natural number n, a group $(D_n, *)$ generated by two elements a and b satisfying the conditions $a^n = b^2 = e$, $b * a * b^{-1} = a^{-1}$ is called n^{th} dihedral group.

Let $((a), *)$ and $((b), *)$ be two cyclic groups of order n and 2 respectively. Let $H = (a)$ and $K = (b)$. Now we consider the set $G = H \times K = \{(a^u, b^v) | u, v \in Z)$. For any (a^u, b^v), $(a^x, b^y), \in G$ we defined $(a^u, b^v).(a^x, b^y) = (a^{u+x}, b^y)$ if $v = 0$ and $(a^u, b^v), (a^x, b^y) = (a^{u+x}, b^{u+y})$ if $b = 1$. Then G will be generated by $a_1 = (a, b^o)$, $b_1 = (a^o, b)$ with $a_1^n = e$, $b_1^2 = e$, $b_1 . a_1 . b_1^{-1} = a_1^{-1}$. Hence $(G, .)$ is an nth dihedral group. It is clear that $(G, .) \cong (D_n, *)$

Definition 5.3.2: A group $(G, .)$ generated by two elements a and b satisfying the relations $a^4 = e$, $b^2 = a^2$, $b.\ a.\ b^{-1} = a^{-1}$ is called the group of Quaternions.

The set H consisting of the following matrices:

$$\begin{bmatrix} 1 & 0 \\ 0 & 1 \end{bmatrix}, -\begin{bmatrix} 1 & 0 \\ 0 & 1 \end{bmatrix}, \begin{bmatrix} 0 & 1 \\ -1 & 0 \end{bmatrix}, -\begin{bmatrix} 0 & 1 \\ -1 & 0 \end{bmatrix}, -\begin{bmatrix} i & 0 \\ 0 & i \end{bmatrix}, \begin{bmatrix} i & 0 \\ 0 & -i \end{bmatrix}$$

CHAPTER 5

$$\begin{bmatrix} 0 & i \\ i & 0 \end{bmatrix}, -\begin{bmatrix} 0 & i \\ i & 0 \end{bmatrix},$$

together with the operation of matrix multiplication is a group of order 8. If we take $a = \begin{bmatrix} 0 & 1 \\ -1 & 0 \end{bmatrix}, b = \begin{bmatrix} i & 0 \\ 0 & -i \end{bmatrix}$, then we can easily see that $a^4 = \begin{bmatrix} 1 & 0 \\ 0 & 1 \end{bmatrix}$, $b^2 = a^2$, $b.\ a.\ b^{-1} = a^{-1}$. Hence $(H, .)$ is a group of Quaternions. We can also varify that any grqup of Quaternions will be isomorphic to $(H, .)$.

Theorem 5.3.2: *Any non-abelian group $(G, *)$ of order 8 is either $(D_4, *)$ or Quaternion group.*

Proof: Since $(G, *)$ is a non-abelian group, G cannot be cyclic and so G cannot have any element of order 8.

If each element of G is of order 2, then $(G, *)$ is again abelian. Thus there exists an element $a \in G$, $a \neq$ e such that a is not of order 2.

Since $o\ (a) | o\ (G)$ and $o\ (a) \neq 2$, then $o\ (a) = 4$. Thus sub-group $(H, *)$, where H $= \{a\ a^2\ a^3, a^4 = e\}$, is of index 2. So $(H, *)$ is a normal sub-group of (G, o). Let $b \in G$ and b $\notin H$. Then

$$G = H \cup H * b = \{e, a, a^2, a^3, b, a * b, a^2 * b, a^3 * b\}.$$

It is clear that G is generated by a and b. If $a * b = b * a$ then $(G, *)$ becomes abelian. So $a * b \neq b * a$.

Since $(H, *)$ is a normal sub-group of $(G, *)$ and $b \in G$, $a \in H$, then $b * a * b^{-1} \in H$, and $o\ (b * a * b^{-1}) = o\ (a) = 4$. But in H there is only one element a^3 of order 4 other than a.

So $b * a * b^{-1} = a^3 = a^{-1}$. Since $(a^4 = e \Rightarrow a^3 = a^{-1})$

Since $(H, *)$ is normal sub-group of order 4 of a group $(G, *)$ of order 8, then $(G/H, *)$ will be group of order 2, so $(H * b)^2 = H \Rightarrow b^2 \in H$. That implies $b^2 = e$, $b^2 = a$, $b^2 = a^2$, or $b^2 = a^3$. Since

$$b^2 = a \Rightarrow b^8 = a^4 = e \Rightarrow o\ (b) = 8$$

$$b^2 = a^3 \Rightarrow b^8 = a^{12} = (a^4)^3 = e \Rightarrow o\ (b) = 8.$$

Hence $b^2 \neq a$, $b^2 \neq a^3$ since $(G, *)$ is not abelian.

Hence $b^2 = e$, $b^2 = a^2$.

When $b^2 = e$, then $(G, *)$ is $(D_4, *)$ and when $b^2 = a^2$, then $(G, *)$ is a group of Quaternions.

Theorem 5.3.3: *Let p and q be two distinct prime numbers.*

To be definite let $p < q$. If (G, o) is a group of order $p\ q$, then one of the following holds.

(i) G is cyclic (Abelian).

(ii) It has two generators a and b such that $a^p = e$, $b^q = e$, $a^{-1}\ o\ b\ o\ a = b^r$, for some r such that $r \not\equiv 1 \pmod q$, $r^p \equiv 1 \pmod q$ and p divides $(q - 1)$.

Proof: Since $o(G) = p.q.$ then G has at least one sylow q-sub-group of order q. By Theorem 4.4.4 the number of sylow q–sub-groups is $1 + m\,q$ for some $m \geq 0$ and $1 + mq$ divides p. As $q > p$ we have $m = 0$. Thus G has only one sylow q-sub-group Q which is normal of prime order q and is cyclic, say, $Q = (b)$. Again G has at least one sylow p-sub-group of order p and the number of such sylow p-sub-groups of G is $1 + kp$ for some $k \geq 0$ and $1 + kp$ divides q. Some q of prime, $1 + xp = 1$ or $1 + kp = q$. In the former case G has only one sylow p-sub-group P which is normal and cyclic, say, $P = (a)$.

Then $(a) \cap (b) = (e)$ gives $a \ o \ b = b \ o \ a$. Since if (H, o) and (K, o) are normal sub-groups of a group G such that $H \cap K = (e)$, then $h \ o \ k = k \ o \ h$, for all $h \in H, k \in K$.

By theorem 4.1.3 $H \ o \ K = G$. and $o\ (HoK) = o\ (G)$

$\Rightarrow o\ (aob) = o\ (a)\ o\ (b) = p.q = o\,(G)$. Hence $G = (a\ b)$ is a cyclic group of order pq.

Again, $1 + x\,p = q \Rightarrow p$ divides $q - 1$. As G has at least one sub-group, say (a) of order p, $o\ (a) = p$. Now $a^{-1}\ o\ b\ o\ a \in (b)$, as (b) is normal sub-group of G. Consequently, $a^{-1}\ o\ b\ o\ a = b^{\,r}$ for some r with $1 \leq r \leq q$.

If $r = 1$, we get $G = (a\ b)$ a cyclic group of order pq. So if G is non-abelian, $r \neq 1$. Thus $r \neq 1\ (mod\ q)$.

Now $a^{-2}\ o\ b\ o\ a^2 = a^{-1}\ o\ (a^{-1}\ o\ b\ o\ a)\ o\ a = a^{-1}\ o\ b^r\ o\ a = (a^{-1}\ o\ b\ o\ a)^r = (b^r)^2 = b^{r^2}$

Similarly, $a^{-3}\ o\ b\ o\ a^3 = b^{r^3}$ and so on. Finally $a^{-p}\ o\ b\ o\ a^{\,p} = b^{r^p}$. However $a^{\,p} = e$. Hence $b = b^{r^p}$.

This yields $r^{\,p} \equiv 1 \pmod q$. Further notice that $(a) \cap (b) = e$ and hence $o\,[(a)\,(b)] = o\ (a)\ o\ (b) = p.q = o\ (G)$. Hence $G = (a)\ (b)$. Thus each element of G is of the forms $a^n\ b^n$. This shows that G is a group generated by two elements a and b such that $a^{\,p} = e$, $b^{\,q} = e$, $a^{-1}\ o\ b\ o\ a = b^{\,r}$, $r \not\equiv (1\ mod\ q)$, $r^{\,p} \equiv 1\ (mod\ q)$ and $p \mid (q - 1)$.

Conversely let r, p, q be such that $p < q$ are primes and they satisfy the above conditions.

Consider two cyclic group $H = (a)$ and $K = (b)$ of order p and q respectively. Let $G = H \times K = \{(a^u, b^v)\}$ where u, v are integers. $o\,(G) = o\ (H \times K) = o\,(K)\ o\ (K) = p.q.$ that is, G has pq elements. For any $(a^u, b^v), (a^x, b^y) \in G$.

We define a binary operation by

$$(a^u, b^v)\ .\ (a^x, b^y) = (a^{u+x}, b^{vr^x+y}) \qquad ...(1)$$

This operation is well-defined since

$$\left(a^u, b^v\right).\left(a^{u_1}, b^{v_1}\right) \text{ and } \left(a^x, b^y\right)\left(a^{x_1}, b^{y_1}\right)$$

$$\Rightarrow \qquad a^u = a^{u_1}, b^u = b^{u_1} \text{ and } a^x = a^{x_1}, b^y = b^{y_1}$$

$$\Rightarrow \qquad p \mid (u - u_1),\ p \mid (x - x_1),\ q \mid (u - v_1),\ q \mid (y - y_1) \qquad ...(2)$$

$$\Rightarrow \qquad p \mid u - u_1 + x - x_1,$$

$$\Rightarrow \qquad p \mid (u + x) - (u_1 + x_1),$$

CHAPTER 5

$\Rightarrow \qquad a^{u+x} = a^{u_1+x},$

Since $o(a) = p$

Again, consider $v\, r^x + y - v_1 r^{x_1} - y_1$ ....(3)

$$= v\left(r^x - r^{x_1}\right) + r^{x_1}(v - v_1) + y - y_1 \qquad ...(4)$$

since $q \mid v - v_1$ and $q \mid y - y_1$, and if $x = x_1$,

Then $q \mid v.o + r^{x_1}(v - u_1) + (y - y_1)$, from (4), we get

$\Rightarrow \qquad q \mid (vr^x + y) - (v_1 r^{x_1} + y_1)$

$\Rightarrow \qquad b^{vr^x+y} = b^{v_1 r^{x_1}+y_1}$

Suppose $x \neq x_1$. To be definite. Let $x < x_1$, then

$$r^x - r^{x_1} = r^x\left(1 - r^{x_1-x}\right).$$

As $r^p \equiv 1 \ (mod\ q)$ and $p \mid (x_1 - x)$, we have

$$r^{x_1-x_1} = 1 \ (\text{mod } q) \Rightarrow q \mid \left(1 - r^{x_1-x}\right) \Rightarrow q \mid \left(r^x - r^{x_1}\right)$$

Then Again from (4) we get

$$q \mid (vx^r + y) - \left(v_1\, r^{x_1} + y_1\right)$$

$\Rightarrow \qquad b^{vr^x+y} = b^{v_1 re^{x_1}+y_1}$

Hence $\left(a^{u+x}, b^{vr^x+y}\right) = \left(a^{u_1+x_1}, b^{v_1 r^{x_1}+y_1}\right)$

Thus the operation is well-defined.

We see that the system $(H \times K, \cdot)$ under this operation is a group. For all (a^x, b^y), $\left(a^{x_1}, b^{y_1}\right)\left(a^{x_2}, b^{y_2}\right) \in H \times K = G,$

To prove that (G, o) is a group, we consider (a^u, b^v), (a^x, b^y), $\left(a^{x_1}, b^{y_1}\right) \in G$. We see that

(1) Associatively: $\left[\left(a^u, b^v\right) \cdot \left(a^x, b^y\right)\right] \cdot \left(a^{x_1}, b^{y_1}\right)$

$$= \left(a^{u+x}, b^{ux+y}\right) \cdot \left(a^{x_1}, b^{y_1}\right)$$

$$= \left(a^{(u+x)+x_1}, b^{(v\gamma^x+y)v^{x_1}+y_1}\right) \qquad ...(1)$$

Similarly,

$$\left(a^u, b^v\right) \cdot \left[\left(a^x, b^y\right) \cdot \left(a^{x_1}, b^{y_1}\right)\right] = \left(a^u, b^v\right) \cdot \left(a^{x+x_1}, b^{y\gamma^x+y_1}\right)$$

$$= \left(a^{u+(x+x_1)}, b^{(vr^x+y)r^{x_1}+y_1}\right)$$

$$= \left(a^{a+x+x_1}, b^{(ur^x+y)r^{x_1}+y_1}\right) \quad ...(2)$$

From (1) and (2) the law of associativity holds good.

(2) Existence of Identity: For $(a^u, b^v) \in G$, there exists an element $(a^o, b^o) \in G$ such that

$$\left(a^o, b^o\right).\left(a^u, b^v\right) = \left(a^{o+u}, b^{o.\gamma^u+v}\right) = \left(a^u, b^v\right)$$

and
$$\left(a^u, b^v\right).\left(a^o, b^o\right) = \left(a^{u+o}, b^{ur^0+o}\right) = \left(a^u, b^v\right)$$

Hence $(a^o, b^o) \in G$ is the identity in G.

(3) Existence of Inverse: Considering $x = p - u$ and y to be such that $u\ r^x + y$ is divisibly by q. That is, there exists a positive integer l such that

$$v\ r^x + y = lq \text{ and } y = lq - vr^x > 0$$

then
$$(a^u, b^v)\ .\ (a^x, b^y) = \left(a^{u+x}, b^{vr^x+y}\right)$$

$$= \left(a^p, b^{lq}\right) = \left(a^o, b^o\right)$$

Hence $(G, \cdot)$ is a group.

Again, let $a_1 = (a, b^o)$, $b_1 = (a^o, b)$, then

$$a_1^u = \left(a^u, b^o\right), b_1^v = \left(a^o, b^v\right);$$

$$\text{So } a_1^u b_1^v = \left(a^u, b^o\right).\left(a^o, b^v\right) = \left(a^u, b^v\right)$$

That G is generated by a_1 and b_1 Further

$$a_1^P = \left(a^p, b^o\right) = \left(a^o, b^o\right) = e \text{ and } b_1^q = \left(a^o, b^q\right) = \left(a^o, b^o\right) = e$$

The identity of G.

Now
$$a_1^{-1} . b_1 . a_1 = \left(a^{p-1}, b\right).\left(a^o, b\right).\left(a, b^o\right)$$

$$= \left(a^o, b^r\right) = b_1^r$$

Hence, G is a group of order pq generated by elements a_1, b_1 of orders p and q respectively, $r \not\equiv 1\ (mod\ q)$, $r^p \equiv 1\ (mod\ q)$ and $p \mid (q - 1)$.

Theorem 5.3.4: *Any two non-abelian groups G and G′, of order pq are isomorphic.*

Proof: Assume that G and G' having generating sets $\{a, b\}$ and $\{a', b'\}$ such that $a^p = e$, $b^q = e$, $(a')^p = e'$, $(b')^q = e'$, for some integer r, r',

$a^{-1} * b * a = b^r$, $(a')^{-1} \circ (b') \circ a' = (b')^{r'}$ with $r \not\equiv 1$ (mod q) $r' \neq 1$ (mod q), $r^p = 1$ (mod q), $(r')^p \equiv 1$ (mod q).

CHAPTER 5

Now all solutions of the congruence $x^p \equiv 1 \pmod q$ which are not congruent to 1 (mod q) are congruent to some one of $r, r^2, r^3, \ldots . r^{p-1}$ *modulo* q. Thus we have $r' = r^i$ (mod q) for some i, $1 \le i \le p-1$.

Then $$b^{r^i} = b^{r^i} = a^{-i} * b * a^i$$

If we put $a_1 = a^i$, $b_1 = b$ we see that a_1, b_1 also generate G and $a_1^{-1} * b_1 * a_1 = b_1^{r'}$.

Now we define a mapping $\phi : G \to G'$ by

$$\phi\left(a_1^u, b_1^v\right) = \left((a^1)^u, (b')^v\right) \text{ for all integers } u, v.$$

For $\left(a_1^u, b_1^v\right), \left(a_1^{u_1}, b_1^{u_1}\right) \in G$

$$\phi\left[\left(a_1^u, b_1^v\right) * \left(a_1^{u_1}, b_1^{v_1}\right)\right] = \phi\left(a_1^{u+u_1}, b_1^{vr^x+v_1}\right)$$

$$= \left((a')^{u+u_1}, (b')^{vr^x+v_1}\right)$$

$$= \left(a'^u, b'^v\right) o \left(a'^{u_1}, (b')^{v_1}\right)$$

$$= \phi\left(a_1^u, b_1^v\right) o\, \phi\left(a_1^{u_1}, b_1^v\right)$$

$\therefore$ ϕ is a homorphism.

Again, Assume that

$$\phi\left(a_1^u, b_1^v\right) = \phi\left(a_1^{u_1}, b_1^{v_1}\right)$$

$$\Rightarrow \quad \left(a'^u, b'^v\right) = \left(a'^{u_1}, b'^{v_1}\right)$$

$$\Rightarrow \quad a'^u = a'^{u_1} \text{ and } b'^v = b'^{v_1}$$

$$\Rightarrow \quad u = u_1 \text{ and } v = v_1$$

$$\Rightarrow \quad \left(a_1^u, b_1^v\right) = \left(a_1^{v_1}, b_1^{v_1}\right)$$

Since G and G' are of the same order pq.

$\therefore$ ϕ is an isomorphism.

Remark: (1) If $p \dagger (q-1)$, then any group of order pq is cyclic.

(2) If $p \mid (q-1)$, then there are only two non-isomorphic groups of order pq one of which is abelian that is again a cycle group as p and q different primes and other is non-abelian.

Example 5.3.1: Here, consider $p = 2$ and $q = 5$ $p \mid (q-1)$. Thus there are two non-isomorphic groups of order 10 one of them is cyclic and other in generated by two elements a and b such that $a^2 = e$, $b^5 = e$, $a^{-1}\, b \cdot a = b^4$ since $b^4 = 1 \pmod 5$.

Example 5.3.2: Consider $p = 3$, $q = 5$, here $p \dagger (q-1)$ thus any group of order 15 is cyclic.

PROBLEMS

1. Prove that any non-abelian group of order 6 is isomorphic to the symmetric group S_3.
2. Prove that in (S_n, o), the sum of lengths of any set of non-identity pair-wise disjoint cycles cannot exceed n using the fact that any group of order 15 is cyclic, conclude that $(A_5\ o)$, (A_6, o), and (A_7, o) have no sub-groups of order 15, although 15 divides their orders.
3. If a group $(G, *)$ of order 12 is not isomorphic to (A_4, o), prove that $(G, *)$ must contain an element of order 6.
4. If a non-abelian group $(G, *)$ of order 12 is not isomorphic to (A_4, o) and contain at least two elements of order 2, then $(G, *)$ is the dihedral group (D_6, o).
5. Show that the direct product $S_3 \times H$, where $(H, *)$ is a cyclic group of order 2, is isomorphic to D_6.
6. Show that the number of non-isomorphic abelian groups of order 12 is two.

CHAPTER 6 RING THEORY

6.1 RINGS

In this chapter, as shall study algebraic systems (mathematical systems) with two operations. There are many mathematical systems with two binary operations. But here we shall define an algebraic system known as rings. A ring is basically a combination of a commutative group and *a* semigroup. In this system many of the important notions in group theory will be extended to systems with two operations.

Definition 6.1.1: (*a*) A ring is a mathematical system $(R, *, o)$ consisting of a non-empty set R and two binary operations $*$ and o defined on R such that

(1) $(R, *)$ is an abelian group.

(2) (R, o) is a semigroup.

(3) The semigroup operation o is distributive over the group operation $*$.

Here it is convenient and customary to use + for group operation $*$ and . for semigroup operation o. We should note that + and . are specified operations and not necessarily ordinary addition and multiplication.

For convenience, however, we shall refer + as addition and . as multiplication. Thus, making use of these + and . operations we speak $(R, +)$ the additive group of the ring and $(R, .)$ the multiplicative semigroup of the ring.

Thus, we have a unique identity element for addition called the zero element of the ring and is denoted by 0. Similarly for an element $a \in R$ there exists a unique additive inverse $-a \in R$. Now in the light of this terminology we give below Ihe definition of the ring.

Definition 6.1.2: Let $(R, +,.)$ be a ring. Then $(R, * .)$ is a commutative ring if, and only if:

(Commutative) for all $a, b \in R$,

$$a. b = b. a.$$

That is, the operation of multiplication is commutative.

Definition 6.1.3: Let $(R, +, .)$ be a ring. Then $(R, +,.)$ is ring with identity if and only if.

(Identity) there exists an element $I \in R$ such that $a. 1 = 1 . a$ for all $a \in R$.

Definition 6.1.4: An element a of the ring $(R, +, .)$ with identity is called invertible if there exists an inverse $a^{-1} \in R$ of the element $a \in R$ under multiplication such that $a \,.\, a^{-1} = a^{-1}, a = 1$.

The set of all invertible elements is denoted by R^*.

Theorem 6.1.1: *Let* $(R, +, .)$ *be a ring with identity. Then the pair* $(R,^*,.)$ *is a group of invertible elements under multiplication.*

Proof: It is clear that the set $R^* \neq \phi$, for $I \in R^*$. Further, we observe that:

(1) If $a, b \in R^*$, the equations

$$(a \,.\, b) \,.\, (b^{-1} \,,\, a^{-1}) = a \,.\, (b \,.\, b^{-1})\, a^{-1}$$
$$= a \,.\, (1) \,.\, a^{-1}$$
$$= 1$$

and

$$(b^{-1} \,.\, a^{-1})\,(a \,.\, b) = b^{-1} \,.\, (a^{-1} \,.\, a) \,.\, b$$
$$= b^{-1} \,.\, (1) \,.\, b$$
$$= 1$$

which show that $a \,.\, b \in R^*$. That is, R^* is closed under multiplication.

(2) Since $R^* \subseteq R$ and the operation of multiplication is associative in R, it is also associative in R^*.

(3) I is the identity element of R^*.

(4) Since $a \in R^*$ is invertible, then there exists an element $a^{-1} \in R$ such that $a.a^{-1} = a^{-1} \,.\, a = 1$. Thus, the inverse of a^{-1} is a. That is $(a^{-1})^{-1} = a \in R^*$ which shows $a^{-1} \in R^*$.

Hence $(R^*, .)$ is a group. If the ring $(R, +, .)$ is commutative, then $(R^*, .)$ is a communitative group.

Example 6.1.1: The system $(Z, +, .)$ is a ring of integers under ordinary addition and ordinary multiplication.

Note: In the ring $(Z, +, .)$ there are two invertible elements, namely, I and -1, and $(\{1, -1\},.)$ is a group.

Similar to this example we can verify other systems.

Example 6.1.2: The system $(R, +, .)$ forms a ring of real numbers under ordinary addition and multiplication and $(R - \{0\}, .)$ is a group of invertible elements of the ring.

Example 6.1.3: The system $(Q, +, .)$ forms a ring of rational numbers under ordinary addition and multiplication and the system $(Q - \{0\}, .)$ is a multiplicative group of the invertible elements of the ring.

Example 6.1.4: The system $(E, +, .)$ is a ring of even integers under ordinary addition and multiplication.

Example 6.1.5: Let $S = \{a + b\sqrt{2} \mid a, b$ *are* integers$\}$.

Then the ordered triple $(S, +, .)$ is a ring of real numbers of the form $a + b\sqrt{2}$, where a, b are integers, under usual addition and multiplication.

Example 6.1.6: The system $(Z_n, +_n, \odot_n)$ is a commutative ring with identity, known as the ring of integers modulo n, where Z_n is the set of residue classes modulo n, namely, $\{0, 1, 2, 3........., n-1\}$ and $+_n$ denotes the addition of resdue classes modulo n, and $\odot_n$ denotes the multiplication modulo n.

Example 6.1.7: Show the ordered triple $(P(S), \Delta, \cap)$ is a ring, where $P(S)$ is the set of all subsets of a non-empty set S and Δ is the symmetric difference of two sets defined by

$$A \Delta B = \{x \mid x \in A \cup B, x \notin A \cap B\},$$

or

$$A \Delta B = (A \cup B) - (A \cap B)$$

$$= (A - B) \cup (B - A).$$

and $\cap$ denotes the usual intersection of two sets.

Solution: In example 2.1.5 we have shown that $(P(S), \Delta)$ is a commutative group. Now we verify that $(P(S). \cap)$ is a semigroup.

(1) (Closure) for all $A \subseteq S$, $B \subseteq S$, $A \cap B \subseteq S$.

(2) (Associativity). For all $A \subseteq S$, $B \subseteq S$, $C \subseteq S$, $(A \cap B) \cap C = A \cap (B \cap C)$.

(3) (Commutativity). For all $A \subseteq S$, $B \subseteq S$, $A \cap B = B \cap A$.

which proves that $(P(S), \cap)$ is a commutative semigroup.

Now, for proving $(P(S), \Delta, \cap)$ a ring we verify the peroperty of left distributivity of $\cap$ over Δ, we choose $A, B, C \subseteq S$, and obtain

$$A \cap (B \Delta C) = A \cap [(B - C) \cup (C - B)]$$

$$= \{A \cap (B - C\} \cup \{(A \cap (C - B)\}$$

$$= (A \cap B) - (A \cap C) \cup (A \cap C) - (A \cap B)$$

$$= (A \cap B) \Delta (A \cup C).$$

Hence $(P(S), \Delta, \cap)$ is a commutative ring.

Example 6.1.8: Let $(G, *)$ be an arbitarary commutative group and Hom G be the set of all homomorphisms from $(G, *)$ onto itself. Then show that (Hom G, +, o) is a ring with identity, where the operation + is defined by

$$(f + g)(a) = f(a) * g(a), \; a \in G, \forall f g \in \text{Hom } G,$$

and o denotes the functional composition.

Solution: We first check up the group axioms for the pair (Hom G, +).

Closure property: $\forall f g \in$ Hom G, and $\forall a, b \in G$,

$$(f + g)(a * b) = f(a * b) * g(a * b)$$

$$= (f(a) * f(b)) * (g(a) * g(b))$$

$$= (f(a) * g(a) * (f(b) * g(b)) \text{ since } (G, *)$$

is commutation group.

$$= (f + g)(a) * (f + g)(b),$$

So that the sum $f + g \in$ Hom G.

Associative property: $\forall f\, g,\, h \in$ Hom G, $a \in G$, we have

$$\begin{aligned}((f + g) + h)(a) &= (f + g)(a) * h(a)\\ &= ((f(a) * g(a)) * h(a)\\ &= f(a) * (g(a) * h(a))\\ &= f(a) * (g + h)(a)\\ &= (f + (g + h))(a).\end{aligned}$$

Thus $(f + g) + h = f + (g + h)$.

Existance of identity: $\forall f \in$ Hom G, there exist constant mapping Z which map all elements of G on e, the identity of $(G, *)$ such that

$$\begin{aligned}(f + Z)(a) &= f(a) * Z(a)\\ &= f(a) * e = f(a).\end{aligned}$$

Thus $f + Z = f \Rightarrow Z$ is an identity in Hom G, that is, the mapping Z in Hom G is the zero element.

Existance of inverse: $\forall f \in$ Hom G, $\exists -f \in$ Hom G defined by $(-f)(a) = f(a)^{-1}$, such that, $\forall\, a \in G$

$$\begin{aligned}(f + (-f))(a) &= f(a) * f(a)^{-1}\\ &= e = Z(a)\end{aligned}$$

which implies that $f + (-f) = Z$, that is, $f + (-f)$ maps all elements of G on the identity element of G which is the characteristic of the zero-element of Hom G.

Commutative property: $\forall\, f, g \in$ Hom G, $a \in G$, we have

$$\begin{aligned}(f + g)(a) &= f(a) * g(a)\\ &= g(a) * f(a) \text{ since } (G, *) \text{ is commutative}\\ &= (g + f)(a).\end{aligned}$$

Thus $$f + g = g + f.$$

Hence (Hom G, +) is a commutative group. We also know that (Hom G, o) is a semigroup with identity by theorem 3.3.1.

Now for proving that (Hom G, +, o) is a ring with identity there remains to show that o is distributive over +. For this, we have, for any $a \in G$,

$$\begin{aligned}f\, o\, (g + h)(a) &= f((g + h)\, a)\\ &= f(g(a) * h(a)\\ &= f(g(a)) * f(h(a))\\ &= (fog)(a) * (foh)(a)\end{aligned}$$

Therefore, $f\, o\, (g + h) = f o\, g + foh$. Similarly, we can prove right distributive law.

Thus (Hom G, +, o) is a ring with identity.

6.2 ELEMENTARY PROPERTIES OF RINGS

There are two properties of rings which are 0. a = 0 and (– a). (–b) = a. b. If they are established as a theorem for any arbitrary ring then we often explain these properties as a "rule" or "law". Here we shall establish several results. Since (R, +, .) is a commutative group under addition, the cancellation law for addition is valid for arbitrary rings. If a, b, $c \in R$, (R, +,.) is a ring, $a + b = a + c$ then $b = c$, for if $a \in R$, $\exists -a \in R$ such that $a + (-a) = 0$.

$$(-a) + (a + b) = (-a) + (a + c)$$

$$\Rightarrow \quad [(-a) + a] + b = [(-a) + a] + c$$

$$\Rightarrow \quad 0 + b = 0 + c$$

$$\Rightarrow \quad b = c.$$

Thus the results which are true for group (R, +) will also be, valied for ring (R, +, .).

Theorem 6.2.1: *In any ring (R, +, .), if a ∈ R, then a. 0 = 0. a = 0.*

Proof: we know

$$a + 0 = a \text{ for } a \in R,$$

and 0 is the additive identity of R.

By distributive property, we have

$a \,.\, a + 0 = a.\ (a + 0) = a \,.\, a + a \,.\, 0.$

By left cancellation law under addition, we have.

$a \,.\, 0 = 0.$

Similarly, we can prove $0 \,.\, a = a$.

It is clear if R = {0} the ((0), +, .) is a ring.

Remark: The ring ({0}, +, .) with one element 0 called zero ring.

Theorem 6.2.2: *Let (R, +,.) be a ring with identity such that that R ≠ {0}. Then the elements 0 and I are distinct.*

Proof: Since $R \neq \{0\}$, some non-zero element $a \in R$. Now, if 1 = 0, then it would follow that

$a = a \,.\, I = a \,.\, 0 = 0, \forall\, a \in R$

which contradicts that $R \neq \{0\}$.

Therefore any ring with identity element contains more than one element and 1 and 0 elements are different.

We have seen in a ring (R, +,.), if $a \in R$ the a 0 = 0. Which implies the product of two elements in a ring is zero whenever either factor is zero. But the converse is not true. There are examples where $a \neq 0$, $b \neq 0$, $a \,.\, b = 0$. (See the example of (Z_4, $+_4$, $\cdot_4$) residue classes modulo 4).

Definition 6.2.1: A ring $(R, +, .)$ is said to have divisors of zero, if there exists two non-zero elements $a, b \in R$ such that $a \cdot b = 0$.

Example 6.2.1: Let R' be set of ordered pairs of real numbers, *i.e.*, $R' = R \times R$ and addition and multiplication be defined by

$$(a, b) + (c, d) = (a + c, b + d],$$

$$(a, b)(c. d) = (ac, bd).$$

The straightforward calculations will show that $(R, +, .)$ is a commutative ring with identity (1, 1). The pair (0, 0) is the additive identity of $(R, +, .)$.

We observe that

$(1, 0) \cdot (0, 1) = (0, 0)$, where $(1, 0) \neq (0, 0)$, $(0, 1) \neq (0, 0)$.

Thus, (1, 0) and (0, 1) are zero divisors of the ring $(R', +, .)$.

Theorem 6.2.3: *Let $(R, +, .)$ be a ring and $a, b \in R$, then $-(a \cdot b) = a \cdot (-b) = (-a) \cdot b$*

Proof: From the definitions of additive inverse, we have

$$b + (-b) = 0$$

$$\Rightarrow \quad a \cdot b + a \cdot (-b) = a\,(b + (-b)) = a.\ 0 = 0.$$

$$\Rightarrow \quad a \cdot (-b) = -(a \cdot b) \text{ by the definition of } -(a \cdot b).$$

Similarly, we can prove

$$(-a)\,b = -(a \cdot b).$$

Corollary 1: For any element $a, b \in R$

Proof: We have

$$\begin{aligned}(-a) \cdot (-b) + a \cdot (-b) &= (-a + a) \cdot (-b)\\ &= 0 \cdot (-b)\\ &= 0\\ &= a.\ (b + (-b))\\ &= a \cdot b + a\,(-b).\end{aligned}$$

by right cancellation law, we have

$$(-a) \cdot (-b) = a.\ b.$$

Second Method.

By the theorem $(a)\,(-b) = -(a \cdot b)$.

We have

$$(-a) \cdot (-b) = -((-a) \cdot b) = -(-(a \cdot b)) = a \cdot b.$$

Corollary 2: If $a, b, \in R$. then

$a \cdot (b - c) = a \cdot b - a \cdot c,$

$(b - c) \cdot a = b \cdot a - c \cdot a$

CHAPTER 6

that is, multiplication is distributive over differences.

Proof: Since multiplication is left distributive over addition,

$$\begin{aligned} a \,.\, (b - c) &= a.\, (b + (-c)) \\ &= a \,.\, b + a \,.\, (-c) \\ &= a \,.\, b + (-(a \,.\, c)) \\ &= a \,.\, b - a \,.\, c. \end{aligned}$$

Similarly, we can show that

$$(b - c) \,.\, a = b \,.\, a - c \,.\, a.$$

Corollary 3: If $(R, +, .)$ is a ring with identity element I, then for all $a \in R$

(*i*) $(-1) \,.\, a = -a$,

(*ii*) $(-1) \,.\, (-1) = 1$

Proof: (*i*) $(-1) \,.\, a = -(1 \,.\, a) = -a$.

(*ii*) Let $I \in R$, $1 \,.\, a = a$ for all $a \in R$.

$$\begin{aligned} (1)\ a + (-1) \,.\, a &= 1 \,.\, a + (-1) \,.\, a \\ &= (1 + (-1)) \,.\, a \\ &= (0) \,.\, a = 0. \end{aligned}$$

Hence $$(-1) \,.\, a = -a,$$

(2) If $a = -1$, then

$$(-1) \,.\, (-1) = -(-1) = 1.$$

Theorem 6.2.4: *A ring $(R, +, .)$ is without zero divisors if and only if the cancellation law holds for multiplication.*

Proof: First, let $(R, +, .)$ be a ring without divisors of zero. Let $a, b, c \in R$ such that $a \neq 0$ and

$$\begin{aligned} a \,.\, b &= a \,.\, c, \text{ then} \\ a \,.\, b - a \,.\, c &= 0 \\ \Rightarrow \quad a \,.\, b + a \,.\, (-c) &= 0 \\ \Rightarrow \quad a \,.\, (b + (-c)) &= 0 \end{aligned}$$

Since $a \neq 0$, and $(R, +, .)$ has no zero divisors then

$$b + (-c) = 0 \Rightarrow b = c.$$

Similarly, we can prove that $b \,.\, a = c \,.\, a$, with $a \neq 0$, gives $b = c$.

Conversely, the cancellation law holds for multiplication in $(R, +, .)$ let $a \neq 0, b \neq 0 \in R$ and $a \,.\, b = 0$, then

$a \,.\, b = 0 = a \,.\, 0$. Since left cancellation law holds, then $b = 0$.

which is contrary to our assumption that $b \neq 0$. Hence, if cancellation law holds, then $(R, +, .)$ does not have zero divisors.

Corollary 4: Let $(R, +, .)$ be a ring. Let $a, b \in R$. Then the equation $a + x = b$ has unique solution.

Proof: This follows from the fact that $(R, +)$ is a group.

Since $a + x = b$, then

$$(-a) + (a + x) = -a + b$$

$$\Rightarrow \quad (-a + a) + x = -a + b$$

$$\Rightarrow \quad x = -a + b.$$

which shows that if such an element x exists, then $x = -a + b$

But we verify that

$$a + ((-a) + b) = (a + (-a)) + b$$
$$= 0 + b$$
$$= b.$$

which prove that $x = -a + b$ is the unique solution of the equation $a + x = b$.

Some Special Classes of Rings

We have study the system $(R, +, .)$ in which $(R, +)$ is an abelian group and $(R, .)$ is a semigroup and multiplication is distributive over addition. That is, the system $(R, .)$ satisfies only two properties, namely, closure property and associatively. As it satisfies other properties, on the basis of the properties satisfied by $(R, .)$ we name the system.

Definition 6.2.2: A ring $(R, +, .)$ is said to be ring without zero divisors if for any $a \neq 0, b \in 0 \in R, a \,.\, b \neq 0$.

Example 6.2.2: The set Z together with addition and multiplication is ring $(Z, +, .)$ without zero divisors since for any two non-zero integers

$a, b \in Z, a \,.\, b \neq 0$.

Definition 6.2.3: A commutative ring with identity is a integral domain if it has no zero divisors. That is, the system $(D, +, .)$ is called an integral domain, if

(1) $(D, +)$ is an abelian group.

(2) $(D, .)$ is a commutative semigroup with identity.

(3) Multiplication (.) is distibutive over addition + .

(4) For any $a \neq 0, b \neq 0 \in D, a \,.\, b \neq 0$.

Example 6.2.3: The set Z together with addition and multiplication is an integral domain.

Definition 6.2.4: A ring is said to be a division ring if its non-zero elements form a group under multiplication, that is, the system $(F, +, .)$ is called a division ring or skew-field if

(1) $(F, +)$ is an abelian group.

(2) $(F - \{0\}, .)$ is a group.

(3) Multiplication (.) is distibutive over addition (+).

CHAPTER 6

Example 6.2.4: The set $M_{n \times n}$ of all square matrices of order n with real numbers as its entries under matrix addition and matrix multiplication is a divisor ring because matrix multiplication is not commutative.

Finally, we make the definition of the ultra-important object known as field.

Definition 6.2.5: A field is a commutative division ring, that is, the system $(F, +, .)$, where F has at least two elements, 0 and 1, is called a field if

(1) $(F, +)$ is an abelian group.

(2) $F - \{0\})$ is an abelian group.

(3) Multiplication (.) is distributive over addition '+'.

Example 6.2.5: (1) The set Q of rational numbers under addition and multiplication is a field $(Q, +, .)$.

(2) The set R of real numbers under addition and multiplication is a field $(R, +,.)$.

(3) The set C of complex numbers under addition and multiplication is a field $(C, +,.)$

(4) The set Z_p of integers of modulo p under addition modulo p and multiplication modulo p is a field $(Z_p, +_p, o_p)$.

Here we also define a special system known as ordered integral domain or ordered field.

Definition 6.2.6: An integral domain $(D, +,.)$ is said to be ordered if it contains a subset $D+$, such that

(*i*) $D+$ is closed under addition and multiplication, *i.e.*, For all $a, b \in D_+$, $a + b \in D_+$ $a \,.\, b \in D_+$.

(*ii*) For any $a \in D_+$, one and only one of the following holds:

$$a = 0 \text{ or } a \in D_+ \text{ or } -a \in D_+.$$

D_+ is the set of positive elements of D. Also the set $D - D_+$ is called the set of negative elements of D.

Definition 6.2.7: A field $(F, +, .)$ is said to be an ordered field if it is an ordered integral domain.

Example 6.2.6: (*i*) The set of integers under addition and multiplication is an ordered integral domain.

(*ii*) The set of rational (Real) numbers under addition and multiplication is an ordered field (integral domain).

(*iii*) The set of complex numbers under addition and multiplication is not an ordered field.

Theorem 6.2.5: *Any field is an integral domain.*

Proof: Let $(F, +, .)$ be a field. To show that $(F, +,.)$ is an integral domain, here we have to show that F is with out zero divisors.

For any $a \neq 0,\ b \neq 0 \in F$, we assume that

$$a \cdot b = 0.$$

$$\Rightarrow \quad a^{-1} \cdot (a \cdot b) = a^{-1}\, 0 \text{ since } a^{-1} \in F$$

$$\Rightarrow \quad (a^{-1} \cdot a) \cdot b = 0$$

$$\Rightarrow \quad b = 0$$

which is contradiction to $b \neq 0$.

Similarly, it can be proved that $a \neq 0$.

Hence any field is an integral domain.

But every integral domain is not a field. We know that the system $(Z, +, .)$ of integers under addition and multiplication is not a field. But if we make the set finite, it becomes a field as shown in the following theorem.

Theorem 6.2.6: *A finite integral domain is a field.*

Proof: Let $(D, +, .)$ be a finite integral with n elements $a_1, a_2, a_3, \ldots, a_n$. In order to prove that $(D, +, .)$ is a field we must

(1) Produce an element $1 \in D$ such that $a \cdot 1 = 1 \cdot a = a$ for all $a \in D$.

(2) For every element $a \neq 0 \in D$ we produce an element $b \in D$ such that $ab = 1$.

Let $a \neq 0 \in D$, then $a \cdot a_1, a \cdot a_2, \ldots, a \cdot a_n$ all belong to D. We claim that they are all distinct. For suppose that $a \cdot a_i = a \cdot a_j$ for $i \neq j$.

$\Rightarrow a \cdot a_i - a \cdot a_j = 0$

$\Rightarrow a \cdot (a_i - a_j) = 0$

$\Rightarrow a_j - a_j = 0$ since $a \neq 0$

$\Rightarrow a_i = a_j$

Contradicting to $i \neq j$.

Thus $a \cdot a_1, a \cdot a_2, \ldots, a \cdot a_n$ are n district elements lying in D, which has exactly n elements.

Since D is an integral domain, it has multiplicative identity 1. So in particular for some integer r.

$$a \cdot a_r = 1.$$

$$\Rightarrow \quad a \cdot a_r = a_r\, a = 1 \text{ since . is commutative in } D$$

$$\Rightarrow \quad a^{-1} = a_r \in D.$$

Thus every non-zero element of D has its inverse in D.

Hence $(D, +, .)$ is a field.

Theorem 6.2.7: *If p is prime, then the ring $(Z_p, +_p) \odot_p)$ is a field.*

Proof: Hence $Z_p = \{0, 1, 2, 3, \ldots, p-1\}$ and

For $a, b \in Z_p$,

CHAPTER 6

$$a +_p b = \begin{cases} a+b \text{ if } a+b < p \\ r \quad \text{if } a+b \geq p \end{cases}$$

where r is the remainder obtained by dividing $a + b$ by p.

and $$a \odot_p b = \begin{cases} ab \text{ if } a.b < p \\ r_1 \quad \text{if } a.b \geq p \end{cases}$$

where r_1 is the remainder obtained by dividing ab by p.

If $a, b \in Z_p$ and $a \odot_p b = 0$

$\Rightarrow$ $p \mid a\,b$

$\Rightarrow$ $p \mid a$ or $p \mid b$ since p is prime

$\Rightarrow$ $a \equiv 0 \pmod p$ or $b \equiv o \pmod p$.

Hence in Z_p one of them zero. This shows that Z_p, has no zero-divisor and consequently $(Z_p, + p, \odot_p)$ is a finite integral domain and by theorem 6.2.6 it is a field.

Example 6.2.7: $(Z_5 +_5, \odot_5)$ is a field where 5 is prime and the system $(Z_4, +_4, \odot_4)$ is not an integral domain because $2 \in Z_4$ and $2 \odot_4 2 = 0$. So Z_4 has zero-divisor.

Now, we consider ordered field.

Theorem 6.2.8: *The unity element of an ordered integral domain (D, +,.) is a positive elemant of D.*

Proof: Let D^+ be the set of positive elements of an integral domain $(D, +, .)$

Suppose that $1 \notin D^+$,

$$1 \neq 0,\ 1 \notin D^+ \Rightarrow -1 \in D^+$$
$$\Rightarrow (-1)\,.\,(-1) \in D^+$$
$$\Rightarrow 1 \in D^+$$

So our supposition is wrong. Hence $1 \in D^+$.

Theorem 6.2.9: *The field $(Z_p, + p, \odot_p)$ is not ordered field, where p is prime.*

Proof: Let $(Z_p, +_p, \odot_p)$ be an ordered field. Then there is subset P of Z_p which contains the positive elements of Z_p.

By definition we know that for an element $a \in Z_p$

$$a = 0 \text{ or } a \in P \text{ or } -a \in P.$$

Suppose $1 \in P$, then $1 \neq 0$ and since

$$1 +_p (p-1) = 0 \Rightarrow -1 = p - 1$$

Now $1 \in P \Rightarrow 1 +_p 1 = 2 \in P \Rightarrow 2 +_p 1 = 3 \in P \ldots$

$$\Rightarrow p - 1 \in P. \text{ by repetition.}$$
$$\Rightarrow -1 \in P$$

which is contradiction to the definition of ordered field.

Hence $(Z_p, +_p, \odot_p)$ is not ordered field.

Theorem 6.2.10: *The field of complex numbers under addition and multiplication is not an ordered field.*

Proof: Suppose $(C, +, .)$ is a field of complex numbers.

Let P be the set of positive elements of C.

Clearly $i \in C, i \neq 0$.

Hence either $i \notin P$ or $-i \in P$.

$$i \in P \Rightarrow i.\, i \in P$$

$$\Rightarrow -1 \in P \text{ as } i^2 = -1.$$

which is contradiction to $1 \in P$.

Therefore $i \notin P$.

Thus $\quad i \neq 0, i \notin P \Rightarrow -i \in P.$

$$\Rightarrow (-i)\,.\,(-i) \in P$$

$$\Rightarrow i^2 \in P$$

$$\Rightarrow -1 \in P.$$

Again, we get a contradiction.

So $\quad -i \notin P.$

Thus $i \neq o, i \notin P, -i \notin P$, *i.e.*, the condition

$i = o$ or $i \in P$ or $-i \in P$ does not hold.

Here $(C, +, .)$ is not an ordered integral domain.

PROBLEMS

1. Let $(R, +, .)$ be a ring. Prove the following:

(a) $a\,.\,(b + (-c)) = a\,.\,b + (-a\,.\,c), \forall\, a, b, c \in R$.

(b) $-(a + b) = -a + (-b), \forall\, a, b \in R$.

(c) $-(-a) = a, \forall\, a \in R$.

(d) $(a + (-b)) + (b + (-c)) = a + (-c), \forall\, a, b, c \in R$.

2. Let $(R, +, .)$ be a ring with identity and without zero divisors. Does the equation $ax = b$ have the solution?

3. Let $(R, +, .)$ be a ring with identity, prove that if $R \neq \{0\}$, then $1 \neq 0$.

4. In a ring $(R, +, .)$ with identity, prove that no divisor of zero can possess a multiplicative inverse.

5. Given that a, b, c, d are elements of the ring $(R, +, .)$ prove that

(a) $(a + b) . (c + d) = a . c + b . c + a . d + b . d$,

(b) $-(a . b . c) = (-a) . (-b) . (-c)$,

(c) If $a . b = b . a = 0$, then $(a + b)^n = a^n + b^n$ for $n \in N$.

6. If the set A contains more than one element, prove that every non-zero element of the power set $P(A)$ is a zero divisor in the ring $(P(A), \Delta, \cap)$.

7. Let R be the set of ordered pairs of non-zero real numbers. In the following cases, determine whether $(R, +, .)$ is a commutative ring with identity. For those systems failing to be so, indicate which axioms are not satisfied :

(a) $(a, b) + (c, d) = (ac, be + d), (a, b) . (c, d) = (ac, bd)$.

(b) $(a, b) + (c, d) = (a + c, b + d), (a, b) . (c, d) = (ac + bd, ad + bd)$

(c) $(a, b) + (c, d) = (ad + bd, bd), (a, b) .(c, d) = (ac, bd)$,

(d) $(a, b) + (c, d) = (a + c, b + d)$ $(a, b) . (c, d) = (ac, ad + be)$.

8. Define two binary operation * and o on the set Z of integers as follows :

$a * b = a + b-1$,

$a \, o \, b = a + b - ab$.

Draw that the system $(Z, *,.)$ is a commutative ring with identity.

9. Let $(R, +, .)$ be an arbitrary ring. In R define a new binary operation o by the rule

$a \, o \, b = a . b + b . a$ for all $a, b \in R$.

Establish that $(R, +, o)$ is a commutative ring.

6.3 SUB-RINGS

In chapter 1, we have discussed subsets of sets, in chapter 2 sub-group of groups. In this chapter, we shall study subsystems of rings, so-called subrings of rings. In addition it is possible to discuss another subsystem of a ring, called an ideal.

Definition 6.3.1: Let $(R, +, .)$ be a ring and $\phi \neq S \subseteq R$. Then the ordered triple $(S, +, .)$ is a sub-ring of $(R, +, .)$ if and only if, $(S, +, .)$ is a ring.

From the definition of a sub-ring it is clear that $(S, +)$ is sub-group of the additive group $(R, +)$, of the ring $(R, +, .)$. Since $(S, +)$ is a sub-group of $(R, +)$, then $a + (-b) \in S$ whenver $a, b \in S$, since $(S, .)$ is a semigroup, thus, if

$a, b \in S$, then $a . b \in S$.

In view of this observation we can define the triple $(S, +,.)$ is a subring of the ring $(R, +, .)$ whenever

(1) S is non-empty subset of R.

(2) $(S, +)$ is a sub-group of $(R, +)$.

(3) S is closed under multiplication.

Definition 6.3.2: Let $(R, +, .)$ be a ring and $(\phi \neq S \subseteq R$. Then the system $(S, +, .)$ is a sub-ring of the ring $(R, +, .)$ if and only if

(1) $a - b \in S$ whenever $a, b \in S$ (closed under difference).

(2) $a \cdot b \in S$ whenever $a, b \in$ S (closed under multiplication).

Example 6.3.1: Let $(Z, +, .)$ be a ring of integers. Then the system $((m), +, .)$ is a subring of the ring $(Z, +, .)$. We know any set (m) generated by m under addition is a subgroup $((m), +)$ of the group $(Z, +)$. Let $mk_1, mk_2 \in (m)$, then $mk_1, mk_2 = m\,(k_1\, mk_2) \in (m)$. Which shows (m) is closed under multiplication.

Example 6.3.2: Every ring $(R, +, .)$ has two trivial subrings $(\{0\}, +, .)$ and $(R, +, .)$ where 0 *is* the additive identity of $(R, +, .)$.

Example 6.3.3: In the ring $(Z, +, .)$ of integers, the system (special case of $((m), +, .)$ $(E, +, .)$ is a subring. Of course, the ring $(Z, +, .)$ is a ring with identity but the ring $(E, +, .)$ of even integers is a subring of $(Z, +, .)$ without identity element.

Now here we shall note a very important and interesting thing about rings and its subrings which are given below:

Let $(R', +, .)$ be a subring of the ring $(R, +, .)$. Then (1) if the ring $(R, +, .)$ has an identity 1 and if this element 1 belongs to R', then 1 is the identity of the subring $(R\ , +, .)$. In this case, we shall call $(R', +, .)$ a *unitary subring of* $(R, +, .)$ *or* $(R, +, .)$ *unitary over ring of* $(R', +, .)$.

Example 6.3.4: The ring of integers $(Z, +,.)$ is a subring of the ring of rational numbers $(Q, +, .)$. The ring $(Q, +, .)$ has an identity element 1 which belongs to Z. This element 1 is also the identity of the ring $(Z, +, .)$. So $(Z, +, .)$ is a unitary subring of $(Q, +, .)$.

(2) The ring $(R, +, .)$ has an identity while $(R', +, .)$ does not have an identity.

Example 6.3.5: The ring of even integers $(E, +, .)$ is a subring of the ring of integers $(Z, +, .)$. The ring $(Z, +, .)$ possesses an identity I but the subring $(E, +, .)$ has no identity element.

(3) Both the Ring $(R, +, .)$ and the subring $(R', +, .)$ possess an identity, but the identity of $(R, +, .)$ does not belong to R. That is, the identities of $(R, +, .)$ and $(R' +, .)$ are different.

Example 6.3.6: *Let* $(A, +, .)$ and $(B, +, .)$ be two rings with identities e_A and e_B respectively. Let R be the set of all ordered pairs (a, b), where $a \in A$ and $b \in B$. If we define addition of multiplication in R by

$$(a, b) + (a', b') = (a + a', b + b'),$$

$$(a, b). (a', b') = (aa', bb').$$

Then $(R, +, .)$ is a ring with identity (e_A, e_B). Now let R' be the subset of R consisting of all elements of the (a, o). Then $(R', +, .)$ is a subring of $(R, +, .)$ which has an identity element $(e_A, 0)$. Thus the identity (e_A, e_B) of R does not belong to R' and both identities (e_A, e_B) and $(e_A, 0)$ are different.

(4) The subring $(R', +, .)$ has an identity but the ring $(R, +, .)$ has no identity.

Example 6.3.7: Let $(A, +, .)$ be a ring with identity e_A and let $(B, +, .)$ be a ring without identity then the ring $(R, +, .)$ and subring $(R', +, .)$ exist as it is seen in the example 6.3.6. The ring $(R, +, .)$ does not have identity but the subring $(R', +, .)$ has an identity $(e_A, 0)$.

Theorem 6.3.1: *In cases (3) and (4) the identity of the subring must be a divisor of zero in the parent ring.*

Proof: Let $(S, +, .)$ be a subring of the ring $(R, +, .)$. Let $1'$ be the identity of the subring $(S, +,.)$. We assume that $1'$ is not the identity of the ring $(R, +, .)$.

Since $S \subseteq R$, then $1' \in S \Rightarrow 1' \in R$. Then there exists an element $a \in R$ such that $a \,.\, 1' \neq a$. We have

$$(a \,.\, 1') \,.\, 1' = a \,.\, (1' \,.\, 1') = a \,.\, 1'$$

$$\Rightarrow \quad (a \,.\, 1') \,.\, 1' - a \,.\, 1' = 0$$

$$\Rightarrow \quad (a \,.\, 1' - a) \,.\, 1' = 0$$

Since $1' \neq 0$, $a.\, 1' - a \neq 0$, (since $1' \neq 1$) which shows $1'$ is the zero divisor.

Example 6.3.8: We considered the ring $(R \times R, +,.)$ of the ordered pair of real numbers under addition and multiplication defined by

$(a, b) + (c, d) = (a + c, b + d)$.

$(a, b) \,.\, (c, d) = (ac, bd)$.

Now we shall consider the subring $(R \times \{0\}, +,.)$. The element $(1,0)$ is the identity of the subring $(R \times \{0\}, +, .)$ while $(1, 1)$ is the identity *of* the ring $(R \times R, +, .)$ we have also seen that $(1, 0)$ is the zero divisor. For, $(1, 0). (0, 1) = (0, 0)$, $(0, 1) \in R \times R$.

6.4 COEFFICIENTS AND EXPONENTS

Since addition and multiplication in rings are binary operations with associative property, sums and products of more than two elements must be defined recursively. For example, if k is a positive integer such that

$a_1 + a_2 + \ldots + a_k$ is defined, we define

$a_1 + a_2 + \ldots + a_{k+1} = (a_1 + a_2 + \ldots + a_k) + a_{k+1}$.

It is then possible to prove generalized associative and distributive laws. The generalized associative law of addition may be stated as follows :

Let n be a positive integer. If r is a positive integer such that $1 \leq r < n$, we have

$$(a_1 + a_2 + \ldots + a_r) + (a_{r+1} + \ldots + a_n) = a_1 + a_2 + \ldots + a_n.$$

In view of this results, it is customary to write sums without use of parentheses since the way parentheses might be inserted is immaterial; of course, similar results hold for products as well.

It $a \in R$ and m is a positive integer, a^m is defined in the usual way. That is,

$$a^m = a \,.\, a \,.\, a \ldots a \text{ (} m \text{ times)}.$$

Moreover, for all positive integers m and n we have

$$\begin{aligned} a^m \,.\, a^n &= (a \,.\, a \,.\, a \ldots a \; m \text{ times}) \,.\, (a \,.\, a \ldots a \; n \text{ times}) \\ &= (a \,.\, a \,.\, a \ldots a \; (m + n) \text{ times}) \\ &= a^{m+n}, \end{aligned}$$

and

$$(a^m)^n = a^m\, a^m \ldots. a^m \ (n \text{ times})$$
$$= a^{m+m\, +\ldots\ldots+\, m} \ (n \text{ summands})$$
$$= a^{mn}.$$

These are often known as the laws of exponents. If R is commutative, that is, $a \,.\, b = b \,.\, a$ for all $a, b \in R$, then

$$(a \,.\, b)^m = a^m \,.\, b^m,$$

for each integer m. In general, this familiar law of exponents does not hold for non-commutative rings.

If $a \in R$, and m is a positive integer, we define $ma = a + a + \ldots\ldots + a$ (m summands).

We also define $0 \,.\, a = 0$ and $(-m)\, a = -(ma)$, and observe that $(-m)\, a = m\,(-a)$. If, now, $a, b \in R$, and m and n are arbitrary integers, the following hold:

$$ma + na = (m + n)\, a,$$
$$m\,(na) = (mn)\, a,$$
$$(m(a + b) = ma + mb$$
$$m\,(ab) = (ma)\, b = a\,(mb),$$
$$(ma)\,(nb) = (mn)\,(ab).$$

If a and b are elements of a commutative ring $(R, +, .)$ and n is a positive integer, the familiar binomial expansion of $(a + b)^n$ holds as in elementary algebra. That is, if for each positive integer $r < n$, we define the positive number $C\,(n, r)$ by

$$C\,(n, r) = \frac{n\,!}{r!\,(n-r)\,!}$$

then

$$(a + b)^n = a^n + C\,(n, 1)\, a^{n-1}\, b + C\,(n, 2)\, a^{n-2}\, b^2 + \ldots.. + C(n, n-1)\, ab^{n-1} + b^n.$$

Definition 6.4.1: Let $(R, +, .)$ be a ring. If there exists a positive integer n such that $n \,.\, a = 0$ for all $a \in R$, then such smallest integer n is called the characteristic of R. If no such positive integer exists, then $(R, +, .)$ is said to be of characteristic zero.

Example 6.4.1: In the rings $(Z, +, .)$, $(Q, +, .)$ and $(R, +, .)$ no such positive integer n is possible for which $n \,.\, a = 0$ for all $a \in Z$, or $a \in Q$, or $a \in R$. Thus these systems are said to have characteristic zero.

Example 6.4.2: In the rings $(P(S), \Delta, \cap)$ of subsets of a set under the operation of symmetric difference of two sets and intersection of two sets, we have

$A \,\Delta\, A = 0$, for all $A \in P\,(S)$.

Thus the ring has characteristic 2.

Example 6.4.3: In the ring $(Z_6, +_6, \odot_6)$ of integers modulo 6, we have

$$1 +_6 1 +_6 1 +_6 1 +_6 1 = 0,$$

CHAPTER 6

$$2 +_6 2 +_6 2 +_6 2 +_6 2 = 0,$$
$$3 +_6 3 +_6 3 +_6 3 +_6 3 = 0,$$
$$4 +_6 4 +_6 4 +_6 4 +_6 4 = 0,$$
$$5 +_6 5 +_6 5 +_6 5 +_6 5 = 0,$$

Thus, the ring $(Z_{6} +_{6}, ._{6})$ has characteristic 6.

Example 6.4.4: In all commutative rings, of characteristic 2, $(a + b)^2 = a^2 + b^2$, for all a, b in the ring.

Solution: Let $(R, +, .)$ be a commutative ring of characteristic two, then $\forall a, b \in R$, $a + a = 0$, or $2a = 0$. For $a, b \in R$,

$$\begin{aligned}(a + b)^2 &= (a + b) . (a + b) \\ &= a . (a + b) + b . (a + b) \\ &= a . a + (a . b + b . a) + b . b \\ &= a^2 + 2a\,b + b^2 \text{ (since } a . b = b . a) \\ &= a^2 + b^2 \text{ (since } 2a = 0).\end{aligned}$$

Definition 6.4.2: An element a of the ring $(R, +, .)$ is said to be idempotent if $a . a = a^2 = a$.

Example 6.4.5: In the ring $(Z_6, +_6, \odot_6)$ of integers modulo 6, we have

$$4 . \odot_6{}^4 = 4$$
$$3 . \odot_6{}^3 = 3$$

Thus 0, 1 , 3 and 4 are idempotent elements.

Definition 6.4.3: An element a of the Ring $(R, +, .)$ is said to be nilpotent if there exists a positive integer n such that $a^n = 0$.

It is obvious that zero of the ring is nilpotent.

Theorem 6.4.1: *Every non-zero nilpotent element is neccssarily a divisor of zero.*

Proof: Since a is nilpotent element, then there exists a positive integer n such that

$$a^n = 0$$
$$\Rightarrow a . (a^{n-1}) = 0$$

where $a^{n-1} \neq 0$ which shows a is a divisor of zero.

Theorem 6.4.2: *Let* $(R_1, +, .)$ *and* $(R_2 +, .)$ *be two subrings of the ring* $(R, +, .)$*. Then* $(R_1 \cap R_2, +, .)$ *is a suhring of* $(R, +, .)$.

Proof: Since $R_1 \subseteq R$ and $R_2 \subseteq R$, then

$R_1 \cap R_2 \subseteq R$ and it is non-empty since $0 \in R_1$ and $0 \in R_2$.

Let $a, b \in R_1 \cap R_2$, then

$a \in R_1, b \in R_1$, and

$a \in R_2$, $b \in R_2$.

Since (R_1 +, .) and (R_2 +, .) are subrings, then $a - b \in R_1$, $a \,.\, b \in R_1$ and $a - b \in R_2$, $a \,.\, b \in R_2$

Therefore,

$a - b \in R_1 \cap R_2$, and a, $b \in R_1 \cap R_2$.

This proves that ($R_1 \cap R_2$ +, .) is a subring.

Corollary : The intersection of any collection of subrings under the operations of addition and multiplication of the parent ring (R, +, .) is a subring.

Definition 6.4.4: Let (R, +, .) be a ring such that $a \,.\, a = a$ for all $a \in R$, then the ring (R, +, .) is said to be Boolean ring.

Example 6.4.6: The ring ($P(S)$, Δ, ∩) of power set of S is a Boolean ring. Since $A \cap A = A$, for all $A \in P(S)$.

Theorem 6.4.3: *If* (R, +, .) *is a Boolean ring, then* $a + a = 0$, $\forall\, a \in R$.

Proof: Since $a \in R$, then $a + a \in R$.

Since (R, + , .) is a Boolean ring, then

$$
\begin{aligned}
(a + a) + 0 &= (a + a) \\
&= (a + a) \,.\, (a + a) \\
&= (a \,.\, a + a \,.\, a) + (a \,.\, a + a \,.\, a) \\
&= (a + a) + (a + a).
\end{aligned}
$$

By cancellation law we have

$$a + a = 0$$

That is, each element of the Boolean ring is the additive inverse of itself.

Example 6.4.7: Let a and b be the elements of a Boolean ring (R, +, .) with $a + b = 0$. Prove that $a = b$.

Solution: Since $a \in R$, then

$$a + a \in R.$$

Since (R, +, .) is a Boolean ring, then

$$a + a = 0, \text{ by the theorem 6.4.3}$$

and we have $a + b = 0$

So $$a + a = 0 = a + b.$$

By cancellation law we obtain

$$a = b.$$

CHAPTER 6

Theorem 6.4.4: *Every Boolean ring* $(R, +, .)$ *is commutative.*

Proof: Let $a, b \in R$, then

$$a + b \in R.$$

Since $(R, +, .)$ is a Boolean ring, then

$$(a + b) = (a + b)^2 = (a + b).(a + b).$$

or $$= a . (a + b) + b . (a + b).$$

or $$= a . a + a . b + b . a + b . b$$

or $$= a^2 + a . b + b . a + b^2$$

or $$= (a + b) + (a . b + b . a).$$

Thus $a + b + 0 = a + b + (b.a + a.b)$

By cancellation law we obtain

$$0 = b . a + a . b.$$

By the example 6.4.3 we have

$$b . a = a . b.$$

Definition 6.4.5: Let $(R, +,.)$ be a ring. The set $\{c \in R \mid c . x = x . c, \forall x \in R\}$, denoted by Cent R, is called the centre of the ring $(R, +,.)$. In other words, Cent R is the set of all those elements of R which commute with every element of R with respect to multiplication.

Theorem 6.4.5: (*Cent* $R, +, .$) *is a subring of* $(R, +,.)$.

Proof: Let $a, b \in$ Cent R, then

$$a . x = x . a, \forall x \in R,$$

$$b . x = x . b, \forall x \in R.$$

Now $(a - b). x = a . x - b . x, \quad \forall x \in R$

$$= x . a - x . b, \forall x \in R$$

$$= x . (a - b), \forall x \in R$$

which shows $a - b \in$ Cent R.

Again

$$(a . b) . x = a . (b . x)$$

$$= a . (x . b)$$

$$= (a . x) . b$$

$$= (x . a) . b$$

$$= x . (a . b).$$

and it shows that

$$a. b \in \text{Cent } R$$

This proves the theorem.

PROBLEMS

1. Let ({m , n, p, q}, +, .) be a ring and addition and multiplication be defined by the following tables:

+	m	n	p	q
m	m	n	p	q
n	n	m	q	p
p	p	q	m	n
q	q	p	n	m

.	m	n	p	q
m	m	m	m	m
n	m	n	m	n
p	m	p	m	p
q	m	q	m	q

Check associative law, distributive law, and commutative law. Does the relation $(p + q)^2 = p^2 + 2\ p\ .\ q + g^2$ hold ?

2. Prove that ($P(S)$, U, ∩) is not ring, where addition and multiplication are defined by set union and set intersection.

3. Find the characteristic of each of the following rings:

(*a*) The ring of integers of modulo 7, with addition and multiplication modulo 7.

(*b*) The ring of real numbers, with ordinary addition and multiplication.

(*c*) The ring of even integers, with ordinary addition and multiplication.

(*d*) The ring of all 2 × 2 matrices with addition and multiplication of matrices.

4. Let (R, +,.) be a ring. Then show that (S, +,.) is a subring of (R, +,.) if and only if:

(*i*) S is a non-empty subset of R,

(*ii*) $a, b \in S \Rightarrow a - b \in S$,

(*iii*) $a, b \in S \Rightarrow a\ .\ b \in S$.

5. Prove that the system ({0, 3, 6, 9}, $+_{12}$, $\odot_{12}$) is a subring of (Z_{12}, $+_{12}$, $\odot_{12}$), the ring of integers modulo 12.

6. Let (R, +, .) be a ring such that $x^2 = x, \forall x \in R$. The prove that

(1) (R, +, .) is commutative.

(2) (R, +, .) has characteristic 2.

(3) $(a + b)^2 = a^2 + b^2 = (a - b)^2$

7. Let (R, +, .) be a ring with identity. Then prove that (R, +, .) has characteristic $n > o$ if and only if n is the least positive integers for which $n.\ 1 = 0$.

8. Prove that the set of (2 × 2) matrices over the integers with addition and multiplication of matrices is a non-commutative ring.

9. Define the characteristic of a ring and deduce the characteristic of (Z/ (4), $+_4$, $\odot_4$).

CHAPTER 6

6.5 IDEALS AND QUOTIENT RINGS

In this section, we introduce an important class of subrings which are more special than subrings, known as ideals.

Definition 6.5.1: $(R, +, .)$ be a ring and $\phi \neq I \subseteq R$. Then the triple $(I, +, .)$ is an ideal of the ring $(R, +,.)$ if, and only if :

(1) $a, b \in I \Rightarrow a - b \in I$,

(2) $a \in I, r \in R \Rightarrow a . r \in I$ and $r . a \in I$.

In a commutative ring we need only $r . a \in I$.

Definition 6.5.2: If $a, b \in I, a - b \in I$ and if $a \in I, r \in R, a . r \in I$, then $(I, +, .)$ is called right ideal of the Ring $(R, +, .)$.

Definition 6.5.3: If $a, b \in I, a - b \in I$, and if $a \in I, r \in R, r . a \in I$, then $(I, +, .)$ is called leftideal of the ring $(R, +, .)$.

Theorem 6.5.1: *Let $(R, +, .)$ be a ring and $(I, +, .)$ be an ideal in $(R, +, .)$ then $(I, +,.)$ is a subring of $(R, +, .)$.*

Proof: Let $(I, +, .)$ be an ideal. Then by the definition of an ideal, I is non-empty set, and if $a, b \in I$, then $a - b \in I$. Thus, $(I, +, .)$ is sub-group of the additive group $(R, +, .)$.

Let $a, b \in I$, then $b \in R$ since $I \subset R$. By the definition of an ideal $a . b \in I$, which implies I is closed under multiplication, this completes the proof of the theorem.

But the converse of the theorem is not true, that is, some rings have subrings which are not ideals. Some examples are given below:

Example 6.5.1: Eet $(Q, +,.)$ be a ring of rational numbers with the usual operation of addition and multiplication. The system $(Z, +, .)$ is subring of integers of the ring $(Q, +, .)$. In order to establish the fact, it is sufficient to find an integer a and one rational number r such that $a . r \notin Z$, we observe that $1 \in Z$, $1/2 \in Q$ but $1. 1/2 \notin Z$, which shows $(Z, +, .)$ is not an ideal.

Example 6.5.2: Let $(Q, +, .)$ be a subring of the ring $(R, +, .)$ of real numbers. We see that $1/2 \in Q$, $\sqrt{2} \in R$ but $1/2 . \sqrt{2} \notin Q$, which shows that the subring $(Q, +, .)$ of rational numbers of the ring $(R, +,.)$ of real numbers is not an ideal.

Example 6.5.3: In any ring $(R, +, .)$ the trivial subrings $(R, +,.)$ and $(\{0\}, +, .)$ are both ideals.

Definition 6.5.4: A ring $(R, +, .)$ is called a simple ring if it does not contain proper ideals.

Example 6.5.4: $(\{0, 3, 6, 9), +_{12}, \odot_{12})$ is an ideal of the ring $(Z_{12}, +_{12}, \odot_{12})$ the ring of integers modulo 12.

Example 6.5.5: For a fixed integer $a \in Z$, the set $(a) = [na \mid n \in Z\}$. We see that the triple $((a), +, .)$ is an ideal of the ring $(Z, +, .)$. For, if $n\ a \in (a)$, $m\ a \in (a)$ then $na - ma = (n - m) . a \in (a)$ and $(m) (na) = (mn)\ a \in (a)$.

In particular $a = 2$, the ring of even integers $(E, +, .)$ is an ideal of $(Z, +, .)$, the ring of integers.

Example 6.5.6: Let $(K_2, +,.)$ be a ring of all matrices of order 2 over the real numbers. Let U be the set of all matrices of K_2 of the form

$$\begin{bmatrix} a & b \\ 0 & 0 \end{bmatrix}, \text{ then } (U, +,.) \text{ is right ideal.}$$

For, if $\begin{bmatrix} c & d \\ p & q \end{bmatrix} \in K_2$, then

$$\begin{bmatrix} a & b \\ 0 & 0 \end{bmatrix} . \begin{bmatrix} c & d \\ p & q \end{bmatrix} = \begin{bmatrix} ac+dp & ad+bq \\ 0+0 & 0+0 \end{bmatrix}$$

which belongs to U.

Let V be the set of all matrices of K_2, of the form $\begin{bmatrix} a & 0 \\ b & 0 \end{bmatrix}$ then $(V, + .)$ is a left ideal.

For, if $\begin{bmatrix} c & d \\ p & q \end{bmatrix} \in K_2$, *then*

$$\begin{bmatrix} c & d \\ p & q \end{bmatrix} . \begin{bmatrix} a & o \\ b & o \end{bmatrix} = \begin{bmatrix} ac+dp & o+o \\ ap+bq & o+o \end{bmatrix} \in V.$$

And let W be the set of all matrices of K_2 of the form $\begin{bmatrix} a & b \\ o & c \end{bmatrix}$, then $(W, +,.)$ is a subring of $(K_2, +,.)$ but neither a right ideal nor a left ideal.

Theorem 6.5.2: *Let $(I_1 +,.)$ and $(I_2, +,.)$ be two ideals of the ring $(R, +,.)$, the $(I_1 \cap I_2, +,.)$ is also an ideal.*

Proof: We observe that $I_1 \cap I_2$ is non-empty since $0 \in I_1$, and $0 \in I_2$.

Let $a, b \in I_1 \cap I_2$, then

$a, b \in I_1$ and $a, b \in I_2$.

Since $(I_1, +,.)$ and $(I_2, +, .)$ are ideals,

$a, b \in I_1, \Rightarrow a - b \in I_1$ and $a . r \in I_1, b . r \in I_1$.

$a . r \in I_1, b . r \in I_1 \Rightarrow a . r - b . r \in I_1$, thus, if $a - b \in I_1$, then $(a - b). r \in I_1$.

Again, $a\ b \in I_2 \Rightarrow a - b \in I_2, a . r$ and $b . r \in I_2$

Thus $(a - b) . r \in I_2$.

Therefore $a - b \in I_1 \cap I_2$ and

$ar, b . r \in I_1 \cap I_2$.

Hence $(I_1 \cap I_2 +, .)$ is an ideal.

CHAPTER 6

Corollary: If $(I_i, +, .)$ is an arbitrary indexed collection of ideals of the ring $(R, +, .)$ then so is also $(\cap I_i, +, .)$.

Theorem 6.5.3: *If (I, +, .) is a proper ideal of the ring (R, +, .) with identity, then no element of I has a multiplicative inverse.*

Proof: Let $a \in I$ such that there exists $a^{-1} \in R$. Since I is closed under multiplication by the elements of R, then

$$a \cdot a^{-1} = 1 \in I.$$

Let $r \in R$, then

$$1 \cdot r = r \in I,$$

which shows that $R \subseteq I$ and we have $I \subseteq R$. This proves $R = I$ which is contradicting the hypothesis that I is a proper subset of R. This proves the theorem.

Definition 6.5.5: The intersection of all ideals in a ring $(R, +, .)$ which contain a given non-empty set K of elements of R is called the ideal generated by K.

Now have a special case, if $K = \{a\}$.

Definition 6.5.6: An ideal in a commutative ring $(R, +, .)$ with identity generated by one element of R is called a Principle ideal. The ideal generated by the element a is denoted by $((a), +, .)$.

Definition 6.5.7: A commutative ring $(R, +, .)$ with identity is called a Principal ideal ring if every ideal in the ring $(R, +, .)$ is a principal ideal.

Theorem 6.5.4: *The ring (Z, +, .) of integers is a principal ideal ring.*

Proof: Let $(A, +,.)$ be an ideal of the ring $(Z, +, .)$ of integers. If $A = (0)$, then $(\{0\}, +, .)$ is a principal ideal generated by the element 0. If $A \neq (0)$, A contains positive integers, and let us assume that n is the smallest positive integer such that $n \in A$. Certainly, $(n) \subseteq A$, and we only need to prove that $A \subseteq (n)$. Let $a \in A$, then there exist integers q and r such that

$$a = q\,n + r,\ 0 \leq r < n.$$

Since $a \in A$, $n \in A$, and $qn \in A$, and $r = a - qn \in A$.

If $r > 0$, we have a contradiction to the assumption that n is the smallest positive integer. That is, $r = 0$ and $a = qn$. It follow, that $a \in (n)$, so $A \subseteq (n)$ which implies $(n) = A$.

Thus the proof is completed.

There are many rings which have ideals that are not principal ideals. The ring $(K_2, +,.)$ of all matrices of order 2 has the ideal $(U, +, .)$ of all matrices of K_2 of the form $\begin{bmatrix} a & b \\ 0 & 0 \end{bmatrix}$ which is not a principal ideal.

Theorem 6.5.5: *Let $a_1, a_2, \ldots\ldots, a_n$ be non-zero elements of a principal ideal of ring (R, +,.). Then $(\cap (a, .), +, .) = ((a), +, .)$,*

where a is the least common multiple of $a_1, a_2, \ldots\ldots a_n$.

Proof: Let $a_1, a_2, ...,a_n$ be the elements of the ring $(R, +,.)$. By the common multiple a of $a_1, a_2, ...a_n$ we mean that a is common multiple of $a_1, a_2, ...a_n$ under the operation of ring multiplication. Thus $a \in R$. Now the element a is called the least common multiple of $a_1, a_2,....,a_n$ if a is common multiple of $a_1, a_2, ...,a_n$ and every other common multiple of $a_1, a_2, ...a_n$ is a multiple of a as well.

Since the ring $(R, +, .)$ is a principal ideal ring, then the principal ideal $((a), +, .)$ generated by the least common multiple a of $a_1, a_2,....a_n$ exists, and we have the principal ideals $((a_i), +,.)$ $i = 1, 2, ...n$. Therefore $(\cap (a_i), +, .)$ is an ideal of $(R, +, .)$ by corollary of theorem 6.5.2. But the ideal $(\cap (a_i), +, .)$ is a principal ideal because $(R, +, .)$ is the principal ideal ring. Therefore, there exists an element $a \in R$ such that $(a) = \cap (a_i)$ which implies $(a) \subseteq (a_i)$, $i = 1, 2,.....,n$. Therefore for some $r_i \in R$, $a = r_i a_i \Rightarrow a$ is common multiple of $a_1, a_2,.......a_n$.

Now we assume that b is common multiple of $a_1, a_2, ...a_n$ Then $b = s. \ . \ a_i$, $i = 1, 2, ...n$, for some $s_i \in R$.

If $r \in R$, then

$r \ . \ b \ . = r \ . \ (s_i \ . \ a_i) = (r \ . \ s_i) \ . \ a_i \in (a_i), \ i = 1, 2, ...n$

Which implies $(b) \subseteq (a_i)$ for all i, that is,

$$(b) \subseteq \cap (a_i) = (a).$$

Since a is the least common multiple, b must be a multiple of a.

Example 6.5.7: The ring of integers $(Z, +, .)$ is a principal ideal ring. We consider two principal ideal $((2), +, .)$ and $((3), +, .)$ generated by 2 and 3 respectively. Then we see that

$$(2) \cap (3) = (6)$$

and $$((2) \cap (3), +, .) = ((6), +, .).$$

Theorem 6.5.6: *A division ring is a simple ring.*

Proof: Let A [$\neq 0$] be an ideal of a division ring $(R_1 + ,.)$. Since $A \neq (0)$, then there exists a non-zero element $a \in A$. Since $(R, +, .)$ is a division ring, a is a unit in R. Then there is an element $b \in R$ such that $a.b = 1 \Rightarrow 1 \in A$, since $a \in A$. For any $r \in R$, $r = I. \ r \in A$. Consequently $R \subset A$. But we have $A \subset R$. Hence $A = R$. Hence $(R, +, .)$ has no proper ideal.

Theorem 6.5.7: *Let $(R, +, .)$ be a ring with unity. It R has no. right ideals except* (0) *and R itself, then R is division ring.*

Proof: In order to prove this theorem we prove that every non-zero element of R has its right multiplication inverse. Let $a \neq 0 \in R$. Consider $aR = \{a \ x | x \in R\}$. Clearly $a = a. \ 1 \in R. \ R \neq \phi$.

Further $a.b - ac = a\ (b - c) \in a\ R$ and for any $\beta \in R$ $(ar)\ s = a\ (rs) \in a\ R \Rightarrow (a\ R, +, .)$ is a right ideal since $a \neq 0 \in R$, $a\ R = R$.

Since $1 \in R$, there is an element $b \in R$ such that $ab = 1 \Rightarrow a$ has its right multiplicative inverse.

Hence $(R, +, .)$ is a division ring.

Theorem 6.5.8: *A field has no proper ideal.*

Proof: Let $(U, +, .)$ be an ideal of the field $(F, +,.)$. Hence we have to show that either $U = (0)$ or $u = F$.

It $U = (0)$, the theorem is proved.

If $U \neq (0)$. Then there exists an element $a \neq 0 \in U$.

Since $U \subset F$, and $a \in U$, then $a \in F$ and because it is the element of the field, so $a^{-1} \in F$.

Now $a \in U, a^{-1} \in F \Rightarrow a. a^{-1} = 1 \in U$

and for any $x \in F, 1 \in U, x.1 = x \in U$

Hence $F \subset U$. But $U \subset F$.

Hence $U = F$. This proves the theorem.

Theorem 6.5.9: *Let $(R, +, .)$ be a commutative ring with identity whose only ideals are (0) and R it self. Then $(R, +, .)$ is field.*

Proof: In order to prove this theorem we have to show that each non-zero element of R has its inverse in R.

So, suppose that $a \neq 0 \in R$. We consider the set $R. a = \{xa \mid x \in R\}$. We claim that $(Ra, +, .)$ is an ideal.

Let $u, v \in Ra$, then for some $r_1, r_2 \in R$ $u = r_1 a$ and $v = r_2 a$. Now $u - v = r_1 a - r_2 a = (r_1 - r_2) a \in Ra$ since $r_1 - r_2 \in R$. So $(Ra, +)$ is a additive sub-group of $(R, +)$.

Moreover, if $r \in R$, $ru = r (r_1 a) = (r r_1) a \in Ra$ since $rr_1 \in R$. Hence $(Ra, +, .)$ is an ideal of R.

By our assumption on R, $Ra = (0)$ or $Ra = R$. Since $0 \neq a = 1.a \in Ra, Ra \neq (0)$.

Therefore, we are left with the only other possibility $Ra = R$. This implies that every element of R is a multiple of a by some element of R.

In particular $1 \in R$ so it can be written as multiple of a by some element of R. That is, there exists an element $b \in R$ such that

$$1 = a \,.\, b \Rightarrow a^{-1} = b.$$

This complete the proof of the theorem.

Theorem 6.5.10: *If $(R, +, .)$ be a commutative ring and $a \in R$, then $Ra = \{ra \mid r \in R\}$ is ideal of R.*

Proof: Hence $Ra = \{ra \mid r \in R\}$. In order to prove that $(Ra, +, .)$ is an ideal, we shall show that is additive sub-group of R and closed under multiplication by the elements of R. Now let $x, y \in Ra$. Then for some element $r_1, r_2 \in R$

$$x = r_1 a \text{ and } y = r_2 a.$$

Now $$x - y = r_1 a - r_2 a = (r_1 - r_2) a \in Ra, \text{ since } r_1 - r_2 \in R.$$

Again for $x \in Ra$ and $r \in R$.

$$rx = r(r_1 a) = (rr_1) a \in Ra \text{ since } rr_1 \in R$$

and

$$\begin{aligned} xr &= (r_1 a) r = r_1 (ar) \\ &= r_1 (ra) \text{ since } R \text{ is commutative} \\ &= (r_1 r) a \in R \text{ since } r_1 r \in R \end{aligned}$$

Hence $(Ra, +, .)$ is an ideal.

Theorem 6.5.11: If $(R, +, .)$ *is a commutative ring with identity and* $a \in R$, *then* $Ra = \{r a \mid r \in R\}$ *is a principal ideal of R generated by a.*

Proof: In above theorem we have seen that $(Ra, +, .)$ is an ideal of $(R, +, .)$.

Let $(S, +, .)$ be ideal generated by $a \in$ R. So, $S = (a)$.

$$S = \{(a) = \{ra + as + ma \mid r, s \in R, m \in Z\}.$$

Now $ra + a s + ma = ra + sa + ma$ since • 1s commutative

$$\begin{aligned} &= ra + sa + m(ea) \text{ since } R \text{ has an identity} \\ &= ra + sa + (me) a \\ &= ra + sa + r^1 a \text{ since } me = r^1 \in R \\ &= (r + s + r^1) a \\ &= xa \text{ where } x = r + s + r^1. \end{aligned}$$

Hence $S = \{xa \mid x \in R\} = Ra$. But $S = (a)$.

Hence $Ra = (a)$.

We have shown that $(Ra, +, .)$ is an ideal generated by (a). So it is principal ideal of R.

Hence, we shall see a very interesting result.

Theorem 6.5.12: *Let* $(R, +, .)$ *be a commutative ring with unity and let* $a \neq 0, b \neq 0 \in R$. *Then* $(a) = (b)$ *iff* a/b *and* b/a *(i.e., a and b are associates).*

Proof: Let $(R, +, .)$ be a commutative ring with unity.

Let $a \neq 0, b \neq 0 \in R$. Then

$$(a) = \{ax \mid x \in R\} \text{ and } (b) = \{bx \mid x \in R\}$$

Let $(a) = (b)$ then

$$\begin{aligned} (a) = (b) &\Rightarrow (a) \subset (b) \text{ and } (b) \subset (a) \\ &\Rightarrow a \in (b) \text{ and } b \in (a) \\ &\Rightarrow a = br \text{ and } b = as \text{ for some } r, s \in R. \\ &\Rightarrow b/a \text{ and } a/b. \end{aligned}$$

That is, a and b are associates.

Conversely, $a/b, b/a \Rightarrow b = am$ and $a = bn$ for some $m, n \in R$

$$\begin{aligned} &\Rightarrow b \in (a) \text{ and } a \in (b) \\ &\Rightarrow (b) \subset (a) \text{ and } (a) \subset (b) \\ &\Rightarrow (a) = (b). \end{aligned}$$

This proves the theorem.

CHAPTER 6

6.6 COSETS OF A RING

Now we study the cosets in a ring. The ideals play the same role in ring theory as the normal sub-groups do in the group theory. If $(U, +, .)$ is an ideal of the ring $(R, +, .)$ then since addition is commutative, the system $(U, +)$ is a normal sub-group of the additive group $(R, +)$ of the ring $(R, +, .)$. Thus we can construct quotient group of R by U. Thus cosets of U assume the form $a + U = (a + i | i \in U\}$, when $a \in R$.

We have seen that cosets of a normal sub-group in a group are identical or disjoint. Two cosets $a + U$ and $b + U$ are equal if $a - b \in U$. The set of cosets is denoted by R/U, where $(U, +)$ is the normal sub-group of $(R, +)$. We see that with two operations R/U forms a ring.

First, we have to define sum and product of two cosets of U in R, so that the operation of addition and multiplication are well defined.

Let addition and multiplication of cosets be defined by

$$(a + U) + (b + U) = (a + b) + U, \text{ and}$$

$$(a + U) \,.\, (b - U) = (a \,.\, b) + U$$

To see that the addition and multiplication are well-defined in the sense that they are independent of the representative of cosets of U. To obtain the sum and product of the cosets,

let $\quad a + U = a' + U$, where $a, a' \notin U$,

and $\quad b + U = b' + U$, where $b, b' \notin U$.

Then we shall show that

$$(a + b) + U = (a' + b') + U,$$

and $\quad (a \,.\, b) + U = (a' \,.\, b') + U$.

If

$$a + U = a' + U \Rightarrow a - a' \in U,$$

$$b + U = b' + U \Rightarrow b - b' \in U.$$

Since $(U, +, .)$ is an ideal, then.

$$(a - a') \in U, b - b' \in U \Rightarrow (a - a') + (b - b') \in U$$
$$\Rightarrow (a + b) - (a' + b') \in U$$
$$\Rightarrow (a + b) + U = (a' + b') + U.$$

Again, $a - a' \in U, b - b' \in U$, then there exist two elements $\underline{x} \in U, y \in U$ such that

$$a - a' = \underline{x} \text{ and } b - b' = y$$

or $\quad a = a' + x, b = b' + y.$

Now $\quad a \,.\, b = (a' + x) \,.\, (b' + y)$

$$= a' \,.\, b' + a' y + x \,.\, b' + xy.$$

Since $(U, +, .)$ is an ideal, then

$$a \,.\, y + x \,.\, b' + x \,.\, y \in U$$

which follows that

$$ab - a' \,.\, b' = a' \,.\, y + x \,.\, b' + x \,.\, y \in U$$

which implies

$$a \,.\, b + U = a' \,.\, b' + U.$$

Now we proceed to obtain that the system $(R/I, +, .)$ is a ring.

Theorem 6.6.1: *Let $(I, +, .)$ be an ideal of the ring $(R, +, .)$ then the system $(R/I, +, .)$ is a ring, known as the quotient ring of R by I.*

(1) *(Closure).* We have seen that the sum of two cosets of I is again a coset of I, that is, $(a + I) + (b + I) = (a + b) + I \in R/I$.

(2) *(Associativity).* Let $a + I, b + 1 , c + I \in R/I$,

$$\begin{aligned}[(a + I) + (b + I)] + (c + I) &= [(a + b) + I] + (c + I)\\ &= ((a + b) + c) + I\\ &= (a + b + c)) + I\\ &= (a + I) + (b + c) + I\\ &= (a + I) + [(b + I) + (c + I)].\end{aligned}$$

(3) *(Commutativity).* Let $a + I, b + I \in R/I$

$$\begin{aligned}(a + I) + (b + I) &= (a + b) + I\\ &= (b + a) + I\\ &= (b + I) + (a + I).\end{aligned}$$

(4) *(Existence of additive identity).* For all $a + I \in R/I$, $\exists\ 0 + I \in R/I$ such that

$$\begin{aligned}(a + I) + (0 + I) = (a + 0) + I &= a + I = (0 + a) + I\\ &= (0 + I) + (a + I).\end{aligned}$$

Thus $0 + I = I$ is the identity in R/I.

(5) *(Existence of additive inverse).* For $a + I$, there exists $-a + I$ such that

$$\begin{aligned}(a + 1) + (- a + I) &= a + (-a) + I\\ &= 0 + I\\ &= ((-a) + a) + I\\ &= (-a + I) + (a + I),\end{aligned}$$

which proves that R/I *is* an abelian group with addition. Now we prove that $(R/I, .)$ is a semigroup.

(6) (*Closure*}. We have defined that if $a + I \in R/I$, $b + I \in R/I$, then $(a + I) \,.\, (b + I) = a \,.\, b + I \in R/\mathrm{I}$

CHAPTER 6

(7) (Associativity). Let $a + I,\ b + 1,\ c + I \in R/I$,

$$\begin{aligned}[(a + I)\,.\,(b + I)]\,.\,(c + I) &= (a\,.\,b + I)\,.\,(c + I)\\ &= (a\,.\,b)\,.\,c + I\\ &= a\,.\,(b\,.\,c) + I\\ &= (a + I)\,.\,((b\,.\,c) + I)\\ &= (a + I)\,.\,[(b + I)\,.\,(c + I)].\end{aligned}$$

(8) *(Distributivity).* $(a + I)\,.\,[(b + I) + (c + I)] = (a + I)\,.\,[b + c + I] = a\,.\,(b + c) + I = (a\,.\,b + I) + (a\,.\,c + I)$.

Thus, we have seen that $(R/I, +, .)$ is a ring, called factor ring or residue classes ring modulo I or quotient ring.

Example 6.6.1: In the ring $(Z, +, .)$ of integers, we consider the principal ideal $((n), +, .)$, where n is a non-negative integer. The cosets of (n) in Z will assume the form

$$a + (n) = \{a + nk \mid k \in Z\} = [a].$$

Thus the cosets are precisely the congruence classes modulo n. It is clear from the definition of addition and multiplication of cosets, $(Z/(n), +, .)$ is a ring, which is merely the ring of integers modulo n.

$$(Z_n, +_n, \odot_n) = (Z/(n), +, .).$$

(*i*) Let $(Z, +, .)$ be the ring of integers and let $(5Z, +, .)$ be an ideal generated by 5. Then the system $(Z/5Z, +, .)$ is a quotient ring.

The quotient set $Z/5Z$ contains the elements $5Z$, $5Z + 1$, $5Z + 2$, $5Z + 3$, $5Z + 4$.

From the operation tables it is quite obvious that $(Z/5(Z), +, .)$ is a ring.

+	5Z	5Z+1	5Z+2	5Z+3	5Z+4
5Z	5Z	5Z+1	5Z+2	5Z+3	5Z+4
5Z+1	5Z+1	5Z+2	5Z+3	5Z+4	5Z
5Z+2	5Z+2	5Z+3	5Z+4	5Z	5Z+1
5Z+3	5Z+3	5Z+4	5Z	5Z+1	5Z+2
5Z+4	5Z+4	5Z	5Z+1	5Z+2	5Z+3

⊙	5Z	5Z+1	5Z+2	5Z+3	6Z+4
5Z	5Z	5Z	5Z	5Z	5Z
5Z+1	5Z	5Z+1	5Z+2	5Z+3	5Z+4
5Z+2	5Z	5Z+2	5Z+4	5Z+1	5Z+3
5Z+3	5Z	5Z+3	5Z+1	5Z+4	5Z+2
5Z+4	5Z	5Z+4	5Z+3	5Z+2	5Z+1

Example 6.6.2: Given $(I_1, +, .)$ and $(I_2, +, .)$ are ideals of the ring $(R, +, .)$

Let $I_1 + I_2 = \{(a + b) \mid a \in I_1, b \in I_2\}$

Show that $(I_1 + I_2 +, .)$ is also an ideal of $(R, +, .)$.

Solution: Let $a + b \in I_1 + I_2$, then $a \in I_1$, $b \in I_2$ and $a_1 + b_1 \in I_1, + I_2$, then $a_1 \in I_1$, $b_1 \in I_2$,. Since $(I_1, +,.)$ and $(I_2, +, .)$ are ideals,

$$a \in I_1,\ a_1 \in I_1 \Rightarrow a - a_1 \in I_1$$

and $$r \in R,\ a \in I_1 \Rightarrow ar \in I_1 \text{ and } ra \in I_1,$$

again $$b \in I_2,\ b_1 \in I_2 \Rightarrow b - b_1 \in I_2,$$

and $$r \in R,\ b \in I_2 \Rightarrow rb \in I_2 \text{ and } br \in I_2.$$

$a - a_1 \in I_1$ and $b - b_1 \in I_2$,

$$\Rightarrow a - a_1 + b - b_1 \in I_1 + I_2,$$

$$\Rightarrow (a + b) - (a_1 + b_1) \in I_1 + I_2,$$

and $ra \in I_1$, $rb \in I_2$.

$$\Rightarrow ra + rb \in I_1 + I_2.$$

$$\Rightarrow r\,(a + b) \in I_1 + I_2.$$

Hence $(I_1 + I_2, +,.)$ is an ideal.

Exercise 6.6.1: Show by example that if $(I_1, +, .)$ and $(I_2, +, .)$ are both ideals of the ring $(R +, .)$, then $(I_1\ U\ I_2, +,)$ is not necessarily an ideal.

Solution: Let $(Z, +, .)$ be a ring of integers and ((2) . +, .) and ((3), +, .) be ideals. We observe that (2) U (3) contain all integers which are multiples of 2 or 3. Thus $2 \in (2)$ and $3 \in (3)$. We see that $2 + 3 \notin (2)\ U\ (3)$. For ((2) U (3), +, .) to be an ideal it should be additive sub-group. But the sum of 2, $3 \in (2)\ U\ (3)$ does not belong to (2) U (3). Hence ((2) U (3), +, .) is not an ideal.

But we observe that $(2) + (3) = \{a + b \mid (a) \in (2) \text{ and } b \in (3)\}$, contains the sum of all elements of (2) and (3) which is clearly the set generated by {(2) U (3)}.

Hence ((2) + (3), +) is an ideal.

Exercise 6.6.2: Let $(I_1, + .)$ and $(I_2, +, .)$ be two ideals of the ring $(R, +, .)$ such that $I_1 \cap I_2 = \{0\}$. Prove that $a.b = 0$ for every $a \in I_1$, $b \in I_2$.

Solution: Since $(I_1, +, .)$ and $(I_2, +, .)$ are ideals, then, If $a \in I_1$, $b \in I_2 \subset R.$,

then $a \,.\, b \in I_1$

and $b \in I_2$, $a \in I_1, \subseteq R.$

then $a.b \in I_2$.

That is, $a.b \in I_1 \cap I_2 = \{0\}$

Hence $a.b = 0$, for all $a \in I_1$, $b \in I_2$.

Exercise 6.6.3: Let $(I_1, +, .)$ be a commutative ring and $a \in R$, the set I is defined by

$$I = \{x \in R \mid x.a = 0\}$$

Prove that $(I, +, .)$ is an ideal of $(R, +, .)$.

Solution: Let $x, y \in I$. By the definition of I, we have $x \,.\, a = 0$, $y \,.\, a = 0$

or $x \,.\, a - y \,.\, a = 0 - 0 = 0 \Rightarrow (x - y).\ a = 0 \Rightarrow x - y \in I$ Again, if $r \in R$, $x \in I$, then $x \,.\, a = 0$

CHAPTER 6

and $(x \,.\, r) \,.\, a = (r \,.\, x) \,.\, a = r\,(x \,.\, a) = 0 \Rightarrow x \,.\, r \in I$ and $r \,.\, x \in I$.

since $(R, +, .)$ is commutative,

Hence $(I, +, .)$ is an ideal of $(R, +, .)$.

Definition 6.6.1: Let $(R, +, .)$ be a commutative ring, than an ideal P of R is called a prime ideal if for all $a, b \in R$,

$$a \,.\, b \in P \Rightarrow a \in P \text{ or } b \in P.$$

Example 6.6.3: In an integral domain $(D, +, .)$, (0) is prime ideal. Since for all $a, b \in D$, $a.b \in (0) \Rightarrow a.b = 0 \Rightarrow a = 0$ or $b = 0$ since D is with out zero divisors.

Example 6.6.4: In $(Z, +, .)$ the ideal $(5) = \{5\,n \mid n \in Z\}$ is prime ideal, since $ab \in (5) \Rightarrow 5 \mid ab \Rightarrow 5 \mid a$ or $5 \mid b$.

Theorem 6.6.2: *An ideal $(P, +, .)$ of commutative ring $(R, +, .)$ is prime if and only if $(R/P, +, .)$ is an integral domain.*

Proof: First, Let $(P, +, .)$ be a prime ideal of the ring $(R, +, .)$. Since $(R, +, .)$ is a commutative ring with identity, (R/P, +, .) is also a commutative with identity $(1 + p)$. Now we have to show that R/P is with out zero divisors.

Let $a + p,\ b + P \in R/P$ such that

$$\begin{aligned}(a + p) \,.\, (b + p) &= P \\ &\Rightarrow a.b + P = P \\ &\Rightarrow a.b \in P \\ &\Rightarrow a \in P \text{ or } b \in P \text{ since } P \text{ is prime ideal} \\ &\Rightarrow a + P = P \text{ or } b + P = P.\end{aligned}$$

R/P is with out zero divisors.

Hence $(R/P, +, .)$ is an integral domain.

Conversely, let $(R/P, +, .)$ be an integral domain and let $a, b \in R$ such that $a.b \in P$.

Now $(a + p) \,.\, (b + p) = a.b + P = P$ since $ab \in P$

Since R/P does not have zero divisions,

$$\begin{aligned}(a + p) \,.\, (b + p) = P &\Rightarrow a + P = P \text{ or } b + P = P \\ &\Rightarrow a \in P \text{ or } b \in P.\end{aligned}$$

Hence $(P, +, .)$ is a prime ideal.

Definition 6.6.2: An ideal $(M, +, .)$ of the ring $(R, +, .)$ is said to be maximal ideal if

(*i*) $M \neq R$

(*ii*) There exists no ideal $(U, +, .)$ of R such that $M \subset U \subset R$.

This if M is maximal ideal of R, then for any ideal U, $M \subset U \subset R \Rightarrow M = U$ or $U = R$.

Example 6.6.5: In a division ring $(D, + ,.)$, (0) is the maximal ideal. Clearly, $(0) \neq D$ as $1 \in D$, $1 \neq 0$. Let $(U, +, .)$ be any non-zero ideal of $(D, +, .)$. Let $a \in U \subset D$, then $a^{-1} \in D$.

So $1 = a \cdot a^{-1} \in U$ and consequently $U = R$. Hence ((0), +, .) is maximal ideal in division ring.

Example 6.6.6: In the ring (*E*, +, .) of even integers, the ideal ((4), +, .) is maximal ideal. $2 \notin (4)$, $(4) \neq E$ let (*U*, +, .) be an ideal such that $(4) \subset U$. Then there exists an element $a \in U$ and $a \notin (4)$.

This implies that *a* is an even integer which is not divisible by 4. Consequently $a = un + 2$ for some integer *n*. Now $2 = a - 4n \in U$. Since $a \in U$ and $(4) \subset U$. This means that every even integral multiple of 2 belongs to *U*. Hence $E = U$. Hence ((4), +, .) is maximal idea.

Theorem 6.6.3: *An ideal (M, +, .) of the commutative ring (R, +, .) is maximal ideal if and only if (R/M, +, .) is a field.*

Proof: First, let (*M*, + , .) be a maximal ideal of the commutative ring (*R*, +, .). Since (*R*, +, .) is a commutative ring with identely, so is (*R*/*M*, +, .). To prove that (*R*/*M*, +, .) is a field, it remains to verify that non-zero element of *R*/*M* has its multiplicative inverse.

Since (*M*, +, .) is maximal, $a \in R$, $a \notin M$,

$M + (a) = R = (1)$, That is, there exists $b \in R$, $x \in M$ such that $x + ab = 1$

$\Rightarrow \quad ab - 1 = -x \in M$ as $x \in M$

$\Rightarrow \quad ab - 1 \in M$

$\Rightarrow \quad ab + M = 1 + M$

$\Rightarrow \quad (a + M) \cdot (b + M) = 1 + M.$

$\Rightarrow \quad (a + M)^{-1} = b^{+}/M \in R/M.$

Hence (*R*/*M*, +, .) is a field

Conversely, let (*R*/*M*, +, .) be a field. For $a \in R$, $a \in M$. $a + M$ is invertible.

For some $b + M \in R/M$

$(a + M) \cdot (b + M) = 1 + M.$

$\Rightarrow \quad a.b + M = 1 + M.$

$\Rightarrow \quad ab - 1 \in M.$

So for some $x \in M$

$ab - 1 = -x$

$\Rightarrow \quad ab + x = 1$

$\Rightarrow \quad (ab) R + x R = 1.R$

$\Rightarrow \quad (a) + M + R.$

$\Rightarrow \quad$ (*M*, +, .) is a maximal ideal.

Example 6.6.7: Consider the ideal (0) of *Z*, since (*Z*, +, .) is an integral domain, (0) is a prime ideal, but (0) is not a maximal ideal since $(0) \subset (2) \subset Z$.

Example 6.6.8: The ideal ((4), +, .) in the ring (*E*, +, .) is maximal ideal. But it is not a prime ideal sense $2.2 \in (4)$ bet $2 \notin 4$. Hence in a commutative ring with out unity a maximal ideal need not be prime.

Theorem 6.6.4: *Let $(U, +, .)$ be an ideal of the ring of integers $(Z, +, .)$ then $(U, +, .)$ is maximal f and only if $U = (p)$, where p is prime.*

Proof: We first assert that if p is prime number, then $P = (p)$ and $P \subset U$, then $U = (a)$ for some integer $a \in Z$. Since $p \in P \subseteq U$, $p = ma$ for some integer m. Since p is prime, then

$$p = ma \Rightarrow a = 1 \text{ or } a = p.$$

If $a = p$, $P \subset U$ and $(a) \subset P \Rightarrow U = P$.

Again if $a = 1$, then $1 \in U$ and $r = 1.r \in U$ for all $r \in Z$ and $U = Z$.

Thus no ideal, other than Z or P itself, can be put between P and Z from which we deduce that P is maximal.

Conversely, let $M = (a)$ be a maximal ideal of Z. We claim that a must be a prime number. For if $a = bc$, where b and c are positive integers, then $U = (b) \supset M$, hence $U = Z$ or $U = M$. If $U = Z$, then $b = 1$ is an easy consequence; if $U = M$, then $b \in M$, and so $b = ra$ for some integer r, since every element of M is a multiple of a. But then $a = bc = rac \Rightarrow rc = 1$, so that $c = 1$, $a = b$. Thus a is a prime number.

Exercise 6.6.4: Let $(R, +, .)$ be the ring of all real-valued continues functions on the closed internal $[0, 1]$. Let $M = \{f \in R \mid f(1/2) = 0\}$. Show that $(M, +, .)$ is a maximal ideal of R.

Solution: To the function $f : [0, 1] \to R$, the set of real numbers, defined by $f(x) = 0$, for all $x \in R$ belongs to M.

Hence $M \neq Q$.

Let $f, g \in M$, then $f\left(\frac{1}{2}\right) = 0$, $g\left(\frac{1}{2}\right) = 0$

Now $(f - g)\left(\frac{1}{2}\right) = f\left(\frac{1}{2}\right) - g\left(\frac{1}{2}\right) = 0 \quad \Rightarrow \quad f - g \in M$

and if $f \in M$ and $h \in R$, $(hf)\left(\frac{1}{2}\right) = h\left(\frac{1}{2}\right) f\left(\frac{1}{2}\right) = h\left(\frac{1}{2}\right) . 0 = 0$

$\Rightarrow h f \in M$.

$\therefore$ $(M, +, .)$ is an ideal.

Now we claim that $(M, +, .)$ is maximal ideal of R, for if the ideal U contains M and $U \neq M$, then there is a function $g(x) \notin U$ such that $g(x) \notin M$. Since $g(x) \notin M$, $g\left(\frac{1}{2}\right) = \alpha \neq 0$. Now $h(x) = g(x) - \alpha$ is such that $h\left(\frac{1}{2}\right) = g\left(\frac{1}{2}\right) - \alpha = 0$, so that $h(x) \in M \subset U$. But $g(x) \subset U$. Therefore $\alpha = g(x) - h(x) \in U$ and so $1 = \alpha\alpha^{-1} \in U$. Thus for every function $t(x) \in R$, $t(x) = 1.\ t(x) \in U$ and consequently $U = R$. Therefore M is a maximal ideal of R. Similarly, if r is a real number $0 \leq r < 1$, then $M_r = \{f(x) \in R \mid f(r) = 0\}$ is maximal ideal of R.

Exercise 6.6.5: For any two ideals A and B of ring $(R, +, .)$ such that $B \subseteq A$. Prove that A/B and B are nilpotent imply A is nilpotent.

Solution. Let A/B and B be nilpotent $\left(\frac{A}{B}\right)^n = B$, $B^m = (0)$ for some integer $n, m > 0$.

$$\begin{aligned}\left(\frac{A}{B}\right)^n = B &\Rightarrow A^n \subseteq B \\ &\Rightarrow (A^n)^m \subseteq B^m \\ &\Rightarrow A^{mn} = 0 \\ &\Rightarrow A \text{ is nilpotent.}\end{aligned}$$

Now suppose A/B and B are nilpotent. Let $a \in A$, then $a + B \in A/B$. Since A/B is nil, there exists are integer $n > 0$ such that $(a + B)^n = B \Rightarrow a^n \in B$.

Again as B is nil, there exists an integer $m > 0$ such that $(a^n)^m = 0 \Rightarrow a^{mn} = 0$. Thus shows that a is nilpotent.

Hence A is nilpotent ideal of R.

Exercise 6.6.6: Let $(R, +, .)$ be commutative ring with unity.

Then prove that every maximal ideal is a prime ideal.

Proof: Let $(U, +, .)$ be an ideal of a commutative ring $(R, +, .)$ with unity. Then $(R/U, +, .)$ is an integral domain if and only if $(U, +, .)$ is a prime ideal by theorem 6.6.2.

Let $(U, +, .)$ be a maximal ideal of commutative ring with identity $(R, +, .)$, the by theorem 6.6.3 $(R/U, +, .)$ is a field. Every field is an integral domain. Hence $(R/U, +, .)$ is an integral domain and consequently $(U, +, .)$ is prime ideal by Theorem 6.6.2.

This concludes the Theorem.

Remark: The converse of the theorem is no true, because every prime ideal is not necessary a maximal ideal.

Exercise 6.6.7: If $(R, +, .)$ is a finite commutative ring with identity, then prove that every prime ideal of $(R, +, .)$ is a maximal ideal of $(R, +, .)$.

Proof: Let $(U, +, .)$ be a prime ideal of a commutative ring with identity $(R, +, .)$. So $(R/U, +, .)$ is an integral domain by theorem 6.6.2 since $(R, +, .)$ is finite, so is R/U.

Thus $(R/U, +, .)$ is a finite integral domain.

$\Rightarrow (R/U, +, .)$ is a field

$\Rightarrow (U, +, .)$ is maximal ideal of $(R, +, .)$.

Hence this proves the theorem.

PROBLEMS

1. Is the subring $(\{0, 2), +_4, \cdot_4)$ an ideal of the integers modulo 4?
2. Is the subring $(\{0, 3, 6\}, +_4, \cdot_4)$ an ideal of the integers modulo 9?
3. Prove that every subring of the integers is an ideal.

4. Prove that every subring of the integers modulo n is an ideal.

5. Determine the quotient rings for the ideals in question nos. 1 and 2.

6. Suppose that $(I, +, .)$ is an ideal of the ring $(R, +, .)$ Prove that if $(R, +, .)$ is commutative, then the quotient ring $(R/I, +, .)$ is commutative.

7. Let $(S, +, .)$ be an ideal and $(T, +, .)$ be a subring of the ring $(R. +, .)$. Then prove that $(S, +, .)$ is an ideal of $(S + I, +, .)$.

8. Determine all ideals of $(Z_{12} + _{12}, \odot_{12})$, the ring of integers modulo 12.

9. Let $(I, +, .)$ be an ideal of the ring $(R, +, .)$ and let $C(I)$ be the set defined by $C(I) = \{r \in R \mid r . a - a . r \in 1, \forall a \in R\}$.

Determine whether $(C(I), +, .)$ forms a subing of $(R, +, .)$.

10. Show that the ring $(R, +, .)$ of real number is a simple ring.

11. Show that the ring $(Z_n, +_n, \odot_n)$ of integers modulo n is a principal ideal ring.

12. Let $(I, +, .)$ be an ideal of the ring $(R, +, .)$ and define

$ann\ I = \{r \in R \mid r . a = 0, a \in 1]$

Prove that the system $(a_{nn}\ I, +, .)$ is an ideal of $(R, +, .)$, called *annihilator ideal of I.*

13. Let $(I, +, .)$ be an ideal of $(R, +, .)$, a commutative ring with identity. For an arbitrary element $a \in R$, the ideal generated by $I\, U\, \{a\}$ is denoted by $((I . a), +,)$. Assuming $a \notin I$, show that

$(I, a) = \{i + r . a \mid i \in I, r \in R\}$.

[**Hint:** The set generated by $I\ U\ \{a\}$ is the set of all elements of I and of those elements which are of the forms $i + a$ *or* $j . a$ for all $j \in I$. So the set (I, a) generated by $I\ U\ [a]$ is

$(I, a) = \{j + a \mid j \in 1\}$

$= \{r(j + a) \mid j \in I, r \in R\}$, Since $((I, a), +, .)$ is an ideal

$= [i + ra \mid i \in I, r \in R\}$, where $i = r . j \in I$.]

14. In the ring of integers, let us consider two principal ideals $((n), +,.)$ and $((m), +,.)$ generated by two non negative integers n and m respectively. Then show that

$((n), m) = ((m), n) = (n) + (m) = (\{m, n\}) = (d)$,

where d is the greatest common divisor of n and m.

15. Let $(I_1, +, .)$ and $(I_2, +, .)$ be two ideals of the ring $(R, +, .)$. Define the set $I_1 . I_2$ by

$I_1 . I_2 = \{\Sigma a_i b_i \mid a_i \in I_1, b_i \in I_2,\}$.

where S denotes a finite sum with one or more terms. Prove that $(I_1, I_2, +, .)$ is an ideal of $(R, +, .)$

16. Let $(I, +, .)$ be an ideal of the ring $(R, +, .)$, show that

(*a*) the ring $(R/I, +, .)$ may have divisor of zeros, even though $(R, +, .)$ does not have any.

(*b*) If $(R, +, .)$ is a principal ideal ring, then so is the quotient ring $(R/I, +, .)$

17. Let $(R, +, .)$ be a commutative ring with identity, and let N denote the set of all nilpotent elements of R.

(*a*) Prove that $(N, +, .)$ is an ideal of $(R, +, .)$

(*b*) Show that the quotient ring $(R/N, +, .)$ has no nilpotent elements. The set N is non-empty for $0^n = 0, 0 \in N$.

[**Hint:** Let a and b be two nilpotent elements of R. Then there exist positive integers m and n such that $a^m = 0, b^n = 0$.

Now we consider $(a - b)^{m+n}$. By bionomial theorem we have $(a - b)^{m+n}$ $= a^{m+n} - {}^{m+n}C_1\, a^{m+n-1}\, b + ...+ (-1)^{m+n}\, {}^{m+n}C_{m+n-1}\, a.b^{m+n-1} + (-1)^{m+n}\, b^{m+n}$

$= a^m . a^n - {}^{m+n}C_1, a^m, a^{n-1}, b + ...+ (-1)^{m+n-1}\, C_{m+n-1}\, a.b^{m-1}. b^n + (-1)^{m+n}\, b^m. b^n$

$= 0$ since every term contains either a^m or b^n .

Thus $(a - b) \in N$

Hence the pair $(N, +)$ is sub-group of the additive group $(R, +)$ of the ring $(R, +, .)$.

Now for every $r \in R$, we have

$= (r.a)^m$, where m is the positive integer for which $a^m = 0$

$= r^m. a^n$, since $(R, +, .)$. is commutative

$= 0$, since $a^m = 0$.

Similarly, $a . r = 0$. This shows $(N, +, .)$ is an ideal of the ring $(R, +, .)$

(*b*) The set R/N is the set of cosets $a + N$ *of* N, the ideal of nilpotent elements. If $a \in N$, then $a + N = N$, so we consider $a \notin N$, that is, there does not exist any positive integer n for which $a^n = 0$.

So $a^n \neq 0$, for any integer $n > 0$.

Now $(a + n)^n = a^n + N \neq N$ for any integer,

since $a^n \notin N$. Hence the result.]

18. Let $(R, +, .)$ be a ring with the property $a^2 + a \in$ cent R for every $a \in R$. Show that $(R, +, .)$ is commutative.

19. For two ideals $(A, +, .)$ and $(B, +, .)$ of a ring $(R, +, .)$, $(A \cup B, +, .)$ is an ideal if and only if either $A \subseteq B$ or $B \subseteq A$.

6.7 HOMOMORPHISMS OF RINGS

In chapter 3 we have studied, the homomorphism from one group to another, which is a function from one set to another that preserves the group operations. This notion is extended to homomorphisms of rings. Since a ring has two operations, the functions will be considered which will preserve both operations of the rings.

Definition 6.7.1: Let $(R, +, .\odot)$ and $(S, \oplus, .)$ be two rings and f a functtion from R into S, $f. R \to S$. Then f is a homomorphism (or operation preserving function) from $(R, +, .)$ into $(S, \oplus, .\odot)$ if and only if

CHAPTER 6

$f(a + b) = f(a) \oplus f(b)$, for all $a, b \in R$,

$f(a.b) = f(a) \odot f(b)$, for all $a, b \in R$.

The ring $(S, \oplus, .\odot)$ is a homomorphic image of $(R, +, .)$ if and only if there exists a homomorphism from R onto S.

Example 6.7.1: Let $(Z_4, +_4, \odot_4)$ and $(Z_2, +_2, \odot_2)$ be two rings of integers modulo 4 and modulo 2 respectively. The function f is defined from Z_4 to Z_2 by $f(0) = f(2) = 0$ and $f(1) = f(3) = 1$. It is clear that this function is not one to one, since two elements 0 and 2 of Z_4 are mapped on $0 \in Z_2$. Now we verify whether f is an operation preserving function.

$f(0 +_4 0) = f(0) = 0, f(0 +_4 0) = f(0) +_2 f(0) = 0$

$f(0 +_4 2) = f(2) = 0, f(0 +_4 2) = f(0) +_2 f(2) = 0$

$f(1 +_4 1) = f(2) = 0, f(1 +_4 1) = f(1) +_2 f(1) = 0$

$f(0 \odot_4 x) = f(0) = 0, f(0 \odot_4 x) = f(0) \odot_2 f(x) = 0$

$f(1 \odot_4 x) = f(x), f(1 \odot_4 x) = f(1) \odot_2 f(x) = f(x)$

which shows f is a homomorphism from $(Z_4, +_4, \odot_4)$ to $(Z_2, +_2, \odot_2)$.

Example 6.7.2: Let $(R_1, +, .)$ and $(R_2, \oplus, \odot)$ be two arbitrary rings and let $f : R_1 \to R_2$ be the function that sends every element of R_1, onto the zero element 0 of $(R_2, \odot, .\odot)$. Such a mapping is a homomorphism, for

$f(a + b) = 0 = 0 \oplus 0 = f(a) \oplus f(b)$.

$f(a . b) = 0 = 0 \odot 0 = f(a) f(b)$.

Example 6.7.3: Let $(Z, +, .)$ be the ring of integers, and $(Z_n, +_n, \odot_n)$ be the ring of integers modulo n.

Define $f: Z \to Z_n$ by $f(a) = [a]$; that is, f maps each integer into the congruence class containing it. Then

$f(a + b) = (a + b) = [a] +_n [b] = f[a] +_n f(b)$.

$f(a . b) = [a . b] = [a] \odot_n [b] = f(a) \odot_n f(b)$.

Thus, f is a homomorphism from $(Z, +, .)$ onto $(Z_n, +_n, \odot_n)$

Example 6.7.4: The mapping $f : Z \to E$ deformed by $f(a) = 2a$ is not a homomorphism from $(Z, +, .)$ into $(E, +, .)$ For,

let $a \neq 0, b \neq 0$,

$f(a + b) = 2(a + b) = 2a + 2b = f(a) + (b)$.

but $f(a . b) = 2(a . b) \neq 2a . 2b = f(a) . f(b)$.

Thus f preserves addition operation and does not preserve multiplication of rings.

Theorem 6.7.1: *Let f be a homomorphism from the ring $(R, +, .)$ into the ring $(R', \oplus, \odot)$. Then the following statements hold:*

(1) $f(0) = 0'$, *where $0'$ is the additive identity of $(R', \oplus, \odot)$,*

(2) $f(-a) = -f(a)$,

(3) $(f(R), \oplus, \odot)$ *is a subring of* $(R', \oplus, \odot)$

If, in addition, $(R, +, .)$ *and* $(R', \oplus, \odot)$ *are rings with identity elements* I *and* I', *respectively, and* $f(R) = R'$, *then*

(4) $f(1) = 1'$,

(5) $f(a^{-1}) = f(a)^{-1}$, *wherever* a^{-1} *exists in* R.

Proof: **(1)** Let $a, 0 \in R$, and f be a homomorphism of the ring $(R, +, .)$ into the ring $(R', \oplus, \odot)$.

$$f(a) = f(a + 0) = f(a) \oplus f(0), \quad ...(1)$$

and

$$f(a) \oplus 0' = f(a) \quad ...(2)$$

from (1) and (2) we have

$f(a) \oplus f(0) = f(a) \oplus 0'$.

By left cancellation law under addition,

$$f(0) = 0'.$$

(2) Let $a \in R$, then $-a \in R$, (since $(R, +)$ is an abelian group. We know $a + (-a) = 0 = (-a) + a$

$$\Rightarrow \quad f(0) = f(a + (-a)) = f(a) \oplus f(-a) = 0'$$

and

$$f(0) = f((-a) + a) = f(-a) \oplus f(a) = 0'$$

which implies

$$f(-a) \oplus f(a) = 0' = f(a) \oplus f(-a).$$

This shows $f(-a)$ is the additive inverse of $f(a) \in R'$,

Hence $f(-a) = -f(a)$.

From this, $f(a - b) = f(a + (-b)) = f(a) \oplus f(-b) = f(a) - f(b)$.

(3) Let $a, b, \in R$, then $f(a), f(b) \in f(R)$

Now $f(a) \oplus f(b) = f(a + b) \in f(R)$

which implies $f(a) \oplus f(b) \in f(R)$

and $f(a) \oplus f(-b) = f(a - b) = f(a) - f(b) \in f(R)$.

Hence $f(a) - f(b) \in f(R)$,

Again $f(a) \odot f(b) = f(a \,.\, b) \in f(R)$.

That is, if $f(a), f(b) \in f(R)$ $f(a) \odot f(b) \in f(R)$

Hence $(f(R), \oplus \odot)$ is a subring of $(R', \oplus, \odot)$

(4) Let c be an element of $f(R)$, then there exist an element a of R such that $f(a) = c$.

$c = f(a) = f(a.1) = f(a) \odot f(1) = c \odot f(1)$

$c = c \odot 1' = c \odot f(1)$

Since the multiplicative identities *of* R and R' are unique, it follows that $f(1) = 1'$.

CHAPTER 6

Note: (1) To the expression $c \odot 1' = c \odot f(1)$ we cannot apply cancellation law, because R' may have divisors of zero or $c \in R'$ may not have its multiplicative inverse.

(2) If f maps into the set R' (f is not onto mapping), then $f(1)$ is the identity of $f(R)$, $\oplus$, $\odot$). The element $f(1)$ may not be an identity of the entire ring $(R', \oplus, \odot)$. In fact, it may very well happen that $f(1) \neq 1'$.

(3) Let $a \in R$ have an inverse $a^{-1} \in R$.

$$f(a) \odot f(a^{-1}) = f(a \,.\, a^{-1}) = f(1)\, 1'$$
$$= f(1) = f(a^{-1}.\, a) = f(a^{-1}) \,.\, f(a)$$

which shows that

$$(f(a))^{-1} = f(a^{-1}).$$

Definition 6.7.2: Let f be a homomprphism from the ring $(R, +, .)$ into the ring $(S, \oplus, \odot)$. Let Ker (f) be the set of all elements $x \in R$ such that $f(x) = 0'$, the zero element of $(S, \oplus, \odot)$. The set Ker (f) is called the kernel of the homomorphism f. Symbolically we can write Ker $(f) = \{a \in R \mid f(a) = 0'\}$.

Theorem 6.7.2: *Let f be a homomorphism from the ring $(R, +, .)$ into the ring $(S, \oplus, \odot)$, then the system* (Ker (f), +, .) *is an ideal of* $(R, +, .)$.

Proof: The Ker (f) is not empty set since $0 \in$ Ker (f), let $a, b \in$ Ker (f), then $f(a) = 0'$, $f(b) = 0'$, where $0'$ is the zero element of $(S, \oplus, \odot)$. Now

$$f(a + b) = f(a) \oplus f(b) = 0' \oplus 0' = 0'$$
$$f(a - b) = f(a) - f(b) = 0' - 0' = 0'$$

Thus if $a, b \in$ Ker (f), then $a - b \in$ Ker (f). Again, for all $r \in R$, $a \in$ Ker (f), $f(r \,.\, a) = f(r) \odot f(a) = f(r) \odot 0' = 0'$,

which shows $r \,.\, a \in$ Ker (f).

In a like manner we conclude that $a \,.\, r \in$ Ker (f).

Hence (Ker (f), +, .) is an ideal of $(R, +, .)$

Example 6.7.5: Let $(Z, +, .)$ and $(Z_n, +_n, \odot_n)$. be two rings and the mapping $f : Z \to Z_n$ defined by $f(a) = [a]$.

It is established that f is a homomorphism from $(Z, +, .)$ onto $(Z_n, +_n, \odot_n)$. Hence

$$\text{Ker}(f) = \{a \in Z \mid f(a) = [0])$$
$$= \{a \in Z \mid a = kn \text{ for some } k \in Z)$$
$$= (n).$$

In other words, (Ker (f), + , .) is just the principal ideal generated by the integer n.

Definition 6.7.3: Let $(R, +, .)$ and $(S, \oplus, \odot)$ be two rings. Let f be a one to one function from R onto S. Then f is an isomorphism from $(R, +, .)$ onto $(S, \oplus, \odot)$ if and only if

$$f(a + b) = f(a) \oplus f(b),$$
$$f(a \,.\, b) = f(a) \odot f(b),$$

for each pair $a, b \in R$.

The ring $(S, \oplus, \odot)$ is an isomorphic image if and only if f is one to one from R onto S, denoted by

$(R, +, .) \cong (S, \oplus, \odot)$.

Theorem 6.7.3: *Let f be a homomorphism from the ring (R, +,) onto the ring (S, ⊕, ⊙). Then f is an isomorphism if, and only if, the kernel f = {0}.*

Proof: The theorem has two parts. Let f be an isomorphism. We shall prove that the Ker f contains only $0 \in R$. We know that $f(0) = 0'$. So Ker f must contain the element $0 \in R$. Since f is an isomorphism, then the only element in the Ker f is 0.

For the other part of the theorem, let f be a homomorphism with Ker $f = \{0\}$, let $a . b \in R$.

Then

$$
\begin{aligned}
& & f(a) &= f(b) \\
&\Rightarrow & f(a) - f(b) &= f(b) - f(b) \\
&\Rightarrow & f(a - b) &= 0' \\
&\Rightarrow & a - b &\in \text{Ker } f \\
&\Rightarrow & a - b &\in \{0\} \text{ Since Ker } f = \{0\} \\
&\Rightarrow & a - b &= 0 \\
&\Rightarrow & a &= b.
\end{aligned}
$$

Hence f is an isomorphism. This completes the proof.

Example 6.7.6: Show that the mapping $f : C \to M_2$, defined by $f(a + ib) = \begin{bmatrix} a & b \\ -b & a \end{bmatrix}$ is an isomorphism of the set C, the set of complex number, into the ring of 2×2 matrices over the real numbers.

Solution: First we have to prove that f is an operation preserving function. For,

let $a + ib \in C$, $c + id \in C$,

$$
\begin{aligned}
f((a + ib) + (c + id) &= f((a + c) + i(b + d)) \\
&= \begin{bmatrix} a+c & b+d \\ -b+d & a+c \end{bmatrix} \\
&= \begin{bmatrix} a & b \\ -b & a \end{bmatrix} + \begin{bmatrix} c & d \\ -d & c \end{bmatrix} \\
&= f(a + ib) + f(c + id),
\end{aligned}
$$

and $\quad f(a + ib) . (c + id) = f((ac - bd) + i(bc + ad))$

$$= \begin{bmatrix} ac-bd & bc+ad \\ -(bc+ad) & ac-bd \end{bmatrix}$$

$$= \begin{bmatrix} a & b \\ -b & a \end{bmatrix} \cdot \begin{bmatrix} c & d \\ -d & c \end{bmatrix}$$

$$= f(a + ib) \cdot f(c + id).$$

Now we show that f is one to one.

If $f(a + ib) = f(c + id)$, then

$$= \begin{bmatrix} a & b \\ -b & a \end{bmatrix} = \begin{bmatrix} c & d \\ -d & c \end{bmatrix}$$

which means $a = c$, $b = d$ by the definition of equal matrices.

Thus

$a + ib = c + id.$

which proves that f is an isomorphism from the ring $(C, +, .)$ to $(M_2, +, .)$

Second Proof: For showing that f is one to one, let us determine the Ker f.

Let $a + ib \in$ Ker f, then $f(a + ib) = \begin{bmatrix} a & b \\ -b & a \end{bmatrix} = \begin{bmatrix} 0 & 0 \\ 0 & 0 \end{bmatrix} \Rightarrow a = 0, b = 0$

which shows Ker f contains 0 element of the complex number C. Hence f is an isomorphism.

Let $(R, +, .)$ and $(R', +, .')$ be two rings and let the ring $(R, +, .)$ be with identity element. Then there exists an isomorphism between the subring $(S, +, .)$ of $(R, +, .)$ and the ring $(R', +', .')$. That is, we can say that every ring is isomorphic to a subring of a ring with identity. This we shall show in the following theorem.

Definition 6.7.4: A ring $(R, +, .)$ is *imbedded* in a ring $(R', +', .')$ if there exists a subring $(S, +', .')$ of $(R', +', .')$ such that $(S, +' . ') \cong (R, +, .)$.

Theorem 6.7.4: *Any ring can be imbedded in a ring with identity.*

Proof: Let $(R, +, .)$ be any ring. We define

$R \times Z = \{(a, n) \mid a \in R, n \in Z\}$, where Z is the set of all integers. On this set the addition and multiplication are given by

$$(a, n) + (b, m) = (a + b. n + m), \text{ and}$$

$$(a, n) . (b, m) = (a.b + am + bn, nm).$$

Under this operations the triple $(R \times Z, +, .)$ is a ring with identity. It can be easily verified. It has $(0, 1)$ which is the identity in $(R \times Z, +, .)$. For

$(a, n). (0, 1) = (a . 0 + a. 1 + n . 0, n .1) = (a, n),$

and similarly, $(0 . 1) . (a, n) . = (a, n)$

But the set $R \times \{0\} \subset R \times Z$ is the set of all element

$(a, 0), a \in R, 0 \in Z$. If $(a, 0), (b, 0) \in R \times \{0\}$, then

$(a, 0) - (b, 0) = (a - b, 0) \in R \times \{0\}$,

and $(a, 0), (b, 0) = (a . b, 0) \in R \times \{0\}$, This shows $(R \times \{0\}, +, .)$ is a subring of the ring $(R \times Z, +, .)$

Now we define the mapping $f : R \to R \times \{0\}$ by

$f(a) = (a, 0), \forall a \in R$.

It is clear that f is a onto.

Let $a, b \in R$. Then

$$f(a) = f(b) \Rightarrow (a, 0) = (b, 0)$$
$$\Rightarrow a = b,$$

Hence f is one to one. It is also operation preserving function. For $a, b \in R$.

$$f(a + b) = (a + b, 0)$$
$$= (a, 0) + (b, 0) = f(a) + f(b),$$

and
$$f(a . b) = (a . b, 0) = (a, 0) . (b, 0)$$
$$= f(a), . f(b)$$

This proves that f is an isomorphism and

$(R, +, .) \cong (R \times \{0\}, +, .)$ Hence $(R, +, .)$ is imbedded in a ring $(R \times Z, +, .)$ with identity.

Theorem 6.7.5: *Let $(I, +, .)$ be an ideal of the ring $(R, +, .)$. Then the mapping $f: R \to R/I$, defined by $f(a) = a + I$, is a homomorphism from $(R, +, .)$ onto the quotient ring $(R/I, +, .)$ having I as its kernel.*

Proof: First, we shall show that f preserves the ring operations. It follows directly from the definition of quotient ring that

$$f(a + b) = (a + b) + I = (a + I) + (b + I)$$
$$= f(a) + f(b),$$

and
$$f(a . b) = (a.b) + I = (a + I), (b + I)$$
$$= f(a) . f(b).$$

For showing f to be onto mapping, $a + I$ is an element of R/I where $a \in R$ and $f(a) = a + I$.

In as much as the coset I is the identity of the ring $(R/I, +, .)$, we have

$$\text{Ker}(f) = \{a \in R \mid f(a) = I\}$$
$$= \{a \in R \mid a + I = I\} = \{a \mid a \in I\} = I,$$

which completes the proof of the theorem.

Theorem 6.7.6: *If f is a homomorphism from the ring $(R +, .)$ onto the ring $(R', +', .')$, then $(R/Ker(f), +, .) \cong (R', +', .')$.*

CHAPTER 6

Proof: To show that $(R/\text{Ker}(f), +, .) \cong (R', +', .')$ we define a mapping $\phi: R/\text{Ker}(f) \to R'$ by $\phi(a + \text{Ker}(f)) = f(a)$, where $f(a)$ is the image of a under f. It is homomorphism from the ring $(R/\text{Ker}(f), +, .)$ onto the ring $(R', +', .')$ which follows directly from

$\phi\ [(a + \text{Ker}\ (f) + (b + \text{Ker}\ (f)]$

$$= \phi\ ((a + b) + \text{Ker}\ (f))$$
$$= f\ (a + b)$$
$$= f\ (a) +'.\ (b)$$
$$= \phi\ (a + \text{Ker}\ f) +'\ \phi\ (b + \text{Ker}\ (f))$$

and $\quad \phi\ ((a + \text{Ker}\ f)\ .\ (b + \ker f) = \phi\ (a\ .\ b + \text{Ker}\ f).$

$$= f\ (a\ .\ b)$$
$$= f\ (a).'\ f\ (b)$$
$$= \phi\ (a + \text{Ker}\ (f)).'\ \phi\ (b + \text{Ker}\ (f))$$

It is also one to one which follows from

$$\phi\ (a + \text{Ker}\ f) = \phi\ (b + \text{Ker}\ f), \text{ for } a,\ b \in R$$

$\Rightarrow \quad f\ (a) = f\ (b)$

$\Rightarrow \quad f\ (a) +'\ f\ (-b) = 0'$

$\Rightarrow \quad f\ (a + (-b) = 0'$

That is, $a - b \in \text{Ker}\ f$ which implies

$$a + \text{Ker}\ f = b + \text{Ref}\ f$$

To show that ϕ is an onto mapping, let

$x \in R'$, $a \in R$ such that

$f\ (a) = x$, since f is onto mapping from R onto R'.

But $a + \text{Ker}\ f \in R/\text{Ker}\ f$ and by the definition of ϕ

$\phi\ (a + \text{Ker}\ (f)) = f\ (a) = x,$

which shows that every element $x \in R'$ is an image under the mapping ϕ. Hence ϕ is an onto mapping. Which proves $(R/\text{Ker}\ f,\ +,\ .) \in (R',\ +',\ .')$.

Example 6.7.7: Consider homomorphism from the ring $(Z_4, +_4, \odot_4)$ onto the ring $(Z_2, +_2, \odot_2)$ defined by $f\ (0) = f\ (2) = 0$, $f\ (1) = f\ (3) = 1$.

It is clear that the Ker (f) of this homomorphism is the set $\{0, 2\}$. Therefore $Z_4/\text{Ker}\ f = Z_4 /\{0, 2\} = \{\{0, 2\}, \{1, 3\}\}$.

Now we prepare the operations tables for the set $\{0, 2\}$, $\{1, 3\}$. Let $\{0, 2\} = H =$ then $H + 1 = \{1, 3\}$.

$+_4$	H	$H+1$
H	H	$H+1$
$H+1$	$H+1$	H

$\odot_n$	H	$H+1$
H	H	H
$H+1$	H	$H+1$

Thus $(Z_4 /\text{Ker}\ f +_4, \odot_4)$ is a ring which is clearly isomorphic to the ring $(Z_2 +_2, \odot_2)$. We can define a function ϕ by $\phi(H+1) = f(1) = 1$, and $\phi(H) = f(0) = 0$ where $H, H + I \in Z_4/$ Ker f and $1, 0 \in Z_2$.

***Lemma 6.7.1:** The only non-trivial homomorphism from the ring (Z, +, .) of integers onto itself is the identity map e_z.*

Proof: Let $f: Z \to Z$ be any mapping which preserve the operations, that is, f is an homomorphhism from the ring (Z, +, .) onto itself.

Now for $n \in Z+$, we have

$$f(n) = f(1 + 1 + ...+ 1), \text{ since } n = 1 + 1 +........+ 1 \; n \text{ times}$$
$$= f(1) + f(1) +......+ f(1) \; n \text{ times}$$
$$= n f(1)$$

When n is negative integer, then $-n$ is positive and we have

$$f(-(-n)) = -f(-n)$$
$$= -(-n f(1))$$
$$= n f(1).$$

Particularly, $f(0) = 0 \,.\, f(1)$.

Thus we conclude that $f(n) = n.\, f(1), \forall n \in Z$

Since f is non-trivial homomorphism, $f(1) = 1$,

Then $f(n) = n = e_z(n) \Rightarrow f = e_z, \forall n \in Z$.

Hence f is an identity function.

Corollary 1: There exists only one isomorphism under which an arbitrary ring (R, +, .) = (Z, +, .)

Proof: Let $f: R \to Z$ and $g: R \to Z$ be two mappings under which $(R, +, .) \cong (Z, +, .)$. Therefore f and g are one to one functions which follows that f^{-1} and g^{-1} exist. Now $g^{-1}: Z \to R \Rightarrow f \circ g^{-1}: Z \to Z$.

Since f and g^{-1} are isomorphism, their composition $f \circ g^{-1}$ is also isomorphism from the ring (Z, +, .) onto itself. By the Lemma 6.7.1 the only non-trivial homomorphism of (Z, +, .) onto itself is the identity map e_z. Therefore $f \circ g^{-1} = e_z \Rightarrow f = g$. This proves the corollary.

***Theorem 6.7.7:** Any homomorphism from an arbitrary ring (R, +, .) onto the ring of integers (Z, +, .) is uniquely determined by its kernel.*

Proof: Let $f: R \to Z$ and $g: R \to Z$ be two homomorphisms from the ring (R, +, .) onto the ring (Z, +, .) such that Ker (f) = Ker (g). Then by the theorem 6.7.6 there exist induced mappings $\bar{f}$ and $\bar{g}$ under which $(R/\text{Ker}(f) +, .) \cong (Z, +, .)$ and $(R/\text{Ker}(g), +, .) \cong (Z, +, .)$. It follows $(R/\text{Ker}(f) +, .) \cong (R/\text{Ker}(g), +, .)$. Since Ker (f) = Ker (g), then R/Ker (f) = R/Ker(g). So the rings (R/Ker (f), +, .) and R/Ker (g), +, .) are equal. Thus $\bar{f}$ and $\bar{g}$ are two one to one functions under which R/Ker (g), +, .) $\cong$ (R/Ker (g), +, .) $\cong$ (Z, +, .). Therefore by the preceding corollary we have

$$\bar{f} = \bar{g}.$$

CHAPTER 6

By factor theroem 3.2.7 we have

$f = \bar{f}\ o\ \phi_1$ and $g = \bar{g}\ o\ \phi_2$, where

ϕ_1: $R \to R/\text{Ker}\ (f)$ and ϕ_2 . $R \to R/\text{Ker}\ (g)$ are natural mapping.

Then $f = g$, since Ker (f) = Ker (g) and $\bar{f} = \bar{g}$.

Theorem 6.7.8: *Let f be a homomorphism of the ring (R , +, .) onto the ring (R′, +′, ′.). Then each of the following is true:*

(i) If (A, +, .) is an ideal in (R, +, .), then (f (A), +′, .′) is an ideal in (R′, + ′, .′).

(ii) If (A′, +′, .′) is an ideal in (R′, +′, .′), then (f^{-1} (A′), + , .) is an ideal in (R, +, .) with Ker (f) $\subseteq f^{-1}$ (A′).

(iii) If (A, +, .) is an ideal in (R, +, .) with Ker (f) $\subset$ A, then f^{-1} (f (A)) = A.

(iv) The mapping A $\to$ f (A) defines a one to one mapping of the set of all ideals in R which contain Ker (f) onto the set of all ideals in (R′, +′ , .′).

(v) If (A, +, .) and (B, +, .) are ideals in (R, +, .) with Ker (f) $\subset$ A, Ker (f) $\subset$ B, then A $\subset$ B if and only if f (A) $\subset$ f (B).

Proof: (1) We have $f(A) = \{f(a) \in R' \mid a \in A\}$

Let $f(a), f(b) \in f(A)$, for $a, b \in A$.

$f(a) - f(b) = f(a - b) \in f(A)$, since (A, +, .) is an ideal, $a, b \in A \Rightarrow a - b \in A$.

For $r \in R$, $a \in A$, we have

$f(a), f(r) = f(a \,.\, r) \in f(A)$ as $a.r \in A$ because (A, +, .) is an ideal.

Similarly, $f(r) \,.\, f(a) = f(r \,.\, a) \in f(A)$.

This shows that (f (A), +′, .′) is an ideal of (R′, +′, .′).

(*ii*) We have

$f^{-1}(A') = \{a \in R \mid f(a) \in A'\}$

Let $a, b \in f^{-1}(A')$, then $f(a), f(b) \in A'$.

Since (A′, +′, .′) is an ideal, then

$f(a - b) = f(a) - f(b) \in A' \Rightarrow a - b \in f^{-1}(A')$

For, $r \in R$, we have

$f(a \,.\, r) = f(a) \,.\, f(r) \in A' \Rightarrow a \,.\, r \in f^{-1}(A')$,

and $f(r \,.\, a) = f(r) \,.\, f(a) \in A' \Rightarrow r \,.\, a \in f^{-1}(A')$.

Since (A′, +′) is the additive sub-group of the additive group (R′, + ′) of the ring (R′, +′ , .′), $0' \in A'$, the additive identity of the ring (R′, +′, .′).

So $\quad \text{Ker}(f) = \{a \in R \mid f(a) = 0'$

$\quad = \{a \in R \mid a = f^{-1}(0)\}$

So $\quad a \in \text{Ker}(f) \Rightarrow a = f^{-1}(0') \subseteq f^{-1}(A')$

$\quad \Rightarrow a \in f^{-1}(A')$.

Hence $(f^{-1}(A'), +' .')$ is an ideal with Ker $(f) \subset f^{-1}(A)$.

(*iii*) $a \in A \Rightarrow f(a) \in f(A) \Rightarrow a \in f^{-1}(f(A))$.

Hence $A \subseteq f^{-1}(f(A))$.

Now again, $c \in f^{-1}(f(A)) \Rightarrow f(c) \in f(A)$.

Then for some element $a \in A$,

$$\begin{aligned} f(a) = f(c) &\Rightarrow f(a) - f(c) = 0' \\ &\Rightarrow f(a - c) = 0' \\ &\Rightarrow a - c \in \text{Ker}(f) \\ &\Rightarrow a - c \in A, \text{ since Ker } (f) \subset A \end{aligned}$$

Since $a \in A$, and $(A, +,.)$ is an ideal.

$a - c \in A \Rightarrow c \in A$.

This shows $f^{-1}(f(A)) \subseteq$ A and it follows

$$f^{-1}(f(A)) = A.$$

(*iv*) By the second part of the theorem 6.7.8 we have $(f^{-1}(A'), +, .)$ is an ideal in $(R, +, .)$ with Ker $(f) \subseteq f^{-1}(A')$, since $(A', +', .')$ is an ideal in $(R', +', .')$. Now we prove that the mapping $A \to f(A)$ is onto. Since $f : (R \to R')$ is onto mapping, then every element of A' is an image element for some element in R. Thus we can take a subset A of R such that $f(A) = A'$, or $A = f^{-1}(A')$. Consequently which implies $f(A) = f(f^{-1}(A')) = A'$. Hence the mapping is onto. To prove that it is one to one let A and B be two ideals with Ker $(f) \subset A$ and Ker $(f) \subset B$. Then

$$\begin{aligned} f(A) = f(B) &\Rightarrow f^{-1}(f(A)) = f^{-1}(f(B)) \\ &\Rightarrow A = B. \text{ by the III part of the theorem 6.7.8.} \end{aligned}$$

This proves that the mapping is one to one and onto.

(*v*) Let $A \subset B$, then

$A \subset B \Rightarrow f(A) \subseteq f(B)$. If $f(A) = f(B)$, then $A = B$, since the mapping $A \to f(A)$ is one to one by part (*iv*) Hence $f(A) \neq f(B)$. So $f(A) \subset f(B)$, conversely, $f(A) \subset f(B) \Rightarrow A = f^{-1}(f(A)) \subseteq f^{-1}(f(B)) = B$ by part (*iii*).

Since $f(A) \subset f(B)$, we cannot have $A = B$, so $A \subset B$. This completes the proof of the theorem.

Theorem 6.7.9: *Let $\phi : R \to R/K$ be a natural homomorphism from the ring $(R, +, .)$ onto the quotient ring $(R/K, +, .)$, where K is the Kernel of f.* If $(A, +, .)$ an ideal in *$(R, +, .)$ containing K, then $(A/K, +, .)$ is ideal in $(R/K, +, .)$. Moreover, there is one to one mapping of the set Z of all ideals in $(R, +, .)$ which contain K onto the set Σ' of all ideals in $(R/K, +, .)$.*

CHAPTER 6

Proof: Let $(A, +, .)$ be an ideal in $(R, +, .)$ with $K \subset A$. Since $(K, +, .)$ is ideal in $(R, +, .)$, it is also ideal in the ring $(A, +, .)$. Thus the quotient ring $(A/K, +, .)$ exists. Now we show that $(A/K, +, .)$ is an ideal in $(R/K, +, .)$ For, A/K is non-empty since $K \in A/K$ and $A/K \subseteq R/K$ since $A \subset R$.

If $a + K$, $b + K$, $\in A/K$, a, $b \in A$, then

$(a + K) - (b + K) = (a - b) + K \in A/K$ as $a - b \in A$ because $(A, +,.)$ is an ideal.

And for $r + K$, $r \in R$,

$(a + k) . (r + K) = (a.r + K) \in. A/K$, as $a.r \in A$ because $(A, +, .)$ is an ideal. This shows $(A/K, +, .)$ is an ideal in $(R/K, +, .)$.

For the next part of the theorem, we define the mapping

$f : \Sigma \to \Sigma'$ by

$f(A) = A/K$, $\forall\, A \in S$.

Let $(A, +, .)$ and $(B, +, .)$ be two ideals containing K. We assume that $f(A) = f(B)$ $\Rightarrow A/K = B/K$.

Then for $a \in A$, $a + K \in B/K \Rightarrow a \in B \Rightarrow A \subseteq B$.

Similarly, for $b \in B$, $b + K \in A/K \Rightarrow b \in A \Rightarrow B \subseteq A$.

Thus we conclude $A = B$. It is clearly onto mapping because for any ideal $(A/K, +, .)$ there exists an ideal $(A, +, .)$ in $(R, +, .)$ containing K such that

$$f(A) = A/K.$$

This completes the proof of the theorem.

Theorem 6.7.10: *If f is a homomorphism of the ring $(R, +, .)$ onto the ring $(R', +', .')$, with kernel K, and $(A, +, .)$ is an ideal in $(R, +, .)$ with $K \subset A$, then*

$(R/A, +, .) = (f(R)/f(A), +', .')$

Proof: We have $f : R \to R'$ given by $r \to f(r)$, $r \in R$, is a homomorphism from $(R, +, .)$ onto $(R', +', .')$. We define another mapping $\phi : R' \to R'/f(A)$ by $\phi(r') = r' +' f(A) = f(r) +' f(A)$, $r \in R$, $f(r) = r' \in R'$, which is clearly a natural homomorphism.

Since $(A, +, .)$ is an ideal, $(f(A), +', .')$ is ideal in $(R', +', .')$

So $(f(R)/f(A), +', .')$ exists as $f(R) = R'$.

Thus we have the mapping $\Psi : R \to \dfrac{f(R)}{f(A)}$ by

$\psi(r) = (\phi \circ f)(r) = \phi(f(r)) = f(r) +' f(A)$.

Since ϕ and f are homomorphism, then their composition $\Psi = \phi \circ f$ is also homomorphism.

Now to compute the proof of the theorem, we find Ker Ψ

$$\begin{aligned} \text{Ker } \Psi &= \{r \in R \mid \psi(r) = 0\} \\ &= \{r \in R \mid \Psi(r) = f(r) + f(A) = f(A)\}, \end{aligned}$$

where $f(A)$ is the zero of $f(R)/f(A)$.

$$\begin{aligned} &= \{r \in R \mid f(r) + f(A) = f(A)\} \\ &= \{r \in R \mid f(r) \in f(A)\} \\ &= \{r \in R \mid r \in A\} = A. \end{aligned}$$

Hence by fundamental theorem, we have

$(R/A, +, .) \cong (f(R)/f(A), +, .)$.

This completes the proof of the theorem.

Theorem 6.7.11: *If* $(A, +, .)$ *and* $(K, +, .)$ *are ideals in* $(R, +, .)$ *such that* $K \subseteq A$, *then* $(R/A, +, .) \cong ((R/K)/(A/K), +, .)$.

Proof: Since $(A, +, .)$ and $(K, +, .)$ are ideals in $(R, +, .)$ such that $K \subseteq A$, then the quotient rings $(R/A, +, .)$, $(R/K, +, .)$ and $(A/K, +, .)$ exist.

Now we define the mapping

$\phi : R/K \to R/A$ by

$\phi(r + K) = r + A,\ r + K \in R/K.$

We see that for $r_1 + K,\ r_2 + K \in R/K$,

$$\begin{aligned}\phi((r_1 + K) + (r_2 + K)) &= \phi((r_1 + r_2 + K) \\ &= (r_1 + r_2) + A \\ &= (r_1 + A) + (r_2 + A) \\ &= \phi((r_1 + K) + \phi(r_2 + K),\end{aligned}$$

and

$$\begin{aligned}\phi((r_1 + K) \,.\, (r_2 + K) &= \phi(r_1 \,.\, r_2 + K) \\ &= r_1 \,.\, r_2 + A \\ &= (r_1 + A).\,(r_2 + A) \\ &= \phi(r_1 + K) \,.\, \phi(r_2 + K),\end{aligned}$$

So that ϕ is a homomorphism and it is clearly onto. Now for completing the proof of the theorem, we find Ker ϕ,

$$\begin{aligned}\text{Ker } \phi &= \{r_1 + K \mid \phi(r_1 + K) = r_1 + A = A\} \\ &= \{r_1 + K \mid r_1 + A = A\} \\ &= \{r_1 + K \mid r_1 \in A\} \\ &= A/K.\end{aligned}$$

Hence by fundamental theorem, we have

$(R/A, +, .) \cong ((R/K) / (A/K), +, .)$.

Hence the theorem.

Theorem 6.7.12: *If* $(A, +, .)$ *is any ideal of a ring* $(R, +, .)$ *and* $(B, +, .)$ *any sub ring of* $(R, +, .)$, *then*

(1) $(A + B, +, .)$ *is sub ring of* $(R, +, .)$

(2) $(A \cap B, +, .)$ *is an ideal of* $(R, +, .)$ in $(B, +, .)$

(3) $((A + B)/A, +, .)$

$\cong (B/A \cap B, +, .)$.

CHAPTER 6

Proof: **(1)** First of all we shall show that $(A + B, +, .)$ is a subring and $(A, +, .)$ is the ideal of it. For this we have

$a + \alpha, b + \beta \in A + B$, where $a, b \in A, \alpha, \beta \in B$

$(a + \alpha) - (b + \beta) = (a - b) + (a - \beta) \in A + B$, since $a - b \in A, \alpha - \beta \in B$ as $(A, +, .)$ is an ideal and $(B, +, .)$ is a subring.

And

$(a + \alpha) . (b + \beta) = a.b + a . \beta + \alpha.b + \alpha . \beta \in A + B.$

Since

$a.b + a . \beta + \alpha . b \in A$ as $(A, +, .)$ is an ideal.

This shows that $(A + B, +, .)$ is a subring of $(R, +, .)$. Since $A = A + 0 \subset A + B$, 0 is the additive identity in $(R, +, .)$ which serves as zero element in A and in B as well.

Therefore $(A, +, .)$ is an ideal in $(A + B, +, .)$.

(2) The set $A \cap B$ is non-empty since $0 \in A$ and $0 \in B$

$$\text{For, } a, b \in A \cap B \Rightarrow a, b \in A \text{ and } a, b \in B.$$

$$\Rightarrow a - b \in A \text{ and } a - b \in B, \text{ since}$$

$(A, +, .)$ is an ideal and $(B, +, .)$ is a ring

$$\Rightarrow a - b \in A \cap B,$$

and for $r \in B \subset R$, $a \in A \cap B$, we have

$a . r$ and $r . a \in A$, since $(A, +, .)$ is an ideal, and

$a . r, r . a \in B$ since $(B, +, .)$ is a subring.

So $a.r, r.a \in A \cap B$, this shows $(A \cap B, +, .)$ is an ideal in $(B,+, .)$.

(3) Since $(A, +, .)$ is an ideal in $((A + B), +, .)$, then $((A + B)/A, +, .)$ exists.

Now we defined the mapping

$\phi : B \to (A + B)/A$ by

$\phi (b) = b + A.$

For $b_1, b_2 \in B$, we have

$$\begin{aligned} \phi (b_1 + b_2) &= (b_1 + b_2) + A \\ &= (b_1 + A) + (b_2 + A) = \phi(b_1) + \phi (b_2). \end{aligned}$$

Hence ϕ is a homomorphism.

Now we see that ϕ is onto. For, $x \in A + B$, $x + A \in (A + B)/A$, then there exist $a \in A$, $b \in B$ such that $x = a + b$:

We have
$$\begin{aligned} x + A &= (a + b) + A \\ &= (b + a) + A \\ &= b + (a + A) \\ &= b + A. \end{aligned}$$

Thus for $b + A \in (A + B)/A$, there exists an element $b \in B$ such that $\phi(b) = b + A$. Hence ϕ is onto.

To complete the proof of the theorem we find the Ker ϕ,

Ker ϕ =

$$= \{b \in B \mid \phi(b) = b + A = A\}$$
$$= \{b \in B \mid b + A = A\}$$
$$= \{b \in B \mid b \in A\}$$
$$= \{b \in B \cap A] = A \cap B\}.$$

Therefore, by fundamental theorem we have

$$(B/A \cap B, +, .) \cong \left(\frac{A+B}{A}, +, .\right).$$

Theorem 6.7.13: *For a fixed element a of (R, +, .), a ring with identity, define the left multiplication function,* $f_a : R \to R$ *by taking*

$$f_a(x) = a \,.\, x, \ \forall\, x \in R.$$

If F_R *is the set of all such functions, prove the ring analog of Cayley's theorem:*

(a) The triple $(F_R, +, 0)$ *forms a ring, where + denotes the usual pointwise addition of functions and o denotes functional composition.*

(b) $(R, +, .) \cong (F_R, +, 0)$.

Proof: ***(a)*** To prove that the pair $(F_R, +, .)$ is a commutative group, we have,

(1) For $f_a, f_b \in F_R$, $a, b \in R$,

$$\begin{aligned}(f_a + f_b)(x) &= f_a(x) + f_b(x) \ \forall\, x \in R\\ &= a.x + b.x\\ &= (a + b).\, x\\ &= f_{a+b}(x),\end{aligned}$$

which means $f_a + f_b = f_{a+b} \in F_R$. Hence F_R is closed under the operation.

(2) For $f_a, f_b, f_c \in F_R$,

$$\begin{aligned}(f_a + f_b) + f_c &= f_{a+b} + f_c\\ &= f_{(a+b)+c} = f_{a+(b+c)}\\ &= f_a + f_{b+c}\\ &= f_a + (f_b + f_c).\end{aligned}$$

Hence this operation is associative.

(3) For all $f_a \in F_R$, $\exists$ an element $f_a \in F_R$, $0 \in R$ such that $f_a + f_0 = f_{a+0} = f_a = f_{0+a} = f_0 + f_a$.

Therefore, f_0 is the identity in F_R.

(4) For every $f_a \in F_R$, $\exists$ an element $f_{-a} \in F_R$, $-a \in R$,

such that

$$f_a + f_{-a} = f_{a+(-a)} = f_0 = f_{-a+a} = f_{(-a)} + f_a.$$

Thus every element $f_a \in F_R$ has its inverse $f_{(-a)} \in F_R$.

(5) For all $f_a, f_b \in F_R$,

$$f_a + f_b = f_{a+b} = f_{b+a} = f_b + f_a$$

Hence the operation + is commutative.

(6) For all $f_a, f_b \in F_R$,

$$\begin{aligned}(f_a \text{ o } f_b)(x) &= f_a(f_b(x)), \forall x \in R \\ &= f_a(b.x) \\ &= a.(b.x) \\ &= (a.b).x = f_{a.b}(x),\end{aligned}$$

which implies $f_a \text{ o } f_b = f_{a.b} \in F_R$.

(7) For all $f_a, f_b, f_c \notin F_R$,

$$\begin{aligned}(f_a \text{ o } f_b) \text{ o } f_c &= f_{a.b} \text{ o } f_c \\ &= f_{(a.b)c} = f_{a(b.c)} \\ &= f_a \text{ o } f_{b.c} \\ &= f_a \text{ o } (f_b \text{ o } f_c).\end{aligned}$$

(8) For all $f_a \in F_R$, $\exists$ an element $f_1 \in F_R$, $I \in R$ is the identity of $(R, +, .)$, such that

$f_a \text{ o } f_1 = f_{a.1} = f_a = f_{1.a} = f_1 \text{ o } f_a$. Hence f_1 is the identity in F_R.

(9) 'o' is distributive over '+'. For $f_a, f_b, f_c \in F_R$,

$$\begin{aligned}f_a \text{ o } (f_b + f_c) &= f_a \text{ o } (f_{b+c}) \\ &= f_{a.(b+c)} \\ &= f_{a.b + a.c} \\ &= f_{a.b} + f_{a.c} \\ &= f_a \text{ o } f_b + f_a \text{ o } f_c.\end{aligned}$$

Similarly, we can show right distributive law.

This shows that $(F_R, +, o)$ is a ring with identity.

(b) We define the mapping $\phi: R \to F_R$ by

$$\phi(a) = f_a, \forall a \in R.$$

For, $a, b \in R$,

$$\phi(a + b) = f_{a+b} = f_a + f_b = \phi(a) + \phi(b),$$

and

$$\phi(a.b) = f_{a.b} = f_a \text{ o } f_b = \phi(a) \ 0 \ \phi(b).$$

Hence ϕ is a homomorphism. For every $f_a \in F_R$, there is an element $a \in R$ such that $\phi(a) = f_a$ which shows ϕ is onto mapping.

Now for $a, b \in R$, we assume that

$$\phi(a) = \phi(b) \Rightarrow f_a = f_b$$
$$\Rightarrow f_a(x) = f_b(x),\ \forall\ x \in R$$
$$\Rightarrow a \,.\, x = b \,.\, x,\ \forall\ x \in R.$$

In particular, for $x = 1$, the identity of $(R, +, .)$

$$a = a.1 = b.1 = b,$$

which shows that ϕ is one to one. Hence ϕ is an isomorphism and $(R, +, .) \cong (F_R, +, .)$. Hence the theorem.

PROBLEMS

1. Let f be the function from the integrs Z onto the even integers given by $f(x) = 2x$ for all $x \in Z$. Prove that f is not a homomorphism.
2. Let f be the function from the integers modulo 4 onto the integers modulo 2 given by $f(0) = f(1) = 0$, $f(2) = f(3) = 1$. Prove that f is not a homomorphism.
3. State and prove the fundamental theorem of homomorphism for rings.
4. Show that the homomorphic image of ring $(R, +, .)$ is a ring.
5. Let C be the set of all pairs (a, b) of real numbers. Define addition and multiplication in C as follows:

   $(a, b) + (c, d) = (a + c, b + d)$,

   $(a, b), (c, d) = (ac - bd, ad + bc)$.

   Prove that this ring $(C, +, .)$ is isomorphic to the ring of complex numbers under the mapping $(a, b) \to a + ib$.
6. Prove that the ring $(\{a + b\}\sqrt{2}, +, .)$ is isomorphic to the ring of integers.
7. Are the integers isomorphic to the rational numbers as rings? Prove your answer.
8. Prove that the only isomorphism from the ring of integers onto the ring of integers is the isomorphism f such that $f(x) = x$. $\forall\ x \in Z$.
9. Prove that the ring $(Z_4, +_4, \odot_4)$ is not isomorphic to the ring $(Z_2\ +_2, \odot_2)$.
10. Let $(\{a + b\sqrt{2}, +, .)$ be a ring and f be defined by $f(a + b\sqrt{2}) = a$, Is f a homomorphism? Prove your answer. If the answer is yes, find the Kernel of the homomorphism.
11. Let f be a homomorphism from the ring $(R, +, .)$ into itself and S be the set of elements that are left fixed by f:

    $S = \{a \in R \mid f(a) = a\}$.

    Show that $(S, +, .)$ is a subring of the ring $(R, +, .)$.
12. Illustrate fundamental theorem by considering the rings $(Z_6, +_6, \odot_5)$, $(Z_3, +_3, \odot_3)$, and the homomorphism $f : Z_6 \to Z_3$, defined by

    $f(0) = f(3) = 0$, $f(1) = f(4) = 1$, $f(2) = f(5) = 2$.
13. (a) Let f be a homomorphism from the ring $(R, +, .)$ into the ring $(S, +', .')$. If $a \in R$ is nilpotent, then show that $f(a)$ is nilpotent in S.

CHAPTER 6

(*b*) Suppose $(R, +, .)$ is a ring which has no non-zero nilpotent elements. Show that all the idempotent elements of R belong to the centre.

[Hint: If $a^2 = a$. then $(a . r . a - a . r)^2 = (a . r . a - r . a)^2 = 0$ for all of $r \in R$.

14. Let $(S, +, .)$ be a subring and $(I, +, .)$ an ideal of the ring $(R, +, .)$. Assuming $S \cap I = (0)$. Prove that $(S, +, .)$ is isomorphic to a subring of the quotient ring $(R/I, +, .)$

[Hint: Since $(S, +, .)$ is subring and $S \cap I = \{0\}$, then $\forall\ a \neq 0 \in S$, $a + I$ is a non-zero elements of R/I. Now we define the mapping $\phi : S \to R/I$ by

$\phi(a) = a + I,\ a \in S.$

For $a, b \in S$, then

$\phi(a + b) = (a + b) + I = (a + I) + (b + I) = \phi(a) + \phi(a),$

and $\phi(a.b) = a.b + I = (a + I) . (b + I) = \phi(a), \phi(b).$

The function ϕ is not onto because there exists some element $x \in R$ which does not belong to S as $S \subset R$. So $x + I \in R/I$, but $x + I$ is not an image element of any element of S under ϕ.

Now we find out Ker ϕ,

$$\begin{aligned} \text{Ker } \phi &= \{a \in S \mid \phi(a) = I\} \\ &= \{a \in S \mid a + I = I\} \\ &= \{a \in S \mid a \in = I\} \\ &= \{a \in S \cap I\} \\ &= \{a \in \{0\}\} = \{0\} \end{aligned}$$

We know that $S/(0) = S$. Hence by fundamental theorem we have

$(S, +, .)$ is isomorphic to some subring of the quotient ring $(R/I, +, .)$.

6.8 THE FIELD OF QUOTIENTS OF AN INTEGRAL DOMAIN

We know that an integral domain $(D, +,.)$ is a commutative ring with identity and without zero-divisors that is, if $a.b = 0$ for some $a, b \in D$, then at least one of a or b must be 0. The ring of integers is an standard example of an integral domain.

The set of integers can be enlarged to the set of rational numbers which is a field. So one can perform a similar construction for any integral domain.

There are main procedures of embedding of one ring into another. Here in this section we confine ourselves to one of these methods: Embedding of a ring in a ring with unity. (II) Embedding of a domain in a field.

Definition 6.8.1: A ring $(R, +,.)$ can be embedded in a ring $(R', +', .')$ if there is an isomorphism of R onto R'. (If R and R' have unit element 1 and 1′, we insist, in addition the image of 1 under this isomorphism is 1′.

Let $(R, +, .)$ be a ring, consider $R \times Z = \{(a, m)\ (a \in R\ m \in Z\}$. The operation of addition and multiplication on $R \times Z$ can be defined by

(*i*) $(a, m) + (b, n) = (a + b, m + n)$, for all $(a, m), (b, n) \in R \times Z$

(*ii*) and $(a, m) . (b, n) = (ab + an + mb, mn)$.

It can be verified easily that $(R \times Z, +, .)$ is a ring with unity $(0, 1)$.

Theorem 6.8.1: *Every ring can be embedded in a ring with identity.*

Proof: Let $(R, +,.)$ be a ring and $R_1 = R \times Z = \{(a, m) \mid a \in R, m \in Z\}$ forms a ring under operations given in (*i*) and (*ii*) above. Now we define a mapping

$$f: R \to R_1 \text{ by}$$
$$f(a) = (a, 0) \text{ for all } a \in R$$

We see that for $a, b \in R$

$$f(a + b) = (a + b, 0).$$
$$= (a, 0) + (b, 0) = f(a) + f(b)$$

and
$$f(a.b) = (a . b, 0) = (a, 0) . (b, 0) = f(a) . f(b).$$

So f is a homomorphism.

Again for $a, b \in R$.

$$f(a) = f(b) \Rightarrow (a, 0) = (b, 0)$$
$$\Rightarrow a = b$$

And f is clearly into.

Hence f is an isomorphism of R onto R_1.

So $(R, +, .) \cong (f(R), +, .) \subset R_1$

Consequently R is embedded in R_1 which has unit element $(o, 1)$.

Remark: Here $(f(R), +, .)$ is an ideal, we see that for $(a, 0) \in f(R)$, $(b, n) \in R_1$, $(a, 0)$. $(b, n) = (ab + 0b + na, 0n)$

$= (ab + na, 0) \in f(R)$ and $(b, n)(a, 0) = (b . a + na + b0, n . 0)$

$= (ba + na, 0) \in f(R)$. Thus $(f(R), +, .)$ is an ideal of $(R_1, +, .)$

Embedding of domain in a field

Let $(D, +, .)$ be an integral domain roughly speaking the field we seek all quotients a/b, where $a, b \in D$, $b \neq 0$. We note three things:

(1) $\frac{a}{b} = \frac{c}{d}$ if and only if $ad = bc$.

(2) We define addition and multiplication by

$$\frac{a}{b} + \frac{c}{d} = \frac{ad + bc}{bd} \text{ and } \frac{a}{b} \cdot \frac{c}{d} = \frac{ac}{bd}$$

which will provide us guidance in making certain field known as quotient field.

Let $(D, +, .)$ be an integral domain with at least two elements and let $D_0 = D - \{0\}$. Then we consider $D \times D_0 = \{(a, b) \mid a, b \in D, b \neq 0\}$. We define a relation $\sim$ on $D \times D_0$ by $(a, b) \sim (c, d)$ iff $ad = bc$. For $(a, b), (c, d) \in D \times D_0$.

CHAPTER 6

Lemma 6.8.1: *The relation ~ on $D \times D_0$ defined by $(a, b) \sim (c, d)$ if and only $ad = bc$, for all $(a, b), (c, d) \in D \times D_0$ is an equivalence relation.*

Proof: For $(a, b), (c, d), (e, f) \in D \times D_0$ we observe that

(1) Reflexivity: For all $(a, b) \in D \times D_0$,

$$ab = ba \Rightarrow (a, b) \sim (a, b).$$

So the relation ~ is reflexive

(2) Symmetry: For any two $(a, b), (c, d) \in D \times D_0$

$$\begin{aligned}(a, b) \sim (c, d) &= ad = bc \\ &= cb = da \\ &\Rightarrow (c, d) \sim (a, b)\end{aligned}$$

So the relation ~ is symmetric.

Transitivity: For $(a, b), (c, d), (e, f) \in D \times D_0$, we see that

$(a, b) \sim (e, d)$ and $(c, d) \sim (e, f)$

$$\begin{aligned}\Rightarrow \quad ad &= bc \text{ and } cf = de \\ \Rightarrow \quad (ad)(cf) &= (bc)(de) \\ \Rightarrow \quad (dc)(af) &= (cd)(be) \\ \Rightarrow \quad af &= be \\ \Rightarrow \quad (a, b) &\sim (e, f)\end{aligned}$$

Hence the relation ~ is transitive

Since $(D, +, .)$ is an integral domain, it is commutative and it is without zero divisors, $cd \neq 0$

We know that the equivalence relation ~ defined on $D \times D_0$ splits the set into disjoint equivalence classes. Let $[a, b]$ be the equivalence class of (a, b) in $D \times D_0$, and let F be the set of all such equivalence classes $[a, b]$, where $a, b \in D, b \neq 0$. To make F field we define addition and multiplication of equivalence classes $[a, b]$, $[c, d]$ as follows:

$$[a, b] +' [c, d] = [ad + bc, bd],$$

and

$$[a, b] .' [c, d] = [ac, bd].$$

Since D is an integral domain and both $b \neq 0$, $d \neq 0$, then $bd \neq 0$. So $[ad + bc, bd] \in F$ and $[ac, bd] \in F$.

We now assert that this addition and multiplication are well defined, that is, if $[a, b] = [a', b']$ and $[c, d] = [c', d']$, then $[a, b] +' [c, d] = [a', b'] +' [c', d']$. To see that this is true,

$$\begin{aligned}[a, b] &= [a', b'] \text{ and } [c, d] = [c', d'] \\ \Rightarrow \quad ab' &= ba' \text{ and } cd' = dc' \qquad \text{...(1)} \\ \Rightarrow \quad ab'\,dd' &= ba'\,dd' \text{ and } bb'\,cd' = bb'\,c'd \\ \Rightarrow \quad ab'\,dd' + bb'\,cd' &= ba'\,dd' + bb'\,c'd \\ \Rightarrow \quad (ad + bc)(b'\,d') &= (a'\,d' + b'\,c')(bd)\end{aligned}$$

$$\Rightarrow \qquad \frac{ad+bc}{bd} = \frac{a'd'+b'c'}{b'd'}$$

Hence addition is well defined

Again

$$ab' = a'b \text{ and } cd' = c'd$$

$$\Rightarrow \qquad a \,.\, b'\, cd' = a'bc'd$$

$$\Rightarrow \qquad (ac)\,(b'd') = (bd)\,(a'c')$$

$$\Rightarrow \qquad \frac{ac}{bd} = \frac{a'c'}{b'd'}$$

So multiplication is well defined.

Theorem 6.8.2: *Every integral domain can be embedded in a field.*

Proof: Now we shall show that $(F, +', .')$ is a field. So we observe the followings:

(1) The set F is closed under addition $+'$ and multiplication.$'$

(2) The addition $+'$ is associative. For

$$\begin{aligned}\{[a, b] + [c, d]\} + [e, f] &= [ad + bc, bd] + [e, f] \\ &= [ad\,f + bc\,f + bde, bdf] \\ &= [adf + b\,(cf + de), bdf] \\ &= [a, b] + [cf + de, df] \\ &= [a, b] + \{[c, d] + [e, f]\}\end{aligned}$$

(3) *The existence of Identity:* For all $[a, b] \in F$, there exists a class $[0, 1] \in F$ such that

$$[a, b]\,[0, 1] = [a \,.\, 1 + b \,.\, 0, b \,.\, 1] = [a, b]$$

Therefore $[0, 1]$ is the additive identity.

(4) *Existence of inverse:* For each $[a, b] \in F$, there is a class $[-a, b] \in F$ such that

$$\begin{aligned}[a, b] + [-a, b] &= [ab + b(-a), bb] \\ &= [0, b^2] = [0, 1]\end{aligned}$$

This shows that $[-a, b]$ is the additive inverse of $[a, b]$

(5) The addition $+'$ is commutative. For all $[a, b], [c, d] \in F$

$$\begin{aligned}[a, b] + [c, d] &= [ad + bc, bd] \\ &= [cb + da, db] = [c, d] + [a, b]\end{aligned}$$

Thus $[F, +')$ is an abelian group. Now we verify the properties with respect to multiplication.

(6) *Associative Property:* For all $[a, b], [c, d], [e, f] \in F$

$$\begin{aligned}\{[a, b] \,.'\, [c, d]\} \,.'\, [e, f] &= [ac, bd] \,.'\, [e, f] \\ &= [(ac)\,e, (bd)\,f] \\ &= [a\,(ce), b(df)] \\ &= [a, b] \,.'\, [ce, df] \\ &= [a, b] \,.'\, \{[c, d] .'\, [e, f]\}.\end{aligned}$$

CHAPTER 6

Hence the multiplications is associative.

(7) *Existence of identity:* For all $[a, b] \in F$ there is a class $[1, 1] \in F$ such that

$$[a, b] \,.'\, [1, 1] = [a \,.\, 1, b \,.\, 1] = [a, b]$$

So [1, 1] is the multiplicative identity in F.

(8) *Existence of inverse:* For each $[a, b] \neq [0, 1] \in F$, we have $[b, a]$ such that

$$[a, b] \,.'\, [b, a] = [ab, ba] = [1, 1]$$

$$\Rightarrow \qquad [a, b]^{-1} = [b, a]$$

(9) The multiplication .′ is commutative. For all $[a, b], [c, d] \in F$

$$[a, b] \,.'\, (c, d] = [ac, bd] = [ca, db] = [c, d] \,.'\, [a, b].$$

(10) Multiplication .′ is distributive over +′.

For $[a, b], [c, d], [e, f] \in F$

$$[a, b] \,.'\, \{[c, d] +' [e, f]\} = [a, b] \,.'\, [cf + de, df]$$
$$= [acf + ade, bdf] \qquad ...(1)$$

and
$$[a, b] \,.'\, [c, d] +' [a, b] \,.'\, [e, f] = [ac, bd] +' [ae, bf]$$
$$= [acbf + bdae, bdbf]$$
$$= [acf + ade, bdf] \qquad ...(2)$$

From (1) and (2) multiplication is left distributive over addition +′. Since the multiplication .′ is commutative, the other distributive law holds.

Consequently $(F, +', .')$ is a field.

Finally we define $F' = \{(a, 1] \mid a \in D, 1 \text{ is the identity of } D\}$. Then $\{F', +', .')$ is a subring of $(F, +', .')$.

and define a mapping $f : F' \to D$ by

$$f\,[a, 1] = a \text{ for all } [a, 1] \in F'.$$

We see that for all $[a, 1], [b, 1] \in F'$

$$f\,([a, 1] +' [b, 1]) = f\,[a \,.\, 1 + b \,.\, 1, 1 \,.\, 1]$$
$$= f\,[a + b, 1]$$
$$= a + b$$
$$= f\,[a, 1] + f\,[b, 1].$$

and
$$f\,([a, 1] \,.'\, [b, 1]) = f\,[a.b, 1] = a \,.'\, b = f\,[a, 1]\, f\,[b, 1]$$

moreover
$$f\,[a, 1] = f\,[b, 1] \Rightarrow a = b.$$

It is clearly onto.

Hence f is an isomorphism.

Thus $(F', +', .') \cong (D, +, .)$

Hence $(D, +, .)$ is embedded in a field $(F, +, .)$.

This completes the proof of the theorem.

The field $(F, +', .')$ is usually called the field of quotient of $(D, +, .)$.

Example: The field of quotient constructed from the integral domain $(Z, +, .)$ of integers is the field $(Q, +, .)$ of rational numbers.

Theorem 6.8.3: *If a field $(K, +, .)$ contains a non-zero integral domain $(D, +, .)$, then K contains a field of quotients of D. (In this sense the field of quotients of D is the smallest field containing D).*

Proof: Let $F = \left\{\frac{a}{b} \middle| a, b \in D, b \neq 0\right\}$. By $\frac{a}{b}$ we mean $a.b^{-1}$. Since D is non-zero, there is an element $a \neq 0 \in D$. Then $1 = \frac{a}{a} \in F$, Now for $x = \frac{a}{b}$, $y = \frac{c}{d} \in F$, $x - y = \frac{a}{b} - \frac{c}{d} = \frac{ad - bc}{bd} \in F$ and $xy = \frac{ac}{bd} \in F$. Thus $(F, +,.)$ is a subring containing 1. Now if $x \neq 0$, then $a \neq 0$ and $x' = \frac{b}{a} \in F$ such that $xx' = \frac{a}{b}.\frac{b}{a} = 1 \Rightarrow x^{-1} = x' \in F$. This proves that $(F, +, .)$ is a field.

Given $z \neq 0 \in D$, we can write $z = \frac{za}{a} \in F$.

Thus $D \subseteq F$. From the definition it is clear that $(F, +.,)$ is field of quotients of D contained in K.

This proves the theorem.

Theorem 6.8.4: *Any two isomorphic integral domains have isomorphic quotient fields.*

Proof: Let $(D, +, .)$ and $(D', +', .')$ be isomorphic integral domains. Then there exists ring isomorphism

$$f : D \xrightarrow{onto} D'.$$

Let $a, b, c, d \in D$ be arbitrary such that $f(a) = a'$, $f(b) = b'$, $f(c) = c'$, $f(d) = d'$. So we have

$$f(a + b) = f(a) + f(b) = a' +' b' \quad \text{...(1)}$$

$$f(a \,.\, b) = f(a) \,.\, f(b) = a' \,.'\, b'. \quad \text{...(2)}$$

Let $(F, +, .)$ and $(F', +', .')$ be quotient fields of D and D' respectively.
Then

$$F = \left\{\frac{a}{b} \middle| a,\ b \in D,\ b \neq 0\right\},$$

$$F' = \left\{\frac{a'}{b'} \middle| a',\ b' \in D',\ b' \neq 0\right\}.$$

Now we have to show that $(F, +,) \cong (F', +', .')$

We consider a mapping $\phi : F \to F'$ defined by

$$\phi\left(\frac{a}{b}\right) = \frac{a'}{b'}, \text{ for all } \frac{a}{b} \in F$$

Since $\frac{a}{b}, \frac{c}{d} \in F$, then $a, b, c, d \in F$ such that $b \neq 0$, $d = 0$.

$\Rightarrow f(a), f(b), f(c), f(d) \in F'$ such that $f(b) \neq 0$, $f(d) \neq 0$

$\Rightarrow a', b', c', d' \in F'$ such that $b' \neq 0$, $d' \neq 0$.

Now
$$\phi\left(\frac{a}{b}+\frac{c}{d}\right) = \phi\left(\frac{ad+bc}{bd}\right)$$
$$= \left(\frac{a'd'+b'c'}{b'd'}\right)$$
$$= \frac{f(a)f(d)+' f(b)f(c)}{f(b) .' f(d)}$$
$$= \frac{f(a)}{f(b)} +' \frac{f(c)}{f(d)} = \frac{a'}{b'} + \frac{c'}{d'} = \phi\left(\frac{a}{b}\right) +' \phi\left(\frac{c}{d}\right)$$

and
$$\phi\left(\frac{a}{b}\cdot\frac{c}{d}\right) = \phi\left(\frac{ac}{bd}\right) = \frac{a'c'}{b'd'} = \frac{f(a) .' f(c)}{f(b) .' f(d)} = \frac{a'}{b'} .' \frac{c'}{d'}$$
$$= \phi\left(\frac{a}{b}\right) .' \phi\left(\frac{c}{d}\right)$$

Hence ϕ is an homomorphism.

Again,
$$\phi\left(\frac{a}{b}\right) = \phi\left(\frac{c}{d}\right)$$
$$\Rightarrow \quad \frac{a'}{b'} = \frac{c'}{d'}$$
$$\Rightarrow \quad a'd' = b'c'$$
$$\Rightarrow \quad f(a)\, f(d) = f(b)\, f(c)$$
$$f(ad) = f(bc)$$
$$\Rightarrow \quad ad = bc \text{ since } f \text{ is one to one.}$$
$$\Rightarrow \quad (a, b) \sim (c, d)$$
$$\Rightarrow \quad \frac{a}{b} = \frac{c}{d}$$

Hence ϕ is an isomorphism.

Further, since f is an onto, then for each $a', b' \in D$, $b' \neq 0$ there exist $a, b \in D$ such that $f(a) = a'$, $f(b) = b'$, $b \neq 0$.

Then every element $\frac{a'}{b'} \in F'$ has a preimage $\frac{a}{b} \in F$ such that $\phi\left(\frac{a}{b}\right) = \frac{a'}{b'}$. Hence ϕ is onto mapping.

Hence ϕ is an isomorphism and $(F, +, .) \cong (F', +', .')$

Exercise 6.8.1: Show that (Hom $(Z^+, Z^+) \cong Z$.

Solution: Define $f : Z \to$ Hom (Z^+, Z^+) by

$$f(n) = T_n \text{ for all } n \in Z, \text{ where}$$

$T_n(x) = n\,x$ for all $x \in Z^+$.

For $n, m \in Z$,

$$f(n + m) = T_{n+m} = T_n + T_m.$$

Since

$$\begin{aligned} T_{n+m}(x) &= nx + mx \\ &= T_n(x) + T_m(x) \\ &= (T_n + T_m)\,x, \end{aligned}$$

Then $\quad T_{n+m} = T_n + T_m.$

Again, since

$$\begin{aligned} T_{n\,.\,m}(x) &= (n\,.\,m)x \\ &= n\,(m\,(x)) \\ &= n\,T_m(x) \\ &= T_n\,(T_m(x) = (T_n \, o \, T_m)\,(x) \end{aligned}$$

$\Rightarrow \quad T_{n\,.\,m} = T_n \, o \, T_m.$

So $\quad f(n\,.\,m) = T_{n\,.\,m} = T_n \, o \, T_m.$

Further $\quad f(n) = f(m)$

$\Rightarrow \quad T_n = T_m$

$\Rightarrow \quad T_n(x) = T_m(x)$

$\Rightarrow \quad nx = mx$

$\Rightarrow \quad n = m.$

Thus f is an isomorphism. Hence

$$(Z, +, .) \cong (\text{Hom}\,(Z^+, Z^+), +, .\,o).$$

Exercise 6.8.2: $(R, +, .)$ is a ring with identity element 1 and f is a homomorphism of $(R, +, .)$ into an integral domain $(D, +, .)$.

If ker $f \neq R$, prove that $f(1)$ is an identity element in D.

Solution: First we see that $f(1) \neq 0$.

For all $x \in R$, $x = 1\,.\,x$ and $f(1\,.\,x) = f(1)\,.\,f(x)$.

Now

$$\begin{aligned} \text{Ker } f &= \{x \in R \mid f(x) = 0\} \\ &= \{x \in R \mid f(1\,.\,x) = 0\} \\ &= \{x \in R \mid f(1)\,.\,f(x) = 0\} = R. \end{aligned}$$

If $f(1) = 0$, for all $x \in R$, $f(x) = 0$,

which is contradiction to Ker $f \neq R$. Hence $f(1) \neq 0$.

CHAPTER 6

Now $[f(1)]^2 = f(1) f(1) = f(1 . 1) = f(1)$.

Let $r \in D$, then $[r f(1) - r] f(1) = r (f(1)^2 - r f(1))$

$$= r f(1) - r f(1)$$
$$= 0$$

$\Rightarrow$ $r f(1) - r = 0$ since $f(1) \neq 0$.

$\Rightarrow$ $r . f(1) = r$.

$\Rightarrow$ $f(1)$ is the identity in D.

Exercise 6.8.3: Prove that the only homomorphisms of $(Q, +, .)$ into itself are identity mappings and the mapping that sends every element into zero.

Solution: Lef $f : Q \to Q$ be a homomorphism

If $(1) = 0$, then Ker $f = Q \Rightarrow f(x) = 0$ for all $x \in Q$.

If $f(1) \neq 0$, then as $(Q, +, .)$ is an integral domain by exercise 6.8.2 $f(1)$ is unity of Q, So $f(1) = 1$

Case I. Let n be a positive integer. Then

$$f(n) = f(1 + 1) + ...+ 1 \ (n \text{ lines}))$$
$$= f(1) + f(1) +...+ f(1) = n f(1) = n.$$

Case II. If $n = 0$, then $f(0) = 0 \Rightarrow f(n) = n$.

Case III. If n is a negative integer, put $n = -m$, where m is a positive integer. Then $f(-m) = -f(m) = -m$ by first case

Hence $f(n) = n$.

Further for any rational number p/q, p, q, $\in Z$, $q \neq 0$,

$$p = q . \frac{p}{q} \Rightarrow f(p) = f(q) f\left(\frac{p}{q}\right)$$

$$\Rightarrow f\left(\frac{p}{q}\right) = \frac{f(p)}{f(q)}$$

$$\Rightarrow f\left(\frac{p}{q}\right) = \frac{p}{q}, \ q \neq 0, f(q) \neq 0.$$

Hence for all $x \in Q$, $f(x) = x = I(x)$, where I is identity mapping

$$\Rightarrow f = I.$$

PROBLEMS

1. Prove that if $[a, b] = [a', b']$ and $[a, d] = [c', d']$, then $[a, b] [c, d] = [a', b'] [c', d']$.
2. Prove that the mapping $\phi : D \to F$ defined by $\phi(a) = [a, 1]$ is an isomorphism of D into F.

3. Let (R, +, .) be a commutative ring with unit element.

A non-empty subset S of R is called a multiplicative system if:

1. $0 \notin S$
2. $s_1, s_2 \in S \Rightarrow s_1 s_2 \in S$.

Let M be the set of all ordered pairs (r, s) where $r \in R$, $s \in S$. In M define $(r, s) \sim (r', s')$ if there exists an element $s'' \in S$ such that

$s'' (rs' - sr') = 0$.

(*a*) Prove that this defines an equivalence relation on M.

Let the equivalence class of (r, s) denoted by $[r, s]$, and Let R_s be set of all the equivalence classes. In R_s define $[r_1, s_1] + (r_2, s_2] = [r_1 s_2 + r_2 s_1, s_1 s_2]$ and $[r_1, s_1] [r_2, s_2] = [r_1 r_2, s_1 s_2]$.

(*b*) Prove that the addition and multiplication described above are well defined and R_S form a ring under there operations.

(*c*) Can R be embedded in R_S?

(*d*) Prove that the mapping $\phi : R \to R_S$ defined by $\phi (a) = [as, s]$ is a homomorphism of R into R_S and find the Kernel of ϕ.

(*e*) Prove that this kernel has no element of S in it.

4. Let $(D, +, .)$ be an integral domain, $a, b \in d$. Suppose that $a^n = b^n$ and $a^m = b^m$ for any two relatively prime positive integers m and n. Prove that $a = b$.

5. Let $(R, +, .)$ be a ring in which $xy = 0 \Rightarrow x = 0$ or $y = 0$. If $a, b \in R$ and $a^n = b^n$ and $a^m = b^m$ for two relatively prime positive integers m and n, prove that $m = n$.

6. Given an example to show that in the above problem if we replace integral domain by commutative ring the result need not be true.

Hint: In $(Z_8, +_8, \odot_8)$, $4^3 = 2^3$ and $4^4 = 2^4$ but $4 \neq 2$.

CHAPTER 6

CHAPTER 7

POLYNOMIAL RINGS

7.1 POLYNOMIALS AND POLYNOMIAL RINGS

There are many ways in which the subject of polynomials may be introduced. The polynomials which appear in elementary algebra always involve numbers from some suitable set. The set may be the set of integers, the set of real numbers, or some other set Each of the sets mentioned here is at least a ring. So we will begain, then, by assuming that the ring *(R, +,.)* has been specified, and from it we will construct a set of polynomials. The set of polynomials will represent an extension of *(R, +,.)* to a new ring (Poly *R, +,.)* called the ring of polynomials over *R*. This is also referred to as the ring of polynomials with coefficients in *R*.

Definition 7.1.1: An infinite sequence of elements of a set S is a function whose domain is the set of non-negative integers and whose range is a subset of S. Such a sequence is represented by the symbol $(a_0, a_1, a_2,....)$, where each $a_i \in S$ for every non-negative integer i. a_i is called the term of the sequence. Thus, in a sequence, there must be zeroth term, first term, a second term etc., each of which is an element of S.

Example 7.1.1: (0, 1/2, 2/3, 3/4,...) is a sequence of rational numbers defined by the function $f : f(n) = n/n + 1$ whose domain is the set of non-negative integers 0, 1, 2, 3.... and their images under f are:

$f(0) = 0, f(1) = 1/2, f(2) = 2/3, f(3) = 3/4$ etc.

Definition 7.1.2: Let $(R, +,.)$ be an arbitrary ring. *A* polynomial over R is an infinite sequence

$$f = (a_0, a_1, a_2,....)$$

of elements of R in which at most a finite number of the elements $a_k \neq 0$; in other words, $(a_0, a_1, a_2,....)$ is a polynomial if and only if there exists some non-negative integer k such that $a_k = 0$ for all $k > n$.

Sometimes we use the notation $(a_0, a_1, a_2,..., a_n, 0, 0..)$ for a polynomial wilh last non-zero term a_n; when $n = 0$, we also assume $a_0 = 0$ in order to get the zero polynomial (0, 0, 0....) each of whose terms is zero.

The set of all polynomials over R is denoted by

Poly $R = \{(a_0, a_1, \ldots, a_n \ldots) \mid a_k \in R, n > 0)$.

Example 7.1.2: Let $(\mathbb{Z}_2, +_2, \odot_2)$ be a ring of integers modulo 2.

The sequence (1, 1, 1, 0, 0,....) is a polynomial while the sequence (0, 1, 0, 1, ...0, 1,.....) would not.

Definition 7.1.3: Let $(R, +, .)$ be an arbitrary ring and let

$$f = (a_0, a_1, a_2, \ldots) \text{ and}$$
$$g = (b_0, b_1, b_2, \ldots)$$

be two elements of Poly R.

(*a*) $f = g$ if and only if their corresponding terms are equal, that is $a_k = b_k$ for every integer $k > 0$

(*b*) $$f + g = (a_0, + b_0, a_1 + b_1, a_2 + b_2, \ldots)$$

(*c*) $$f \cdot g = (a_0 \cdot b_0, a_0, b_1 + a_1 b_0, a_0 \cdot b_2 + a_1 \cdot b_1 + a_2 \cdot b_2, \ldots) = (c_0, c_1, c_2, \ldots)$$

where

$$c_k = \Sigma_{i+j=k} a_i \cdot b_j = a_0 \cdot b_k + a_1 \cdot b_{k-1} + \ldots + a_k \cdot b_0$$

We observe that if $a_i = 0$ for all $i > n$ and $b_j = 0$ for $j > m$, then $c_k = 0$ for all $k > n + m$ $(i + j = k > n + m \Rightarrow i > n$ or $j > m)$.

Thus, we have seen $f + g$ *and* $f \cdot g$ are again polynomials over R.

Example 7.1.3: *Let* $f =$ (0, 2, 3, 0, 0,....) and $g =$ (1, 0, 3, – 2, 0, 0, ...) be polynomials over the ring $(\mathbb{Z}, +, .)$. The sum

$$\begin{aligned} f + g &= (0, 2, 3, 0, 0, \ldots) + (1, 0, 3, -2, 0, 0, \ldots) \\ &= (0 + 1, 2 + 0, 3 + 3, 0 + (-2)\ 0 + 0, \ldots) \\ &= (1, 2, 6, -2, 0, 0, \ldots) \end{aligned}$$

and the product

$$\begin{aligned} fg &= (0, 2, 3, 0, 0, \ldots)(1, 0, 3, -2, 0, 0, \ldots) \\ &= (0 \cdot 1, 0 \cdot 0 + 2 \cdot 1, 3 \cdot 1 + 2 \cdot 0 + 0 \cdot 3, 0 \cdot 1 + 3 \cdot 0 + 3 \cdot 2 + (-2) \cdot 0, \\ &\quad 0 \cdot 1 + 0 \cdot 0 + 3 \cdot 3 + 2 \cdot (-2) + 0 \cdot 0, 0 \cdot 1 + 0 \cdot 0 + 0 \cdot 3 + 3 \cdot (-2) + 2 \cdot 0 + 00, 0 + 0 \cdot 0 \\ &\quad + 0 \cdot 3 +. 0 \cdot (-2) + 3 \cdot 0 + 2 \cdot 0 + 0 \cdot 0, 0, 0, \ldots) \\ &= (0, 2, 3, 6, 5, -6, 0, 0, \ldots). \end{aligned}$$

Definition 7.1.4: If $f = (a_0, a_1, a_2, 0, 0, \ldots)$ is a polynomial over a ring, then a_k is referred to as the kth coefficient of f for each positive integer k. The element a_0, is called either the constant term of f or the zeroth coefficient of f.

Definition 7.1.5: If $f = (a_0, a_1, a_2, \ldots, a_n, 0, 0, \ldots)$ is a polynomial of degree n if and only if $a_n \neq 0$ but $a_m = 0$ for all $m > n$. If all the coefficients of f are zero, then f has no degree.

Example 7.1.4: The polynomial (0, 0, 0,.....) has no degree. The Polynomial (2, 0, 0,.....) has degree zero. The degree of the polynomial (0, 2, 0, 3, 0, 0,.....) is three, and the degree of (1, 2, 3, 4, 5, 6, 0, 0) is five.

Now we shall try to express polynomials in the familiar form usually encountered in elementary algebra. For this, let ax represent the polynomial

$$(0, a, 0, 0,...)$$

that is, ax is the particular element of Poly R which has the element for its first terms and 0 for all other terms.

In general, the symbols $a\ x^n$, $n > 1$ represent the polynomial

$$(0...... 0, a, 0,...)$$

where the element a is the $(n + 1)$th term if we start counting from 1 and if we start counting from 0, that is, zeroth, first term, second term, etc. Then a is the nth term of the polynomials, for example,

$$ax^2 = (0, 0, a, 0, 0,....)$$

and $$ax^3 = (0, 0, 0, a, 0, 0,.....) \text{ and so on}$$

$$ax^n = (0, 0.....0, a, 0, 0,....).$$

Using this convention we have the representation for each polynomial

$$f = (a_0, a_1, a_2,......, a_n, 0,.....)$$

in the form

$$\begin{aligned} f &= (a_0, 0, 0,.....) + (0, a_1, 0, 0....) \\ &\quad + (0, 0, a_2, 0, 0,....) + (0, 0, 0, a_3, 0....) +.... \\ &\quad + (0, 0, 0,...., 0, a_n\ 0,....) \\ &= a_0 + a_1x + a_2x^2 + a_2x^3 +..... + a_nx^n \end{aligned}$$

Therefore, the set of all polynomials Poly R over an arbitrary ring $(R, +, .)$ can be regarded as the set of all formal expressions

$$f = a_0 + a_1x + a_2x^2 +..... + a_nx^n,$$

here the elements $a_0\ a_1 .... a_n$, the coefficients of f, belong to R.

It is obvious that x is simply a new symbol or indeterminant, totally unrelated to the ring $(R, +, .)$ and it is not an element of $[R]$ To indicate the indeterminant x, it is common to write $R\ [x]$ for the set of polynomials Poly R and $f\ (x)$ for any polynomial.

Remark: If the ring $(R, +, .)$ has an identity, I, then Ix with the indeterminant x, can be written as

$$I\ .\ x = (0, 1, 0, 0...) \in \text{Poly } R$$

Thus, $I\ x$ is a member of $R\ [x]$

Keeping this in view ax can be written as the product of element in $R\ [x]$:

$$ax = (a, 0, 0,...). (0, 1, 0, 0,...)$$

Similarly,

$$x \cdot x = (0, 1, 0, 0...) (0, 1, 0,.....),$$
$$x^2 = (0, 0, 1, 0, 0....),$$
$$x^3 = (0, 0, 0, 1,.....)$$

and so on

$$x^n = (0, 0, 0....0, 1, 0,...),$$

where 1 appears at $(n + 1)$th place of x^n.

Consider the polynomial (2, 3, 0, 4, 0, .) belonging to $Z[x]$, where $(Z, +, .)$ is a ring of integers.

By the above remark, $x^2 = (0, 0, 1, 0....)$, $x^3 = (0, 0, 0, 1,...)$ then

$$\begin{aligned} 2 + 3x + 0x^2 + 4x^3 &= (2, 0,...) + (3, 0,....). (0, 1, 0,...) \\ &\quad + (0,...) . (0, 0, 1, 0...) + (4, 0, ...) (0, 0, 0, 1, 0,....) \\ &= (2, 0,.) + (0, 3, 0,...) + (0,...) \\ &\quad + (0, 0, 0, 4, 0,....) \\ &= (2, 3, 0, 4, 0,...). \end{aligned}$$

Theorem 7.1.1: *The triple* (Poly R, +, .) *forms a ring, known as the ring of polynomials over R. Furthermore, the ring (Poly R, +, .) is commutative with identity if and only if (R, +, .) is a commutative ring with identity.*

Proof: It is left as exercise. We shall write $(R[x], +,.)$ for (Poly R, +, .)

Theorem 7.1.2: *If the ring (R, +, .) is an integral domain, so is also its polynomial ring (R [x], +, .).*

Proof: We have observed that whenever $(R, +, .)$ is a commutative ring with identity, $(R[x], +, .)$ is also a commutative ring with identity. To prove $(R[x], . +, .)$ is an integral domain, we have to show that $R[x]$ is free from divisors of zero.

Let $f(x)$ and $g(x)$ be two polynomials of $R[x]$ with deg $f(x) = m$ and deg $g(x) = n$.

$$f(x) = a_0 + a_1 x + a_2 x^2 + ... + a_m x^m, \; a_m \neq 0.$$
$$g(x) = b_0 + b_1 x + b_2 x^2 +.... b_n x^n, \; b_n \neq 0.$$
$$f(x) . g(x) = a_0 b_0 + (a_0 b_1 + b_0 a_1) x + ... + ...+ (a_m b_n) x^{m+n}.$$

Since $(R, +, .)$ is an integral domain,

$$a_m b_n \neq 0$$

which implies $f(x) . g(x) \neq 0$.

That completes the proof of the theorem.

Theorem 7.1.3: Let *(R, +, .) be an integral domain and f (x) and g (x) be two non-zero polynomials of (R [x], +, .). Then*

(1) *deg (f(x) . g (x)) = deg f (x) + deg g (x), and*

(2) *either f (x) + g (x) = 0, or*

deg (f (x)) < ma x {deg (f(x), deg g (x))}

Proof: **(1)** Suppose $f(x), g(x) \in R[x]$ with deg. $f(x) = n$ and deg $g(x) = m$, so that

$$f(x) = a_0 + a, x + ... + a_n x^n,\ a_n \neq 0,$$

$$g(x) = b_0 = b_0 + b_1 x.... + b_m x^m,\ b_m \neq 0$$

Then $$f(x) g(x) = a_0, b_0 + (a_0 . b_1 + a_i . b_0) x + ... + (a_n . b_m) x^{n+m}.$$

Since $(R, +, .)$ is free from devisors of zero, then $a_n b_m \neq 0$; accordingly, $f(x) . g(x) \neq 0$.

Thus, deg $(f(x) . g(x)) = n + m =$ deg $f(x)$ + deg $g(x)$.

(2) From the definition of two polynomials it is obvious that if $f(x) + g(x) \neq 0$, then

$$\deg(f(x) + g(x)) = \{\max(m, n) \text{ if } m \neq n$$

$$= n, \text{ if } m = n \text{ and } a_n + b_m \neq 0$$

and $$\deg(f(x) + g(x)) < n, \text{ if } m = n \text{ and } a_n + b_m = 0.$$

Remark: If $(R, +, .)$ is not an integral domain, it may happen that $a_n b_m = 0$ even if $a_n \neq 0$, $b_m \neq 0$. In this case,

deg $(f(x) . g(x)) <$ deg $f(x)$ + deg $g(x)$.

Example 7.1.5: As an illustration of what might happen if $(R, +, .)$ has zero-divisors, consider the ring $(Z_8, +_8, \odot_8)$, the ring of integers modulo 8.

Taking

$$f(x) = 1 + 2x,$$

$$g(x) = 4 + x + 4x^2,$$

we then have $f(x) g(x) = (1 + 2x) . (4 + x + 4x^2)$

$1 \odot_8 4 + 1 \odot_8 1\ x + 1 ._8 4\ x^2 + 2 \odot_8 4x$

$+ 2 \odot_8 1x^2 + 2 \odot_8 4x^2$

$= 4 + x + 6x^2$, so that

deg $(f(x) . g(x)) = 2 < 1 + 2 =$ deg $f(x)$ + deg $g(x)$,

Theorem 7.1.4: *The polynomial domain $(R[x], +, .)$ is not a field for any integral domain $(R, +, .)$.*

Proof: It is sufficient to prove the theorem that any non-zero polynomial of $R[x]$ which has positive degree does not have multiplicative inverse. For suppose

$f(x) \in R[x]$ with deg $f(x) > 0$; If $f(x) . g(x) = 1$ for some $g(x) \in R[x]$,

we would obtain the contradiction

$0 =$ deg $1 =$ deg $(f(x) . g(x)) =$ deg $f(x)$ + deg $g(x) \neq 0$.

That is, $g(x)$ does not exist such that $f(x) . g(x) = 1$.

Corollary: *No element of polynomial domain $(R[x] +, .)$, for an integral domain $(R, +, .)$, has an inverse if the degree of the element is greater than zero.*

Definition 7.1.6: If $f(x) = a_0 + a_1x + ... + a_n x^n$, $a_n \neq 0$, is a non-zero polynomial in $R[x]$ we call the coefficient a_n the leading coefficient of $f(x)$ and the integer n. the degree of the polynomial.

Definition 7.1.7: If $(R, +, .)$ is a ring with identity, a non-zero polynomial of degree n whose leading coeficient $a_n = 1$ is said to be monic polynomial. Thus we write

$$f(x) = a_0 + a_1 x + ...+ x^n.$$

Definition 7.1.8: The polynomial in $R[x]$ whose coefficients are all zero except for the constant term, which may be either zero or non-zero is called constant polynomial. Thus, $(a, 0, 0,......)$ is a constant polynomial. The set of all constant polynomials in $R[x]$ is denoted by R'.

$$R' = \{(a_0, 0, 0, .) \mid a_0 \in R\}$$

Theorem 7.1.5: *If* $(R, +, .)$ *is an integral domain and* $(R', +, .)$ *is an integral domain of constant polynomials in* $R[x]$. *Then* $(R', +, .)$ *is isomorphic to* $(R, +, .)$.

Proof: To prove this theorem, we need to find a one to one mapping from R' onto R which preserves the operation of addition and multiplication. For this purpose, define the mapping ϕ from R' into R by $\phi(a_0, 0, 0,...) = a_0$, for each constant polynomial $(a_0, 0, 0,....)$ in R'.

The mapping ϕ is one to one from R' onto R, since, for each element $a \in R$, there is one and only one constant polynomial having a as its zeroth coefficient, and for each constant polynomial $(b, 0, 0,....)$, b is a unique element of R.

Now, in order to prove that operations are preserved by ϕ, we consider any two elements in R', say $f = (a, 0,..., 0...)$ and $g = (b, 0...., 0....)$

$$\begin{aligned}\phi(f + g) &= \phi((a, 0,...., 0,...) + (b, 0,.... b,...)) \\ &= \phi(a + b,....-,....) \\ &= a + b \\ &= \phi(a, 0,....., 0,...) + \phi(b, 0,.... 0,...) \\ &= \phi(f) + \phi(g),\end{aligned}$$

and

$$\begin{aligned}\phi(f \,.\, g) &= \phi(a, 0, 0,...., 0,...). (b, 0, 0,...., 0,....) \\ &= \phi(a \,.\, b, 0, 0,....., 0,....) \\ &= a.b \\ &= \phi(a, 0, 0,....., 0,....)\, \phi(b, 0, 0,..., 0,...) \\ &= \phi(f) \,.\, \phi(g).\end{aligned}$$

Thus, ϕ preserves the operation of addition and multiplication. This completes the proof that ϕ is an isomorphism and hence $(R', +,.)$ is isomorphic to $(R, +,.)$

Remark: The product of a constant polynomial $(a, 0, 0,...)$ and any polynomial $f = (b_0, b_1, b_2,.... b_0, 0, 0,.....)$ is given by $(a, 0, 0,....). (b_0, b_1, b_2.....b_n, 0, 0,.....) = (ab_0, ab_1,...., ab_n,...)$ $= a(b_0, b_1,..., b_n, 0, 0,...) = af.$

7.2 DIVISIBILITY OF POLYNOMIALS

Definition 7.2.1: If $f(x), g(x) \in R[x]$ then $f(x)$ is a divisor (or factor) of $g(x)$ if and only if there is a polynomial $h(x) \in R[x]$ such that $g(x) = f(x) . h(x)$. We write $f(x) | g(x)$, and read as $f(x)$ divides $g(x)$.

Definition 7.2.2: If $f(x) | g(x)$, and $g(x) | f(x)$ then $f(x)$ and $g(x)$, are called associates.

Theorem 7.2.1: *If $f(x)$ and $g(x)$ are polynomials in $R[x]$ which are associates then $f(x) = d\, g(x)$ for some element $d \in R$.*

Proof: If two polynomials $f(x)$ and $g(x)$ are associates, then $f(x) | g(x)$ and $g(x) | f(x)$. Therefore there exist two polynomials $h(x)$ and $k(x)$ in $R[x]$ such that

$$f(x) = h(x) . g(x), \text{ and}$$

$$g(x) = k(x) . f(x). \text{ Thus,}$$

$$f(x) = h(x) . k(x) . f(x).$$

By cancellation law in $(R[x], +,)$, we have

$$1 = h(x) . k(x)$$

That is, $h(x)$ and $k(x)$ are inverses of each others.

We have already seen that the only elements in $R[x]$ which have inverses are the constant polynomials, that is, elements of R. These may or may not have inverses depending on R. If $(R, +, .)$ is a field, all non-zero elements have inverses. In any case, $h(x)$ and $k(x)$ must be elements of R.

Therefore, we conclude $h(x) . k(x) \in R$ and $h(x)$ and $k(x)$ are units of integral domain $(R, +, .)$ which have inverses.

Hence the theorem.

From the above theorem, any polynomial is divisible by its associates and by all units. These divisors are referred to as improper divisors and all other divisors are proper divisors.

Example 7.2.1: Consider $3x + 2 \in Z_5[x]$. Then $3x + 2$ and $x + 4$ are associates, since $2 \odot_5 (3x + 2) = x + 4$ and $3 \odot_5 (x + 4) = 3x + 2$.

Hence $$(3x + 2) = 3 \odot_5 (x + 4)$$
$$= 3 \odot_5 (2 \odot_5 (3x + 2)$$

which implies $$3 \odot_5 2 = 1 \in Z_5.$$

And $3x + 2$ is an improper divisor of $x + 4$, and $x + 4$ is an improper divisor of $3x + 2$.

Example 7.2.2: Every number $a \neq 0$ in the field $(R, +, .)$ of real numbers is a divisor of each polynomial $f(x)$ in $R[x]$. Thus $2 | x^2+1$, since $x^2 + 1 = 2(1/2\, x^2 + 1/2)$ and $\sqrt{2} | 3x - 5$, since

$$3x - 5 = \sqrt{2}\left(\frac{3x}{\sqrt{2}} - \frac{5}{\sqrt{2}}\right)$$

Definition 7.2.3: A polynomial $p(x) \in F[x]$ is said to be irreducible over the field $(F, +, .)$ if any only if $p(x) = f(x) \cdot g(x)$, $f(x), g(x) \in F[x]$ implies that one of $f(x)$ or $g(x)$ has degree o (*i.e.*, constant). Thus if $p(x)$ is irreducible in $F(x)$, it has no root in F.

Example 7.2.3: Show that polynomial

$$f(x) = x^3 + x + 1 \text{ is irreducible over } F = Z/(5).$$

Solution: If $f(x)$ is irreducible it must have a linear factor $x - a$, where a is zero of $f(x)$. By checking the five numbres

$$\begin{aligned} f(0) &= 1, \\ f(1) &= 1^3 +_5 1 +_5 1 = 3, \\ f(2) &= 2^3 +_5 2 +_5 1 = 2 \odot_5 2 \odot_5 2 +_5 2 +_5 1 \\ &= 3 +_5 2 +_5 1 = 1, \\ f(3) &= 3^3 +_5 3 +_5 1 = 3 \odot_5 3 \odot_5 3 +_5 3 +_5 1 \\ &= 2 +_5 3 +_5 1 = 1, \\ f(4) &= 4^3 +_5 4 +_5 1 = 4 \odot_5 4 \odot_5 1 +_5 4 +_5 1 \\ &= 4 +_5 4 + 1 = 4, \end{aligned}$$

We see that $f(x)$ has no zero in $Z/(5)$, and so the polynomial is irreducible.

Example 7.2.4: Show that the polynomials.

$$f(x) = x^2 + x + 4$$

is irreducible over the field $(F, +_{11}, \odot_{11})$ of integers modulo 11.

Solution: Here the field $(F, +_{11}, \odot_{11})$ has eleven elements

0, 1, 2, 3, ... 10.

If $f(x)$ is irreducible it must have linear factor $x - a$, where a is zero of $f(x)$. By checking the eleven numbers

$$\begin{aligned} f(0) &= 4, \\ f(1) &= 1^2 +_{11} 1 +_{11} 4 = 6, \\ f(2) &= 2^2 +_{11} 2 +_{11} 4 = 10, \\ f(3) &= 3^2 +_{11} 3 +_{11} 4 = 5, \\ f(4) &= 4^2 +_{11} 4 +_{11} 4 = 4 \odot_{11} 4 +_{11} 4 +_{11} 4 \\ &= 5 +_{11} 4 +_{11} 4 = 2, \end{aligned}$$

Similarly,

$$\begin{aligned} f(5) &= 1, \\ f(6) &= 2, \\ f(7) &= 5, \\ f(8) &= 10, \\ f(9) &= 6, \\ f(10) &= 1. \end{aligned}$$

We see that $f(x)$ has no zeroes in F, and so $f(x)$ is irreducible in F.

Example 7.2.5: $x^2 - 2$ is irreducible in Q [x], where (Q, +, .) is the field of rational numbers, but is reducible in the field (R, +, .) of real numbers, since $x^2 - 2 = (x + \sqrt{2}$ $(x - \sqrt{2})$ where $x + \sqrt{2}$, $x - \sqrt{2} \in R$ [x].

Example 7.2.6: Any linear polynomial $f(x) = a\,x + b$, $a \neq 0$, is irreducible in F [x], where (F, +, .) is a field. Indeed, since the degree of a product of two non-zero polynomials is the sum of the degrees of the factors, it follows that representation

$$a\,x + b = g\,(x)\,.\,h\,(x)$$

with $0 < \deg g\,(x) < 1$, $0 < \deg h\,(x) < 1$ is impossible. Thus every reducible polynomial has degree at least 2.

Example 7.2.7: $3x^2 + 6$ is reducible over the integers, since $3x^2 + 6 = 3\,(x^2 + 2)$. 3 is not a unit in the set of integers, 3 is proper divisor and $3x^2 + 6$ is reducible over integers. But $3x^2 + 6$ is irreducible over the field (R, +, .) of rational numbers. Since 3 is a unit in the set of rational numbers and is therefore improper divisor in R [x], hence $3x^2 + 6$ is irreducible over R [x].

7.3 VALUE OF A POLYNOMIAL AT $x = r$

Let (R, +, .) be a ring with identity and $r \in R$. For each polynomial

$$f(x) = a_0 + a_1\,x + a_2x^2 + ... + a_ax^n$$

in R [x], we define $f(r) \in R$ by

$$f\,(r) = a_0 + a_1r + ... + a_nr^n$$

The element $f(r)$ is said to be the value of $f(x)$ at $x = r$.

Now, suppose $f(x)$, $g\,(x) \in R$ [x] and $r \in R$. If

$$h\,(x) = f\,(x) + g\,(x),\ k\,(x) = f(x)\,g\,(x),$$

then $h\,(r)$ and k (r) are the elements of R, since addition and multiplication hold in (R, +, .)

But, if (R, +, .) is not commutative then some difficulty arises. In this case, the above result need not hold.

Let $$h(x) = (x - a)\,(x - b)$$

$$= x^2 - (a + b)\,x + a\,.\,b.$$

Then $$h\,(r) = r^2 - (a + b)\,r + a\,.\,b.$$

and $$(r - a)\,(r - b) = r^2 - a\,.\,r - r\,.\,b + a\,.\,b \neq h\,(r),$$

Since (R, +, .) is not commutative.

Thus, $h\,(x) = f(x)\,.\,g\,(x)$ does not always apply $h\,(r) = f(r)\,.\,g\,(r)$.

For Example: If $f\,(x) = 2 - 3x + 4x^2$ is considered an element in Z [x] over (Z, +, .), the integral domain of integers, then

$$f\,(2) = 2 - 3\,.\,2 + 4\,.\,2^2 = 12$$

$$f\,(-3) = 2 - 3\,(-3) + 4\,(-3)^2 = 47.$$

Definition 7.3.1: If $f(x) \in R[x]$ over the ring $(R, +, .)$ and $f(c) = 0$ for an element $c \in R$, then c is called a zero of $f(x)$ or the root of $f(x)$.

Example 7.3.1: The polynomial $f(x) = x^2 - 2$ has no zeros when $f(x) \in Z[x]$ over $(Z, +, .)$, the ring of integers. However, if $f(x)$ is considered as an element in $R[x]$ over the field $(R, +,.)$ of real numbers, then both $\sqrt{2}$ and $-\sqrt{2}$ are zeros of $f(x)$. If $f(x) \in Z_7[x]$ over the ring

$(Z_7, +_7, \odot_7,)$ then 3 and 4 are zeros of $f(x)$.

Theorem 7.3.1: *(Division Algorithm) Let $(R, +, .)$ be ring with identity and $f(x)$, $g(x) \neq 0$ be polynomials in $R[x]$, with leading coefficient of $g(x)$ an invertible element. Then there exist unique polynomials $q(x)$, $r(x) \in R[x)$ such that*

$$f(x) = q(x)\, g(x) + r(x),$$

where either $r(x) = 0$ or $\deg r(x) < \deg (x)$.

Proof: The proof proceeds by induction on the degree of $f(x)$. First, we note that if $f(x) = 0$ or $\deg f(x) < \deg g(x)$, then a representation meeting the requirements of the theorem exists with $q(x) = 0$, $r(x) = f(x)$. Furthermore, if $\deg f(x) = \deg g(x) = 0$, $f(x)$ and $g(x)$ are both elements of R, and it suffices to take $q(x) = f(x) \,.\, g(x)^{-1}$, $r(x) = 0$.

Now, suppose the theorem is true for the polynomials $f(x)$ of degree less than n and $\deg f(x) = n$, $\deg g(x) = m$, where $n > m$; that is,

$$f(x) = a_0 + a_1 x + \ldots\ldots + a_n x^n,\ a_n \neq 0$$

$$g(x) = b_0 + b_1 + \ldots. + b_m x^m,\ b_m \neq 0\ (n \geq m).$$

The polynomial

$$f_1(x) = f(x) - (a_n \,.\, b_m^{-1})\, x^{n-m}\, g(x)$$

lies in $R[x]$ and, since the coefficients of x^n is $a_n - (a_n \,.\, b_m^{-1}) \,.\, b_m = 0$, $f_1(x)$ has degree less than n. By induction, there are polynomials $q_1(x)$, $r(x) \in R[x]$ such that

$$f_1(x) = q_1(x)\, g(x) + r(x),$$

where $r(x) = 0$ or $\deg r(x) < \deg g(x)$.

Substituting, we obtain the equation

$$\begin{aligned} f(x) &= (q_1(x) + ((a_n \,.\, b_m^{-1})\, x^{n-m}) \,.\, g(x) + r(x) \\ &= q(x) \,.\, g(x) + r(x) \end{aligned}$$

which shows that the desired representation also exists when $\deg f(x) = n$.

As for uniqueness, suppose

$$f(x) = q(x) \,.\, g(x) + r(x) = q'(x) \,.\, g(x) + r'(x),$$

where $r(x)$ and $r'(x)$, satisfy the requirement of the theorem.

Then

$$r(x) - r'(x) = (q'(x) - q(x)) \,.\, g(x).$$

If $r(x) - r'(x) \neq 0$, we would have

$\deg (r(x) - r'(x)) = \deg (q'(x) - q(x)) + \deg g(x) \geq \deg g(x)$.

On the other hand, since the degrees of $r(x)$ and $r'(x)$ are both less than that of $g(x)$, it follows that

$\deg (r(x) - r'(x)) < \deg g(x)$

This contradiction implies $r(x) = r'(x)$ which in turn yields $q(x) = q'(x)$. The polynomials $q(x)$ and $r(x)$ in the expression

$$f(x) = q(x) \cdot g(x) + r(x)$$

given by the Division Algorithm are called the quotient and remainder on dividing $f(x)$ by $g(x)$, respecitvely. We also observe that if $g(x)$ is a monic polynomial, or if $(R, +, .)$ is a field, it is not necessary to assume the leading coefficient of $g(x)$ to be invertible.

***Example* 7.3.2:** Let $f(x) = 2x^3 - 4x^2 + 5$ and $g(x) = 2x^2 - 3x + 2$ be considered polynomials in $R[x]$ over the field $(R, +, .)$ of rational numbers. We shall find the quotient and remainder of the division of $f(x)$ by $g(x)$ and express $f(x)$ in the form $f(x) = q(x) \cdot g(x) + r(x)$,

where $r(x) = 0$ or $\deg r(x) < \deg g(x)$.

$$\begin{array}{l} \qquad\qquad\qquad\quad x - 1/2 \\ \hline 2x^2 - 3x + 2 \,.\, 2x^3 - 4x^2 + 5 \\ \\ \qquad\qquad\qquad 2x^3 - 3x^2 + 2x \\ \qquad\qquad\qquad - \quad + \quad - \\ \hline \qquad\qquad\qquad -x^2 - 2x + 5 \\ \\ \qquad\qquad\qquad -x^2 + \frac{3}{2}x - 1 \\ \qquad\qquad\qquad + \quad - \quad + \\ \hline \qquad\qquad\qquad \frac{-7}{2}x + 6 \end{array}$$

Then $2x^3 - 4x^2 + 5 = \left(x - \frac{1}{2}\right)(2x^2 - 3x + 2) + \left(\frac{-7}{2}x + 6\right)$.

Here $q(x) = x - \frac{1}{2} \in R[x]$

and $r(x) = \frac{-7}{2}x + 6 \in R[x]$.

***Theorem* 7.3.2:** *(Remainder Theorem) Let $(R, +, .)$ be a commutative ring with identity. If $f(x) \in R[x]$ and $a \in R$, then there is a unique polynomial $q(x)$ in $R[x]$ such that*

$$f(x) = (x - a) \cdot q(x) + f(a)$$

Proof: Applying the division algorithm to $f(x)$ and $x - a$, we obtain $f(x) = (x - a), q(x) + r(x)$

where $r(x) = 0$ or deg $r(x) <$ deg $(x - a) = 1$. It follows in either case that $r(x)$ is a constant polynomial $r \in R$.

Substituting a for x, we have the desired result

$$f(a) = (a - a) . q(a) + r(a) = 0 + r = r.$$

Thus, we have

$$f(x) = (x - a) . q(x) + f(a).$$

Corollary: *(Factorization theorem) The polynomial $f(x) \in R[x]$ is divisible by $x - a$ if and only if a is a root of $f(x)$.*

Proof: This result follows immediately from the theorem 7.3.2 since

$$f(x) = (x - a) . q(x)$$

if and only if $f(a) = 0$.

Exercise 7.3.1: Determine whether $x - 3$ and $x + 2$ are divisors of $f(x) = x^5 + 32$, considered as polynomials in $R[x]$ over the field $(R, +,.)$ of rational numbers.

Since $f(3) = 275 \neq 0$, $x - 3$ does not divide $x^5 + 32$. However $f(-2) = 0$, so that $x + 2$ divides $x^5 + 32$.

7.4 THE GREATEST COMMON DIVISOR OF TWO POLYNOMIALS

Definition 7.4.1: A greatest common divisor *(GCD)* of two polynomials $f(x), g(x) \in F[x]$ is a polynomial $d(x) \in F[x]$ such that

(a) $d(x)$ is monic.

(b) $d(x) | f(x)$ and $d(x) | g(x)$.

(c) If $c(x)$ is a polynomial such that $c(x) | f(x)$ and $c(x) | g(x)$, then $c(x) | d(x)$.

It is denoted by $d(x) = (f(x), g(x))$

Example 7.4.1: The *GCD* of $3x^3 - 24$ and $3x^2 - 12$, considered as an element of $R[x]$ for the rational numbers R, is $x - 2$. It is true, of course, that $3x - 6$ is not monic.

Example 7.4.2: The *GCD* of $x^2 - 4$ and $x^2 + 6x + 9$, considered as elements of $R[x]$, is 1, since the two polynomials have no common divisors other than constant polynomial. The only monic constant polynomial is 1.

Theorem 7.4.1: *Let $(F, +, .)$ be a field and $f(x), g(x) \in F[x]$, not both of which are zero. Then $f(x)$ and $g(x)$ have a unique greatest common divisor $d(x)$ which can be expressed in the from*

$$d(x) = m(x) f(x) + n(x) g(x)$$

for two polynomials $m(x), n(x) \in F[x]$.

Proof: We consider the set S of all polynomials of the form

$m(x) f(x) + n(x) . g(x)$. Thus,

$S = \{(m(x) . f(x) + n(x) . g(x)) \mid m(x), n(x) \in F[x]\}$

The set S is non-empty since

$f(x) = 1 . f(x) + 0 . g(x) \in S.$

It is obvious that S contains monic polynomials for if $h(x) \in S$ with leading coefficient b, $b^{-1} h(x) \in S$ is a monic polynomial, because

$b^{-1} h(x) = (b^{-1} m(x)) . f(x) + (b^{-1} n(x)) . g(x).$

Now we choose $m(x)$ and $n(x) \in F[x]$ such that $m(x) . f(x) + n(x). g(x)$ is a monic polynomial $d(x)$ in S of least degree. Thus $d(x) = m(x) . f(x) + n(x) . g(x)$.

This $d(x)$ is unique. For, if $d_1(x)$ is also a monic polynomial of least degree in S, then $d(x) - d_1(x)$ is in S which is either 0 or of degree lower than that of $d(x)$.

Since $d(x) \in S$ is of least degree, it follows $d(x) = d_1(x)$.

Next, we shall show that $d(x)$ is GCD of $f(x)$ and $g(x)$. By division algorithm, we have

$$f(x) = q(x) . d(x) + r(x),$$

where $$r(x) = 0, \text{ or } \deg r(x) < \deg d(x).$$

But since $r(x) = f(x) - q(x) . d(x) \in S$, and by assumption than $d(x)$ is of least degree in S, which implies $r(x) = 0$

Thus, $d(x) \mid f(x)$.

Similarly $d(x) \mid g(x)$.

Now, if $d_1(x)$ is any common divisor of $f(x)$ and $g(x)$, then we have

$$f(x) = f_1(x) . d_1(x), \text{ and}$$

$$g(x) = g_1(x) . d_1(x)$$

Since $d(x) \in S$, there exist $m_1(x)$ and $n_1(x)$ polynomials in S such that

$$\begin{aligned} d(x) &= m_1(x) . f(x) + n_1(x) . g(x) \\ &= m_1(x) . f_1(x) d_1(x) + n_1(x) . g_1(x) . d_1(x) \\ &= [m_1(x) . f_1(x) + n_1(x) . g_1(x)] . d_1(x), \end{aligned}$$

which shows $d_1(x) \mid d(x)$. This is also true for other common divisors of $f(x)$ andg (x).

Hence $d(x)$ is GCD of $f(x)$ and $g(x)$.

This completes the proof of the theorem.

Exercise 7.4.1: Find the GCD, $d(x)$, of $f(x) = x^3 - 2x^2 + 3x - 7$ and $g(x) = x^2 + 2$, and express it in the form

$$d(x) = m(x) . f(x) + n(x) . g(x)$$

Solution: By applying division algorithm on $f(x)$ and $g(x)$, we have

$x^3 - 2x^2 + 3x - 7 = (x - 2)(x^2 + 2) + x - 3.$ ....(1)

Again applying division algorithm on divisor and remainder and so on, we have

$$x^2 + 2 = (x + 3)(x - 3) + 11, \quad ....(2)$$

and

$$x - 3 = \left(\frac{x}{11} - \frac{3}{11}\right)(11) + 0. \quad ....(3)$$

Hence 11 is the polynomial which divides $f(x)$ and $g(x)$ Hence 1 is *GCD* of $f(x)$ and $g(x)$.

From the second equation, we have

$$11 = (x^2 + 2) - (x + 3)(x - 3). \quad ...(4)$$

Substituting the value of $x - 3$ from (1) in (4) we obtain

$$11 = (x^2 + 2) - (x + 3)[x^3 - 2x^2 + 3x - 7 - (x - 2)(x^2 + 2)]$$

$$= (x^2 + 2) - (x + 3)(x^3 - 2x^2 + 3x - 7) + (x^2 - x - 6)(x^2 + 2)$$

$$= (x^2 - x - 5)(x^2 + 2) + (-x - 3)(x^3 - 2x^2 + 3x - 7).$$

Dividing by 11, we have the *GCD*, 1 expressed as

$$= \left(\frac{x^2}{11} - \frac{x}{11} - \frac{5}{11}\right)(x^2 + 2) + \left(-\frac{x}{11} - \frac{3}{11}\right)(x^3 - 7x^2 + 3x - 7)$$

Exercise 7.4.2: Find the *GCD* of $f(x) = x^3 - 3x^2 + 2x - 6$ and $g(x) = x^3 - 4x^2 + 4x - 3$, and express it in the form $m(x) f(x) + n(x) g(x)$.

Solution: Applying the procedure of division algorithm, we have

$$x^3 - 3x^2 + x - 6 = 1(x^3 - 4x^2 + 4x - 3) + x^2 - 2x - 3$$

$$x^3 - 4x^2 + 4x - 3 = (x - 2)(x^2 - 2x - 3) + 3x - 9$$

$$x^2 - 2x - 3 = \left(\frac{1}{3}x + \frac{1}{3}\right)(3x - 9) + 0.$$

Hence the *GCD* of $f(x)$ and $g(x)$ is $(3x - 9)$, or $x - 3$. Then

$$3x - 9 = (x^3 - 4x^2 + 4x - 3) - (x - 2)(x^2 - 2x - 3)$$

$$= (x^3 - 4x^2 + 4x - 3) - (x - 2)[x^3 - 3x^2 + 2x - 6 - (x^3 - 4x^2 + 4x - 3)]$$

$$= (x - 1)(x^3 - 4x^2 + 4x - 3) + (-x + 2)(x^3 - 3x^2 + 2x - 6).$$

Hence $x - 3$, the *GCD* of $f(x)$ and $g(x)$ can be expressed as $(x - 3) = \left(\frac{x}{3} - \frac{1}{3}\right)(x^3 - 4x^2 + 4x - 3) + \left(-\frac{x}{3} + \frac{2}{3}\right)(x^3 - 3x^2 + 2x - 6)$.

Theorem 7.4.2: *Let $f(x)$, $g(x)$, and $h(x)$ be polynomials in $F(x)$ for a field $(F, +, .)$. If $f(x) \mid h(x) \,.\, g(x)$ and $(f(x), g(x)) = 1$, then $f(x) \mid h(x)$.*

Proof: Since $(f(x), g(x)) = 1$, then 1 can be expressed as

$1 = m(x).f(x) + n(x).g(x)$ for some $m(x), n(x) \in F[x]$.

Multiplying this equation by $h(x)$, we have

$h(x) = m(x).f(x).h(x) + n(x).g(x).h(x)$.

But $f(x) | g(x).h(x)$, so that $g(x).h(x) = q(x).f(x)$ for some polynomial $q(x) \in F[x]$ Thus

$$h(x) = m(x) f(x) h(x) + n(x) f(x) q(x)$$
$$= f(x) [m(x).h(x) + n(x).q(x)],$$ which shows $f(x) | h(x)$.

Theorem 7.4.3: *If $f(x)$ is an irreducible element in $F[x]$ for a field $(F, +, .)$ and $f(x) | g_1(x).g_2(x) ..... g_k(x)$, $g_i(x) \in F[x]$ then $f(x) | g_i(x)$ for some i, $1 \leq i \leq k$.*

The proof of the theorem is simple and is left to the reader.

Theorem 7.4.4: *(Unique Factorization Theorem). Each polynomial $f(x) \in F[x]$ of positive degree is the product of a non-zero element of field F and irreducible monic polynomials of $F[x]$ and this factorization is unique.*

Proof: We shall prove the theorem by induction on degree n of $f(x)$. Let $n = 1$, then $f(x) = ax + b = a.(x + a^{-1}b)$, where $x + a^{-1}b$ is a monic irreducible polynomial. Next, let $n > 1$ and let the theorem hold for all polynomials of degree less than n. If $f(x)$ is irreducible in $F(x)$, we are through after factoring out its leading coefficient, otherwise, $f(x)$ is reducible and it is possible to write $f(x) = g(x).h(x)$ with $g(x) \in F[x]$, $0 < \deg g(x) < n$, $0 < \deg h(x) < n$. Therefore, by induction hypothesis,

$$g(x) = a_1 g_1(x).g_2(x) .... g_r(x)$$
$$h(x) = a_2.h_1(x).h_2(x) ... h_s(x),$$

where a_1, a_2 are non-zero elements of F (the leading coefficient of $g(x)$ and $h(x)$, and $g_i(x)$ and $h_i(x)$ are irreducible polynomials.

Thus, $f(x) = (a_1.a_2).g_1(x) ... g_r(x).h_1(x) ... h_s(x)$ is a factorization satisfying the conditions of the theorem.

Now we shall show that this factorization is unique.

We again prove by induction on $n = \deg f(x)$. For $n = 1$, $f(x) = a_1.(x + b_1) = a_2(x + b_2)$, $a_1, a_2 \in F$.

Then $a_1 = a_2$ and $a_1.b_1 = a_2.b_2$ which implies $b_1 = b_2$ by the cancellation law for multiplication. Now, let $n > 1$ and $f(x) = cp_1(x).p_2(x)... p_r(x) = d.q_1(x).q_2(x) ... q_s(x)$ be any two factorization of $f(x)$ into irreducible monic polynomials. Surely $c = d$, for each is the leading coefficient of $f(x)$. Furthermore, $p_1(x) | q_1(x).q_2(x).... q_r(x)$ which implies $p_1(x) | q_i(x)$ for some $1 \leq i \leq s$. For instance, $p_1(x) | q_1(x)$. Since these are irreducible polynomials, it follows that $p_1(x) = q_1(x)$ hence, we may cancel to conclude that

$$f_1(x) = p_2(x).... p_r(x) = q_2(x).... q_s(x),$$

def $f_1(x) < n$, its factorization is unique by the induction assumption, so that the sequence $q_2(x),\dots q_s(x)$ is simply a rearrangement of $p_2(x),\dots, p_r(x)$. Together with the observation that $c \,.\, p_1(x) = d_1\, q_1(x)$, this completes the proof.

Theorem 7.4.5: If *(F, +, .) is afield, then the ring (F [x], +, .) is a principal ideal domain.*

Proof: We have already shown that the system (F [x], +, .) is an integral domain over the field (F, +, .) . For showing (F [x], +, .) principal ideal domain we shall show that every ideal of (F [x], +, .) is principal.

Let I be any ideal of (F [x], +, .) and let $I = (0)$ then, the result is trivially true since $0.\, f(x) = 0 \in (0)$ for all $f(x) \in F[x]$.

Again, let $I \neq (0)$, contain a non-zero polynomial $p(x)$ of lowest degree.

For each $f(x) \in I$, there exists $q(x)$ and $r(x)$ in $F[x]$ such that $f(x) = q(x)\,.\,p(x) + r(x)$, where $r(x) = 0$ or $\deg r(x) < \deg p(x)$.

Now

$r(x) = f(x) - q(x)\,.\,p(x) \in I$

Since $p(x) \in I$ is a polynomial of lowest degree, the deg $r(x)$ cannot be less than the deg $p(x)$. Thus $r(x) = 0$ and we have

$f(x) = q(x)\,.\,p(x) \in (p(x))$, the ideal generated by $p(x)$ which shows

$I \subseteq (p(x))$.

Conversely, if $g(x) \in (p(x))$, there exists $q_1(x) \in F[x]$ such that

$g(x) = q_1(x)\,.\,p(x) \in I$ since I is an ideal, which follows that

$(p(x)) \subseteq I$.

Hence $I = (p(x))$ which completes the theorem.

Corollary: *A non-trivial ideal of (F [x], +,.) is maximal if and only if it is prime ideal.*

Example 7.4.3: Show that the polynomial ring (Z [x], +, .) over the ring (Z, +, .) of integers is a not a principal ideal ring.

Solution: To prove this result we must have at least one ideal of (Z [x] , +, .) which is not principal. For this we consider the ideal ((x, 2), +, .) generated by two elements x and 2 and we shall show that this ideal ((x, 2), +, .) is not principal ideal.

Let ((x, 2), +, .) be a principal ideal generated by $a(x) \in Z[x]$. So $(x, 2) = (a(x))$ and consequently $2, x \in (a(x))$. Here for some $b(x)$ and $c(x)$ of $Z(x)$ we have

$2 = a(x)\,.\,b(x)$,

and $x = a(x)\,.\,c(x)$

This implies $\deg (a(x)\,.\,b(x)) = \deg a(x) + \deg b(x) = \deg 2 = 0$,

and $\deg a(x)\,.\,c(x)) = \deg a(x) + \deg c(x) = \deg = 1$.

So $\deg a(x) + \deg b(x) = 0 \Rightarrow a(x)$ and $b(x)$ are non-zero constant of Z.

Consequently $a(x)\,.\,b(x) = 2 \Rightarrow a(x) = 1,\ b(x) = 2$, or $a(x) = -1,\ b(x) = -2,$

or $a(x) = 2,\ b(x) = 1,$

or $a(x) = -2,\ b(x) = -1$

If $a(x) = \pm 1$, then $(a(x)) = Z[x]$ which is contradiction to $(a(x)) = (x, 2) \neq Z[x]$.

Now suppose that

$$a(x) = \pm 2,$$

$$x = a(x) \,.\, c(x) = \pm 2 \,.\, c(x)$$

$$= \pm 2\,(c_0 + c_1 x + c_2 x^2 + ..), \text{ where}$$

$$c(x) = c_0 + c_1 x + c_2 x^2 +,.$$

On comparing the like powers of x, we have

$$1 = \pm 2\, c_1, \text{ where } c_1, \in Z.$$

This is again a contradiction to that there exist no integer c_1 such that $1 = \pm 2\, c_1$

This shows that no polynomial $a(x)$ is possible which generate ideal $(x, 2)$.

Hence $(x, 2)$ is not principal ideal and consequently $(Z[x], +, .)$ is not principal ideal ring.

Theorem 7.4.6: *The mapping*

$f(x) \to (f(x)) = \{g(x) \,.\, f(x) \mid g(x) \in F[x]\}$

is one to one mapping from the set of monic polynomials of $F[x]$ onto the set of non-zero ideals of $F[x]$.

Proof: Let $f(x)$ be a monic polynomial of minimum degree n, then we have a non-zero ideal $(I, +, .)$, where $I = (f(x))$. Suppose $g(x)$ is another monic polynomial of minimal degree n which belong to I. So we have

$$g(x) = q(x) \,.\, f(x).$$

Since $\deg g(x) = \deg f(x) = n$, then $q(x)$ must be constant, say, q. Since $g(x)$ and $f(x)$ are monic polynomials, $q = 1$ which in turn implies $g(x) = f(x)$. This shows that mapping is one to one and onto. It completes the proof.

Now we see that corresponding to any monic polynomial of minimal degree we have non-zero ideal $((f(x)), +, .)$

So, we can have the quotient ring $(F(x) / (f(x)), +, .)$ with the ideal $(f(x)), +, .)$ as their zero element.

Theorem 7.4.7: *The elements $b(x)$ and $c(x)$ are in the same coset of $T = (f(x))$ in $F[x]$ if and only if they leave the 'same remainder when divided by $f(x)$.*

Proof: Let

$$b(x) = q_1(x) \,.\, f(x) + r(x)$$

and

$$c(x) = q_2(x) \,.\, f(x) + r(x).$$

Then

$$b(x) - c(x) = (q_1(x) - q_2(x)) \cdot f(x).$$

Since $(T, +, .)$ is an ideal,

$(q_1(x) - q_2(x)) \cdot f(x) \in T$, conseqnently

$b(x) - c(x) \in T \Rightarrow b(x) + T = c(x) + T.$

Conversely, let

$b(x) + T = c(x) + T.$ Then there exists an element

$q(x) \cdot f(x) \in T$ such that

$b(x) = c(x) + q(x) \cdot f(x).$

Dividing $c(x)$ by $f(x)$, we have

$c(x) = q_1(x) \cdot f(x) + r(x)$, $\deg r(x) < \deg f(x)$.

So that

$$\begin{aligned} b(x) &= q_1(x) \cdot f(x) + q(x) \cdot f(x) + r(x) \\ &= (q_1(x) + q(x)) \cdot f(x) + r(x) \end{aligned}$$

This shows $b(x)$ and $c(x)$ have the same remainder $r(x)$. This proves the theorem.

Theorem 7.4.8: If $T = (f(x))$ *is a proper ideal of* $(F[x], +, .)$, *then the set of elements* $r + T$ *of* $F[x]/T$, *where* $r \in F$, *forms a field isomorphic with* $(F, +, .)$ *and the natural mapping* $r \to r + T$, $\forall r \in F$, *is an isomorphism.*

Proof: Since $(T, +, .)$ is a proper ideal, there $\deg f(x) > 0$. Hence the element $r \in F$ can be obtained as remainder on dividing some polynomials of $F[x]$ by $f(x)$. So

For $r_1, r_2 \in F$, we assume $r_1 + T = r_2 + T$.

Since r_1 and r_2 are remainders on dividing some polynomials $b(x)$ and $c(x)$ by $f(x)$, the there exists $q_1(x)$ and $q_2(x) \in F[x]$ such that

$$b(x) = q_1(x) \cdot f(x) + r_1,$$

and

$$c(x) = q_2(x) \cdot f(x) + r_2.$$

Then

$$\begin{aligned} r_1 + T = r_2 + T &\Rightarrow [b(x) - q_1(x) f(x)] + T = [c(x) - q_2(x) \cdot f(x)] + T \\ &\Rightarrow b(x) + T = c(x) + T \\ &\Rightarrow r_1 = r_2. \end{aligned}$$

by the theorem 7.4.7 that if $b(x) + T = c(x) + T$, then $b(x)$ and $c(x)$ have the same remainders.

This proves the mapping is one to one and it is clearly onto.

Now, For $r, s \in F$, then

$$r + s \to (r + s) + T = (r + T) + (s + T),$$

and $$r \,.\, s \to (r \,.\, s) + T = (r + T) \,.\, (s + T).$$

This means the set $\{r + T \mid r \in F\}$ is a subring under addition and multiplications of two cosets. Since it is isomorphic with $(F, +, .)$, then the subring $(\{r + T \mid r \in F\}, +, .)$ is a field.

Theorem 7.4.9: *Let $(F, +, .)$ be a field, then the following statements are equivalent:*

(1) *$p(x)$ is irreducible in $F[x]$.*

(2) *The principal ideal $((p(x)), +, .)$ is maximal in $(F[x], +, .)$.*

(3) *The quotient ring $(F[x]/(p(x)), +, .)$ is a field,*

Proof: (*a*) From (1) we prove (3). Let $p(x)$ be an irreducible polynomial. Then we have to show that every non-zero elements of $F(x) / (p)(x))$ has a multiplicative inverse.

Let $b(x) + (p(x)) \neq (p(x))$. Then $b(x) \notin (p(x))$. So $b(x)$ does not have $p(x)$ as a factor. Since $p(x)$ is irreducible, then $b(x)$ and $p(x)$ do not have any common factor except the identity of the field. So for some $f(x)$ and $g(x)$ in $F[x]$, we have

$$b(x) \,.\, f(x) + g(x) \,.\, p(x) = 1$$

or
$$\begin{aligned} 1 + (p(x)) &= [b(x) \,.\, f(x) + g(x) \,.\, p(x)] + (p(x)) \\ &= b(x) \,.\, f(x) + (p(x)), \text{ since } g(x) \,.\, p(x) \in (p(x)) \\ &= [b(x) + (p(x))] \,.\, [f(x) + (p(x))]. \end{aligned}$$

Therefore $f(x) + (p(x))$ is the inverse of $b(x) + (p(x))$ since $1 + (p(x),$ is the multiplicative identity in $F[x]/(p(x))$.

Hence $(F[x] / (p(x)), +, .)$ is a field.

(b) From (3) we prove (2).

Let the quotient ring $(F[x] / (p(x)), +, .)$ be a field. Then the ideal $((p(x)), +, .)$ is maximal. Consequently from the part (*c*) below $f(x)$ irreducible.

(*c*) from (2) we prove (1)

Let the ideal $(p(x)), +, .)$ be maximal ideal generated by $p(x)$. Then we have to show that $p(x)$ is irreducible. Since the ideal $((p(x)), +, .)$ is maximal, then for any $g(x) \notin (p(x))$ we have

$$(p(x)) + (g(x)) = F[x]$$

Therefore, for some $q_1(x)$ and $q_2(x)$, $p(x) \,.\, q_1(x) + q_2(x) \,.\, g(x) = 1$, 1 is the constant polynomial of $F[x]$ which is the identity in $F[x]$.

This implies $p(x)$ and all $g(x) \notin (p(x))$ are prime. If $g(x) \in (p(x))$, then $g(x)$ is multiple of $p(x)$ and $p(x)\, q_1(x) + q_2(x) \,.\, g(x) = 1 \in (p(x)) \Rightarrow (p(x)) = F(x)$. This contradiction shows $p(x)$ is irreducible.

(d) From (1) we prove (2), Let $p(x)$ be an irreducible polynomial. Then we have to show that the ideal $((p(x), +, .)$ is maximal.

Any $g(x) \in F[x]$ is prime to $p(x)$. So there are some

$h(x)$ and $q(x)$ for which $p(x) \,.\, h(x) + g(x) \,.\, q(x) = 1$.

Therefore every $f(x) \in F[x]$ can be written as

$p(x) \cdot h(x) \cdot f(x) + g(x) \cdot q(x) \cdot f(x) = 1 \cdot f(x) = f(x)$

Since $p(x) \cdot h(x) \cdot f(x) \in (p(x))$, $g(x) \cdot q(x) \cdot f(x) \in (g(x))$,

Then $(p(x)) + (g(x)) = F[x]$, where $g(x) \notin (p(x))$.

Hence $(p(x))$ is maximal.

Theorem 7.4.10: *Let $f(x)$ be a polynomial in $F(x)$ for a field $(F, +, .)$ with leading coefficient c and degree n. Then if $d_1, d_2, ..., d_n$ are n distinct zeros of $f(x)$,*

$$f(x) = c(x - d_1)(x - d_2)...(x - d_n).$$

Conversely, if $(x - d_1)(x - d_2) ... (x - d_n)$ are divisors of $f(x)$, then each d_i for $i = 1, 2, ..., n$ is a zero of $f(x)$.

Proof: We shall prove the theorem by induction on n. If $f(x)$ has degree one, $f(x) = cx + d$ for $c, d \in F$. If d_1 is a zero of $f(x)$, $f(d_1) = 0$, and hence $c\,d_1 + d = 0$, or $d = -c\,d_1$, then we have

$f(x) = cx + d = cx - c\,d_1 = c(x - d_1)$ as required.

Now, suppose that the theorem holds for all polynomials of degree k, and let $f(x)$ be of degree $k + 1$, with leading coefficient c, and distinct zeros $d_1, d_2, ... d_{k+1}$.

Then, since d_{k+1}, is zero of $f(x)$, $f(x) = (x - d_{k+1}) \cdot g(x)$ for some $g(x) \in F(x)$, by the factor theorem. We note that the degree of $g(x)$ must be k and that $g(x)$ has leading coefficient c. Further, for any zero d_i of $f(x)$, with $i \leq k$,

$f(d_i) = (d_i - d_{k+1})\, g(d_i)$.

Since the zeros of $f(x)$ are distinct

$d_i - d_k \neq 0$.

and therefore $g(d_i) = 0$, so that d_i is a zero of $g(x)$. Thus $g(x)$ has k distinct zeros $d_1, d_2, ...d_k$. By induction hypothesis,

$g(x) = c(x - d_1)(x - d_2) ... (x - d_k)$.

Thus, we have

$f(x) = c(x - d_1)(x - d_2) ... (x - d_k)(x - d_{k+1})$ and the proof is complete.

For the second part of the theorem, if $(x - d_1), (x - d_2), ..., (x - d_n)$ are divisors of $f(x)$, then each d_i for $i = 1, 2, ..., n$ is a zero of $f(x)$.

Definition 7.4.2: Let $f(x)$ be a polynomial of degree n over a field $(F, +, .)$. We say that the equation $f(x) = 0$ is an equation over the field $(F, +, .)$, and n is the degree of the equation.

Example 7.4.4: Write an equation over $R[x]$ having $\frac{1}{2}$, -2 and 3 as roots. By the theorem 7.4.10.

$$f(x) = \left(x - \frac{1}{2}\right)(x+2)(x-3) = x^3 - \frac{3}{2}x^2 - \frac{11}{2}x + 3$$

has $\frac{1}{2}$, –2, and 3 as zeros. Hence the equation

$$x^3 - \frac{3}{2}x^2 - \frac{11}{2}x + 3 = 0 \text{ has roots } \frac{1}{2}, -2, \text{ and } 3.$$

We can write this equation as $2x^3 - 3x^2 - 11x + 6 = 0$.

Theorem 7.4.11: *If $f(x)$ is a polynomial of degree n over a field $(F, +, .)$, the equation $f(x) = 0$ has at most n distinct roots in F.*

Proof: Let $r_1, r_2, \ldots, r_n$ be distinct roots of the equation $f(x) = 0$ in F. We shall show that $f(x)$ can have no other roots in F, and hence the greatest possible number of distinct roots is n.

Suppose that r is any root of $f(x) = 0$, $f(r) = 0$. By factor theorem, we have

$f(x) = c(x - r_1)(x - r_2) \ldots (x - r_n)$,

where c is the leading coefficient of $f(x)$. Since r is the root of $f(x)$,

$0 = f(r) = c(r - r_1)(r - r_2) \ldots\ldots (r - r_n)$

Since $(F, +, .)$ has no zero divisors, one of the factors $r - r_i = 0$, where $1 \le i \le n$. Hence $r = r_i$. That is, any root r of $f(x) = 0$ must be equal to one of the n roots $r_1, r_2, \ldots r_n$. This completes the proof.

Corollary: *Let $f(x)$ and $g(x)$ be non-zero polynomials of degrees $\le n$ over the integral domain $(R, +, .)$. If there exists $n + 1$ distinct elements $a_k \in R$ $(k = 1, 2, \ldots, (n+1))$ for which $f(a_k) = g(a_k)$, then $f(x) = g(x)$.*

Proof: The polynomial $h(x) = f(x) - g(x)$ is such that deg $h(x) \le n$ and has at least $n + 1$ distinct roots in R. By the theorem 7.4.11 this is impossible unless

$$h(x) = f(x) - g(x) = 0, \text{ or } f(x) = g(x).$$

Theorem 7.4.12: *(Fundamental Theorem of Algebra). Let $(C, +, .)$ be a field of complex numbers. If $f(x) \in C[x]$ is a polynomial of positive degree, then $f(x)$ has at least one root in C.*

Proof: Many proofs of this theorem are available, none is strictly algebraic in nature. Thus, we shall assume the validity of the theorem without proof.

Theorem 7.4.13: *If $f(x) \in C[x]$ is a polynomial of degree $n > 0$, then $f(x)$ can be expressed in $C[x]$ as a product of n linear factors.*

Proof: Let $f(x) \in C[x]$ of degree n with leading coefficient a_n, then $f(x)$ has at least one root b_1 *in C* such that $f(b_1) = 0$. Then $f(x)$ is divisible by $(x - b_1)$ and $f(x)$ can be expressed as

$$f(x) = (x - b_1) f_1(x),$$

where

$f_1(x) \in C[x]$ of degree $n - 1$ with leading coefficient a_n. Again by the same reasoning $f_1(x)$ has at least one root b_2 in C, such that $f_1(b_2) = 0$ and $f_1(x)$ is divisible by $x - b_2$. Thus,

$f_1(x) = (x - b_2) . f_2(x) \in C[x]$. This process can be continued till we come to a polynomial $f_{n-1}(x)$ whose leading coefficient is a_n and whose zero is b_n. Thus,

$$f_{n-1}(x) = a_n(x - b_n).$$

Therefore, we can write

$$f(x) = a_n(x - b_1) . (x - b_2)...(x - b_n).$$

Thus, $f(x)$ is the product of n linear factors

$x - b_1,\ x - b_2 ...,x - b_n,$

or $b_1, b_2,... b_n$. are the n zeros of $f(x)$.

Theorem 7.4.14: *If a polynomial $f(x) \in F(x)$, where $(F, +, .)$ is a field of real numbers, has a zero $a + ib$, $b \neq 0$, it has also the conjugate zero $a - ib$.*

Proof: We observe the product

$[x - (a + i\,b)]\,[x - (a - ib)] = (x - a)^2 + b^2.$

Let $g(x) = (x - a)^2 + b^2$ be a monic equation of degree 2.

By division algorithm, there exist quotient $q(x)$ and remainder $r(x)$ on dividing $f(x)$ by $g(x)$. Thus, we have

$f(x) = g(x) . q(x) + r(x),$

where $r(x) = 0$, or $\deg r(x) = 1$.

Let $r(x) = mx + n$, then

$f(x) = [(x - a)^2 + b^2] . q(x) + mx + n.$

Since $a + ib$ is a zero of $f(x)$, $f(a + ib) = 0$, then $0 = f(a + ib) = 0. q(x) + m(a + ib) + n$

or $0 = ma + n + imb,$

which implies $ma + n = 0$, and $imb = 0$

since $b \neq 0$, $m = 0$, and consequently $n = 0$.

Therefore $r(x) = 0$.

Hence $f(x) = (x - a)^2 + b^2 . q(x)$ which means $f(a - ib) = 0$

This completes the proof of the theorem.

Theorem 7.4.15: *Let $f(x) = a_0 + a_1 x + a_2 x^2 +... + a_n x^n$ be a polynomial with integral coefficients. Let p/q be a rational number in lowest terms (That is, the integers p and q are relatively prime). If p/q is a root of the equation $f(x) = 0$, then $p \mid a_0$ and $q \mid a_n$.*

Proof: If p/q is a root of $f(x) = 0$, then

$a_0 + a_1(p/q) +...., a_n(p/q)^n = 0.$ ...(1)

Multiplying the equation (1) by q^n, we obtain

$$a_0 q^n + a_1 pq^{n-1} + ...+ a_{n-1} p^{n-1} q + a_n p_n = 0. \quad ...(2)$$

Solving equation (2) for $a_0 q^n$, we have

$$a_0 q^n = - a_1 pq^{n-1} - ...- a_{n-1} p^{n-1} q - a_n p^n$$
$$= p (-a_1 q^{n-1}..... a_{n-1} p^{n-2} q - a_n p^{n-1})$$

which shows $p \mid a_0 q^n$ and since p and q are relatively prime, $p \mid a_0$ as desired.

Now we solve equation (2) for $a_n p^n$, we obtain

$$a_n p^n = -a_0 q^n - a_1 pq^{n-1} -.... -a_{n-1} p^{n-1} q$$
$$= q (-a_0 q^{n-1} - a_1 pq^{n-2} -.... - a_{n-1} p^{n-1})$$

which shows $q \mid a_n p^n$, and since p and q are relatively prime, $q \mid a_n$. This completes the proof of the theorem.

Example 7.4.5: Find the rational zeros and decomposition of $6x^4 - 7x^3 + 6x^2 - 1$ over the field $(Q, +, .)$ of retional numbers,

Solution: If the rational number $p \mid q$, with $(p, q) = 1$ and with q *positive, is a zero of the polynomial, then* p must divide -1 and q must divide 6. It follows that

$$p = 1, \text{ and } q = +1, +2, +3, +6.$$

Then only possible rational zeros of $f(x)$ are,

$$\frac{1}{2}, \frac{1}{3}, \frac{1}{6}$$

Since $f(\pm 1) \neq 0$, hence ± 1 are not zeros of $f(x)$.

Next, $f\left(\frac{1}{2}\right) = 0$ which means $\frac{1}{2}$ is a zero of $f(x)$. That is, $x - \frac{1}{2}$ must divide $f(x)$. This

$$f(x) = \left(x - \frac{1}{2}\right)(x^3 - 4x^2 + 4x + 2)$$

Now the possible rational zeros of

$$g(x) = 3x^3 - 2x^2 + 2x + 1 \text{ are } \pm 1, \pm \frac{1}{3}.$$

$g(\pm 1) \neq 0$, hence ± 1 are not zeros of $g(x)$. But we see that $g\left(\frac{1}{3}\right) = 0$, which implies $\frac{1}{3}$ is a zero of $g(x)$ and $x - \frac{1}{3} \mid g(x)$. Thus $g(x) = \left(x - \frac{1}{3}\right)(3x^2 - 3x - 3)$.

Therefore, $f(x) = 6\left(x - \frac{1}{2}\right)\left(x - \frac{1}{3}\right)(x^2 - x - 1)$

The only rational zeros are $\frac{1}{2}$ and $\frac{1}{3}$.

Definition 7.4.3: Let $a_1, a_2, ..., a_n$ be a set of elements of the field $(F, +, .)$. Then we define $S_i(a) = S_i(a_1, a_2,...., a_n)$ for $i = 1, 2,...., n$ by

$S_0(a) = e$ (identity of the field $(F, +, .)$),

$S_1(a) = a_1 + a_2 +....+ a_n = \Sigma a_i,$

$S_2(a) = a_1 a_2 + a_2 a_3 +... = \Sigma a_i a_j,$

$S_3(a) = a_1 a_2 a_3 + a_2 a_3 a_4 +.... = \Sigma a_i a_j a_k,$

$S_n(a) = a_1 a_2 .... a_n$

where, $S_i(a)$ in general is a sum of all products of the elements $a_1, a_2, ...,a_n$ taken i at a time.

$S_i(a)$ are called elementary symmetric functions of $a_1, a_2, ..., a_n$.

Theorem 7.4.16: *Let $(F, +,.)$ be a field, and*

$f(x) = a_0 + a_1x + ... + a_{n-1} x^{n-1} + a_n x^n$ in $F[x]$ has n zeros $r_1, r_2,..., r_n$, then

$$S_j = (-1)^j \frac{a_{n-j}}{a_n},$$

where S_j is the jth elementary symmetric function of the zeros $r_1, r_2,.., r_n$.

Proof: Since $r_1, r_2,..., r_n$, are the zeros of $f(x)$, $(x - r_1), (x - r_2),... (x - r_n)$ divide $f(x)$, and $f(x) = a_n (x - r_1)(x - r_2) ... (x - r_n)$

or $a_n x^n + a_{n-1} x^{n-1} + a_{n-2} x^{n-2} +....+ a_1 x + a_0$

$= a_n [x^n - (\Sigma r_1) x^{n-1} + (\Sigma r_1 r_2) x^{n-2} +...+ (-1)^n r_1 r_2 . r_n]$

$= a_n [x^n - S_1 x^{n-1} + (-1)^2 S_2 x^{n-2} +....+ (-1)^n S_n].$

Comparing the coefficients of like powers of x on both sides, we obtain

$a_{n-1} = -a_n S_1, S_1 = \dfrac{-a_{n-1}}{a_n}$

$a_{n-2} = (-1)^2, S_2, a_n S_2 = (-1)^2 a_{n-2}/a_n,$

$a_{n-j} = (-1)^j a_n S_j, S_j = (-1)^j a_{n-j}/a_n.$

This completes the proof of the theorem.

Exercise 7.4.3: If the sum of two zeros of the polynomial $x^3 + px^2 + qx + r$ is equal to the third zero, prove that

$$p^3 - 4pq + 8r = 0.$$

Solution: Let r_1, r_2, r_3, be zeros of $x^3 + px^2 + qx + r = 0$, such that $r_1 + r_2 = r_3$, then

$$r_1 + r_2 + r_3 = -p \quad ...(1)$$

$$r_1 r_2 + r_2 r_3 + r_1 r_3 = q \quad ...(2)$$

$$r_1, r_2, r_3 = -r \quad ...(3)$$

Since $r_1 + r_2 = r_3$, the identity (1) gives

$2r_3 = -p$ or $r_3 = -p/2$.

From (2), $r_1 r_2 + (r_1 + r_2) r_3 = q$

or $$r_1 r_2 + r_3 \cdot r_3 = q$$

or $$r_1 r_2 + \frac{p^2}{4} = q$$

$$r_1 r_2 = q - p^2/4.$$

From (3), we get

$$r_1 r_2 r_3 = -r$$

$$(q - p^2/4)(-p/2) = -r$$

or $$(4q - p^2)(-p) = -8r$$

$$p^3 - 4pq + 8r = 0.$$

PROBLEMS

1. Let $f = (-1, 0, 2, 0,...)$ and $g = (0, 2, -3, 0,...)$ be two polynomials over the integral domain $(Z, +, .)$ of integers. Find

   (a) $f + g$; (b) $f . g$; (c) f^2; (d) $f^2 . g^2$; (e) g^3.

2. Let $U = (1, 2, 0,...)$ and $V = (3, 4, 0...)$ be two polynomials over the field $(Z_5, +_5, \odot_5)$. Find

   (a) $u + v$; (b) $u . v$; (c) u^2;

   (d) v^2; (e) $u^2 . v^2$; (f) $= u^3$.

3. Prove that the system $(T, +, .)$ where T is the set of all polynomials in $Z[x]$ which have zero for all coefficients other than constant term, + and. denote addition and multiplication of polynomials, forms a ring. Note the constant term may be zero or non-zero. That is,

   $T = \{(a, 0, 0,....) \mid a \in Z\}$.

4. Let $U = (3, 0, 0, 2, 0,...)$, $V = (0, 1, 2, 0,...)$, and $W = (0, 0, 4, 0,...)$ be polynomials in $Z_5[x]$ for the field $(Z_5, +_5, \odot_5)$ of integers modulo 5.

   Check by direct computation that these polynomials satisfy each of the following identities;

   (a) $u + (v + w) = (u + v) + w$; (b) $u (v . w) = (u . v) . w$;

   (c) $u. (v + w) = u . v + u . w$; (d) $(u + v) . w = u . w + v . w$.

5. Let $(D, +, .)$ be a field and let S be the subset of $D[x]$ consisting of all polynomials which have degree zero or no degree. That is, $S = \{(a, 0, 0,..., 0,...)\ a \in D\}$. Prove that each non-zero polynomial in S has an inverse in S.

Note that these are the only polynomials in $D[x]$ which have inverses. And the system $(S, +, .)$ is a field isomorphic to the field $(D, +, .)$.

6. Prove that $f(x) = x^2 + 6$ is reducible over Z_7 and Z_{11}, and over the real numbers. To prove this, merely show a correct factorization of $f(x)$.

7. Decide whether $f(x) = 2x^2 + 2x - 2$ is reducible or irreducible over each of the following integral domains. Do not show proof, but give the factors in each case where $f(x)$ is reducible.

(a) $(Z, +, .)$, the domain of integers.

(b) $(K, +, .)$, the field of real numbers.

(c) $(Q, +, .)$, the field of rational numbers.

(d) $(C, +, .)$, the field of complex numbers.

8. Factor (a) $6x^2 + 2x + 6$,

(b) $3x^2 + 2x + 2$,

(c) $x^2 + 3$

over the field $(Z_7, +_7, \odot_7)$.

9. Find the *GCD.* $(f(x), g(x))$ of $f(x)$ and $g(x)$ in each of the following cases, and express it in the form $m(x) \, . \, f(x) + n(x) \, . \, g(x)$, when $f(x), g(x) \in R[x]$, where $(R, +, .)$ is the field of rational numbers:

(a) $f(x) = 6x^3 + 5x^2 - 2x + 35$ and $g(x) = 2x^2 - 3x + 5$.

(b) $f(x) = x^3 + 1$ and $g(x) = x^2 + 3x - 5$

when $f(x)$ and $g(x) \in Z_5[x]$, where $(Z_5, +_5, \odot_5)$ is the field of integers module 5.

(c) $f(x) = 3x^3 - 2x^2 + 4$ and $g(x) = x^4 + 3x^2 + 1$.

(d) $f(x) = 2x^2 + x + 2$ and $g(x) = x^2 + x + 4$.

10. Find the rational zeros and the decomposition of the polynomials $f(x) = 4x^4 + 8x^3 + 7x^2 + 3 \in Q[x]$ over the field $(Q, +, .)$ of rational numbers.

11. Let $(F, +, .)$ be a subfield of the field $(K, +, .)$, and suppose $f(x), g(x) \in F[x]$ are relatively prime in $F[x]$. Prove that they are relatively prime in $K[x]$.

12. Prove that $(F[x]/(x^2 + x + 4), +, .)$ over the field $(Z_{11}\ +_{11}, \odot_{11})$ of integers modulo 11 is a field.

13. Let $(F, +, .)$ be a field of real numbers. Prove that $(F[x]/(x^2 + 1), +, .)$ is a field isomorphic to the field of complex numbers.

14. If $f(x)$ is in $F[x]$, where $(F, +, .)$ is the field of integers modulo p, p *a* prime, and $f(x)$ is irreducible over F of degree n prove that $(F[x]/(f(x)), +, .)$ is a field with p^n elements.

15. For an arbitrary ring $(R, +, .)$ with identity, prove that

(a) the polynomial $1 \, . \, x^n \in$ Cent $R[x]$,

(b) If $(A, +, .)$ is an ideal of $(R, +, .)$, then $(A[x], +, .)$ is an ideal of the polynomial ring $(R[x], +, .)$

(c) If $(R, +, .) \cong (R', +', .')$, then $(R[x], +, .) \cong (R'[x], +', .')$

16. Let $(D, +, .)$ be an integral domain, show that the characteristic of $(D[x], +, .)$ is equal to the characteristic of $(D, +, .)$.

17. Prove that no monic polynomial can be a zero divisor in $(R[x], +, .)$.

18. Show that the relation $\sim$ defined by taking $f(x) \sim g(x)$ if and only if

$\deg f(x) = \deg g(x)$.

is an equivalence relation in the set of non-zero polynomial *of* $R[x]$.

19. *(a)* Let P be the set of all polynomials in $Z[x]$ with constant term 0:

$p = \{a_1 x + a_2 x^2 + ... + a_n x^n \mid a_k \in Z, n \geq 1\}$.

Show that the triple $(P, +, .)$ is a prime ideal of $(Z(x), +, .)$.

(b) Show that principal ideal $((1.x), +, .)$ is a prime ideal of the ring $(Z[x], +, .)$ but not a maximal ideal.

20. If $(R, +, .)$ is a commutative ring with identity, then $1 + ax$ is invertible in $R[x]$ if and only if a is nilpotent.

21. Let an element r of the commutative ring $(R, +, .)$ with identity be fixed.

(a) If $R(r)$ denotes the set

$$R(r) = \{f(r) \mid f(x) \in R[x]\},$$

Prove that the triple $(R(r), +, .)$ is a sub ring of $(R, +, .)$.

(b) Show that the mapping $\phi(f(x)) = f(r)$ is a homomorphism of $(R[x], +, .)$ onto $(R(r), +, .)$.

22. Define the derivative $f'(x)$ of the polynomial

$$f(x) = a_0 + a_1 x + ... + a_n x^n, \; a_n \neq 0.$$

as $$f'(x) = a_1 + 2a_2 x + 3a_3 x^2 + ... + na_n x^{n-1}.$$

Prove that if $f(x) \in F[x]$, where $(F, +, .)$ is a field of rational numbers, then $f(x)$ is divisible by the square of a polynomial if and only if $f(x)$ and $f'(x)$ have a greatest common divisor $d(x)$ of positive degree.

CHAPTER 8

FACTORIZATION IN INTEGRAL DOMAINS

In this chapter we shall confine our attention on the factorization of elements of integral domains only. If instead of integral domains we choose the elements of some field for the problem of factorization, then we observe that since all elements of a field have multiplicative inverse, every non-zero element divides other element. If $a \neq 0$, $b \neq 0$ are elements of some field, then $b = a \,.\, (a^{-1} b)$ which shows a divides b. But it partly happens when the elements a, b are of some integral domain because it has no zero divisors and we do not assume the multiplicative inverses. Thus the theory of arithmatic play an important role in the study of its structure.

8.1 DIVISIBILITY

Definition 8.1.1: Let $(D, +, .)$ be an integral domain with $a, b \in D$. Then we say that a divides b, denoted by $a \mid b$, or b is a multiple of a, or a is a factor of b, if $a \neq 0$, and if there exists $c \in D$ such that $b = ac$. Wc write $a \mid b$ if a does not divide b.

Theorem 8.1.1: *Let $(D, +, .)$ be an integral domain with $a, b, c \in D$.*

Then

(i) $a \mid a$, *for all* $a \in D$, *(Reflexive),*

(ii) $a \mid b$, $b \mid c \Rightarrow a \mid c$, *(Transitive),*

(iii) $a \mid b \Rightarrow a \mid b\, x,\ \forall\, x \in D$,

(iv) $a \mid b$, $a \mid c \Rightarrow a \mid b \pm c$.

The proofs of above results are very simple and left to the reader.

Remarks: *(i)* Since $0 \,.\, a = 0$, then $a \mid 0$. Thus in integral domain $(D, +, .)$ every element $a \in D$ divides 0.

Definition 8.1.2: If $(D, +, .)$ is an integral domain, $a \in D$, and $a \mid I$ we say that a is unit of D. That is, those elements of integral domain are called units which have multiplicative inverses. Clearly in any integral domain the multiplicative identity 1 and its additive inverse -1, are units.

Example 8.1.1: In the integral domain $(Z, +, .)$ of integers there are only two units 1 and -1.

Example 8.1.2: In the integral domain $(D, +, .)$, where $D = \{a + ib] \; a, b \in Z\}$, $i^2 = -1$, the units are $1, -1, i, -i$.

Example 8.1.3: In the integral domain $(D, +, .)$ of rational numbers every non zero rational has its inverse a rational number. So every non-zero element of this domain is a unit in *D*.

Example 8.1.4: In the integral domain whose elements are $\{a + b\sqrt{2} \mid a, b \in Z\}$, the units are $1, -1, 1 + \sqrt{2}, -1 + \sqrt{2}, (1 - \sqrt{2}), -(1 - \sqrt{2})$. The units consists of elements of the form $a + b\sqrt{2}$ where $a^2 - 2b^2 = \pm 1$.

For example $(1 + \sqrt{2})(-1 + \sqrt{2}) = 1$

Theorem 8.1.2: *If $a \mid b$ and $b \mid a$, then $a = b\,u$, where u is a unit.*

Proof: $a \mid b \Rightarrow b = a \,.\, c$, for some $c \in D$,

$b \mid a \Rightarrow a = b \,.\, d$, for some $d \in D$,

Hence $b \,.\, 1 = b = a \,.\, c. = (b \,.\, d) \,.\, c = b\,(dc)$.

Since $b \neq 0$, then by cancellation Law we have

$1 = dc \Rightarrow d \mid 1$. This show *d* is unit, say, *u*. Hence $a = b \,.\, u$.

Definition 8.1.3: Let $(D, +,.)$ be an integral domain with $a, b \in D$. An element *a* is called an *associate of b* if $a = b.\, e$, where *e* is unit in *D*.

For Example: In the set of integers $-n$ and n are associate of any $n \in Z$.

Example 8.1.5: In the integral domain of example 8.1.2, 5, –5, $5i$ and $-5i$ are associates of 5.

Example 8.1.6: In the integral domain of example 8.1.3, every number except 0 is a unit and therefore any two non-zero numbers are associates.

Example 8.1.7: In the integral domain of example 8.1.4, the associate of $3 + 5\sqrt{2}$ are $\pm(3 + 5\sqrt{2})$, $\pm(3 + 5\sqrt{2})(1 + \sqrt{2})$, and $\pm(3 + 5\sqrt{2})(1 - \sqrt{2})$.

Theorem 8.1.3: *If u and u, are units of an integral domain $(D, +,.)$, then u,. u′ is a unit in D.*

Proof: since u and u' are units, then

$u \mid 1$ and $u' \mid 1$. Then there exists $c, c' \in D$

such that $uc = 1$ and $u'c' = 1$.

Hence $(uc)(u'c') = (uu')(cc') = 1 \Rightarrow uu' \mid 1$.

This proves the theorem.

Theorem 8.1.4: *The relation $(\sim)$ on D, defined by $a \sim b$ if and only if a and b are associates, is an equivalence relation.*

Proof: (1) Since $1 = 1.a$, $a \sim a$, where 1 is the unit of any integral domain *D*. Hence $\sim$ is reflexive.

(2) $a \sim b \Rightarrow a = b \,.\, e$ where e is unit in D. Therefore there exists $c \in D$ such that $e \,.\, c = c \,.\, e = 1$. Thus $c \mid 1$ and c is also a unit of D. But

$a \,.\, c = (b\ e) \,.\, c = b. (e \,.\, c) = b \,.\, 1 = b \Rightarrow b \sim a$ Hence $\sim$ is symmetric.

(3) $a \sim b,\ b \sim c \Rightarrow a = bu,\ b = cd,$ where u and d are units of D.

$\Rightarrow a = (cd) \,.\, u = c\ (du)$

$\Rightarrow a \sim c$ since du is a unit by theorem 8.1.3.

Theorem 8.1.5: *The positive integer n is a factor of the positive integer m if and only if $(n) \supseteq (m)$.*

Proof: Let $(n) \supseteq (m)$. then the set (n) generated by n contains all elements which are multiple of m. Particularly, $m \in (n)$, that is, m is a multiple of n.

Conversely, Let n divide m. then there exists an element $q \in Z$ such that

Then $$m = n \,.\, q.$$

So that $$am = a \,.\, (n.q) = (a \,.\, q) \,.\, n$$

Example 8.1.8: $(5) = \{0, \pm 5, \pm 10, \pm 15, \pm 20, \pm 25, \pm 30, , .\}$ and $(10) = \{0, \pm 10 \pm 20, \pm 30,.....\}$.

It is clear $(5) \supset (10)$, and $5 \mid 10$.

Remark: If $a \mid b$, then $(a) \supseteq (b)$, and if $b \mid a$, then $(b) \supseteq (a)$. Hence if a and b are associates, then $(a) = (b)$.

Definition 8.1.4: Let $(D, +, .)$ be an integral domain. Then $b \in D$ is called irreducible if $b \neq 0$ and b is not a unit, but $a \mid b \Rightarrow a$ is a unit or an associate of b. Otherwise b is called reducible.

Definition 8.1.5: In an integral domain an element $a \neq 0$ is called prime if it is not unit and if when ever $a \mid bc$, then $a \mid b$ or $a \mid c$.

Example 8.1.9: Let $(R, +, .)$ be an integral domain, then the polynomial ring $(R\ [x], +, .)$ is also an integral domain. Now if we consider $x^2 + x + 1 \in R\ [x]$, $2x + 5 \in R\ [x]$ and $7 \in R\ [x]$, then these are all prime elements in $R\ [x]$. But $x^2 + 5x + 6 = (x + 2)\ (x + 3)$, $2x + 6 = 2\ (x + 3)$, and $8 = 2 \,.\, 4$, so $x^2 + 5\,x + 6$, $2x + 6$, and $8 \in R\ [x]$ are not prime elements in $R\ [x]$.

Example 8.1.10: Let $(D, +, .)$ be an integral domain, where $D = \{a + b\ \sqrt{5} \mid a, b \in Z\}$. Now first of all we determine the units of D.

To do this we define the norm of an element $\alpha \in D$ by

$N(\alpha) = a^2 - 5\ b^2$, where $\alpha = a + b\ \sqrt{5} \in D$.

If $\beta = c + d\ \sqrt{5} \in D$, then we can see that

$N\ (\alpha \,.\, \beta) = N\ (\alpha) \,.\, N\ (\beta)$.

Now if $\alpha = (a + b\ \sqrt{5})$ is unit then there exists $\beta = c + d\ \sqrt{5}$ such that

$$\alpha \,.\, \beta = (a + b\ \sqrt{5}) \,.\, (c + d\ \sqrt{5}) = 1.$$

Hence $N\ (\alpha \,.\, \beta) = N\ (1) = 1$. Thus, we have

$N(\alpha . \beta) = N(\alpha) . N(\beta) = (a^2 - 5b^2) . (c^2 - 5b^2) = 1$. This means that if α is a unit, then $(a^2 - 5b^2) . (c^2 - 5d^2) = 1$.

Therefore if $\alpha = a + b\sqrt{5}$ is a unit then $N(\alpha) = \pm 1$, if $N(\alpha) = 1$, then $(a + b\sqrt{5})(a - b\sqrt{5}) = a^2 - 5b^2 = 1$, and if $N(\alpha) = -1$, then $(a + b\sqrt{5}) . (-a + b\sqrt{5}) = -(a^2 - 5b^2) = 1$ and if $N(\alpha) = 1$, $\alpha | 1$ so that α is a unit of D.

Thus we have drawn the result that if

$\alpha = a + b\sqrt{5}$ is unit of D, then $N(\alpha) = \pm 1$. From this we conclude that

(1) $1, -1, 2 - \sqrt{5}, 2 + \sqrt{5}, -2 - \sqrt{5}, -2 + \sqrt{5}$ are all units in D.

(2) $1 + \sqrt{5}$ and $-3 + \sqrt{5}$ are not units, since

$(2 - \sqrt{5})(1 + \sqrt{5}) = -3 + \sqrt{5}$.

(3) $1 + \sqrt{5}$ and $-3 + \sqrt{5}$ are associates, we now prove that

(4) $2, 1 + \sqrt{5}$, and $-1 + \sqrt{5}$ are all primes of D. Suppose 2 is not prime in D, then $2 = \alpha . \beta$, where α and β are not units. So $4 = N(2) = N(\alpha . \beta) = N(\alpha) . N(\beta) \Rightarrow N(\alpha) = \pm 2$ and $N(\beta) = \pm 2$. But if $N(\alpha) = a^2 - 5b^2 = \pm 2$, then $a^2 \cong \pm 2 \pmod 5$ and this congruence equation does not have any solution. This shows 2 is prime. Similarly we can show that $1 + \sqrt{5}$ and $1 - \sqrt{5}$ are primes of D.

Further more 2 is not associate of $1 + \sqrt{5}$ since if $2 = (a + b\sqrt{5})(1 + \sqrt{5})$ where $a + b\sqrt{5}$ is a unit of D, we have $2 = (a + 5b) + (a + b)\sqrt{5}$. Hence $a + 5b = 2$ and $a + b = 0$. Thus $a = -b$ and $4b = 2$ or $b = 1/2$. This means that $a + b\sqrt{5} \notin D$. Hence 2 is not associate of $1 + \sqrt{5}$.

We observe that

(5) $4 = 2.2 = (1 + \sqrt{5})(-1 + \sqrt{5})$. Thus we have two factorization of 4 in D into primes that are not associates. It means that the factorization into primes is not unique in every integral domain.

Example 8.1.11: In the integral domain $(D, +, .)$ of Gaussian integers $\{a + ib | a, b \in Z\}$, $1 + i$ is irreducible element. If $a + ib | 1 + i$, then $(1 + i) = (a + ib)(c + id)$, for proving $1 + i$ irreducible we must show that either $a + ib$ is a unit or $c + id$ is a unit. So we have $1 - i = (a - ib)(c - id)$. Multiplying $1 + i$ and $1 - i$

we obtain $(1 + i)(1 - i) = (a + ib)(a - ib)(c + id)(c - id)$

or $2 = (a^2 + b^2)(c^2 + d^2)$ which implies either $a^2 + b^2 = 2$ or $a^2 + b^2 = 1$. Similarly either $c^2 + d^2 = 1$ or $c^2 + d^2 = 2$. If $a^2 + b^2 = 1$, then $(a + ib)(a - ib) = 1 \Rightarrow a + ib$ is a unit. If $c^2 + d^2 = 1$, then $(c + id)(c - id) = 1 \Rightarrow c + id$ is a unit. This shows that $1 + i$ is irreducible.

Lemma 8.1.1: *In an integral domain $(D, +, .)$ with unity every prime p is irreducible.*

Proof: Let p be a prime element. Let $a \in D$ divide p,

$a | p \Rightarrow p = a . b \Rightarrow p | a$ or $p | b$.

If $p | a$, then a is an associate of p,

and if $p | b$, then $b = cp$.

So $1 \,.\, p = p = a \,.\, b = a\,(cp) = (a \,.\, c) \,.\, p$. Since $p \neq 0$, by cancellation law $1 = a\,c \Rightarrow$ a is unit. Hence $p = b \,.\, a$. Hence p is irreducible.

The converse of the lemma is not true.

It is shown by an example given below:

Example 8.1.12: We consider the integral domain $(D, +, .)$ where $D = \{a + b\sqrt{5} \mid a, b \in Z\}$. We have seen that $1 + \sqrt{5}$ is a prime element of D.

But $1 + \sqrt{5}$ is not irreducible, for if $1 + \sqrt{5} = \alpha \,.\, \beta$, where neither α nor β is unit.

Then we have to prove that $N(1 + \sqrt{5}) = N(\alpha).\,N(\beta) \Rightarrow -4 = N(\alpha)\,.\,N(\beta) \Rightarrow$ either $N(\alpha) = +2$ or $N(\alpha) = -2$.

If $N(\alpha) = +2$, then $a^2 - 5b^2 = 2 \Rightarrow \alpha = a + b\sqrt{5}$ is not unit, if $N(\alpha) = -2 = a^2 - 5b^2 = -2 \Rightarrow \alpha$ is not unit, because α is a unit if $a^2 - 5b^2 = \pm 1$.

Similarly β is not a unit. This shows $1 + \sqrt{5}$ is not irreducible.

Lemma 8.1.2: *Every irreducible element of principal ideal domain (PID) is prime.*

Proof: Let p be an irreducible element of a *PID* $(R, +, .)$. Since p is irreducible, $p \neq 0$, p is not unit, then we shall show that if $p \mid a.b$ and if $p \nmid a$, then $p \mid b$. Since $(R, +, .)$ is principal ideal domain the ideal $(p)+(b) = (d)$ for some $d \in R$. This implies $(p) \subseteq (d)$ and $p = d \,.\, r$ for some $r \in R$. But p is irreducible, so either d is a unit or r is a unit. Suppose d is a unit, there exists x and y in R such that

$px + by = 1 \Rightarrow a = ap \,.\, x + a\,by \Rightarrow p \mid a$, since $p \mid apx$ and $p \mid ab$, which is a contradiction to our supposition. This means that d is not a unit but r is a unit. Then $p = d \,.\, r \Rightarrow d = p \,.\, r^{-1} \Rightarrow (d) = (p)$. Therefore $(p) + (b) = (p) \Rightarrow (b) \subseteq (p) \Rightarrow p \mid b$. This proves the lemma.

Theorem 8.1.6: *Let $(R, +, .)$ be a PID. Then any proper ideal $((m), +, .)$ is maximal if and only if m is irreducible.*

Proof: By the lemma 8.1.2 we have every irreducible element in *PID* is prime. So our problem is that $((m), +, .)$ is maximal if and only if m is prime.

Let $((m), +, .)$ be maximal ideal in *PID* $(R, +,.)$. If m is not prime, then $m = 1 \Rightarrow (m) = R$, which is ruled out by the fact that a maximal ideal is proper, or

$m = q \,.\, n$, where

$1 < q \leq n < m \Rightarrow (1) \supset (q) \supset (n) \supset (m)$

this is a contradiction to the maximality of (m). So clearly this implies that m is prime.

Conversely, suppose m is prime then if (m) is not maximal, either $(m) = (1)$ which is ruled out since 1 is not multiple of m, or there is a proper ideal (n) such that

$(1) \supset (n) \supset (m)$.

Clearly this implies that

$m = n \,.\, q$, where $q > 1$,

which contradicts the assumption that m is prime. Hence (m) is maximal.

Theorem 8.1.7: *Let $(D, +, .)$ be an integral domain and let $a \in D$. $(D/(a), +, .)$ has zero divisors if a is not irreducible. If a is a prime, the $D/(a)$ is an integral domain.*

Proof: If a is not irreducible, then $a = x \,.\, y$, where x and y are not units and they are not also associates of a. This means that $x \notin (a)$ and $y \notin (a)$.

Now we define the natural homomorphism

$\phi: D \to D/(a)$ by

$\phi(x) = x + (a), \forall\, x \in D.$

Then

$a + (a) = xy + (a) = (x + (a)) \,.\, (y + (a))$

$\Rightarrow \bar{0} = (a) = \bar{x} \,.\, \bar{y}$, and since $x \notin (a)$ and

$y \notin (a)$, $\bar{x} \neq \bar{0}$, $\bar{y} \neq \bar{0}$, where $\bar{x} = x + (a)$ and

$\bar{y} = y + (a)$

Hence $\bar{x}$ is a zero divisor.

Now suppose that a is prime and

$\bar{u} \,.\, \bar{v} = (u + (a)), (v + (a)) = (a) = \bar{0}$, then

$\bar{0} = \bar{u} \,.\, \bar{v} = \overline{u \,.\, v},$, that is $u \,.\, v + (a) = (a) \Rightarrow u \,.\, v \in (a)$

$\Rightarrow a \mid u \,.\, v$ and a is prime, then $a \mid u$ or $a \mid v$. Hence either $u \in (a)$ or $v \in (a)$ which implies either $\bar{u} = \bar{0}$ or $\bar{v} = \bar{0}$. Hence if a is prime, then $(D/(a), +, .)$ is an integral domain. Hence the theorem.

Definition 8.1.6: Let $(D, +,.)$ be an integral domain. Then we call $d \in D$ a **greatest common divisor** *(GCD)* of two elements $a, b \in D$ $(a, b \neq 0)$ if

(1) $d \mid a$ and $d \mid b$

and

(2) if $c \mid a$ and $c \mid b$, then $c \mid d$.

The greatest common divisor of a and b is denoted by (a, b).

Definition 8.1.7: *Let $(D, +, .)$* be an integral domain and let a and b be non zero elements of D. Then we call an element $l \in D$ a **least common multiple** *(LCM)* of a and b if

(1) $a \mid l$ and $b \mid l$, and

(2) if $a \mid m$ and $b \mid m$, then $l \mid m$.

The least common multiple of a and b is denoted by $[a, b]$.

Lemma 8.1.3: *If b and c have a greatest common divisor or a least common multiple, each is unique to within multiplication by unit.*

Proof: Let d and d' be two greatest common divisor of b and c then we have $d|b$, $d|c$, $d'|d$ when d is a greatest common divisor and $d'|b$, $d'|c$, $d|d'$, when d' is a greatest common divisor. Thus $d|d'$ and $d'|d \Rightarrow d$ and d' are associates. Hence $d' = de$, where e is a unit.

A similar proof can be given for least common multiple of b and c.

Theorem 8.1.8: *Let a and $b \neq 0$ belong to an integral domain $(D, +, .)$. Let $a = bq + r$. If b and r have a greatest common divisor d, then d is a greatest common divisor of a and b.*

Proof: Since d is a greatest common divisor of b and r, $d|b$ and $d|r$.

From $a = bq + r$ it is clear that $d|a$. Thus d is a common divisor of a and b.

From $r = a - b \,.\, q$ we have that any common divisor, c, of a and b must divide r which implies c is a common divisor of b and r. So c will divide the greatest common divisor d of b and r. This means that $c|d$, the common divisor of a and b. This proves d is the greatest common divisor of a and b.

CHAPTER 8

Theorem 8.1.9: *Let $(D, +,.)$ be a principal ideal domain (PID). Then every pair of non-zero elements has (1) a greatest common divisor, (2) a least common multiple.*

***Proof* (1):** Let $a, b \in D$. Since $(D, +,.)$ is principal ideal domain, $((a), +, .)$, $((b), +, .)$ and $((a) + (b), +, .)$ are principal ideals. So there exists an element d such that

$((a) + (b), +, .) = ((d), +, .)$.

Now we shall show that $d = (a, b)$.

Now we have that

$$(a) + (b) = (d) \Rightarrow (d) \supseteq (a) \text{ and } (d) \supseteq (b)$$
$$\Rightarrow d|a \text{ and } d|b \Rightarrow d \text{ is a common divisor of } a \text{ and } b.$$

Let h be another divisor of a and b, that is,

$$h|a \Rightarrow (h) \supseteq (a),$$

and

$$h|b \Rightarrow (h) \supseteq (b).$$

This means that $(h) \supseteq (a) \cup (b)$. Since $((h), +, .)$ is an ideal containing $(a) \cup (b)$, the ideal $((h), +, .)$ will contain the ideal $((a) + (b), +, .)$ because it is the smallest ideal containing $(a) \cup (b)$. Thus

$(h) \supseteq (a) + (b) = (d) \Rightarrow h|d$.

This shows $d = (a, b)$ is the greatest common divisor of a, *and* b.

***Proof* (2):** Since $(D, +, .)$ is *PID*, then $((a), +, .)$ and $((b), +, .)$ and $((a) \cap (b), +, .)$ are principal ideals.

So there exists some elements $l \in D$ such that

$(a) \cap (b) = (l)$, where $((l), +,.)$ is also principal ideal.

This implies (l), $\subset (a)$ and $(l) \subset (b) \Rightarrow a|l$ and $b|l \Rightarrow l$ is a common multiple.

Let m be another common multiple of a and b. That is,

$$a|m \Rightarrow (m) \subset (a)$$

$$b \mid m \Rightarrow (m) \subset (b) \Rightarrow (m) \subset (a) \cap (b) = (l)$$
$$\Rightarrow l \mid m$$

This shows that l is the least common multiple.

Remark: If $d = (a, b)$, then we have

$(d) = (a) + (b)$. Then for some $x, y, \in D$

We have

$$d = ax + by$$

If the greatest common divisor (a, b) of a and b is 1. Then a and b are called co-prime and

$1 = ax + by$ for some $x, y \in D$.

PROBLEMS

1. In commutative ring show that no unit is a zero divisor.

2. Find the units in the integral domain $(Z_p, +_p, \odot_p)$, prime.

3. Let $(D, +, .)$ be an integral domain. Show that

(*i*) If $b \mid c$ then b is a greatest common divisor and c is a least common multiple of b and c,

(*ii*) If $1 = (a, b)$, then $(a\, e, b\, e) = e$.

4. In an integral domain $(D, +,.)$, where $D = [a + b\sqrt{2} \mid a, b \in Z\}$, Let $g\,(a + b\sqrt{2}) = \lvert a^2 - 2b^2 \rvert$.

(1) Show that if $\alpha, \beta \in D$, then $g\,(\alpha\,.\,\beta) = g\,(a), g\,(\beta)$.

(2) Show that $1 + 2\sqrt{2}$ is an irreducible element in D.

(3) Show that there are infinite number of units in D.

5. Prove that in an integral domain with unity, every pair of non-zero elements has *GCD* if and only if every pair of non-zero elements has *LCM*.

6. In an integral domain if $a\,b \mid a\,c$, then prove that $b \mid c$.

7. If non-zero elements b and c have *GCD* (b, c) and an *LCM* $[b, c]$, then b, c and $[b, c]$ (b, c) are associates. Moreover if b and c have an *LCM*, then $[b, c] \mid bc$ and $bc / [b, c]$ is a greatest common divisor of b and c.

8. Prove that an integral domain $(R, +, .)$ with unity is a field if and only if $(R\,[x], +, .)$ is *PID*.

9. If $(A, +, .)$. $(B, +, .)$, and $(C, +, .)$ are non-zero ideals in a ring $(R, +, .)$, show that if $(R, +, .)$ is *PID* then

(1) $A \cap (B + C) = A \cap B + A \cap C$

(2) $A + (B \cap C) = (A + B) \cap (A + C)$

(3) $A\,(B \cap C) = AB \cap AC$

(4) $A = (a)$ and $B = (b) \Rightarrow AB = (ab)$.

10. Two non-zero ideals $(A, +, .)$ and $(B, +, .)$ of a ring $(R, +, .)$ are said to be co-maximal if $A + B = R$, If $(R, +, .)$ is PID, then show that $AB = A \cap B$ if and only if $(A, +, .)$ and $(B, +, .)$ are co-maximal.

11. In an integral domain if prime p divides another prime q, then show that p and q are associates.

12. In a commutative ring with identity show that an associative of a prime (irreducible) elements is prime (irreducible).

8.2 EUCLIDEAN DOMAINS

Let $a, b \in Z$, the set of integers. If $b \neq 0$, then there exists two unique integers q and r such that $a = bq + r$ and $0 \leq r < |b|$. This is known as Division Algorithm. Here we generalize this concept.

Definition 8.2.1: An Euclidean Domain $(D, +, ., d)$ is an integral domain $(D, +, .)$ and a function d from $D - \{0\}$ into the set Z_0 of non-negative integers such that

1. For all pairs $a, b \in D$ with $b \neq 0$ there exists q and r in D such that $a = bq + r$ and either $r = 0$ or $d(r) < d(b)$.
2. For all $a \neq 0, b \neq 0 \in D$,
   $d(a) \leq d(a \,.\, b)$.
3. $d(a) \geq 0$ for all $a \in D - \{0\}$.

We do not assign any value for $d(0)$. The condition 1 is known as Euclidean Algorithm.

Example 8.2.1: The integral domain $(Z, +, .)$ of integers is an Euclidean domain $(Z, +, ., d)$, where the function d is defined by $d(a) = |a|, \forall\, a \in Z - \{0\}$. This function d satisfies all the three conditions.

Example 8.2.2: The every field $(F, +, .)$ is an Euclidean domain $(F, +, ., d)$, where the function d is defined, by

$d(a) = 1$ for all $a \in F - \{0\}$. For any $a \neq 0, b \neq 0 \in F$,

$ab \neq 0$. So $d(a \,.\, b) = 1 = d(a)$. The condition 2 is satisfied.

For any two non-zero elements a, b of F, we have

$a = (a \,.\, b^{-1}) \,.\, b + 0$ which implies that the condition (1) is satisfied.

Example 8.2.3: The field of rational numbers $(Q, +, .)$ is not an Euclidean domain $(Q, +, ., d)$ if d is defined by

$d(a) = |a|$.

If it is an Euclidean domain, then

For $3/2 \in Q$, we have

$$d\left(\frac{3}{2}\right) = \left(\frac{3}{2}\right) \leq d\left(\frac{3}{2} \cdot \frac{2}{3}\right) = \left|\frac{3}{2} \cdot \frac{2}{3}\right| = 1, \text{ which is not true.}$$

This shows that the system $(Q, +, ., d)$ is not Euclidean domain.

Example 8.2.4: The integral domain $(Z_i), +, .)$ of Gaussian integers is an Euclidean domain $(Z(i); +, ., d)$, where $Z(i) = (a + ib \mid a, b \in Z\}$ and the function d is defined by

$$d(\alpha) = \alpha \,.\, \bar{\alpha}$$

or $d(a + ib) = a^2 + b^2 = (a + ib)(a - ib), \forall\ a + ib \in Z(i)$.

For this we certainly have $d(0) = 0$ if and only if

$\alpha = a + ib = 0 + i\,0 = 0$. Hence if $\alpha \neq 0$, $d(\alpha) \geq 1$.

For all $\alpha = a + ib$ and $\beta = c + id$, we have

$$d(\alpha \,.\, \beta) = (\alpha \,.\, \beta)\left(\overline{\alpha \,.\, \beta}\right)$$

$$= (\alpha \,.\, \beta)\left(\bar{\alpha}\,,\bar{\beta}\right)$$

$$= (\alpha \,.\, \bar{\alpha}) \,.\, \left(\beta \,.\, \bar{\beta}\right) = \left(a^2 + b^2\right) . \left(c^2 + d^2\right) = d(\alpha) \,.\, d(\beta).$$

It is clear that $(a^2 + b^2) < (a^2 + b^2) \,.\, (c^2 + d^2)$, as $c^2 + d^2 \geq 1$ which implies $d(\alpha) \leq d(\alpha) \,.\, d(\beta) = d(\alpha \,.\, \beta)$. Hence condition (2) is satisfied.

Now we prove condition (1) as follows.

For given α and $\beta \neq 0$, there must exist σ and ρ such that

$\alpha = \beta.\ \sigma + \rho$, where either $\rho = 0$ or $d(\rho) < d(\beta)$.

Since $\beta \neq 0$, we have

$\rho = \alpha - \beta \,.\, \sigma = \beta(\alpha/\beta - \sigma)$ and in the field of complex numbers, since $d(\rho) = \rho.\rho = |\rho|^2$,

$|\rho|^2 = |\beta|^2\, |\alpha|\beta - \sigma|^2 \Rightarrow d(\rho) < d(\beta)$ if $|\alpha/\beta - \sigma|^2 < 1$.

Now we have to choose $\sigma \in Z(i)$ so that $|\alpha/\beta - \sigma|^2 < 1$.

Let $\alpha = a + ib$, $\beta = c + id$. So

$$x + iy = \alpha/\beta = \frac{a+ib}{c+id} = \frac{(a+ib)(c-id)}{(c+id)(c-id)}$$

$$= \frac{(ac+bd)+i(bc-ad)}{c^2+d^2}$$

$$= \frac{ac+bd}{c^2+d^2} + i\frac{bc-ad}{c^2+d^2}.$$

which shows x and y are rational numbers. Here we can choose two integers e and f such that

$|x - e| \leq |\dfrac{1}{2}$ and $||y - f| \leq \dfrac{1}{2}$ and let

$\sigma = e + fi$. Then $\sigma \in Z(i)$ and

$|\alpha/\beta - \sigma|^2 = |x + iy - (e + fi)|^2$

$$= |(x - e) + i\,(y - f)|^2$$

$$= (x - e)^2 + (y - f)^2 \leq \frac{1}{4} + \frac{1}{4} = \frac{1}{2} < 1$$

Thus we have found σ for which

$|\alpha|\beta - \sigma|^2 < 1$. Hence $\rho = \alpha - \beta\,\sigma$ holds.

This satisfies the condition 1. Hence the integral domain of Gaussian integers is an Euclidean domain.

Example 8.2.5: If we consider the polynomial ring $(F[x], +, .)$ over the field $(F, +, .)$, $(F[x], +, .)$ is an integral domain. Now if we define the function d from $F[x] - \{0\}$ into Zo, the set of non-zero integers by

$d(f(x)) = \deg(f(x))$. Then we see that

$\deg(f(x)\,.\,g(x)) = \deg(f(x)) + \deg(g(x)) > \deg(f(x))$ which implies $d(f(x)\,.\,g(x)) \geq d(f(x))$. Hence the condition (2) is satisfied. For $f(x), g(x) \in F[x]$ we have $f(x) = g(x)\,.\,q(x) + r(x)$ and either $r(x) = 0$ or $\deg r(x) < \deg g(x)$.

This means that

$f(x) = g(x)\,.\,q(x) + r(x)$, where $r(x) = 0$ or $d(r(x)) < d(g(x))$

Hence the condition (1) is satisfied.

$d(f(x) > 0$ for all $f(x) \in F[x] - \{0\}$.

This shows that the integral domain $(F[x], +, .)$ over the field $(F, +, .)$ is an Euclidean domain $(F[x], +, .\,d)$.

Theorem 8.2.1: *Let $(D; +, ., d)$ be an Euclidean domain and Let $(A, +, .)$ be an ideal of $(D, +, .\,d)$. Then the ideal $(A, +, .)$ is a principal ideal, that is, there exists an element $a_0 \in A$ such $A = \{a_0\,x \mid x \in D\}$.*

Proof: For $0 \in A$, we have $A = \{0\}$ and consequently $(A, +, .)$ is principal ideal.

Now let $A \neq \{0\}$. Then there exists an element $a_0 \in A$ for which $d(a_0)$ is minimal. For any $a \in A$, we have $a = b\,a_0 + r$, where either $r = 0$ or $d(r) < d(a_0)$, since $(D, +, ., d)$ is an Euclidean domain. Thus $r = a - ba_0, \in A$, as $a \in A$, $ba_0 \in A$ because $(A, +, .)$ is an ideal.

If $r \neq 0$, then $d(r) < d(a_0)$, that is, there exists an element $r \in A$ whose d-value $d(r)$ is less than the minimal d value $d(a_0)$ of an element $a_0 \in A$. This contradiction shows that r must be zero and consequently we have $a = ba_0$. This shows that every element of A is a multiple of a_0. Hence $(A, +, .)$ is a principal ideal.

Corollary 1: *A Euclidean domain $(D, +, ., d)$ has an unit element.*

Proof: Since $(D, +, ., d)$ is an Euclidean domain, the ideal $(D, +, .)$ is a principal ideal. So there exists an element $a_0 \in D$ such that every element of D is a multiple of a_0, that is, $D = (a_0)$.

In particular, there exists an element c for which $a_0 = a_0\,c$.

For $a \in D$, there exists an element $x \in D$ such that $a = x\, a_0$.

Now $a\, c = (x\, a_0)\, c = x\, (a_0\, c) = xa_0 = a$. By cancellation law we have $c = 1$, the unit element, this proves the corollary.

Remark: From the preceding theorem and its corollary we conclude that every Euclidean domain is a principal ideal domain. But every principal ideal domain is not a Euclidean domain.

Lemma 8.2.1: *(i) Let $(D, +, ., d)$ be an Euclidean domain. Then if $a \in D$, $a \neq 0$ then $d(a) \geq d(1)$.*

Proof: We have $1.a = a$. By the second condition of Euclidean domain

$$d(1) \leq d(1 \,.\, a)$$

$$\Rightarrow \qquad d(1) \leq d(a).$$

Lemma 8.2.2: *If $O \neq a \in D$, then $d(a) = d(1)$ if and only if a is a unit.*

Proof: Let a be a unit. Then we have to prove $d(a) = d(1)$. Since a is a unit, then there exists an element $c \in D$ for which $a \,.\, c = 1$. Thus by the condition (2) of Euclidean domain.

We have

$$d(a) \leq d(a\, c) = d(1)$$

But by the preceding lemma 8.2.1 $d(1) \leq d(a)$, this implies $d(a) = d(1)$.

Conversely, if $d(a) = (1)$, we consider

$1 = a \,.\, q + r$ where $r = 0$ or $d(r) < d(a) = d(1)$.

Since $d(r) < d(1)$ is impossible, $r = 0$ and consequently $1 = a \,.\, q \Rightarrow a \mid 1 \Rightarrow a$ is a unit

Lemma 8.2.3: *If $a, b \in D$ and if b is not a unit in D then $d(a) < d(a \,.\, b)$.*

Proof: If $b \in D$ is a unit, then there exists an element $c \in D$ such that $b \,.\, c = 1$. Now by the condition (2) we have $d(a) \leq d(a \,.\, b)$,

and $d(a \,.\, b) \leq d(a \,.\, b \,.\, c) = d(a \,.\, 1) \leq d(a)$.

Thus $d(a) \leq d(a \,.\, b)$ and $d(a \,.\, b) \leq d(a)$ give us

$d(a) = d(a \,.\, b)$, when b is a unit.

If b is not a unit then we clearly have

$d(a) < d(a \,.\, b)$.

Theorem 8.2.2: *Let $(D, +, .\, v)$ be Euclidean domain. If $a, b \in D$ and one of them is non-zero, then a and b have a greatest common divisors, and more over, there exists $u, t \in D$ such that $d = au + bt$.*

Proof: Let A be the set of all elements of the form $ra + sb$, where $r, s \in D$. *i.e.,* $A = \{ra + sb \mid r, s \in D\}$

For $r_1\, a + s_1\, b$ and $r_2\, a + s_2\, b \in A$, we have

$(r_1\, a + s_1\, b) - (r_2\, a + s_2\, b) = (r_1 - r_2)\, a + (s_1 - s_2)\, b \in A$ as

$r_1 - r_2$ and $s_1 - s_2 \in D$,

and for all $u \in D$, we have

$u\ (r_1\ a + s_1\ b) = (ur_1)\ a + (us_1)\ b \in A$ as ur_1 and $us_1 \in D$.

This shows $(A, +, .)$ is an ideal.

Since $(D, +, ., v)$ is a Euclidean domain, it is a principal ideal domain and consequently $(A, +, .)$ is a principal ideal. So the every element of A is a multiple of some element $d \in A$. Since every element of A is of the form $ra + sb$, then for some u and $t \in D$, $d = au + bt$. Since the Euclidean domain $(D, +, .\ v)$ has a unit element 1, then

$a = 1\ .\ a + 0\, b \in A$, and

$b = 0\ a + 1\ .\ b \in A$.

Since $a, b \in A$, a and b are multiples of d. So $d \mid a$ and $d \mid b$, that is, d is a common divisor of a and b. Suppose $c \mid a$ and $c \mid b$, then $c \mid au$ and $c \mid bt$ which implies $c \mid au + bt = d$. Hence any common divisor of a and b divides d. This proves that d is the greatest common divisor of a and b which can be expressed as $d = au + bt$.

CHAPTER 8

Definition 8.2.2: $(D, +, .\ v)$ be an Euclidean domain. If $GCD\ (a, b) = 1$ we say a and b are co-prime or relatively prime.

Theorem 8.2.3: *In an Euclidean domain $(D, +, ., v)$, $GCD\ (a, b) = 1$ if and only if there are elements x and y in D such that $ax + by = 1$.*

Proof: If $GCD\ (a, b) = 1$, then by theorem 8.2.2 we have $ax + by = 1$, for some $x, y \in D$.

Conversely, if $ax + by = 1$ and $d \mid a$, and $d \mid b$, then $d \mid 1$, that is, d is a unit and hence 1 is a GCD of a and b.

Theorem 8.2.4: *i (f $a \mid bc$ and $(a, b) = 1$, then $a \mid c$.*

Proof: From $(a, b) = 1$, we have

$$ax + by = 1 \text{ for some } x, y \in D.$$

or
$$a\, c\, x + b\, c\, y = c.$$

$a \mid b\ c$ and $a\ \mid a\ c \Rightarrow a \mid c$.

Theorem 8.2.5: *In an Euclidean domain any irreducible element is a prime.*

Proof: Let a be irreducible and let $a \mid bc$ and suppose that $a \nmid b$. Then we have to prove $\underline{a} \mid \underline{c}$. From the Theorem 8.2.4 we conclude that we have to show that $(a, b) = 1$.

Let x be a common divisor of a and b.

$x \mid a \Rightarrow x$ is a unit or an associate of a, since a is irreducible.

$x \mid b$, and x is a unit or is an associate of $a \Rightarrow a \mid b$ which is contradiction to assumption that $a \nmid b$. Thus every common divisor of a and b is a unit.

So $(a, b) = 1$. Hence the result

Theorem 8.2.6: *If $a \mid c$ and $b \mid c$ and $(a, b) = 1$, then $a\ b \mid c$.*

Proof: Since $a \mid c$ and $b \mid c$, $(a, b) = 1$, then

$c = ac_1$ and $c = bc_2$ and $ax + by = 1$ for some

$c_1, c_2, x, y \in D$.

Then

$$c = acx + bcy = abc_2 x + bac_1 y$$
$$= ab\,(c_2 x + c_1 y) \Rightarrow ab \mid c.$$

Theorem 8.2.7: *If $a, b_1, b_2.., b_n$ are elements of an Euclidean domain, and if $(a, b_i) = 1$ for all i, then $(a, (b_1 b_2 ... b_n) = 1$.*

Proof: To prove the theorem we shall apply induction on n. If $n = 1$, then $(a, b_1) = 1$, the theorem holds. By induction we assume that

$$(a, (b_2...b_n)) = 1.$$

Therefore $ax + (b_2...b_n)\,y = 1$ and $au + b_1\,v = 1$

Multiplying we have

$1 = a\,(au\,x + (b_2....b_n)\,uy + b_1\,xv) + b_1\,(b_2....b_n)\,yv,$

or $1 = ar + (b_1,...b_n)\,s$, where

$r = aux + (b_2...b_n)\,uy + b_1\,xv$, and $s = yv$.

This follows the result.

Theorem 8.2.8: *In an Euclidean domain $(D, +, ., d)$ if $a \mid b$, $b \neq 0$, and a is neither a unit nor an associate of b, then $d(a) < d(b)$.*

Proof: Since a is not an associate of b, $b \mid a$. Hence $a = bq + r$, where $r = 0$ or $d(r) < d(b)$, for $q, r \in D$. But we can write $b = ac$. Now $r = a - bq = a - acq = a\,(1 - cq)$.

So $d(r) = d\,(a\,(1 - cq))$ and $d(a) \leq d\,(a\,(1 - cq) = d(r)$

Thus we have $d(a) < d(b)$.

Theorem 8.2.9: *In an Euclidean domain $(D, +, ., d)$ each non-zero element is either a unit or can be written as a finite product of primes, which is unique in the following sense.*

If $a = p_1 p_2..p_n = q_1 . q_2 \dots q_m$ where each p_i, and q_j is prime, then $n = m$ and each p_i, $1 \leq i \leq n$ is an associate of some q_j, $1 \leq j \leq m$.

Proof: We shall prove the first part of the theorem by mathematical induction on $d(a)$.

Let $d(a) = d(1)$. Then a is a unit of D. Hence the theorem holds for $d(a) = d(1)$.

Now we assume that the theorem is true for all elements $x \in D$ such that $d(x) < d(a)$. By applying this assumption we have to show that a is either a unit or can be written as a finite product of primes.

If $a \in D$ is a prime, then there is nothing to prove. So suppose that a is not prime and $a = b\,.\,c$, where neither b nor c is a unit of D. We have

$d(b) \leq d(bc) = d(a)$, and

$d(c) \leq d(bc) = d(a)$.

Thus by induction, b and c can be written as a finite product of primes of D:

$b = p_1 \,.\, p_2...p_n$, and

$c = q_1\,q_2 \dots q_m$, where p_i and q_j are prime elements of D. Thus

$a = bc = p_1 \cdot p_2 \ldots p_n\, q_1\, q_2 \ldots q_m$. Hence in this way a can be written as a product of finite number of prime elements. This completes the first part of the theorem.

For uniqueness of the theorem we suppose that $a \neq 0$ and it is not a unit of D.

Let

$a = p_1 \cdot p_2, \ldots p_n, = q_1 \cdot q_2 \ldots q_m$, where p_i and q_j are primes. Since $p_1 | p_1\, p_2 \ldots p_n$, then $p_1 | q_1 \cdot q_2 \ldots q_m \Rightarrow p_1$ must divide some q_j.

Since p_1 and q_j are both primes of D and $p_1 | q_j$, then p_1 must be an associate of q_j and so we have

$q_j = u_1 \cdot p_1$, u_1, is a unit of D.

Thus

$$p_1\, p_2 \ldots p_n = q_1 \cdot q_2 \ldots q_m = u, p_1\, q_1\, q_2 \ldots q_{j-1} \ldots q_{j+1} \ldots q_m$$

$$\Rightarrow p_2 \cdot p_3 \ldots p_n = u_1\, q_1\, q_2 \ldots q_{j-1}\, q_{j+1} \ldots q_m$$

Again we repeat the above argument for p_2 and similarly for p_3, p_4 and so on.... After n steps the left side is reduced to 1, and the right hand side a product of a certain number of q' s the excess of m over n. This implies $n \leq m$ since q' s are not units. Similarly, we have $m \leq n$. Hence $m = n$.

In the process we have also proved that every p_i is an associate of some q_j and conversely. This completes the proof of the theorem.

CHAPTER 8

PROBLEMS

1. Show that the system $(H, +, ., d)$ is a Euclidean domain where $H = \{a + b\sqrt{2} \mid a, b \in Z\}$, and $d(\alpha) = a^2 + 2b^2$ if $\alpha = a + b\sqrt{2}$.
2. Let $(D, +, ., d)$ be a Euclidean domain. Show that $a \sim b$, then $d(a) = d(b)$.
3. In the domain of Gaussian integers find the greatest common divisor of $3 + 4i$ and $7 - i$.
4. Let $\alpha = 1 + 2i$ and $\beta = 3 + 4i$. In the Gaussian integers final σ and ρ so that $\alpha = \beta\sigma + \rho$ with $d(\rho) < d(\beta)$.
5. Let $(D, +, ., d)$ be a Euclidean domain. Let k be an integer such that $d(1) + k \geq 0$. Define a new function d' on D by

   $d'(a) = d(a) + k$. Show that $(D, +, ., d')$ is Euclidean domain. If s is a positive integer and a function d'' is defined by $d''(a) = sd(a)$, show that $(D, +, ., d'')$ is a Euclidean domain.
6. Prove that if d is a common divisor of a and b, then $(a/d, b/d) = 1$ if and only if $(a, b) = d$.
7. Let a and b be non-zero elements of a Euclidean domain. Then show that, $ax + by = d$ if and only if $(a, b) | d$.
8. If $a_1, a_2, \ldots, a_n$ are elements of a Euclidean domain which are mutually co-prime, and if $a_i | c$ for all i, then show that $a_1 \,.\, a_2\; a_n | c$.

9. Let a, b, c belong to a Euclidean domain $(D, +, ., d)$. Then prove that the equation $c = ax + by$ has solution in D if and only if $(a, b) | c$.

10. In Euclidean domain the congruence $ax \equiv b \ (mod\ (n)$ has solution if and only if $(a, n) | b$.

11. Let $(D, +, ., d)$ be a Euclidean domain and let p be a prime in D. Show that $(D/(p), +, .)$ is a field and $((p), +, .)$ is a maximal ideal.

12. Prove that the only units of $Z(\sqrt{-5})$ are + 1 and –1.

13. Show that $(1 + 2\sqrt{-5})(1 - 2\sqrt{-5}) = 3 . 7$.

14. Show that $(1 + 2\sqrt{-5})$ and 3 are irreducible in $Z(\sqrt{-5})$.

15. Show that 3 is not a prime in $Z(\sqrt{-5})$.

16. Let $a = 3$, $b = 1 + 2\sqrt{-5}$ and $c = 7. (1 - 2\sqrt{-5})$. Prove that a and b are co-prime but that ac and bc have no greatest common divisor and also least common multiple in $Z(\sqrt{-5})$.

17. Prove in a Euclidean domain (a, b) can be found as follows:

$b = q_0 a + r_1$ where $d(r_1) < d(a)$

$a = q_1 r_1 + r_2$ where $d(r_2) < d(r_1)$

$r_1 = q_2 r_2 + r_3$ where $d(r_3) < d(r_2)$

.

.

.

$r_{n-1} = q_n r_n$

and $r_n = (a, b)$.

8.3 UNIQUE FACTORIZATION DOMAINS

Is the integral domain of integers we have seen that any integer $n > 1$ can be written as the product of finite number of primes and this product is unique. Now here we have to study this property in other rings which are integral domains. There are some rings in which this property of uniqueness does not hold. For this we classify the rings:

Definition 8.3.1: An integral domain $(D, +, .)$ is called a *unique factorization domain* (UFD) if

(1) every non-zero element $r \in D$ is either a unit or can be written as a finite product of irreducibles.

(2) If $r = p_1 \dots p_n = q_1 \dots q_m$, where p_i and q_j are irreducibles in D, then $m = n$, and each p_i is an associate of some g_j.

By unique factorization theorem in a Euclidean domain and in Polynomial domain, every Euclidean domain and every polynomial domain over a field are unique factorization domains.

Example 8.3.1: Every field $(F, +, .)$ is a unique factorization domain because every non-zero element of a field is a unit.

Example 8.3.2: The integral domain $(Z, +, .)$ of integers is an unique factorization domain, because except 1, 0, and –1, every integer is expressible as a product of finite number of primes. In integers every prime is an irreducible element.

Example 8.3.3: The integral domain $(D, +, .)$ where $D = \{a + b\sqrt{-5} \mid a, b \in Z]$ is not a unique factorization domain. Because $4 = 2.2 = (1 + \sqrt{5})(-1 + \sqrt{5})$. Since 2 is a prime of D and $1 + \sqrt{5}$ and $-1 + \sqrt{5}$ are also prime of D and 4 can be written in two ways as above, the domain is not a UFD.

Remark: Every Euclidean domain is necessarily a UFD, but every UFD is not a Euclidean domain.

Lemma 8.3.1: *Every irreducible element of UFD $(D, +, .)$ is a prime.*

Proof: Let $(D, +, .)$ be a UFD and let r be an irreducible element of D. Let $r \mid a \,.\, b$ while $r \nmid a$. Then we have to show that $r \mid b$. Since a and b are element of a UFD, then $a = p_1 \dots p_n$ and $b = q_1 \dots q_m$ are unique factorization of a and b respectively. Thus $a \,.\, b = p_1 \dots p_n\, q_1 \dots q_m$ is a unique factorization of ab.

Since $r \mid ab$, $ab = rc$ for some $c \in D$. Let us suppose that c has a factorization $t_1 \,.\, t_2 \dots t_u$ into irreducibles. Thus $a \,.\, b = p_1 \dots p_n \,.\, q_1 \dots q_m = r \,.\, t_1 \dots t_u$.

Since the factorization is unique, $r c$ is an associate of r and it must appear among $p_1 \dots p_n \,.\, q_1 \dots q_m$. Since $r \mid a$, then rc must be among $q_1\, q_2 \dots q_m$. This implies $r \mid rc$, $rc \mid q_1 \dots q_m \Rightarrow r \mid q_1 \dots q_m$ or $r \mid b$.

Lemma 8.3.2: *In a principal ideal domain $(D, +, .)$, for every ascanding chain of ideals*
$(a_1) \subseteq (a_2) \subseteq (a_3) \subseteq \dots \subseteq (a_n) \subseteq \dots$
there exists an integer n such that $(a_t) = (a_n)$, $\forall\ t \geq n$.

Proof: Let $(a_1) \subseteq (a_2) \subseteq (a_3) \subseteq \dots$ be an infinite ascending chain of ideals in D.

Let $A = U\,(a_i)$ be the union of the sets (a_i).

Since every $(a_i) \subseteq A$, $\forall\ i$, A is non-empty. Now we shall prove that $(A, +, .)$ is an ideal of $(D, +, .)$.

$b_1, b_2 \in A \Rightarrow b_1 \in (a_k),\ b_2 \in (a_l)$. We can suppose $k \leq l$.

Then $b_1, b_2 \in (a_l)$. Since $((a_l), +, .)$ is an ideal, $b_1 - b_2 \in (a_l)$ and $b_1\, x$ and $x \,.\, b_1$ for any $x \in D$ are in (a_l).

This implies that $b_1, b_2, b_1\, x$, and $x\, b_1$ are in A. Hence $(A, +, .)$ is an ideal.

Since $(D, +, .)$ is a principal ideal domain, then $A = (d)$, where $d \in A$. Since $d \in A$, then $d \in (a_n)$ for some integer n. Hence $A = (d) = (a_n)$, consequently if $t \geq n$, then $A = (a_n) \subseteq (a_t)$. Since $A = U\,(a_i)$, then $(a_t) \subseteq A = (a_n)$ this implies $(a_n) = (a_t)$, this proves the lemma.

Lemma 8.3.3: *For every non-zero unit element a in a PID $(D, +, .)$ there exists an irreducible element p such that $p \mid a$.*

Proof: Let $a \neq 0$ be a non unit element of D. Since $(D, +, .)$ is PID, then $((a), +, .)$ is a principal ideal.

If the ideal $((a), +, .)$ is maximal, then a is an irreducible element and we have to prove nothing. If $((a) +, .)$ is not maximal, there exists an ideal $((p), +, .)$ such that

$(a) \subset (p) \subset D.$

$(a) \subseteq (p) \Rightarrow p \mid a$. If $((p)), +, .)$ is maximal, then p is irreducible and divides a. Again if $((p), +, .)$ is not maximal, there exists an other ideal $((p_1), +, .)$ such that $(a) \subset (p) \subset (p_1) \subset D$ which implies $p_1 \mid a$.

If $((p_1), +, .)$ is maximal, p_1, is irreducible and divides a. Since $(D, +, .)$ is a PID, then D contains no properly ascending infinite sequence of ideals. That is, there exists a maximal ideal (p_n) such that $(p) \subset (p_1) \subset (p_2) \subset ...\subset (p_n) \subset D$. Since $((p_n), +, .)$ is maximal, p_n is irreducible and consequently $p_n \mid a$. Hence the lemma follows.

Theorem 8.3.1: *Every principal ideal domain $(D, +, .)$ is a unique factorization domain (UFD).*

Proof: Let a *be* a non zero non unit element of D. Then we have to show that a can be factorized into irreducibles of D. Irreducibles of PID are primes of PID. There exists some prime p_1 which divides a . $p_1 \mid a$, $a = a_1 . p_1$, for some $a_1 \in D$. Thus we have $a_1 \mid a \Rightarrow (a) \subseteq (a_1)$.

If $(a) = (a_1)$, then $a_1 = a.x$ for some $x \in D \Rightarrow a = a_1 p_1 = ax . p_1 \Rightarrow x . p_1 = 1 \Rightarrow p_1$, is a unit which is absurd because p_1, is an irreducible element. Hence $(a) \subset (a_1)$.

If a_1 is a unit, a is an associate of p_1 and hence a is irreducible and there is nothing to prove.

So we assume a_1 is not a unit. Then there is another irreducible element p_2 such that $p_2 \mid a_1 \Rightarrow a_1 = a_2 p_2$ for some $a_2 \in D$. Thus we have $a = a_1 p_1 = a_2 p_1 p_2$. If a_2 is a unit, then p_2 is an associate of a_1 and hence a_1 is irreducible. So $a = a_1 p_1$ is required factorization. If a_2 is not unit, then $a_2 \mid a_1 \Rightarrow (a_1) \subset (a_2)$ and there exists an irreducible element p_3 such that $p_3 \mid a_2$, *i.e.*, $a_2 = a_3 p_3$ for some $a_3 \in D$ which is not unit. So we have $(a_2) \subset (a_3)$. If we continue this process we have a ascending chain of ideals

$(a) \subset (a_1) \subset (a_2) \subset (a_3) \subset ....$

But this ascending chain of ideals must be finite because this chain is holding in PID $(D, +, .)$, Therefore for some integer n we must have the ideal $((a_n), +, .)$ maximal and hence a_n is irreducible, say p_{a+1} such that

$(a) \subset (a_1) \subset (a_2) \subset (a_3) \subset ... \subset (a_n) \subset D.$ Then

$a = a_1 p_1$, $a_1 = a_2 p_2$, $a_2 = a_3 p_3$,...., $a_{n-1} = a_n p_n$ give us $a = a_n p_1 p_2 p_3... p_n = p_1 p_2 p_3...p_n p_{n+1}$; where $a_n = p_{n+1}$, a product of finite number of irreducible elements. Hence every non-zero non-unit element of D can be expressed as product of finite number of irreducibles.

For uniquences Let

$a = p_1 p_2 ... p_{n+1} = a_1 q_2 ... q_m.$

We shall prove it by induction on $n + 1$. When $n = 0$ the result is trivially true.

Now $p_1 . p_2 ... p_{n+1} = q_1 q_2 ... q_m \Rightarrow q_1 \mid p_1 p_2 ... p_{n+1}$

Since $p_1, p_2...p_{n+1}$ are all primes, $q_1 | p_i$ for some $i = 1, 2,..n +1$.

$p_i = u_1 q_1$, u_1 is a unit. Thus

$\Rightarrow u_1 q_1 p_1 \cdot p_2...p_{i-1} \cdot p_{i+1}...p_{n+1} = q_2 \cdot q_2...q_m$

$\Rightarrow u_1 p_1 \cdot p_2...p_{i-1} \cdot p_{i+1}...p_{n+1} = q_2 \cdot q_2...q_m$

$\Rightarrow p'_1 \cdot p'_2...p_{i-1}...p'_{i+1}...p_{n+1} - = q_2 q_3...q_m,$

where $u_1 p_1 = p_1$, $p_j = p_j$, $j \geq 2$

If we put $b = q_2 q_3...q_m$, then

$b = p'_1 p'_2....p'_{i-1} \cdot p'_{i+1} \cdots p'_{n+1}$.

To apply induction we suppose that the rosult holds for all those non-zero non-unit elements which are written as a finite product of less than $(n + 1)$ irreducible elements. Hence the factorization of b is unique and consequently $n = m - 1 \Rightarrow n + 1 = m$. This shows that factorizations of a have the same number of primes and each p_i is an associate of some q_j. This completes the theorem.

Lemma 8.3.4: *In a UFD (D, +, .) any two elements a, b have a GCD (greatest common divisor (a, b) and LCM (least common multiple) [a, b].*

Proof: Since the elements a and b belong to a unique factorization domain, then.

$a = e\, p_1^{a_1} .... p_n^{a_n}$, where e is a unit, and $a_i \geq 0$, p_i are primes,

and by permitting the exponents on primes p_i to be non-zero, we have

$b = f\, p_1^{b_1} p_2^{b_2} .... p_n^{b_n}$, where f is a unit, $b_i \geq 0$, and each p_i is prime.

Now

c_i = minimum of a_i and b_i,

and d_i = maximum of a_i and b_i,

Then $p_1^{c_1} p_2^{c_2} ... p_n^{c_n}$

is the greatest common divisor of a and b and $p_1^{d_1} p_2^{d_2} .... p_n^{d_n}$ is the least common multiple of a and b.

Example 8.3.4: The GCD and LCM of 15 and –12 can be found as follows:

$$-12 = (-1)2^2, 3^1, 5^0,$$

$$15 = (1)2^0, 3^1, 5^1.$$

So GCD $(-12, 15) = 2^0\, 3^1 \,.\, 5^0 = 3,$

and LCM $[-12, 15] = 2^2\, 3^1 \,.\, 5^1 = 60.$

Corollary: *Any finite number of non-zero elements $a_1,...., a_n$ of a UFD have an GCD and LCM.*

Lemma 8.3.5: If $d = (a_0, a_1...a_n)$, *then*

$$(a_0 | d, a_1 | d,...a_n | d) = 1$$

Now here we shall pay attention to the polynomial ring over a UFD.

Definition 8.3.2: Let $(D, +, .)$ be a UFD and let $f(x) = a_0 + a_1 x +... + a_n x^n \in D[x]$. Then the greatest common divisor $d = (a_0\, a_1, ... a_n,$ of $a_0\, a_1,... a_n$ is called the content of $f(x)$ denoted by $c(f)$.

CHAPTER 8

If d and d' are two greatest common divisors of $a_0, a_1 ...., a_n$, then d is an associate of d' and vice-versa. So $c\ (f)$ is unique within units of D.

Definition 8.3.3: A non-zero polynomial $f\ (x) = a_0 + a_1\ x + ... + a_n\ x^n \in D\ [x]$ over the unique factorization domain $(D, +, .)$ is called *primitive* if the greatest common divisor d of a_0, a_1, a_n is 1, *i.e.*, $c\ (f) = 1$.

Example 8.3.5: $F\ (x) = 2 + 6x + 10x^2 + 18x^3 \in Z\ [x]$ is not primitive because GCD (2, 6, 10, 18) = 2 ≠ 1.

The polynomial $f\ (x) = x^3 + 4x^2 - 3x + 1 \in Z\ [x]$ is primitive since GCD (1, 4, –3, 1) = 1.

Lemma 8.3.6: *Let $f\ (x) \in D\ [x]$ over a UFD $(D, +, .)$. Then there exist a primitive polynomial $g\ (x) \in D\ [x]$ and an element $d \in D$) such that $f\ (x) = d\ .\ g\ (x)$. Moreover, if $f\ (x) = dg\ (x) = bh\ (x)$, where $g\ (x)$ and $h\ (x)$ are primitive polynomial in $D\ [x]$ and $d, b \in D$, then d is an associate of b in D and $g\ (x)$ is an associate of $h\ (x)$ in $D\ [x]$.*

Proof: Let $f\ (x) = a_0 + a_1 x + ... + a_n\ x^n \in D\ [x]$. Let d is a GCD $(a_0, a_1, ..., a_n)$. Then

$$f(x) = d\left(a_0/d + a_1/dx + ... + \frac{a_n}{d}\ x^n\right)$$

$= d\ (g(x))$, where $g(x) \in D\ [x]$ is primitive polynomial since $(a_0/d, a_1/d, ..., a_n/d) = 1$.

$= c\ (f)\ .\ g\ (x)$

For uniqueness let us suppose that

$$f(x) = dg\ (x) = bh\ (x), \qquad .....(1)$$

Where

$g\ (x) = g_0 + g_1\ x + ... + g_n\ x^n$, and

$h\ (x) = h_0 + h_1\ x + ... + h_n\ x^n$ are primitive.

Let $c = (d, b)$. It follows from (1) that

$$dg_i = b\ h_i \Rightarrow (d/c)\ g_i = (b/c)\ h_i, \ \forall\ i$$

$\Rightarrow (d/c) | h_i$ and $(b/c) | g_i$ since $(d/c, b/c) = 1$.

$\Rightarrow (d/c) |$ GCD $(h_0, ... h_n)$,

and $(b/c) |$ GCD $(g_0, g_1 ..., g_n)$

$\Rightarrow (d/c) | 1$ and $(b/c | 1$, since

GCD $(h_0, ... h_n) = 1$ and

GCD $(g_0, ... g_n) = 1$ as $h(x)$ and $g(x)$ are primitive

$\Rightarrow d/c = u,\ b/c = v$, u and v are units of D.

$\Rightarrow d = cu,\ b = c\ v$.

$\Rightarrow d \sim c,\ b \sim c \Rightarrow d \sim b$.

Hence d and b are associates.

It now follows that $ug_i = v\ h_i \Rightarrow g\ (x)$ and $h\ (x)$ are associates.

Lemma 8.3.7: *If $f(x)$ and $g(x)$ are two primitive polynomials over UFD, $f(x) \cdot g(x)$ is a primitive polynomial over UFD.*

Proof: Let $f(x) = a_0 + a_1, x + ... + a_n x^n$

and $g(x) = b_0 + b_1 x + ... + b_m x^m$ be primitive polynomials.

Let $f(x) \cdot g(x)$ be not a primitive. Then the coefficients of $f(x)$ and $g(x)$ would be divided by some element of D which is greater than 1, say, some prime element p. Since $f(x)$ and $g(x)$ are primitive, p does not divide some coefficient a_i of $f(x)$ and some coefficients of b_j of $g(x)$.

Let a_i and b_j *be* the first coefficients of $f(x)$ and $g(x)$ respectively which are not divisible by p.

Now the coefficient of x^{i+j} in the product $f(x)\, g(x)$ is

$c_{i+j} = a_i b_j + (a_{i+1}\, b_{j-1} + a_{i+2}\, b_{j-2} + ... + a_{i+j}, b_0) + (a_{i-1}\, b_{j+1} + a_{i-2}\, b_{j+2} + ... + a_0\, b_{j+1})$

$\Rightarrow a_i\, b_j = c_{i+j} - (a_{i+1}\, b_{j-1} + ... + a_{i+j}\, b_0) - (a_{i-1}\, b_{j+1} + ... + a_0\, b_{j+1})$

$\Rightarrow p \mid a_i\, b_j$, since p divides each term on the right side

$\Rightarrow p \mid a_i$ or $p \mid b_j$.

which is a contradiction. This proves the lemma.

Corollary: *For two non-zero polynomial $f(x)$ and $g(x)$ over UFD, $c(f \cdot g) = c(f) \cdot c(g)$ (up to units)*

Proof: Let $f(x), g(x) \in D[x]$. Then

$f(x) = c(f) \cdot f_1(x)$, *and*

$g(x) = c(g) \cdot g_1(x)$. Where $f_1(x)$ and $g_1(x)$ are primitive polynomials and $c(f)$ and $c(g)$ are contents of $f(x)$ and $g(x)$ respectively.

We have $f(x) \cdot g(x) = c(f) \cdot c(g)\, f_1(x) \cdot g_1(x)$.

By the Lemma 8.3.7 $f_1(x) \cdot g_1(x)$ is primitive. Hence this proves that the content of $f(x)\, g(x)$ is $c(f) \cdot c(g)$. Thus

$c(f \cdot g) = c(f) \cdot c(g)$.

Remark: From this corollary it is clear that the product of a primitive polynomial and a polynomial is not primitive.

Let $(D, +, .)$ be a unique factorization domain. Since UFD $(D, +, .)$ is an integral domain, there exists a quotient field $(Q. +, .)$ of the domain $(D, +, .)$. Therefore $(D, +, .)$ is a subring of the field $(Q, +, .)$. So we can consider $(D[x], +, .)$ as a subring of the domain $(F[x], +, .)$. For example, the field $(Q, +, .)$ of rational numbers is the Quotient field of the domain $(Z, +, .)$ of integers and $(Z[x], +, .)$ is a subring of the ring $(Q[x], +, .)$ If $f(x) \in Q[x]$, that is, $f(x) = a_0 + a_1 x + ... + a_n, x^n$, $a_0, a_1, ... a_n \in Q$, then

$f(x) = \dfrac{1}{a} f_1(x)$, where a is an integer and a is LCM of all denominators of $a_0, a_1, ..., a_n$. Thus the coefficient of $f_1(x)$ belong to Z, the set of integers, that is, $f_1(x) \in Z(x)$. Hence we can conclude that if $(Q, +, .)$ is the quotient field of the integral Domain $(D, +, .)$, then $f(x) \in Q[x]$ is written as

$$f(x) = \frac{1}{a} f_1(x),\ a \in D,\ f_1(x) \in D[x].$$

***Lemma 8.3.8:** Let (F, +, .) be the quotient field of UFD (D, +, .). Then an irreducible primitive polynomial f(x) in D[x] is also irreducible polynomial in F[x].*

Proof: Let $f(x)$ be an irreducible and primitive polynomial in $D[x]$ and let $f(x)$ be reducible in $F[x]$. Then there exist two polynomial $g(x)$ and $h(x)$ in $F[x]$ of positive degrees such that $f(x) = g(x) . h(x)$. But $g(x)$ can be written as $\frac{1}{a} g_0(x)$, where $a \in D$, $g_0(x) \in D[x]$ and

$$h(x) = \frac{1}{b} h_0(x), \text{ where } b \in D \text{ and } h_0(x) \in D[x].$$

Hence

$$f(x) = \frac{1}{ab} g_0(x) . h_0(x).$$

We can also write $g_0(x) = c(g_0)\, g_1(x)$, $h_0(x) = c(h_0)\, h_1(x)$, where $c(g_0)$ and $c(h_0)$ are contents of $(g), (x)$ and $h_0(x)$, and $g_1(x)$, $h_1(x)$ are primitive in $D[x]$. If we put $\alpha = c(g_0)$, $\beta = c(h_0)$, then

$$f(x) = \frac{1}{ab}(\alpha\beta) g_1(x) . h_1(x),$$

$$\Rightarrow a.b\, f(x) = (\alpha . \beta) g_1(x) . h_1(x).$$

Since $h_1(x)$ and $g_1(x)$ are primitive, $g_1(x) . h_1(x)$ is also primitive and consequently $f(x)$ is primitive. Thus it follows that $c(f) = c(g_1 . h_1) \Rightarrow a.b = \alpha . \beta \Rightarrow f(x) = g_1(x) . h_1(x)$. This shows $f(x)$ is reducible in $D[x]$ which is contradiction that $f(x)$ is irreducible in $D[x]$. Hence $f(x)$ is irreducible in $F(x)$.

***Lemma 8.3.9:** If the primitive element f(x) in D[x] is irreducible as an element of F[x], it is also irreducible as an element of D[x].*

Proof: Let $f(x)$ be a primitive polynomial in $D[x]$ which is irreducible in $F[x]$.

Since the ring $(D[x], +, .)$ is a subring of the domain $(F[x], +, .)$, then if $f(x)$ is irreducible in $F(x)$ it is a fortiori, irreducible in $F[x]$.

***Lemma 8.3.10:** If $f_1(x)$ and $f_2(x)$ are primitive in D[x] and are associates in F[x], where (F,+,.) is the quotient field of the domain (D,+, .), then $f_1(x)$ and $f_2(x)$ are associates in D[x].*

Proof: Let $f_1(x)$ and $f_2(x)$ be two primitive polynomials in $D[x]$ which are associates in $F[x]$. Since the units of $F[x]$ are non-zero elements of F,

$f_1(x) = a . f_2(x),\ a \neq 0 \in F.$

We can write $a = b_1 . b_2^{-1},\ b_1, b_2 \in D.$

Then $f_1(x) = b_1 . b_2^{-1} f_2(x)$

or $b_2 f_1(x) = b_1 f_2(x)$.

Since $f_1(x)$ and $f_2(x)$ are primitive, $c(f_1) = c(f_2)$, *i.e.*, $b_1 = b_2$ within units. It follows that $f_1(x)$ differs from $f_2(x)$ by a unit in $D[x]$. Hence $f_1(x)$ is an associate of $f_2(x)$ in $D[x]$.

Lemma 8.3.11: *(Gauss) Let (D, +, .) be a UFD and let (Q, +, .) be its field of quotients. Let f (x) be a non scalar primitive polynomial in D [x], f (x) is reducible in D [x] if and only if f (x) reducbile in Q [x].*

Proof: Since $(D\,[x], +, .)$ is a subring of $(Q\,[x], +, .)$, then if $f(x)$ is reducible in $D\,[x]$, it is, a fortiori, reducible in $Q\,[x]$.

Conversely, if $f(x)$ reducible in $Q\,[x]$, then there exist two polynomials $g\,(x)$ and $h\,(x)$ in $Q\,[x]$ of positive degrees such that

$$f(x) = g\,(x)\,.\,h\,(x).$$

But we can write $g\,(x) = \dfrac{1}{a}\, g_0\,(x)$, and $h\,(x) = \dfrac{1}{b}\, h_0\,(x)$, where $a, b \in D$ and $g_0\,(x)$ and $h_0\,(x) \in D\,[x]$. We also can write

$g_0\,(x) = c\,(g_0)\, g_1\,(x)$, $h_0\,(x) = c\,(h_0)\, h_1\,(x)$, where $c\,(g_0)$ and $c\,(h_0)$ are contents of $g_0\,(x)$ and $h_0\,(x)$ respectively, and $g_1\,(x)$ and $h_1\,(x)$ are primitive in $D\,[x]$.

If we put $c\,(g_0) = \alpha$, $c\,(h_0) = \beta$. Then

$$f(x) = \frac{\alpha.\beta}{ab}\, g_1\,(x)\,.\,h_1\,(x).$$

Since $g_1\,(x)$ and h_1, (x) are primitive, $g_1\,(x)\,.\,h_1\,(x)$ is primitive. It follows that $c\,(f) = c\,(g_1\,h_1)$, that is, $a\,b = \alpha\,.\,\beta \Rightarrow f\,(x) = g_1\,(x)\,.\,h_1\,(x)$. Hence $f(x)$ is reducible in $Q\,[x]$.

Example 8.3.6: $x^4 + 3x + 1$ is a primitive polynomial over integers. It is irreducible over the rational numbers since it has no linear factor as it has no rational zero. Thus if it factors, it must factor into the product of two quadratic polynomials with rational coefficients. By preceding theorem it must have quadratic factors with integral coefficients. Since it is monic, then

$$\begin{aligned} r^4 + 3x + 1 &= (x^2 + ax + b)\,(x^2 + cx + d) \\ &= x^4 + (a + c)\,x^3 + (b + d + a\,c)\,x^2 + (bc + ad)\,x + bd. \end{aligned}$$

comparing the coefficient of like powers of x we obtain,

$$\begin{aligned} a + c &= 0 \\ b + d + ac &= 0 \\ bc + ad &= 3 \\ bd &= 1 \end{aligned}$$

Thus, $b = d = 1$ or $b = d = -1$, while $a = -c$, so $a\,c = -a^2 = -(b + d) = \pm 2$. No integral solution exists.

Corollary: *Let (D, +, .) be a UFD. If f (x) is an irreducible polynomial of D [x], then f (x) is a prime in D [x].*

Proof: Let $f\,(x) \in D\,[x]$ be an irreducible element. If $\deg f\,(x) = 0$, then $f\,(x) \in D$ and it is prime by Lemma 8.3.1 So let $\deg f\,(x) \geq 1$ and let $f\,(x) \mid g\,(x)\,.\,h\,(x)$ in $D\,[x]$. Then we have to show that either $f\,(x) \mid g\,(x)$ or $f(x) \mid h\,(x)$. Let $(Q, +, .)$ be the field of quotients of D. Then clearly $f\,(x) \mid g\,(x)\,.\,h\,(x)$ in $Q\,[x]$.

Since $f(x)$ is irreducible in $D[x]$, it is also irreducible in $Q[x]$ by Lemma 8.3.8. Hence $f(x)$ is prime in $Q[x]$ and so $f(x) | g(x)$ or $f(x) | h(x)$ in $Q[x]$.

Now let $f(x) | g(x) . h(x)$ in $D[x]$, then

$f(x) | g(x) . h(x)$ in $Q[x]$

Since $f(x)$ is prime in $Q[x]$, then

either $f(x) | g(x)$ in $Q[x]$

or $f(x) | h(x)$ in $Q(x)$.

This implies either $f(x) | g(x)$ in $D[x]$ or $f(x) | h(x)$ in $D[x]$. Hence $f(x)$ is prime in $D[x]$.

Now we shall prove very important.

Theorem 8.3.2: *If (D, +, .) is a UFD, then (D [x], +, .) is also a UFD.*

Proof: We have the triple $(D[x], +, .)$ is an integral domain. We have to show that every non-zero and non-unit element $f(x)$ of $D[x]$ can be written as product of finite irreducibles in $D[x]$ and this product of irreducibles is unique.

We shall prove the theorem by induction on deg $f(x)$. If deg $f(x) = 0$, then $f(x) \in D$ and so $f(x)$ has a unique factorization.

If deg $f(x) > 0$, then we can write

$f(x) = dg(x)$, $d \in D$ and $g(x)$ is primitive of $D[x]$. Since $d \in D$, d has a unique factorization. If $g(x)$ is reducible, $g(x) = g_1(x) . g_2(x)$, deg $g_1(x) <$ deg $g(x)$, deg $g_2(x) <$ deg $g(x)$. Since $g(x)$ is primitive, $g_1(x)$ and $g_2(x)$ are both primitive.

So by induction $g_1(x)$ and $g_2(x)$ must have unique factorizations. That is, $g_1(x)$ and $g_2(x)$ can be expressed as the product of finite number of irreducibles in $D[x]$ and irreducibles in $D[x]$ are primes in $D[x]$ by the corrollary of Lemma 8.3.11. Hence $f(x) = dg(x)$ can be written as the product of finite number of irreducibles (primes) in $D[x]$.

We shall prove uniqueness by induction on deg $f(x)$. If deg $f(x) = 1$, $p_1 p_1(x) = q_1 q_1(x) \Rightarrow p_1 = q_1 \Rightarrow p_1(x) = q_1(x)$. Hence the theorem is true.

Now suppose that deg $f(x) = n > 1$ and

$$\begin{aligned} f(x) &= p_1 . p_2 ... p_n \, p_1(x) . p_2(x) .... p_n(x) \\ &= q_1 . q_2 ... q_m \, q_1(x) . q_2(x) ... q_m(x) \end{aligned} \quad (1)$$

are two factorization of $f(x)$ into primitive irreducibles factors $p_i(x)$ and $q_i(x)$ with deg $p_i(x) > 0$, deg $q_j(x) > 0$, and $p_i, q_j \in D$ are also irreducibles.

Since $p_i(x)$ and $q_j(x)$ are primitive, $p_1(x) ... p_n(x)$ and $q_1(x) ... q_m(x)$ are primitive. It follows that the content $c(p_1 . p_2 .... p_n) = c(q_1 . q_2 ... q_m)$, that is

$p_1 . p_2 ... p_n = q_1 . q_2 ... q_m$.

So

$p_1(x) . p_2(x) ... p_n(x) = q_1(x) . q_2(x) ... q_m(x)$

$\Rightarrow p_1(x) | q_j(x)$ for some $j = 1, 2, ... m$

Hence $p_1(x)$ and $q_j(x)$ are associates. So

$p_1(x) \ldots p_2(x) \ldots p_n(x) = p_1(x)\, q_i(x) \ldots q_{j-1}(x)\, g_{j+1}(x) \ldots q_m(x).$

or $f_1(x) = p_2(x) \ldots p_n(x) = q_1(x) \ldots q_{i-1}(x) \ldots q_{j+1}(x) \ldots q_m(x)$, $deg\ f_1(x) < n$.

We now make use of induction that the decomposition is unique for polynomials of degree less than n. So the decomposition $f_1(x)$ is unique. Hence $n - 1 = m - 1 \Rightarrow n = m$. Hence in the decomposition of $f(x)$ the number of irreducible factors are same and every $p_i(x)$ is an associate of some $q_j(x)$. Hence the decomposition of $f(x)$ is unique.

Example 8.3.7: *Factor the element*

$f(x) = 2x^3 + 3x^2 + 2x + 3,$

when $(F, +, .) = (Z/(5), +, .)$.

Solution: Using the preceding theorem and zeros of $f(x)$, we have

$f(1) = 2 \odot_5 1^3 +_5 3 \odot_5 1^2 +_5 2 \odot_5 1 +_5 3 = 0$

$f(2) = 2 \odot_5 2^3 +_5 3 \odot_5 2^2 +_5 2 \odot_5 2 +_5 3 = 0$

$f(3) = 2 \odot_5 3^3 +_5 3 \odot_5 3^2 +_5 2 \odot_5 3 +_5 3 = 0$

So that $(x - 1)$, $(x - 2)$, $(x - 3)$ are irreducible factors of $f(x)$.

Since $-1 = 4$, $-2 = 3$, $-3 = 2$, then factors of $f(x)$ are

$(x + 4)$, $(x + 3)$, and $(x + 2)$.

Theorem 8.3.3: *(Eisenstein's Irreducibility criterion). Let $(U, +, .)$ be a UFD and $(Q, +, .)$ its field of quotients. Let $f(x) = a_0 + a_1 x \ldots. + a_n x^n$ in $D[x]$. If there is prime $p \in D$ such that $p \mid a_0, p \mid a_1, \ldots p \mid a_{n-p}, p \mid a_n$ and $p^2 \mid a_0$, then $f(x)$ is irreducible in $Q[x]$.*

Proof: Suppose that $f(x) = b(x) . c(x)$ in $D[x]$, where

$b(x) = b_0 + \ldots. + b_r x^r,\ b_r \neq 0$

and $c(x) = c_0 + \ldots. + c_m x^m,\ c_m \neq 0$

Then

$a_0 = b_0 c_0,$

$a_1 = b_0 c_1 + b_1 c_0,$

.

.

.

$a_n = b_r c_m$ with $r + m = n$.

Since $p \mid a_0$, and $p^2 \mid a_0$, then either $p \mid b_0$ or $p \mid c_0$. We choose $p \mid c_0$ then $p \mid b_0$.

Since $p \mid a_n$, $p \mid b_r$ and $p \mid c_m$.

Now consider the integers $b_0, b_1, \ldots, b_r$, where $p \mid b_0$ and $p \mid b_r$. Hence there exists a first element $b_k \in D$ such that $p \mid b_k$, and $p \mid b_i$, $i = 0, 1, \ldots, k - 1$. Now we consider the coefficient of x_k in the product $b(x) . c(x)$.

$a_k = \Sigma b_i c_i$ with $i + j = k < r < r + m = n$ or $a_k = b_0 c_k + \ldots + b_k C_0$.

$p \mid b_i$, $i = 0, 1, 2, \ldots k - 1$, $p \mid b_0\, c_k$, $p \mid b_1\, c_{k-1} \ldots\, p \mid b_{k-1}\, c_1$

$p \mid c_0$, $p \mid b_k \Rightarrow p \mid b_k\, c_c$ and consequently $p \mid a_k$.

But it is a contradiction to the given that $p \mid a_k$, $k = 0, 1, 2, \ldots, n - 1$. Hence $f(x)$ cannot be factored and $f(x)$ is indeed irreducible.

Example 8.3.8: $x^2 + 2x + 2$, $x^3 + 2x^2 + 2x + 2$, $x^{100} + 3x + 3$, and $x^{10} - 15x^2 \dagger 135$ are irreducible over R by Eisenstein's criterion.

Example 8.3.9: For any prime p, $f = \dfrac{x^p - 1}{x - 1} = x^{p-1} + \ldots + 1$ is irreducible over the rationals.

Here Eisenstein's criterion is not immediately applicable, but a simple trick makes it so. We consider the polynomial.

$$g(x) = f(x + 1) = \frac{(x+1)^p - 1}{(x+1) - 1} = \frac{(x+1)^p - 1}{x} = x^{p-1} + \ldots + p.$$

As a consequence of the binomial theorem, we have

$$f(x + 1) = \frac{1 + px + \dfrac{p(p-1)x^2}{21} + \dfrac{p(p-1)(p-2)x^3}{31} \ldots + x^p - 1}{x}$$

$$= x^{p-1} + \ldots + p$$

So Eisenstein's criterion is satisfied by $f(x + 1)$. Hence $f(x + 1)$ is irreducible in $Q[x]$.

For example in $x^2 + 1 \in Z[x]$. Eisenstein's criterion is not applicable. If we put $x + 1$ in place of x, we have

$f(x + 1) = f(x + 1)^2 + 1 = x^2 + 2x + 2$, then Eisenstein's criterion is applicable. Hence $f(x + 1)$ is irreducible in $Q[x]$. So $x^2 + 1$ is irreducible in $Q[x]$.

PROBLEMS

1. If $(D, +, .)$ is an integral domain with unit element, prove that any unit in $D[x]$ must already be a unit in D.
2. Show that the domain $(D, +, .)$, where $D = \{a + b\sqrt{6} \mid a, b \in Z\}$ is not a *UFD.*
3. Show that $(Z\sqrt{-7} +, .)$ is not a UFD.
4. Let $(R, +, .)$ be commutative ring with no non-zero nil potent elements (that is, $a^n = 0 \Rightarrow a = 0$). If $f(x) = a_0 + a_1 x + \ldots + a_m x^m$ in $R[x]$ is a zero divisor, prove that there is an element $b \neq 0$ in R such that $ba_0 = ba_1 = \ldots = ba_m = 0$
5. Do problem 4 dropping the assumption that R has no non-zero nilpotent elements.
6. Find all the irreducible polynomials of degree ≤ 3 in $Z_2[x]$.
7. Prove that in $R[x]$, where $(R, +, .)$ is a commutative ring, $f(x) \mid g(x) \Rightarrow f(a) \mid g(a)$, $\forall\, a \in R$.

8. Let $f(x) \in Z[x]$. Show that if $f(x)$ is monic and irreducible in $Z_p[x]$, $f(x)$ is irreducible in $Z[x]$, use this to show that $x^4 + 3x + 1$ is irreducible in $Q(x)$ where $(Q, +, .)$ is the fieid of rational numbers.

9. If p is a prime, prove that the polynomial $x^n - p$ is irreducible over the rationals.

10. If a is rational and $(x - a)$ divides a monic polynomial with integral coefficients, prove that a must be an integer.

11. If $(D, +, .)$ is UFD, then prove that any $f(x) \in D[x]$ is an irreducible element of $D(x)$ if and only if either $f(x)$ is an irreducible elements of R or $f(x)$ is an irreducible primitive polynomial in $D[x]$.

12. Let $(K, +, .)$ be a quotient field of a UFD $(R, +, .)$ Then every irreducible element of R is a prime element of $R[x]$.

CHAPTER 9

VECTOR SPACES

9.1 DEFINITION AND EXAMPLES OF VECTOR SPACES

So far we have studied the mathematical systems with one binary operation and with two binary operations known as groups and rings, integral domains, and fields. In the present chapter, we shall discuss the matter of combining two different mathematical systems (algebraic systems) into a single system known as a vector space.

To combine two algebraic systems we have to define two binary operations which are called internal binary operation and external binary operation respectively. That is, if $(V, +)$ and $(F, +, .)$ be two systems, the system $(F, +,.)$ is a field. Then the binary operation f: $VXV \to V$ is called internal binary operation (addition) and the binary operation g: $FXV \to V$ is called external binary operation (scalar multiplication). Now we define the required mathematical systems.

Definition 9.1.1: (Vector Space)–A system $((V, \oplus), (F, +, .), .)$ is called a vector space (linear space) over the field $(F, +, .)$ if and only if

(*a*) $(F, +, .)$ is a field (scalars) whose identity elements with respect to addition and multiplication are denoted by 0 and 1 respectively.

(*b*) $(V, \oplus)$ is a commutative group, whose elements are called vectors, with respect to the operation of addition of two vectors, (called internal operation). That is,

(1) for each pair of vectors $\alpha, \beta \in V$, vector $\alpha \oplus \beta \in V$.

(2) $\oplus$ addition is commutative, $\alpha \oplus \beta = \beta \oplus \alpha$.

(3) $\oplus$ addition is associative, $(\alpha \oplus \beta) \oplus \gamma = \alpha \oplus (\beta \oplus \gamma)$.

(4) there is a unique vector O in V, called the zero vector, such that $\alpha \oplus O = O \oplus \alpha = \alpha$, for all $\alpha \in V$.

(5) for each vector $\alpha \in V$, there is unique vector $- \alpha \in V$ such that $\alpha + (- \alpha) = O$.

(*c*) A rule $\odot$ or (external operation), called scalar multiplication, which associates with each scalar $a \in F$ and a vector $\alpha \in V$, a vector $a \odot \alpha \in V$, called the product of a and α, in such way that for all $a, b \in F$ and all $\alpha, \beta \in V$:

(1) $(a + b) \odot \alpha = (a \odot \alpha) \oplus (b \odot \alpha)$,

(2) $a \odot (\alpha \oplus \beta) = (a \odot \alpha) \oplus (a \odot \beta)$,

(3) $(a \,.\, b) \odot \alpha = a \odot (b \odot \alpha)$,

(4) $1 \odot \alpha = \alpha$, where 1 is the field identity.

Fortunately, the notation used above can be simplified. From the type of letters involved it is always clear whether we are adding two scalars or two vectors, or multiplying two scalars, or a scalar and a vector. Therefore, we can use + to indicate both types of addition and juxtaposition to indicate both type of multiplications.

We shall denote a vector space over the field $(F, +, .)$ by $V(F)$.

We observe that the hypothesis $1x = x$ is quite essential; without it, every field and commutative group would yield a vector space under the trivial scalar multiplication $c\,\alpha = 0$ for all $c \in F$, $\alpha \in V$.

Example 9.1.1: Let $(F, +, .)$ be any field, and let V be the set of all n-tuples $\alpha = (x_1, x_2 ... x_n)$ of scalars $x_i \in F$. If $\beta = (y_1\ y_2 ... y_n)$ of scalars $y_i \in F$, the sum of α and β is defined by $(\alpha + \beta) = (x_1 + y_1, x_2 + y_2, ..., x_n + y_n)$.

The product of scalar c and vector α is defined by $c\,\alpha = (cx_1, cx_2 ...... cx_n)$.

Then $V(F)$ is a vector space over the field $(F, +, .)$.

The fact that this vector addition and scalar multiplication satisfy conditions given in the definition 9.1.1 of vectors space is easy to verify, using the similar properties of addition and multiplication of elements of F the verification of vector space axioms are left to the reader.

This space is called n-tuple space, denoted by $V_n(F)$ or F^n.

Example 9.1.2: Let $(F, +, .)$ be any field and let m and n be positive integers. Let V be the set of all $m \times n$ matrices over the field F.

$A = [a_{ij}]$ of scalars $a_{ij} \in F$,

where $i = 1, 2, ...m$, and $j = 1, 2, ...,n$.

If $B = [b_{ij}]$ of scalars $b_{ij} \in F$ and $i = 1, 2 ......m$. $J = 1, 2, ...,n$,

the sum of A and B is defined by

$$A + B = [a_{ij} + b_{ij}].$$

The product of scalar $k \in F$ and vector A is defined by $kA = [ka_{ij}]$, called scalar multiplication.

Then we observe that the system $(V, +)$ is an abelian group under addition.

Now, to prove $V(F)$ is a vector space, we verify other conditions of vector space.

Let $A, B \in V$, and $k_1, k_2 \in F$, then

$$\begin{aligned}(1)\ (k_1 + k_2) A &= (k_1 + k_2) [a_{ij}] = [(k_1 + k_2) a_{ij}] \\ &= [k_1 a_{ij} + k_2 a_{ij}] \\ &= [k_1 a_{ij}] + [k_2 a_{ij}] = k_1 A + k_2 A.\end{aligned}$$

(2) $k_1 (A + B) = k_1 [a_{ij} + b_{ij}]$
$= [k_1 (a_{ij} + b_{ij})] = [k_1 a_{ij} + k_1 b_{ij}]$
$= [k_1 a_{ij}] + [k_1 b_{ij}] = k_1 A + k_1 B.$

(3) $(k_1 k_2) A = k_1 k_2 [a_{ij}] = k_1 [k_2 a_{ij}]$
$= k_1 (k_2 A).$

(4) $I . A = 1 \{a_{ij}] = [1 . a_{ij}] = [a_{ij}] = A,$

where 1 is the identity element of the field.

Thus $V(F)$ is a vector space.

Example 9.1.3: Let $(R, +, .)$ be the field of real numbers and $(C, +, .)$ be the field of complex numbers then the field $(C, +, .)$ of complex numbers forms a vector space over the field $(R, +, .)$ of real numbers.

Example 9.1.4: Let $(F, +, .)$ be a field and $(F[x], +, .)$ be the ring of polynomials in the indeterminant x with coefficients from F. Then the operations necessary to give $F[x]$ a vector space structure are the followings:

if $f(x) = a_0 + a_1 x + a_2 x^2 + ...+ a_n x^n$ in $F[x]$

and $g(x) = b_0 + b_1 x + b_2 x^2 + ...+ b_m x^m$ in $F[x]$

where $m \leq n$, then

$f(x) + g(x) = (a_0 + b_0) + (a_1 + b_1) x + ...(a_m + b_m) x_m +...+ a_n x^n$ and for $c \in F$,

$cf(x) = (c . a_0) + (c . a_1) x + (c . a_2) x^2 + ...+ (c . a_n) x^n$ *belong to* $F[x]$.

Let us retain symbol $F[x]$ for this vector space, in preference to the correct but awkward $F[x] (F)$.

It is referred to as the space of polynomial functions over the field $(F, +, .)$.

Example 9.1.5: Let $(F, +, .)$ be any field and let S be any non-empty set. Let V be the set of all functions from the set S into F.

The sum of two vectors $f, g \in V$ is the vector $f + g$, *i.e.*, the function from S into F, defined by

$$(f + g) (s) = f(s) + g(s).$$

The product of the scalar c and the function f is the function cf defined by

$$(cf) (s) = cf (s).$$

Now, we shall verify that the operations we have defined satisfy vector space conditions.

(*a*) Since addition in F is commutative,

$$f(s) + g(s) = g(s) + f(s) \Rightarrow (f + g) (s) = (g + f) (s)$$

so
$$f + g = g + f.$$

(*b*) Since addition in F is associative,

$$f(s) + [g(s) + h(s)] = [f(s) + g(s)] + h(s)$$

so for each $s \in S$, $f + (g + h) = (f + g) + h$.

(c) The unique zero vector is the zero function which assigns to each element $s \in S$ the scalar $f(s) = 0 \in F$.

(d) For each f in V, $(-f)$ is the function which is given by

$(-f)(s) = -f(s)$.

Thus, $(V, +)$ is a commutative group.

Now, we see that the scalar multiplication satisfies the properties: For all $f, g \in F$, we have

(i) $((c_1 + c_2) f) = (c_1 + c_2) f(s)$

$$= c_1 f(s) + c_2 f(s) = (c_1 f + c_2 f)(s)$$

for $s \in S$, so $(c_1 + c_2) f = c_1 f + c_2 f$,

(ii) $(c_1 c_2) f(s) = c_1 (c_2 f(s))$

for each $s \in S$, so $(c_1 c_2 f) = c_1 (c_2 f)$,

(iii) $(c_1 (f + g))(s) = c_1 ((f + g)(s))$

$$= c_1 (f(s) + g(s))$$

$$= c_1 (f(s) + g(s))$$

$$= (c_1 f + c_1 g)(s)$$

for each $s \in S$,

so $c_1 (f + g) = c_1 f + c_1 g$,

and

(iv) $(I \,.\, f)(s) = 1.\, f(s) = f(s)$

for each $s \in S$, so $1 \,.\, f = f$.

Which shows that the set of functions from S into F is a vector space over the field $(F, +, .)$

Theorem 9.1.1: *If $V(F)$ is a Vector Space and $x \in V$, $y \in V$, $c \in F$, then*

(1) $0\, x = \bar{0}$

(2) $c\, \bar{0} = \bar{0}$,

(3) $-(c\, x) = (-c)\, x = c\,(-x)$,

(4) $a\,(x - y) = ax - ay$,

(5) $ax = \bar{0}, a \neq 0 \Rightarrow x = \bar{0}$,

(6) $ax = ay,\ a \neq 0 \Rightarrow x = y$.

Proof: **(1)** To establish (1), we see the field result

$$0 + 1 = 1. \text{ Then}$$

$$0x + x = 0x + 1x = (0 + 1)\, x = 1x = x = \bar{0} + x.$$

Since $(V, +)$ is a commutative group, the cancellation law gives

CHAPTER 9

$$0\,x = \bar{0}.$$

(2) The proof of this follows from the group result

$$\bar{0} + x = x, \text{ we have}$$

$$c\,\bar{0} + c\,x = c\,(\bar{0} + x) = cx = \bar{0} + cx.$$

Again the cancellation law gives

$$c\,\bar{0} = \bar{0}$$

(3) To obtain this, we observe that

$$\bar{0} = 0\,x = (c + (-c))\,x = c\,x + (-c)x.$$

This means that $(-c)\,x = -(c\,x)$. Similarly

$$\bar{0} = c\,\bar{0} = c\,[x + (-x)] = c\,(x) + c\,(-x) \text{ which implies}$$

$$c\,[-x] = -\,[cx]$$

(4) We have

$$a\,(x - y) = a[x + (-y)] = ax + a(-y)$$

$$= ax - (a\,y)$$

since by (3) we have $a\,(-y) = -\,(a\,y)$.

(5) We have

$$ax = \bar{0},\; a \neq 0 \Rightarrow a^{-1} \text{ exists such that}$$

$$a\,.\,a^{-1} = a^{-1}\,.\,a = 1.$$

By premultiplying $ax = \bar{0}$ by a^{-1}, we obtain

$$a^{-1}\,(a\,.\,x) = a^{-1}\,.\,\bar{0}$$

$$\Rightarrow \quad (a^{-1}\,.\,a)\,.\,x = \bar{0}$$

$$\Rightarrow \quad 1x = \bar{0}$$

$$\Rightarrow \quad x = \bar{0}$$

(6) We have

$$ax = ay,\; a \neq 0 \Rightarrow a^{-1} \text{ exists such that}$$

$$a^{-1}\,.\,a = a\,.\,a^{-1} = 1.$$

Premultiplying $ax = ay$ by a^{-1}, we have

$$a^{-1}.\,(ax) = a^{-1}\,(ay)$$

$$\Rightarrow \quad (a^{-1}\,.\,a)\,.\,x = (a^{-1}\,.\,a)\,.\,y$$

$$\Rightarrow \quad 1\,.\,x = 1\,.\,y$$

$$\Rightarrow \quad x = y.$$

9.2 SUBSPACES

Whenever, a mathematical system has been considered, the question of sub-systems arose. In the case of vector spaces, the sub-vector spaces are referred to as subspaces.

Definition 9.2.1: Let $V(F)$ be a vector space over the field $(F, +, .)$ and $W \subseteq V$, $W \neq \phi$. Then $W(F)$ is a subspace of $V(F)$ provided that $W(F)$ satisfies the vector space axioms when equipped with the same operations defined on $V(F)$.

Since $W \subseteq V$, much of the algebraic structure of $W(F)$ is inherited from $V(F)$. The minimum conditions that $W(F)$ must satisfy to be space are:

(1) $(W, +)$ is a subgroup of $(V, +)$.

(2) W is closed under scalar multiplication.

From the definition of subspace it is clear that usual criterion for deciding whether $(W, +)$ is sub-group of $(V, +)$ is to see if W is closed under differences. Thus, $\alpha, B \in W$, then $\alpha - \beta \in W$. The second of the above conditions implies that $-\alpha = (-1)\alpha \in W$, whenever $\alpha \in W$. Because $\alpha - \beta = \alpha + (-\beta)$, and condition (2), together with the closure of W under addition is sufficient to guarantee that W be closed under differences.

Example 9.2.1: Every vector space $V(F)$ has two trivial subspaces, namely $V(F)$ itself and the zero subspace $\{0\}(F)$. Subspaces distinct from $V(F)$ are said to be proper subspaces.

Example 9.2.2: Consider the set W of vectors in $V_3(F)$ whose components add up to zero:

$W = \{(a_1, a_2, a_3) \mid a_1 + a_2 + a_3 = 0\}$.

If $(a_1, a_2, a_3), (b_1, b_2, b_3) \in W$, then

their difference

$$(a_1, a_2, a_3) - (b_1, b_2, b_3) = (a_1 - b_1, a_2 - b_2, a_3 - b_3)$$

and

$$\begin{aligned}&(a_1 - b_1 + a_2 - b_2 + a_3 - b_3)\\ &= (a_1 + a_2 + a_3) - (b_1 + b_2 + b_3) = 0 - 0 = 0\end{aligned}$$

which shows W is a sub-group under addition.

And for any $c \in F$,

$$c(a_1, a_2, a_3) = (ca_1, ca_2, ca_3)$$

and

$$\begin{aligned}&ca_1 + ca_2 + ca_3\\ &= c(a_1 + a_2 + a_3) = 0.\end{aligned}$$

Thus, W is closed under scalar multiplication. Hence $W(F)$ is a subspace of $V_3(F)$.

Example 9.2.3: Let W denote the collection of all elements from the space $M_2(F)$ of the form

$$\begin{bmatrix} a & b \\ -b & a \end{bmatrix}$$

Let $\begin{bmatrix} a & b \\ -b & a \end{bmatrix}, \begin{bmatrix} c & d \\ -d & c \end{bmatrix} \in W$, then

$$\begin{bmatrix} a & b \\ -b & a \end{bmatrix} - \begin{bmatrix} c & d \\ -d & c \end{bmatrix} = \begin{bmatrix} a-c & b-d \\ -(b-d) & a-c \end{bmatrix} \in W,$$

$$k\begin{bmatrix} a & b \\ -b & a \end{bmatrix} = \begin{bmatrix} k.a & k.b \\ -(k.b) & k.c \end{bmatrix} \in W,$$

Hence $W(F)$ is subspace of the vector space $M_2(F)$.

Example 9.2.4: Let W be the set of all symmetric matrices of order n taken from the set V of all $m \times n$ matrices over the field $(F, +, .)$ of the form

$$[a_{ij}] = [a_{ij}],\ i, j = 1, 2, 3, \ldots, n \text{ and } a_{ij} \in F.$$

Then $W(F)$ *is a* subspace of the vector space $V(F)$.

Theorem 9.2.1: *A nonvoid subset W of V is a subspace W (F) of the vector space V (F) over the field (F, +, .) if and only if for every* $\alpha, \beta \in W$ *and every* $c \in F$,

(1) $\alpha + \beta \in W$.

(2) $c\,\alpha \in W$.

Proof: The proof of the theorem consists of two parts. In the first part if $W(F)$ is a subspace of the space $V(F)$, then $W(F)$ is closed under addition of vectors and scalar multiplication of vectors by scalars. That is, if $\alpha, \beta \in W$, $c \in F$, then $\alpha + \beta \in W$ and $c\,\alpha \in W$.

Conversely, assume that W is closed under the operations. Then $(-1)\,\alpha = -\alpha \in W$ for every $\alpha \in W$, and $\alpha + (-\alpha) = \bar{0} \in W$. Vectors addition in W is the same as in V, so it is associative and commutative, and W forms a commutative group. The other axioms are satisfied in W because those properties are inherited from V.

Theorem 9.2.2: *W (F) is a subspace of the vector space V (F) if and only if* $\phi \neq W \subseteq V$ *and* $c\,\alpha + d\,\beta \in W$, *whenever* $\alpha, \beta \in W$, $c, d \in F$.

Proof: If $W(F)$ is a subspace, then by definition, W is nonempty and contains $c\,\alpha + d\,\beta$ for all $\alpha, \beta \in W$, $c, d \in F$.

Conversely, if this condition holds, that $c\,\alpha + d\,\beta \in W$ for all $\alpha, \beta \in W$, $c, d \in F$, W must contain $1\,\alpha + 1\beta = \alpha + \beta$ and $c\,\alpha + 0\,\beta = c\,\alpha$ for every $\alpha, \beta \in W$, $c \in F$. Accordingly, W is closed with respect to the vector space operations.

Corollary 1: $W(F)$ is a subspace of the vector space $V(F)$ if and only if $\phi \neq W \subset V$ and $c\,\alpha + \beta \in W$ whenever $\alpha, \beta \in W$, $c \in F$.

Proof: Let $W \neq \phi$ and $c\,\alpha + \beta \in W$, whenever $\alpha, \beta \in W$, $c \in F$.

Since W is non-empty, there is a vector $\alpha \in W$ and hence $(-1)\,\alpha + \alpha = \bar{0} \in W$. Then if α is any vector in W and $c \in F$, the vector

$$c\,\alpha = c\,\alpha + \bar{0} \in W.$$

In particular, $(-1)\,\alpha = -\alpha \in W$, finally if $\alpha\,\beta \in W$, then $\alpha + 1\,\beta = 1\,\alpha + \beta \in W$. Thus $W(F)$ is a subspace. Conversely, if W is a subspace of V, $\alpha, \beta \in W$, $c \in F$ certainly $c\,\alpha + \beta \in W$.

Theorem 9.2.3: *If W_1 (F) and W_2 (F) are subspaces of the vector space V (F), then so is ($W_1 \cap W_2$) (F).*

Proof: The set $W_1 \cap W_2$ is not empty, for $\bar{0} \in W_1 \cap W_2$.

Let $\alpha, \beta \in W_1 \cap W_2$ and $c \in F$, then since both W_1 (F) and W_2 (F) are subspaces,

$$\alpha + \beta \in W_1,\ c\,\alpha \in W_1,$$

and $$\alpha + \beta \in W_2,\ c\ \ \alpha \in W_2,$$

that is, $$\alpha + \beta \in W_1 \cap W_2$$

and $$c\,\alpha \in W_1 \cap W_2.$$

Thus ($W_1 \cap W_2$) (F) is a subspace of V (F), since $W_1 \cap W_2$ is closed under its operations.

Corollary: If W_i (F) is an indexed collection of subspaces of the vector space V (F), then ($\cap W_i$) (F) is a subspace of V (F).

Theorem 9.2.4: *If W_1 (F) and W_2 (F) are subspaces of the vector space V (F), then ($W_1 \cup W_2$) (F) is not necessarily a vector space.*

Proof: ($W_1 \cup W_2$, +) should be a sub-group of the group (V, +) to be a subspace. But we have seen in the chapter of group theory that if (W_1, +) and (W_2, +) are sub-groups of the group (V, +), then ($W_1 \cup W_2$ +) is not necessarily a sub-group of (V, +). By the definition of a subspace, if ($W_1 \cup W_2$, +) is not a sub-group of (V, +), then ($W_1 \cup W_2$) (F) is not a subspace.

CHAPTER 9

Example 9.2.5: Let M_2 (F) be the vector space of all 2-rowed square matrices over the field (F, +, .). Take U to be the set of all scalar matrices

$$\begin{bmatrix} a & 0 \\ 0 & a \end{bmatrix},\ a \in F.$$

and W to be the set of all matrices of the form

$$\begin{bmatrix} 0 & b \\ -b & 0 \end{bmatrix},\ b \in F.$$

U (F) and W (F) are both subspaces of M_2 (F),

For, if $\begin{bmatrix} a & 0 \\ 0 & a \end{bmatrix}, \begin{bmatrix} c & 0 \\ 0 & c \end{bmatrix} \in U,\ c_1, d_1 \in F,$

then $$c_1 \begin{bmatrix} a & 0 \\ 0 & a \end{bmatrix} + d_1 \begin{bmatrix} c & 0 \\ 0 & c \end{bmatrix} = \begin{bmatrix} c_1 a & 0 \\ 0 & c_1 . a \end{bmatrix} + \begin{bmatrix} d_1 . c & 0 \\ 0 & d_1 . c \end{bmatrix}$$

$$= \begin{bmatrix} c_1 a + d_1 c & 0 \\ 0 & c_1 a + d_1 c \end{bmatrix} \in U$$

and if $\begin{bmatrix} 0 & b_1 \\ -b_1 & 0 \end{bmatrix}, \begin{bmatrix} 0 & b \\ -b & 0 \end{bmatrix} \in W,\ c, d \in F,$

then $c\begin{bmatrix} 0 & b_1 \\ -b_1 & 0 \end{bmatrix} + d\begin{bmatrix} 0 & b \\ -b & 0 \end{bmatrix} = \begin{bmatrix} 0 & cb_1 \\ -cb_1 & 0 \end{bmatrix} + \begin{bmatrix} 0 & db \\ -db & 0 \end{bmatrix}$

$$= \begin{bmatrix} 0 & cb_1 + bd \\ -(cb_1 + db) & 0 \end{bmatrix} \in W.$$

But we note that while the matrices

$$\begin{bmatrix} 1 & 0 \\ 0 & 1 \end{bmatrix} \text{ and } \begin{bmatrix} 0 & 1 \\ -1 & 0 \end{bmatrix}$$

belong to U ∪ W, and their sum

$$\begin{bmatrix} 1 & 0 \\ 0 & 1 \end{bmatrix} + \begin{bmatrix} 0 & 1 \\ -1 & 0 \end{bmatrix} = \begin{bmatrix} 1 & 1 \\ -1 & 1 \end{bmatrix} \notin U \cup W.$$

Thus, $(U \cup W)$ (F) is not a subspace of M_2 (F).

Example 9.2.6: Let V_3 (F) be a vector space. Let

U = $\{(a, 0, 0) \mid a \in F\}$ and $W = \{(0, a, 0) \mid a \in F\}$.

We can easily verify that $U(F)$ and $W(F)$ are both subspaces. Further, we note that if $(a, 0, 0)$ and $(0, b, 0)$ belong to $U \cup W$, then their sum $(a, 0, 0) + (0, b, 0) = (a, b, 0)$ belongs to neither U nor W.

Thus, $(U \cup W)$ (F) is not a subspace of the vector space V_3 (F).

Definition 9.2.2: If U and W are non-empty subsets of the vector space V (F), their (linear) sum is defined to be the set

$$U + W = [u + w \mid u \in U, w \in W\}.$$

Theorem 9.2.5: *If $U(F)$ and $W(F)$ are subspaces of the vector space $V(F)$, then $(U + W)$ (F) is also a subspace.*

Proof: The sum $U + W \neq \phi$, since $\bar{0}, \in U$, $\bar{0} \in W$, hence

$$\bar{0} = \bar{0} + \bar{0} \in U + W.$$

Let $\alpha, \beta \in U + W$, then

$\alpha = u_1 + w_1$ and $\beta = u_2 + w_2$, where $u_1, u_2 \in U$, $w_1, w_2 \in W$. For scalars $c, d \in F$,

$$c\,\alpha + d\,\beta = c\,(u_1 + w_1) + d\,(u_2 + w_2)$$
$$= (cu_1 + du_2) + (cw_1 + dw_2).$$

Since U (F) and W (F) are subspaces of the space V (F), it follows $cu_1 + du_2 \in U$ and $cw_1 + dw_2 \in W$. Thus

$$c\,\alpha + d\,\beta \in U + W.$$

Hence $(U + W)$ (F) is a subspace.

Theorem 9.2.6: *If $U(F)$ and $W(F)$ are subspaces of the vector space $V(F)$, then $(U + W)(F)$ is the smallest subspace containing both U and W, that is, $(U + W)(F)$ is the subspace generated by $U \cup W$, $(U + W)(F) = [U \cup W](F)$.*

Proof: We have $U + W = \{u + w \mid u \in U, w \in W\}$. So for any scalars $a, b \in F$, $a\,u \in U$ and $b\,w \in W$ and $a\,u + b\,w \in U + W\}$. Therefore every element of $U + W$ can be written as the linear combination of elements of $U \cup W$. But the set $[U \cup W]$ of all linear combinations of elements of $U \cup W$ forms a subspace $[U \cup W](F)$ of $V(F)$. This implies $U + W \subseteq [U \cup W]$.

Since $\quad U \subset U + W,\ W \subset U + W$, then

$U \cup W \subseteq U + W$. That is, $(U + W)(F)$ is the subspace of the vectorspace $V(F)$ which contains $U \cup W$. As the $[U \cup W](F)$ is the smallest subspace containing $U \cup W$, we have $[U \cup W] \subseteq U + W \Rightarrow U \cup W = [U \cup W]$ since $U + W \subseteq [U \cup W]$.

Theorem 9.2.7: *Let $U(F)$ and $W(F)$ be two subspaces of the vector space $V(F)$. Then following conditions are equivalent:*

(1) $U \cap W = \{\bar{0}\}$.

(2) *Every vector $\alpha \in U \cup W$ is uniquely represented in form $\alpha = u + w$ where $u \in U$, $w \in W$.*

Proof: Let $U \cap W = \{\bar{0}\}$. Let the vector $\alpha \in U \cup W$ be represented in two forms

$$\alpha = u_1 + w_1 = u_2 + w_2,\ u_1, u_2 \in U,\ w_1, w_2 \in W.$$

Then $u_1 - u_2 = w_2 - w_1$, but $u_1 - u_2 \in U$ and $w_2 - w_1 \in W$, since $U(F)$ and $W(F)$ are subspaces.

The equation $u_1 - u_2 = w_2 - w_2$ means that

$$u_1 - u_2 \in U \cap W \text{ and } w_2 - w_1 \in U \cap W,$$

therefore, it follows that

$$u_1 - u_1 = 0,\ w_2 - w_1 = 0$$

or

$$u_1 = u_2,\ w_1 = w_2.$$

In other words, α is uniquely represented in the form $u + w$. Conversely, assume the statement (2) holds and the vector $\beta \in U \cap W$. We may express β in two different ways as the sum of a vector in U and *a* vector in W, namely $\beta = \beta + \bar{0}$, where $\beta \in U$, $\bar{0} \in W$ and $\beta = \bar{0} + \beta$, where $\bar{0} \in U$, $\beta \in W$. The uniqueness gives $\beta = \bar{0}$, that is, $U \cap W = \{\bar{0}\}$.

Definition 9.2.3: The two subspaces $U(F)$ and $W(F)$ of the vector space $V(F)$ are complementary if $U \cap W = \{\bar{0}\}$ and $U + W = V$. The sum $U + W$ is called the direct sum and denoted by $U \oplus W$, if $U + W = V$ and $U \cap W = \{\bar{0}\}$.

Definition 9.2.4: Two subspaces $U(F)$ and $W(F)$ are said to be disjoint if $U \cap W = \{\bar{0}\}$.

Theorem 9.2.8: *Let $U(F)$ and $W(F)$ be two subspaces of the vector space $V(F)$. Every vector $x \in V$ is uniquely expressible in the form $x = u + w$, where $u \in U$, $w \in W$, if and only if $V = U \oplus W$.*

Proof: Let $V = U \oplus W$. Then $U \cap W = \{\bar{0}\}$ and $U + W = V$.

Let $\quad v \in V$ and let $v = x_1 + x_2 = y_1 + y_2,\ x_1, y_1 \in U$

CHAPTER 9

$$x_2, y_2 \in W$$

So $x_1 + x_2 = y_1 + y_2 \Rightarrow x_1 - y_1 = y_2 - x_2$

Since $x_1 - y_1 \in U$, $y_2 - x_2 \in W$, then

$$\begin{aligned} x_1 - y_1 = y_2 - x_2 &\Rightarrow x_1 - y_1 \in W \text{ and } y_2 - x_2 \in U \\ &\Rightarrow x_1 - y_1, y_2 - x_2 \in U \cap W = \{\bar{0}\} \\ &\Rightarrow x_1 - y_1 = \bar{0} \text{ and } y_2 - x_2 = \bar{0} \\ &\Rightarrow x_1 = y_1 \text{ and } x_2 = y_2. \end{aligned}$$

Hence every element of V is uniquely expressed as a sum of an element of U and of an element of W.

Conversely, if $x = u + w \in U + W$, where $x \in V$, then

$$U + W = V.$$

$$\begin{aligned} x \in U \cap W &\Rightarrow x \in U \text{ and } x \in W \\ &\Rightarrow x = u + \bar{0} \text{ and } x = \bar{0} + w \\ &\Rightarrow u + \bar{0} = \bar{0} + w \\ &\Rightarrow u = w = \bar{0}. \end{aligned}$$

Since the representation of x is unique. Hence $x = \bar{0} \Rightarrow U \cap W = \{\bar{0}\}$.

Hence the theorem.

Remark: For given a subspace $U(F)$ of $V(F)$ there exist several complementary subspaces. In case of V_2 (F), if

$$U = \{(a, 0) \mid a \in F\},\ W = \{(0, a) \mid a \in F\},\ W' = \{(a, a) \mid a \in F\}.$$

Then W (F) and W' (F) are complementary subspaces of the subspace U (F).

Theorem 9.2.9: *Let $W(F)$ be a subspace of the vector space $V(F)$ and the cosets $\alpha + W$, $\beta + W \in V/W$. If vector addition and scalar multiplication are given by*

$$(\alpha + W) + (\beta + W) = (\alpha + \beta) + W,$$

$$c\,(\alpha + W) = c\,\alpha + W,\ c \in F,$$

Then (V/W) (F) is itself a vector space, known as the quotient space of V by W.

Proof: Since $(V, +)$ is a commutative group, the sub-group $(W, +)$ of the group $(V, +)$ is also commutative. Hence $(W, +)$ is a normal sub-group of $(V, +)$. Thus, $(V/W, +)$ is a quotient group of V *by* W. The quotient group $(V/W, +)$ is a commutative group since $(V, +)$ is a commutative group.

For other vector space axioms, we note that if $\alpha + W$, $\beta + W \in V/W$, $c_1, c_2 \in F$, then

(*i*)
$$\begin{aligned} (c_1 + c_2)(\alpha + W) &= (c_1 + c_2)\,\alpha + W \\ &= (c_1\,\alpha + c_2\,\alpha) + W \\ &= (c_1\,\alpha + W) + (c_2\,\alpha + W) \\ &= c_1\,(\alpha + W) + c_2\,(\alpha + W). \end{aligned}$$

(ii) $(c_1 \cdot c_2)(\alpha + W) = (c_1 \cdot c_2)\alpha + W)$

$= c_1 \cdot ((c_2 \alpha) + W)$

$= c_1 (c_2 (\alpha + W))$

(iii) $c [(\alpha + W) + (\beta + W)] = c [(\alpha + \beta) + W]$

$= (c (\alpha + \beta)) + W$

$= (c \alpha + c \beta) + W$

$= (c \alpha + W) + (c \beta + W)$

$= c (\alpha + W) + c (\beta + W).$

(iv) $1. (\alpha + W) = \alpha + W.$

this proves the theorem.

Definition 9.2.5: Let A be a set of vectors in a vector space $V(F)$. The subspace spanned by A is defined to be the intersection W of all subspaces of V which contain A. When A is a finite set of vectors, $A = \{a_1, a_2 .... a_n\}$, we shall simply call W the subspace spanned by the vectors $\alpha_1, \alpha_2,....\alpha_n$.

Definition 9.2.6: Suppose $V(F)$ is a vector space and $\alpha_1, \alpha_2,.....\alpha_n \in V$, any finite sum of the form

$$c_1\alpha_1 + c_2\alpha_2 +.....+ c_n \alpha_n,$$

where each $c_1 \in F$, is said to be a linear combination (over F) of the vectors $\alpha_1, \alpha_2,......\alpha_n$. A linear combination is called trivial if all its coefficients $c_i = 0$, and non-trivial if at least one coefficient is different from zero.

If the set $A = \{\alpha_1, \alpha_2,....\alpha_n\}$, the set $[A]$ of all linear combinations of vectors of A is the collection of all finite sums of the form

$$c_1 \alpha_1 + c_2 \alpha_2 +.....+ c_n \alpha_n.$$

where $\qquad c_i \in F, \alpha_i \in A,$ and $i = 1, 2,......n.$

***Theorem 9.2.10:** The set [A] of all linear combinations of any non-void subset A of vectors of V is a subspace [A] (F) of the vector space V (F).*

Proof: It is clear that the set $[A]$ of all linear combinations contains A. So the set $[A]$ is non-empty.

Let $\quad \alpha, \beta \in [A], c, d \in F.$

Since $\alpha, \beta \in [A]$, α and β are linear combinations of $\alpha_1, \alpha_2, \alpha_3.......\alpha_n \in A$. Thus

$$\alpha = \alpha_1 \alpha_1 + \alpha_2 \alpha_2 +.......+ \alpha_n \alpha_n \ (a_i \alpha F),$$

and $\qquad \beta = b_1 \alpha_1 + b_2 \alpha_2 +.......+ b_n \alpha_n. \ (b_i \in F)$

we note that

$$c\alpha + d\beta = c (\Sigma a_i \alpha_i) + d (\Sigma b_i \alpha_i)$$

$$= \Sigma (ca_i) \alpha_i + \Sigma (db_i) \alpha_i$$

$$= \Sigma (ca_i + db_i) \alpha_i \in [A],$$

since $ca_i + db_i \in F$,

which shows $[A]$ (F) is a subspace of the vector space $V(F)$.

Example 9.2.7: Let $(F, +, .)$ be a subfield of the field $(C, +, .)$ of complex numbers. Let $A = \{\alpha_1, \alpha_2, \alpha_3\}$

Suppose

$$\alpha_1 = (1, 2, 0, 3, 0)$$
$$\alpha_2 = (0, 0, 1, 4, 0),$$
$$\alpha_3 = (0, 0, 0, 0, 1).$$

A vector α is in the subspace $[A]$ (F) of F^5 if and only if there exists $c_1, c_2, c_3, \in F$ such that

$$\alpha = c_1 \alpha_1 + c_2 \alpha_1 + c_3 \alpha_3$$
$$= c_1 (1, 2, 0, 3, 0) + c_2 (0, 0, 1, 4, 0) + c_3 (0, 0, 0, 0, 1)$$
$$= (c_1, 2c_1, c_2, 3c_1 + 4c_2, c_3).$$

Thus, $[A]$ consists of the vectors of the form

$$\alpha = (c_1, 2c_1, c_2, 3c_1 + 4c_2, c_3),$$

Thus for example $(-3, -6, -1, -5, 2) \in [A]$, whereas

$$(2, 4, 6, 7, 8) \notin [A].$$

Example 9.2.8: Let A be the subset of V_n (F) (or F^n) whose elements are the n-vectors $e_1 = (1, 0, 0,......0)$, $e_2 = (0, 1,......, 0)$,........, $e_n = (0, 0,......, 1)$; in general, e_k is the vector with 1 in the kth component position and 0 elesewhere :

$$e_k = (\delta_{1k}, \delta_{2k},.... \delta_{nk}),\ k = 1, 2,...., n,\ \delta_{ij} = 1,\ i = j,\ \delta_{ij} = 0,\ i \neq j.$$

Here $[A]$ (F), the subspace spanned by these vectors, is all of V_n (F).

Indeed, for any n-tube $(a_1, a_2.....a_n)$ over F, we have $(a_1, a_2,......a_n) = a_1 (1, 0, 0........0) + a_2 (0, 1, 0.....0) + ...+ a_n (0, 0......1)$

$$= a_1 e_1 + a_2 e_2 +.......+ a_n e_n \in [A].$$

Example 9.2.9: Let V (F) be the space of all polynomial functions over F.

Let S be the subset of V consisting of the polynomial functions f_1, f_2.... defined by

$$f_n (x) = x^n,\ n = 0, 1, 2,.......$$

Then $V(F)$ is the subspace spanned by the set S.

PROBLEMS

In the exercises below, the symbol F stands for an arbitrary field.

1. If $V(F)$ is a vector space, prove that

   (a) $n (cx) = (nc) x = c (nx)$

   for all $x \in V$, $c \in F$ and $n \in Z$.

(b) If $x \in V$ with $x \neq 0$, then $c_1 x = c_2 x \Rightarrow c_1 = c_2$.

(c) verify that

$(x_1 + x_2) + (x_3 + x_4) = [x_2 + (x_1 + x_3)] + x_4$

for all x_1, x_2, x_3, and x_4 in V.

2. Let V be the set of all pairs (x, y) of real numbers and $(F, +, .)$ be the field of real numbers. Define $(x, y) + (x_1, y_1) = (x + x_1, y + y_1)$, $c(x, y) = (cx, y)$.

Is V, with these operations, a vector space over the field of real numbers?

3. Let V be the set of all ordered pairs (x, y) of real numbers and let $(F, +, .)$ be a field of real numbers.

Define

$$(x, y) + (x_1, y_1) = (3y + 3y_1, -x - x_1),$$

$$c\,(x, y) = (3cy, -cx)$$

Verify that V, with these operations, is not a vector space over the field of real numbers.

4. Consider the set of all the triples of real numbers (a_1, a_2, a_3), subject to the conditions stated below. In each case determine whether the system forms a vector space relative to addition of triples and multiplication of a triple by a real number.

(1) $a_1 = 0$; a_2 and a_3 arbitrary.

(2) $a_1 = -a_3$; a_2 arbitrary.

(3) a_1, a_2, arbitrary, $a_3 = 1 + a_1 - a_2$.

(4) a_1, a_2 arbitrary, $a_1 = 3a_1 - 4a_2$

(5) $a_1\, a_2 \geq 0$, a_3 arbitrary.

5. Consider the set S of all solutions of a homogeneous differential equation of order n:

$$y^n + a_1(x)\, y^{n-1} + \ldots. + a_{n-2}(x)\, y'' + a_n(x)\, y' + a_n(x)\, y = 0.$$

Prove that S forms a vector space relative to addition and scalar multiples of functions.

6. If V is the space of all functions from R into R which are continuous on $-1 \leq x \leq 1$, which of the following subsets are subspaces of V?

(1) All differentiable functions.

(2) All polynomials of degree two.

(3) All polynomials of degrees less than five.

(4) All odd functions $f(-x) = -f(x)$.

(5) All even functions $f(-x) = f(x)$.

(6) All functions for which $f(0) = 0$.

CHAPTER 9

(7) All negative functions.

(8) All constant functions.

7. For each of the following sets W, determine whether $W(R)$ is a subspace of the vector space $V_n\ (R)$ over the field of real numbers.

(1) $W = \{(a_1, a_2 ..., a_n) \mid a_1 + a_2 + ... + a_n \neq 0\}$.

(2) $W = \{(a_2, a_2 ....., a_n) \mid a_1 = a_2 = ... = a_n\}$.

(3) $W = \{(a_1, a_2, ..., a_i) \mid a_1 a_2 = 0\}$.

(4) $W = \{(a_1, a_2, ..., a_n) \mid a_k \in Z, \forall k\}$.

8. Consider the set of all triples of real numbers, (a_1, a_2, a_3) subject to conditions stated below. In each case determine the set forms a subspace of $V_3\ (R)$.

(*i*) $a_1 = 5a_2$

(*ii*) $a_1 + a_2 = a_3^{-1}$.

(*iii*) $a_1 \geq 0$.

(*iv*) $a_1 = 1$.

(*v*) a_1 is a rational number.

9. Let $U\ (F)$ and $W\ (F)$ be subspace of the vector space $V\ (F)$. Prove that $(U \cup W)\ (F)$ forms a subspace of $V(F)$ if and only if $U \supseteq W$ or $W \supseteq U$.

10. Find all of the subspace of $V_2\ (Z_2)$ and $V_3\ (Z_2)$.

11. In the vector space $V_3\ (F)$, define the subsets U and W by $U = \{(a, b, a + b) \mid a, b \in F\}$, $W = \{(c, c, c) \mid c \in F\}$ verify $V_3 = U \oplus W$.

12. If S and T are subsets of the vector space $V\ (F)$ over the field F, then

(*a*) $S \subseteq T \Rightarrow [S] \subseteq [T]$,

(*b*) $[S\ U\ T] \Rightarrow [S] + [T]$,

(*c*) $[[S]] \Rightarrow [S]$

(*d*) $S \subseteq [T] \Rightarrow [S] \subseteq [T]$.

13. Let $V\ (R)$ be the vector space of all functions form the set R of real numbers into itself, and let

$U = \{f \in V \mid f(-x) = -f(x)\}$, and

$W = \{f \in V \mid f(-x) = f(x)\}$. Then prove that

$V = U \oplus W$.

14. Determine whether $W(F)$ is a subspace of the indicated vector space:

(*a*) $V_3\ (F)$; for fixed scalars $a_1, a_2, a_3, \in F$,

$W = \{(x_1, x_2, x_3) \mid a_1 x_1 + a_2 x_2 + a_3 x_3 = 0\}$.

(*b*) $V\ (F)$; $W = \{(a_1, a_2, ... a_n, 0, 0, ...) \mid a_k \in F, n \in N\}$.

(*c*) $F_\infty\ [x]$; W consists of all polynomials of degree greater than 4.

(*d*) $V_n\ (F)$; for fixed $n \times n$ matrix $[a_{ij}]$,

$W = \{x \in V_n\ (F) \mid [a_{ij}]\ x = 0\}$.

(e) $M_3(F)$, $W = \{\begin{bmatrix} a & b & 0 \\ 0 & a+b & 0 \\ 0 & 0 & b \end{bmatrix} \in M_3(F) \mid a, b \in F\}$.

15. Suppose $U(F)$, $W(F)$, are subspaces of the vector space $V(F)$. Prove that

(a) If $U \subseteq V$, $W \subseteq V$, then $U + W \subseteq V$,

(b) $(U \cap V) + (W \cap V) \subseteq (U + W) \cap V$,

(c) If $U \subseteq V$, then $U + (W \cap V) = (U + W) \cap V$.

9.3 LINEAR DEPENDENCE AND LINEAR INDEPENDENCE

We now come to one of the most useful concepts in the theory of vector spaces, that of linear dependence and linear independence. We give below some useful definitions.

Definition 9.3.1: Let $S = \{\alpha_1, \alpha_2, \ldots \alpha_n)$ be a finite subset of the vector space $V(F)$. (i) S is said to be linearly dependent (*L.D.*) if there exist n scalars.

$$c_1, c_2, \ldots, c_n \in F, \text{ not all zero, such that}$$

$$c_1 \alpha_1 + c_2 \alpha_2 + \ldots + c_n \alpha_n = \bar{0}$$

(ii) On the other hand, S is said to be linearly independent (*L. I*) if and only if every equation of the form

$$c_1 \alpha_1 + c_2 \alpha_2 + \ldots + c_n \alpha_n = 0$$

implies that $c_1 = c_2 = c_3 = \ldots = c_n = 0$

(iii) A vector β which can be expressed in the form

$$\beta = c_1 \alpha_1 + c_2 \alpha_2 + \ldots + c_n \alpha_n.$$

is said to be a linear combination of vectors $\alpha_1, \alpha_2, \ldots \alpha_n$ where $c_1 \in F$.

Example 9.3.1: Let $(F, +, .)$ be a subfield of the complex numbers. In $V_3(F)$ the vectors

$$\alpha_1 = (3, 0, -3),$$
$$\alpha_2 = (-1, 1, 2),$$
$$\alpha_3 = (4, 2, -2),$$
$$\alpha_4 = (2, 1, 1),$$

are linearly dependent, since

$$c_1 \alpha_1 + c_2 \alpha_2 + c_3 \alpha_3 + c_4 \alpha_4$$
$$= c_1 (3, 0, -3) + c_2 (-1, 1, 2) + c_3 (4, 2, -2) + c_4 (2, 1, 1)$$
$$= (3c_1 - c_2 + 4c_3 + 2c_4, c_2 + 2c_3 + c_4, -3c_1 + 2c_2 - 2c_3 + c_4) = (0, 0, 0)$$

implies $3c_1 - c_2 + 4c_3 + 2c_4 = 0,$

$$c_2 + 2c_3 + c_4 = 0$$
$$-3c_1 + 2c_2 - 2c_3 - 1c_4 = 0,$$

which give $c_1 = 2, c_2 = 2, c_3 = -1, c_4 = 0,$

CHAPTER 9

The vectors $e_1 = (1, 0, 0)$

$e_1 = (0, 1, 0)$

$e_2 = (0, 0, 1)$

are linarly independent, since

$$c_1 e_1 + c_2 e_2 + c_3 e_3 = c_1 (1, 0, 0) + c_2 (0, 1, 0) + c_2 (0, 0, 1)$$
$$= (c_1, c_2, c_3) = (0, 0, 0)$$

implies $c_1 = c_2 = c_3 = 0.$

Example 9.3.2: In R [x], the vector space of polynomials in x over the field of real numbers (R, +, .), consider the three polynomials $1 + x + 2x^2$, $2 - x + x^2$, $-4 + 5x + x^2$ of degree two, since

$2 (1 + x + 2 x^2) + (-3) (2 - x + x^2) + (-1) (-4 + 5x + x^2) = 0$, then

the given polynomials are linearly dependent.

Example 9.3.3: Show that the vectors

(1, 2, 3), (3, –2, 0) form a linearly independent set.

Let $k_1 (1, 2, 3) + k_2 (3, -2, 0) = (0, 0, 0)$

which means $k_1 + 3 k_2 = 0,$

$2 k_1 - 2 k_2 = 0$

$3 k_1 + 0k_2 = 0.$

Therefore, on solving these equations, we have

$$k_1 = 0 = k_2.$$

Hence, the given vectors are linearly independent.

Here we shall consider the vectors taken from the given vector space $V(F)$ over the field (F, +, .).

Theorem 9.3.1: *If* β *is linear combination of the set of vectors* $\alpha_1, \alpha_2,..\alpha_n$ *then the set* $\{\beta, \alpha_1, \alpha_2,...\alpha_n\}$ *is linearly dependent.*

Proof: We have

$$\beta = a_1 \alpha_1 + a_2 \alpha_2 +....+ a_n \alpha_n, a_i, \in F$$

or

$$\beta - (a_1 \alpha_1 + a_2 \alpha_2 +....+ a_n \alpha_n,) = \bar{0}$$

It is clear that the coefficient of β is not zero, which is sufficient to show that the set $(\beta_1 \alpha_1....\alpha_n\}$ is a linearly dependent.

Theorem 9.3.2: *If* $\{\alpha_1, \alpha_2, ..., \alpha_n\}$ *is a linearly independent set and* $\{\beta, \alpha_1, \alpha_2,....., \alpha_n)$ *is linearly dependent set,* β *is a lineary combination of* $\alpha_1, \alpha_2.....\alpha_n$.

Proof: We have

$$a\beta + a_1 \alpha_1 + a_1\alpha_2 +...+ a_n \alpha_n = \bar{0}, a, a_i \in F$$

where the coefficients are not all zero.

Now, if $a \neq 0$,

$$\beta = a^{-1} (-1) (a_1 \alpha_1 + a_1 \alpha_2 + ... + a_n \alpha_n)$$

So that β is a linear combination of $\alpha_1, \alpha_2, ... \alpha_n$.

If $a = 0$,

$$a_1 \alpha_1 + a_2 \alpha_2 + ... a_n \alpha_n = \bar{0}$$

implies $a_1 = a_2 = a_3 = ... = a_n = 0$, since $\alpha_1, \alpha_2, \alpha_3 ... \alpha_n$ are linearly independent and we have a contradiction.

Theorem 9.3.3: *Every super set of a linearly dependent set is linearly dependent.*

Proof: If $S_1 \subseteq S_2$, then S_2 is called the super set of S_1.

Let $S_2 = \{\alpha_1, \alpha_2, ... \alpha_p, \alpha_{p+1}, ... \alpha_n\}$,

and $S_1 = \{\alpha_1, \alpha_2, ... \alpha_p,\}$ be linearly dependent set. Thus, S_2 is a super set of the set S_1.

Since S_1 is linearly dependent, there exists a relation of the form

$$a_1 \alpha_1 + a_2 \alpha_2 + ... + a_p \alpha_p = \bar{0}, \; a_i \in F \quad ....(1)$$

where $a_1, a_2, ... a_p$ are not all zero.

We have relation of the form

$$a_1 \alpha_1 + a_2 \alpha_2 + ... + a_p \alpha_p + a_{p+1} + a_{p+1} + ... + a_n \alpha_n = \bar{0} \quad ....(2)$$

where $a_{p+1} = a_{p+2} = ... = a_n = 0$,

which shows the set $\{\alpha_1, \alpha_2, ... \alpha_{p-1}, \alpha_p, , \alpha_{p+1} ..., \alpha_n\}$ is linearly dependent since all a_i's are not zero in the relation

Theorem 9.3.4: *A set*

$$\{\alpha_1, \alpha_2, ... \alpha_r\}$$

of non-zero vectors is linearly dependent if, and only if, some

$$\alpha_p, \; 2 \leq p \leq r$$

is a linear combination of the preceding vectors.

Proof: Suppose that the set is L.D., being non-zero set consisting only of a_1, is L.I. Now there must exist a number $p \geq 2$ such that

$\{\alpha_1, \alpha_2, ... \alpha_{p-1}\}$ is linearly independent (L.I.) and

$\{\alpha_1, \alpha_2, ... \alpha_{p-1}, \alpha_p\}$

is a linearly dependent set for, at most $p = r$ since $\alpha_1, \alpha_2, ..., \alpha_{p-1}, \alpha_p$ are L.D. and $\alpha_1, \alpha_2, ..., \alpha_{p-1}$ are L.I. then α_p is a linear combination of $\alpha_1, \alpha_2, ... \alpha_{p-1}$ by Theorem 9.32.

Conversely, if α_p is a linear combination of $\alpha_1, \alpha_2, ... \alpha_{p-1}$, then the set $\alpha_1, \alpha_2 ...... \alpha_{p-1}, \alpha_p$ is linearly dependent. And any super set $(\alpha_1, \alpha_2, ..., \alpha_p, ..., \alpha_n\}$ of a linearly dependent set $\{\alpha_1, \alpha_2, ..., \alpha_p\}$ is linearly dependent by the Theorem 9.3.2.

Theorem 9.3.5: *If the set $A = \{\alpha_1, \alpha_2, ..., \alpha_n\}$ is a linearly independent set, then any vector $\alpha \in [A]$ is uniquely expressible as a linear combination of the vectors $\alpha_1, \alpha_2, ... \alpha_n$.*

Proof: Let α have two representations,

$$\alpha = a_1 \alpha_1 + a_2 \alpha_2 + ... + a_n \alpha_n = b_1 \alpha_1 + b_2 \alpha_2 + ... + b_n a_n.$$

This leads to the relation

$$[a_1 - b_1) \alpha_1 + (a_2 - b_2) \alpha_2 + ... + (a_n - b_n) \alpha_n = 0.$$

Since the vectors $\alpha_1, \alpha_2 ....., \alpha_n$ are linearly independent,

$$a_k - b_k = 0, \text{ for all } k = 1, 2, ...n$$

or $$a_k = b_k.$$

Exercise 9.3.1: If α, β, γ are L.I. vectors in $V(F)$,

Prove that $\alpha + \beta$, $\beta + \gamma$, $\gamma + \alpha$ are also L.I.

Solution: To show that the elements $a + \beta$, $\beta + \gamma$, $\gamma + \alpha$ are linearly independent, there are scalars x, y, z such that

$$x(\alpha + \beta) + y(\beta + \gamma) + z(\gamma + \alpha) = \bar{0}$$

or $$(x + z)\alpha + (x + y)\beta + (y + z)\gamma = \bar{0} \quad ...(1)$$

Since α, β, γ are linearly independent, then we contain from the relation (1), that

$$x + y = 0, \quad ...(2)$$

$$x + z = 0, \quad ...(3)$$

$$x + x = 0. \quad ...(4)$$

On adding (2), (3) and (4), we get

$$x + y + z = 0 \quad ...(5)$$

Subtracting (2), (3), and (4) from (5) respectively, we get

$$z = 0, x = 0, y = 0.$$

which shows, $\alpha + \beta$, $\beta + \gamma$, $\gamma + \alpha$ are linearly independent.

Exercise 9.3.2: Show that the set $\{\alpha_1 \alpha_2 \alpha_3\}$ is linearly dependent if the set $\{\alpha_1 + a\alpha_2 + b\alpha_3, \alpha_2, \alpha_3\}$ is linearly dependent, where $\alpha_1, \alpha_2, \alpha_3 \in V(F)$, $a, b \in F$.

Solution: Since the set $(\alpha_1 + a\alpha_2 + b\alpha_3, \alpha_2, \alpha_3\}$ is linearly dependent, there exist scalars $l, m, n, \in F$ not all zero, such that

$$l(\alpha_1 + a\alpha_2 + b\alpha_3) + m(\alpha_2) + n(\alpha_2) = \bar{0}$$

or $$l\alpha_1 + (al + m)\alpha_2 + (bl + n)\alpha_3 = \bar{0} \quad ...(1)$$

If $l \neq 0$, then $\alpha_1, \alpha_2, \alpha_3$ are linearly dependent.

If $l = 0$, the relation (1) reduces to the form

$$m\alpha_2 + n\alpha_3 = \bar{0}.$$

Since l, m, n are not all zero, and $l = 0$, at least one of m, n is non-zero. Which implies α_2, α_3 are linearly dependent and $\{\alpha_2, \alpha_3\} \subset \{\alpha_1, \alpha_2, \alpha_3\}$ which means $\{\alpha_1, \alpha_2, \alpha_3)$ is linearly dependent.

9.4 BASES AND DIMENSION

Theorem 9.4.1: *Let n vectors $v_1, v_2, ..., v_n$ span a vector space containing m linearly independent vectors $u_1, u_2, ..., u_m$, then $n \geq m$.*

Proof: Since $v_1, v_2, ...v_n$ span the space, u_1 can be expressed linear combination of the v_i. Therefore this equation can be solved for some one of the v_i, say v_k in terms of u_1 and the rest of the v_i, consequently the set consisting of u_1, and the rest of the v_i spans the vector space, since any linear combination of the v_i becomes a linear combination of u_1 and all the v_i except v_k when the expression for v_k interms of u_1, and other v_i is used to eliminate v_k. Then u_2 can be expressed as a linear combination of u_1 and all the v_i except v_k. Since u_i are linearly independent, some v_i must have a non-zero coefficient, and therefore this v_i can be expressed in terms of u_1, u_2 and the remaining $(n - 2)$ v_i, and these n vectors span the space. The process can be continued untill all m of the v_i vectors are used, and since at each stage one u_i vector is replaced, the number of vector v_i must have been at least as great as the number of vectors u_i. That is, $n \geq m$.

Theorem 9.4.2: *If two sets of linearly independent vectors span the same space, there are the same number of vectors in each set.*

Proof: If there are m vectors in one set and n in the other, then by Theorem 9.4.1 $m, \geq n$ and $n \geq m$, *and thus* $m = n$.

Definition 9.4.1: A basis of vector space is a linearly independent subset which generates (spans) the whole space. A vector space is finite-dimensional if and only if it has a finite basis.

Example 9.4.1: Let $(F, +, .)$ be a field and in vector space V_n (F) (or F^n) let S be the subset consisting of the vectors

$$e_1, e_2, ..., e_n \text{ defined by}$$

$$e_1 = (1, 0, 0, ...0).$$

$$e_2 = (0, 1, 0 ...... 0),$$

$$e_3 = (0, 0, 1 ...... 0),$$

$$\cdots \quad \cdots$$

$$\cdots \quad \cdots$$

$$e_n = (0, 0, 0, ...1).$$

Let $a_1, a_2, ...a_n$ be scalars in F and put

$$\alpha = a_1 e_1 + a_2 e_2 + ... + a_n e_n$$

Then $\alpha = (a_1, a_2, ...a_n)$.

This shows $e_1, e_2, ..., e_n$ span V_n (F) (or F^n). Since $\alpha = \bar{0}$ if and only if $\alpha_1 = \alpha_2 = ... = a_n = 0$, the vectors $e_1, e_2 ..., e_n$ are linearly independent. Thus the set $S = (e_1, e_2, .., e_n\}$ is a basis for V_n (F). This particular basis is called standard basis of V_n (F).

Example 9.4.2: Let M_n (F) be the vector space of n *x* n matrices over a field $(F, +, .)$. This space has a basis consisting of the n^2 matrices E_{ij}, where E_{ij} is the square matrix of order n having as its $(i, j)^{th}$ entry one and zeros elsewhere, any matrix $(a_{ij}) \in M_n$ (F) can be written as

$(a_{ij}) = a_{11} E_{11} + a_{12} E_{12} ... + a_{nn} E_{nn}$. Moreover $[a_{ij}] = \bar{0}$ if and only if $a_{11} = ...a_{nn} = 0$. Hence $E_{11}, E_{12}, ... E_{nn}$, are linearly independent over F.

Example 9.4.3: Let $F[x]$, the vector space of polynomials in x with coefficients from F. A basis for $F[x]$ is formed by the set

$$S = \{1, x, x^2, ..., x^n, ...\}.$$

By definition, each polynomial $p(x) = a_0 1 + a_1 x + ... + a_n x_n \in F[x]$ is linear combination of the elements of S. The set S is linearly independent. For any finite subset

$\{x^{n_1}, x^{n_2}, ..., x^{n_k}\}$, $(0 \leq n_1 < n_2 < .... < n_k)$, the relation

$c_1 x^{n_1} + c_2 x^{n_2} + ... + c_k x^{n_k} = \bar{0}$

holds if and only if $c_1 = c_2 = ... = c_k = 0$.

Corollary: If $V(F)$ is a finite dimensional vector space, then the two bases have the same (finite) number of elements.

Proof: Since $V(F)$ is a finite dimensional it has a finite basis

$$\beta_1, \beta_2, ...\beta_m$$

Then, by the Theorem 9.4.1 every basis of $V(F)$ is finite and contains no more than m elements. Thus if $\alpha_1, \alpha_2, ...\alpha_n$ is a basis, $n \leq m$. By the same argument $m \leq n$. Hence $m = n$.

Definition 9.4.2: The number of elements in a basis of the finite dimensional vector space $V(F)$ is known as the dimension of $V(F)$ denoted by $\dim V(F)$.

Example 9.4.4: If $(F, +, .)$ be a field, the dimension of $V_n(F)$ is n. For the standard basis of $V_n(F)$ contains n vectors.

Corollary: *Let $V(F)$ be n-dimensional vector space, then*

(*a*) any set of vectors in $V(F)$ which contains more than n vectors is linearly dependent.

(*b*) no set which contains less than n vectors can span $V(F)$.

Theorem 9.4.3: *Let S be a* L.I. *subset of a (finite dimensional) vector space $V(F)$. Suppose $\beta \in V$ which is not in the subspace $[S](F)$, then the set obtained by adjoining β to S is* L.I.

Proof: Let S have distinct vectors $\alpha_1, \alpha_2, ...\alpha_m$, and that $c_1 \alpha_1 + c_2 \alpha_2 + ... + c_m \alpha_m + b\beta = \bar{0}$, $c_i \in F$, then $b = 0$, for otherwise,

$$\beta = \frac{(-c_1)}{b}\alpha_1 + ... + \frac{(c_m)}{b}\alpha_m$$

and $\beta \in [S](F)$. Thus $c_1 . \alpha_1 + c_2 a_2 +, ... + c_m \alpha_m = \bar{0}$ and since S is a L.I. set, each $c_i = 0$, which completes the proof.

Theorem 9.4.4: *If $W(F)$ is a subspace of a finite-dimensional vector space $V(F)$, every linearly independent subset of W is finite and is a part of a (finite) basis for $W(F)$.*

Proof: Let S_n be a linearly independent (L.I.) subset of W. If S is L.I. subset of W containing S_n, then S is also a L.I. subset of V. Since $V(F)$ is a finite dimensional, S contains no more than dim V elements. Thus there is a maximal L.I. subset S of W which contains S_n. Since S is a maximal L.I. subset of W containing S_n, the preceeding lemma shows that

W is the subspace $[S]$ spanned by S. Hence S is a basis for $W(F)$ and S_n is the part of a basis of $W(F)$.

Corollary 2: In a finite dimensional vector space $V(F)$ every non-empty L.I. set of vectors is a part of a basis.

Corollary 2: If $W(F)$ is a proper subspace of a finite dimensional vector space $V(F)$, then W is finite dimensional and dim $W <$ dim V.

Theorem 9.4.5: *Any linearly independent set of vectors in n-dimensional space $V(F)$ can be extended to a basis.*

Proof: *Let $B_k = \{\alpha_1, \alpha_1, \ldots\ldots \alpha_k\}$ be linearly independent and $A = \{\beta_1, \beta_2, \ldots, \beta_n\}$ be a basis of $V(F)$. If $\beta_i \in B_k$, for $i = 1, 2, \ldots n$, then $[B_k] = V$. Otherwise, for some $\beta_i \notin B_k$, $\{B_k\ \beta_i\}$ is linearly independent thus the original set has been extended to a larger independent set and theorem follows by repeating the argument until the enlarged set spans $V(F)$.*

Example 9.4.5: Show that the set

$$(1, 2, 1), (2, 1, 0), (1, -1, 2) \text{ forms a basis for } V_3(F).$$

Solution: Directly let (x, y, z) be any vector in $V_3(F)$. We want to know if there exist a_1, a_2, a_3 such that

$$a_1(1, 2, 1) + a_2(2, 1, 0) + a_3(1, -1, 2) = (x, y, z).$$

For this, we get the equations

$$a_1 + 2a_2 + a_3 = x \quad \ldots(1)$$

$$2a_1 + a_2 - a_3 = y \quad \ldots(2)$$

$$a_1 + 0a_2 + 2a_3 = z \quad \ldots(3)$$

From (1) and (2) $\quad a_1 + a_2 = \frac{1}{3}(x + y),$

From (1) and (3) $\quad a_1 + 2a_3 = 2zx - z.$

$$a_2 = \frac{5x}{9} - \frac{y}{9} - \frac{z}{3} \quad \ldots(4)$$

$$a_1 = \frac{-2x}{9} + \frac{4}{9}y + \frac{z}{3} \quad \ldots(5)$$

Putting the value of a_1 in (3) $a_3 = \frac{x}{9} - \frac{2y}{9} + \frac{z}{3}$ ...(6)

Hence $S = \{(1, 2, 1), (2, -1, 0) (1, -1, 2)$ spans $V_3(F)$. As the dimension of $V_3(F)$ is 3 and S has only 3 vectors, these must be all independent.

Theorem 9.4.6: *If $W(F)$ is a subspace of a finite dimensional vector space $V(F)$, then*

(a) $W(F)$ is finite dimensional with dim $W \leq$ dim V,

(b) dim W = dim V if and only if $W = V$.

Proof: (a) Let $V(F)$ be a finite dimensional vector space with dim $V = n$. Then there is a linear independent set $\{x_1, x_2, \ldots, x_n\}$ of n vectors which spans $V(F)$. So the L.I. set $\{x_1, x_2, \ldots, x_n\}$ is *a* basis of $V(F)$. Since $W(F)$ is a subspace of $V(F)$, it must have basis, that is,

CHAPTER 9

there are L.I. vectors $y_1, y_2, \ldots, y_m$ which span $W(F)$, that is, dim $W = m$. The set of L.I. vectors $y_1, y_2, \ldots y_m$ can be extended to the set of L.I. independent vectors $y_1, y_2, \ldots y_m, y_{m+1}, \ldots, y_n$ which spans $V(F)$. That is, the basis of $W(F)$ is a subset of the basis of $V(F)$ which implies dim $W \le$ dim V.

(b) Let dim W = dim $V = n$. Then there exist a basis $\{y_1, y_2, \ldots, y_n\}$ for $W(F)$ and a basis $\{x_1, x_2, \ldots, x_n]$ for $V(F)$. Two basis of a vector space have the same number of L.I. vectors by corrollary of Theorem 9.4.1 This implies both basis $\{y_1, y_2, \ldots, y_n\}$ and $\{x_1, x_2, \ldots, x_n\}$ span $V(F)$. Hence $W = V$.

Conversely, suppose that $W = V$. We have by (*a*) dim $W \le$ dim V. if the finite number of vectors $x_1, x_2, \ldots, x_n$ span $V = W$ then any set of independent vectors in $W(F)$ contains no more than m elements, that is, $n \le m$ or dim $V \le$ dim W. Thus dim W = dim V.

Theorem 9.4.7: *If W (F) is a subspace of a finite dimensional vector space V(F), then the quotient space (V/W) (F) is also a finite dimensional and*
dim V = dim W + dim V/W.

Proof: Since $W(F)$ is a subspace of a finite dimensional vector space $V(F)$, $W(F)$ is finite dimensional with dim $W \le$ dim V. Let $\{x_1, x_2, \ldots, x_n\}$ be a basis for $W(F)$. But this basis $\{x_1, x_2, \ldots, x_n\}$ *can* be extended to the basis of $V(F)$ by adding vectors $y_1, y_2, \ldots, y_m$. Thus we have the basis $\{x_1, x_2, \ldots, x_n, y_1, y_2, \ldots, y_m\}$ for $V(F)$.

Any vector $x \in V$ can be written as

$$x = a_1 x_1 + a_1 x_2 + \ldots + a_n x_n + b_1 y_1 + \ldots + b_{m}, y_m$$

where a'^s, and b'^s are in F.

Since $W(F)$ is a subspace, $a_1 x_1, + a_2 x_2 + \ldots + a_n x_n \in W$,

and

$$a_1 x_1 + a_2 x_2 + \ldots + a_n x_n = x - (b_1 y_1 + \ldots + b_m y_m) \in W.$$

$$\Rightarrow \quad x + W = b_1 y_1 + b_2 y_2 + \ldots + b_m y_m + W$$

$$= (b_1 y_1 + W) + (b_2 y_2 + W) + \ldots + (b_m y_m + W)$$

$$= b_1 (y_1 + W) + b_2 (y_2 + W) + \ldots + b_m (y_m + W).$$

Thus the coset $x + W \in V/W$ is expressed as the linear combination of the cosets $y_1 + W, y_2 + W_1, \ldots, y_m + W$, that is, they span $(V/W(F)$.

Now we have to show that $y_1 + W, y_2 + W, \ldots, y_m + W$, form a basis for $(V/W)(F)$. For this, we assume

$$c_1 (y_1 + W) + c_2 (y_2 + W) + \ldots + c_m (y_m + W) = \bar{0} + W = W.$$

or $$c_1 y_1 + W + c_2 y_2 + W + \ldots + c_m y_m + W = W$$

or $$(c_1 y_1 + c_2 y_2 + \ldots + c_m y_m) + W = W \text{ which implies}$$

$c_1 y_1 + c_2 y_2 + \ldots + c_m y_m \in W$ which must be a linear combination of the vectors $x_1, x_2, \ldots, x_n$ of W. So

$$c_1 y_2 + c_2 y_2 + \ldots + c_m y_m = d_1 x_1 + d_2 x_2 + \ldots + d_n x_n,$$

for some $d_1, d_2, \ldots, d_n$ of F.

This means that

$$c_1 y_1 + c_2 y_2 + \ldots + c_m y_m - d_1 x_1 - d_2 x_2 \ldots - d_n x_n = \bar{0}$$

Since $x_1, x_2, \ldots, x_n, y_1, \ldots, y_m$, are linearly independent, $c_1 = c_2 = \ldots = c_m = d_1 = d_2 = \ldots = d_2 = 0$.

Thus $y_1 + W, y_2 + W_2, \ldots, y_m + W$ form a basis of (V/W) (F) which has m elements.

Hence $\dim (V/W) = m = (m + n) - n = \dim V - \dim W$.

or $\dim V = \dim W + \dim \dfrac{V}{W}$

Example 9.4.6: We consider the vector space V_3 (F) and subspace W (F), where $W = \{(a, 0, 0) \mid a \in F\}$.

Hence, the quotient space (V_3/W) (F) consists of the cosets of the form $(o, b, c) + W$. W (F) has the basis $\{(1, 0, 0)\}$ which can be extended to the basis of V (F) by adjoining the vector $(0, 1, 0)$ and $(0, 0, 1)$.

The cosets $(0, 1, 0) + W$ and $(0, 0, 1) + W$ will form the basis of (V_3/W) (F).

If $(a, b, c) \in V_3$, then

$(a, b, c) + W = b\,[(0, 1, 0) + W] + c\,[(0, 0, 1) + W]$.

Thus $\dim V/W = 2 = \dim V - \dim W$.

Theorem 9.4.8: *It $U(F)$ and $W(F)$ are subspaces of a finite dimensional vector space $V(F)$, then*

dim $(U + W) = \dim U + \dim W - \dim (U \cap W)$

If $V = U \oplus W$,

dim $(U \oplus W) = \dim U + \dim W$.

CHAPTER 9

Proof: Since $U(F)$ and $W(F)$ are subspaces of $V(F)$, then $(U \cap W)$ (F) and $(U + W)$ (F) are also subspaces of $V(F)$. Since $V(F)$ is finite dimensional, all its subspaces are finite dimensional.

Let $\{x_1, x_2, \ldots, x_n]$ be a basis for $(U \cap W)$ (F).

Since $U \cap W \subset U$ and $U \cap W \subset W$, the basis $\{x_1, \ldots x_n\}$ of $(U \cap W)$ (F) can be extended to a basis $\{x_1, \ldots x_n, u_1, \ldots, u_m\}$ of U (F) by adjoining vectors $u_1, u_2, \ldots, u_m$ and to a basis $\{x_1, x_2, \ldots x_n, w_1, w_2, \ldots, w_r\}$ of W (F) by adjoining vectors $w_1, w_2, \ldots, w_r$, respectively. Thus the basis of $U(F)$ has $n + m$ elements and of W (F) $n + r$ elements.

Since $(U + W)$ (F) is the smallest subspace generated by $U \cup W$, every element of $U + W$ can be represented as a linear combination of the vectors of the set

$\{u_1, u_2, \ldots, u_m, x_1, x_2, \ldots, x_n, w_1, w_2, \ldots, w_r\}$. Therefore this set spans $(U + W)$ (F). If this is linearly independent, then we can say $(U + W)$ (F) has dimension $m + n + r$. For this we assume that

$a_1 u_1 + a_2 u_2 + \ldots + a_m u_m + b_1 x_1 + \ldots + b_n x_n + c_1 w_1 + \ldots + c_r w_r = \bar{0}$ where all $a's$, $b's$, and $c's$ are in F.

Let $z = c_1 w_1 + c_2 w_2 + c_2 w_2 + \ldots + c_r w_r$, $z \in W$ as $W(F)$ is a subspace. Then, we have

$z = -[a_1 u_1 + \ldots + a_m u_m + b_1 x_1 + \ldots + b_n x_n] \in U$ as

$U(F)$ is a subspace with the basis $\{u_1, u_2, \ldots, u_m, x_1, \ldots, x_n\}$.

This means that $z \in U \cap W$. So z can be written as a linear combination of the basis elements $\{x_1, x_2,..., x_n]$ of $(U \cap W)$ (F). Therefore for some $d_1, d_2,..., d_n$ of F, we have

$$z = d_1 x_1 + d_2 x_2 + ... + d_n x_n.$$

Hence

$$z = c_1 w_1 + c_2 w_2 + ... + c_r w_r = d_1 x_1 + d_2 x_2 + ... + d_n x_n$$

$$\Rightarrow \quad c_1 w_1 + c_2 w_2 + ... + c_r w_r - d_1 x_1 - d_2 x_2 .... - d_n x_n = \bar{0}$$

$\Rightarrow \quad c_1 = c_2 = ... c_r = d_1 = d_2 = ... = d_n = 0$ since $w_1, w_2,..., w_r, x_1, x_2,...., x_n$ is a basis for W (F).

Since $\{x_1, x_2,..., x_n, u_1, u_2, ..., u_m\}$ is a basis of U (F), then $a_1 u_1 + a_2 u_2 + ... + a_m u_m + b_1 x_1 + b_2 x_2 + ... + b_n x_n = \bar{0}$

$a_1 = a_2 = ... = a_m = b_1 = ... = b_n = 0.$

Therefore this proves.

$a_1 u_1 + ... + a_m u_m + b_1 x_1 + ... b_n x_n + c_1 w_1 + c_2 w_2 + ... + c_r w_r = \bar{0} \Rightarrow a_1 = a_2 = ... = a_m = b_1 = ... = b_n = c_1 = ... = c_r = 0.$

Hence the set $\{u_1, u_2,..., u_m, x_1, x_2,...., x_n, w_1, w_2,..., w_r\}$ is L.I. and forms a basis of $(U + W)$ (F) with $\dim (U + W) = m + n + r$.

Hence

$$\dim (U + W) = m + n + r = (m + n) + (n + r) - n$$
$$= \dim U + \dim W - \dim (U \cap W).$$

In particular, if $U \oplus W = V$, then $U \cap W = \{\bar{0}\}$, $U + W = V$.

$$\text{So } \dim (U \oplus W) = \dim U + \dim W - \{\bar{0}\}$$
$$= \dim U + \dim W, \text{ since } \dim \{\bar{0}\} = 0.$$

Hence the theorem.

PROBLEMS

1. For each of the following vector spaces, determine whether the sets listed are linearly dependent or independent:

   (1) V_4 (F): $\{(1, 0, 0, 0), (1, 1, 0, 0), (1, 1, 1, 0), (1, 1, 1, 1)\}$.

   (2) $F[x]$: $\{x^2 + x - x, x^2 - x - 2, x^2 + x + 1\}$,

   (3) V_3 (Z_5): $\{(4, 1, 3), (2, 3, 1), (4, 1, 0)\}$,

   (4) V_3 (C): $\{1, 2, + i, 3), (2 - i, i, 1), (i, 2 + 3i, 2)\}$.

2. If F is the field of real, numbers, show the vectors (1, 1, 0, 0), (0, 1, –1, 0), (0, 0, 0, 3) in V_4 (F) are L.I. over F.

3. Show that the system of three vectors

   (1, 3, 2), (1, –7, –8), (2, 1, –1) of V_3 (R) is L.D.

4. In $V(R)$, where R is the field of real numbers, determine whether each of the following set of vectors is L.I. or L.D.

(1) (1, 1, 1), (l, 0, 1), (0, 1, 0,),

(2) (8, 4, 8), (2, 1, 2), (0, 0, 1),

(3) (1, –1, 1) (2, 1, –2), (8, 1, –4).

5. Show that the set

$$S = \{(1, 0, 0), (1, 1, 0), (1, 1, 1), (0, 1,0)\}$$

spans V_3 *(R)* but does not form a basis.

6. Show that the set of all 2 × 2 matrices over real numbers forms a vector space under the usual addition and scalar multiplication and the

set $$\left\{\begin{bmatrix}1 & 0\\0 & 0\end{bmatrix}, \begin{bmatrix}1 & 1\\0 & 0\end{bmatrix}, \begin{bmatrix}1 & 1\\1 & 0\end{bmatrix}, \begin{bmatrix}1 & 1\\1 & 1\end{bmatrix}\right\}$$

forms the basis for M_2 *(F)*.

7. Given

$u_1 = (1, -1, 3)$, $u_2 = (2, 3, 5)$, $u_3 = (-1, 4, -2)$ and $u_4 = (4, 1, -2)$, find c_1, c_2, c_3, c_4 not all zero, such that $c_1 u_1 + c_2 u_2 + c_3 u_3 + c_4 u_4 = \bar{0}$.

8. Prove that a set consisting of a single non-zero vector is L.I.

9. Prove that every sub-set of L.I. set is L.I.

10. Prove that any $n + 1$ vectors from n-dimensional vector space are L.D.

11. Prove that if $x \in [x_1, x_2,..., x_n]$ is uniquely expressible in the form $x = b_1 x_1 + b_2 x_2 +...+ b_n x_n$ for some b's of F, the vectors $x_1, x_2,....x_n$ are linearly independent.

12. (*a*) Find a basis for the vector space *C (R)* and all basis for the space V_3 (Z_2).

(*b*) For what values of a do the vectors $(1 + a, 1, 1)$,

$(1, 1 + a, 1)$, and $(1, 1, 1 + a)$ form a basis of V_3 *(R)*.

13. Assume $\{x_1, x_2, x_3\}$ is a basis for the vector space V_3 *(R)*. Verify that the sets $\{x_1 + x_2, x_2 + x_3, x_3 + x_1\}$ and $\{x_1, x_1 + x_2, x_1 + x_2 + x_3\}$ also serve as bases of V_3 *(R)*. is this situation true in the space V_3 (Z_3) ?

14. If diag *Mn* is the set of all diagonal matrices of order *n,* show that diag *Mn* (*F*) is a subspace of *Mn* (*F*) and determine its dimension.

15. Prove that if *W (F)* is a proper subspace of the finite dimensional vector space *V* (*F*), then dim $W <$ dim *V.*

16. Let *V (F)* be a finte dimensional vector space with bases $[x_1, x_2,...., x_n\}$. If W_k (*F*) is the subspace generated by x_k $(k = 1, 2,...., n)$, verify that

$$V = W_1 \oplus W_2 \oplus.... \oplus W_n .$$

17. Let *U* (*F*) and *W* (*F*) be subspaces of V_n (*F*) such that dim $U > n/2$ and dim $W > n/2$. Show that $U \cap W \neq \{0\}$.

18. Determine the dimension of the quotient space (V_3/W) (*F*), where the set $W = \{(a, b, a + b \mid a, b e F\}$.

19. Prove that every subspace of finite dimensional vector space has a complement.

CHAPTER 9

Hint. Let $U(F)$ be subspace with dim $U = m$ of the finite dimensional vector space $V(F)$ with dim $V = n$.

Let $\{x_1, x_2, ..., x_m\}$ be a basis of $U(F)$. Then this set

$\{x_1, x_2, ..., x_m\}$ can be extended to the basis

$\{x_1, x_2, ..., x_m, x_{m+1}, ..., x_n\}$ of $V(F)$ by adjoining n-m vectors $x_{m+1}, x_{m+2}, ..., x_n$. If we suppose that $W(F)$ is a subspace spanned by $x_{m+1}, x_{m+2}, ..., x_n$,

then $(x_{m+1}, ..., x_n\}$ is a basis for $W(F)$.

For any $x \in V$, we have

$x = a_1 x_1 + a_2 x_2 + ... + a_m x_m + a_{m+1} x_{m+1} + ... + a_n x_n$.

So

$x \in U + W$ since $a_1 x_1 + a_2 x_2 + ... + a_m x_m \in U$

and $a_{m+1} x_{m+1} + ... + a_n x_n \in W$

This means $V = U + W$.

Now

$$y \in U \cap W \Rightarrow y \in U \text{ and } y \in W$$

$\Rightarrow \quad y = b_1 x_1 + b_2 x_2 + ... + b_m x_m = c_{m+1} x_{m+1} + ... + c_n x_n$, since $\{x_1, ... x_n\}$ is a basis of $U(F)$ and $(x_{m+1}, ..., x_n)$ is basis of $W(F)$.

$\Rightarrow \quad b_1 x_1 + b_2 x_2 + ... + b_m x_m - c_{m+1} x_{m+1} ... - c_n x_n = \bar{0}$

$\Rightarrow \quad b_1 = b_2 = ... = b_m = c_{m+1} = ... = c_n = 0,$

since $\{x_1, x_2, ..., x_m, x_{m+1}, ...x_n)$ is L.I. being a basis of $V(F)$. Hence

$$y = U \cap W = \{\bar{0}\}$$

This shows that $U \cap W = \{\bar{0}\}$ and $V = U + W$, that is, $V = U \oplus W$. Hence $U(F)$ has its complement $W(F)$.

9.5 INNER PRODUCT SPACES

In our discussion of vector spaces over the fields $(F, +, .)$ the nature of field F has played virtually no role. In this section we shall consider vector space V over the field of real or complex numbers. The vector space $V(F)$ over the field of Real numbers is a *real vector space* and over the field of complex numbers is called a *complex vector space.*

In real vector space we have studied the idea of length, idea of perpendicularity and angle between two vectors.

Let us recall some properties of dot product in three dimensional real vector space. Given vectors $u = (x_1, x_2, x_3)$, $v = (y_1, y_2, y_3)$, where x's and y's are real numbers, the dot product of u and v denoted by $u . v$, was defined by $u . v = x_1 y_1 + x_2 y_2 + x_3 y_3$. The length of $u = \sqrt{u . u} = \sqrt{x_1^2 + x_2^2 + x_3^2}$ and the angle θ between u and v is determined by

$$\cos\theta = \frac{u . v}{\sqrt{u . u}\sqrt{v . v}} = \frac{x_1 y_1 + x_2 y_2 + x_3 y_3}{\sqrt{x_1^2 + x_2^2 + x_3^2}\sqrt{y_1^2 + y_2^2 + y_3^2}}$$

we list some properties of the dot product.

(1) $u \,.\, v \geq 0$ and $u \,.\, u = 0$ if and only if $u = 0$;

(2) $u \,.\, v = v \,.\, u$; and

(3) $u \,.\, (av + bw) = a\,(u \,.\, v) + b\,(u \,.\, w)$.

For any vectors u, v, w and real numbers a, b.

But everything that has been said cannot be carried over to complex vector space for instance, if $u = (1, i, 0)$, then by above concept $u \,.\, u = 1 \,.\, 1 + i \,.\, i + 0 \,.\, 0 = 0$. $u \,.\, u$ represent the length of the vector and the length of non-zero vector must be non-zero. So the product need modification.

We can achieve this much by altering the definition of dot products slightly. If $\bar{\alpha}$ denotes complex conjugate of a complex number α, we define.

$u \,.\, v = x_1\bar{y}_1 + x_2\bar{y}_2 + x_3\bar{y}_3$. For real numbers this new definition coincides with the old one. We also see that $u \,.\, u$ is real and positive as $u = (1, i, 0)$,

$u \,.\, u = 1 \,.\, \bar{1} + i \,.\, \bar{i} + 0 \,.\, \bar{0} = 1 \,.\, 1 + i \,.\, (i - i) + 0 \,.\, 0 = 1 + 1 + 0 = 2.$

However, we lose something, for instance $u \,.\, v \neq v \,.\, u$.

In fact, the exact relationship between these is $u \,.\, v = \overline{v \,.\, u}$. We observe the following properties of this dot product.

(1) $u \,.\, v = \overline{(v \,.\, u)}$;

(2) $u \,.\, u \geq 0$, and $u \,.\, u = 0$ if and only of $u = 0$;

(3) $(au + bv) \,.\, w = a\,(u \,.\, v) + b\,(v \,.\, w)$;

(4) $u \,.\, (av + bw) = \bar{a}\,(u \,.\, v) + \bar{b}\,(u \,.\, w)$.

for all real or complex numbers a, b and all complex vectors u, v, w.

Now, we define the product called inner product of two vectors over real or complex numbers.

Definition 9.5.1: The scalar product (or dot product, or inner product) of two complex vectors $u = (a_1, a_2, \ldots\ldots a_n)$ and

$v = (b_1, b_2, \ldots\ldots, b_n)$ is defined by the complex scalar as follows:

$$u \,.\, v = a_1\bar{b}_1 + a_2\bar{b}_2 + \ldots\ldots + a_n\bar{b}_n$$

$$v \,.\, u = b_1\bar{a}_1 + b_2\bar{a}_2 + \ldots\ldots + b_n\bar{a}_n \qquad \ldots(1)$$

They are called the Hermitian scalor product. They show that these two products are complex conjugates of each other *i.e.*,

$$u \,.\, v = \overline{v \,.\, u} \qquad \ldots(2)$$

They length of a complex vector is defined by

$$|u| = \left[a_1\bar{a}_1 + a_2\bar{a}_2 + \ldots\ldots + a_n\bar{a}_n\right]^{1/2} \qquad \ldots(3)$$

CHAPTER 9

It is also called the Hermitian length or norm of the vector denoted by $\|u\|$.

From (1) and (3), we get

$$\|u\| = \sqrt{u \cdot u} \quad ...(4)$$

Since $a_n \bar{a}_n$ is real, if follows from (3) that the length is real quantity. In particular case, if we get $|u| = 1$ then u is said to be normalized. The vectors u and v are said to orthogonal if

$$u \cdot v = v \cdot u = 0 \quad ...(5)$$

They with (2) show that if u and v are orthogonal, then real and imaginary parts of the scalar product $(u \cdot v)$ are separately zero.

A real vector space is also called a Euclidean space and a complex vector space is also called a Unitary space.

Definition 9.5.2: Let v be the vector space over the field $(F, +, .)$.

Let $u, v \in V$. Then a scaler complex (real) number, denoted by $(u. v)$ or (u, v), is called the inner product (or scalar product or dot product) of u and v if it satisfies the following axioms:

(1) $(u, v) = \overline{(u, v)}$;

(2) $(u, u) \geq 0$ and $(u, u) = 0$ if and only if $u = 0$;

(3) $(au + bv, w) = a(u, w) + b(v, w)$;

for any $u, v, w \in V$ and $a, b \in F$.

The vector space V over F together with inner product is called the *inner product space.*

If $(F, +,.)$ is the field of complex numbers, then property 1 implies that (u, u) is real.

We see that

$$\begin{aligned} (u, av + bw) &= \overline{(av + bw, u)} \\ &= \overline{a(v, u) + b(w, u)} \\ &= \bar{a}\,\overline{(v, u)} + \bar{b}\,\overline{(w, u)} \\ &= \bar{a}\,(u, v) + \bar{b}\,(u, w). \end{aligned}$$

Example 9.5.1: In $F^{(n)}$ define, for $u = (a_1,....,a_n)$, and $v = (b_1,....,b_n)$, $(u, v) = a_1 \bar{b}_1 + a_2 \bar{b}_2 +....+ a_n \bar{b}_n$. This is an inner product in $F^{(n)}$. We see that.

Solution:

1.

$$\begin{aligned} (u, v) &= a_1\bar{b}_1 + a_2\bar{b}_2 + ... + a_n\bar{b}_n \\ (v, u) &= b_1\bar{a}_1 + b_2\bar{a}_2 + ... + b_n\bar{a}_n \\ \overline{(v, u)} &= \bar{b}_1 a_1 + \bar{b}_2 a_2 + ... + \bar{b}_n a_n \end{aligned}$$

So $(u, v) = \overline{(v, u)}$

2. $$(u, u) = a_1\bar{a}_1 + a_2\bar{a}_2 + \ldots\ldots + a_n\bar{a}_n$$

$$= a_1^2 + a_2^2 + \ldots\ldots. + a_n^2 > 0.$$

3. $$(\alpha u + \beta v) = (\alpha a_1, \alpha a_2, \ldots.., \alpha a_n) + (\beta b_1, \beta b_2, +\ldots..+, \beta b_n)$$

$$= (\alpha a_1 + \beta b_1, \ldots\ldots., \alpha a_n + \beta b_n)$$

If $w = (c_1, c_2, \ldots.., c_n)$

$$(\alpha u + \beta v, w) = (\alpha a_1 + \beta b_1)\bar{c}_1 + \ldots\ldots. + (\alpha a_n + \beta b_n)\bar{c}_n$$

$$= \alpha a_1\bar{c}_1 + \beta b_1\bar{c}_1 + \ldots\ldots. + \alpha a_n\bar{c}_n + \beta b_n\bar{c}_n$$

$$= \alpha(a_1\bar{c}_1 + \ldots\ldots. + a_n\bar{c}_n) + \beta(b_1\bar{c}_1 + \ldots\ldots + b_n\bar{c}_n)$$

$$= \alpha(u, w) + \beta(v, w).$$

Hence, the given product is a inner product.

Example 9.5.2: In F^2 define, $u = (a_1, a_2)$ and $v = (b_1, b_2)$, (u, v)

$= 2a_1\bar{b}_1 + a_1\bar{b}_2 + a_2\bar{b}_1 + a_2\bar{b}_2$. Show that this defines inner product.

Solution: Hence

$$(u, v) = 2a_1\bar{b}_1 + a_1\bar{b}_2 + a_2\bar{b}_1 + a_2\bar{b}_2 \quad \ldots(1)$$

$$(v, u) = 2b_1\bar{a}_1 + b_1\bar{a}_2 + b_2\bar{a}_1 + b_2\bar{a}_2$$

$$\overline{(v, u)} = 2\bar{b}_1 a_1 + a_2\bar{b}_1 + a_1\bar{b}_2 + a_2\bar{b}_2 \quad \ldots(2)$$

Hence $(u, v) = \overline{(v, u)}$.

$$(u, u) = 2a_1\bar{a}_1 + a_1\bar{a}_2 + a_2\bar{a}_1 + a_2\bar{a}_2 > 0$$

Let $w = (c_1, c_2)$

$$(\alpha u + \beta v) = (\alpha a_1 + \beta b_1, \alpha a_2 + \beta b_2)$$

$$(\alpha u + \beta v, w) = 2(\alpha a_1 + \beta b_1)\bar{c}_1 + (\alpha a_1 + \beta b_1)\bar{c}_2$$

$$+ (\alpha a_2 + \beta b_2)\bar{c}_1 + (\alpha a_2 + \beta b_2)\bar{c}_2$$

$$= 2\alpha a_1\bar{c}_1 + 2\beta b_1\bar{c}_1 + \alpha a_1\bar{c}_2 + \beta b_1\bar{c}_2 + \alpha a_2\bar{c}_1$$

$$+ \beta b_2\bar{c}_1 + \alpha a_2\bar{c}_2 + \beta b_2\bar{c}_2$$

$$= \alpha(2a_1\bar{c}_1 + a_1\bar{c}_2 + a_2\bar{c}_1 + a_2\bar{c}_2)$$

$$+ \beta(2b_1\bar{c}_1 + b_1\bar{c}_2 + b_2\bar{c}_1 + b_2\bar{c}_2)$$

$$= \alpha(u, w) + \beta(v, w).$$

Hence defined product is a inner product in F^2.

Example 9.5.3: Let V be the vector space over C of all continuous complex valued functions on the interval [0, 1]. For $f(t), g(t) \in V$, define $(f(t), g(t)) = \int_0^1 f(t)\overline{g(t)}\,dt$.

Solution: We have

$$(f(t), g(t)) = \int_0^1 f(t)\overline{g(t)}\,dt$$

$$\overline{(g(t), f(t))} = \int_0^1 \overline{g(t)\overline{f(t)}}dt$$

$$= \int_0^1 f(t)\overline{g(t)}dt.$$

Hence $$(f(t), g(t)) = \overline{(g(t), f(t))}$$

Again $$(f(t), f(t)) = \int_0^1 f(t)\overline{f(t)}dt = \int_0^1 |f(t)|^2 dt \geq 0$$

and the equality holds if and only if $f(t) \neq 0$.

$$(\alpha f(t) + \beta g(t), h(t)) = \int_0^1 (\alpha f(t) + \beta g(t))\overline{h(t)}\,dt$$

$$= \int_0^1 (\alpha f(t))\bar{h}(t) + \beta g(t))\overline{h(t)}\,dt$$

$$= \alpha\int_0^1 f(t)\,\bar{h}(t)\,dt + \beta\int_0^1 (g(t)\,(\bar{h}(t))\,dt$$

$$= \alpha\,(f(t), h(t)) + \beta\,(g(t), h(t)).$$

Hence given product is inner product.

Example 9.5.4: If $u = (i, 2i, 1)$, $v = (1, 1 + i, 0)$, $w = (i, 1 - i, 2)$, then (1) Find the length of each vector, (2) Find the inner product of (u, v) and (v, u), (3) Find a vector r which is orthogonal to both u and w, and (4) show that $x = (1 - i, -1, 1 - i)$ is orthogonal to both u and v, all taken in the Hermitian sense.

Solution: We have

(1) $$\|u\| = \sqrt{u\,.\,u} = \sqrt{i\,.\,\bar{i} + 2i\,.\,\overline{2i} + 1.1}$$

$$= \sqrt{i\,.\,(-i) + 2i\,.\,2(-i) + 1} = \sqrt{1 + 4 + 1} = \sqrt{6}$$

$$\|v\| = \sqrt{v\,.\,v} = \left[1\,.\,\bar{1} + (1+i)\,\overline{(1+i)} + 0.\bar{0}\right]^{1/2}$$

$$= \left[1 + (1+i)(1-i) + 0\right]^{1/2} = \left[1 + 1 - i^2 + 0\right]^{1/2} = \sqrt{3}$$

$$\|w\| = \sqrt{w\,.\,w} = \left[i\,\bar{i} + (1-i)\,\overline{(1-i)} + 2\,.\,2\right]^{1/2}$$

$$= \left[i(-i) + (1-i)\,(1+i) + 4\right]^{1/2} = [1 + 2 + 4]^{1/2} = \sqrt{7}$$

(2) $$(u, v) = i\,.\,\bar{1} + 2i\,\overline{(1+i)} + 1\,.\,\bar{0}$$

$$= i + 2i\,(1 - i) + 0 = i + 2i - 2i^2 + 0 = 3i + 2$$

and $$(v,\ u) = 1\,.\,\bar{i} + (1+i)\left(\overline{2i}\right) + 0\,.\,1 = -i + (1 + i)\,(-2i) + 0$$

$$= -i - 2i - 2i^2 = 2 - 3i$$

(3) The condition for orthogonality is

$u\,.\,u = 0,\ r\,.\,u = 0,\ w\,.\,r = 0$ and $r\,.\,w = 0$.

If we take $r = (r_1, r_2, r_3)$, than these conditions give

$$-i\,r_1 - 2i\,r_2 + r_3 = 0 \qquad ...(1)$$

$$i\,r_1 + 2i\,r_2 + r_3 = 0 \qquad ...(2)$$

and $$-i\,r_1 + (1 + i)\,r_2 + 2\,r_3 = 0 \qquad ...(3)$$

$$i\,r_1 + (1 - i)\,r_2 + 2\,r_3 = 0 \qquad ...(4)$$

From (1) and (3), we get

$$r_3 + (1 + 3i)\,r_2 = 0 \qquad ...(5)$$

From (2) and (4), we get

$$r_3 + (1 - 3i)\,r_2 = 0 \qquad ...(6)$$

If we take $r_3 = r_{31} + ir_{32}$, $r_2 = r_{21} + ir_{22}$, then from the relation (5) or (6), we get

$$r_{31} + r_{21} - 3r_{22} = 0 \text{ and } r_{32} + r_{22} + 3r_{21} = 0.$$

Thus, we get two equations involving four unknowns, so that two of them are arbitrary. Let us choose $r_{31} = 3$, $r_{32} = -1$, then we get $r_{21} = 0$, $r_{22} = 1$, so that $r_3 = 3 - i$, $r_2 = i$. Then we find $r_1 = -1 - 5i$. Thus, the vector $r = (-1 - 5i,\ i,\ 3 - i)$ is one of many vectors which are orthogonal to both u and v.

(4) We see that

$$(u,\ x) = (i,\ 2i,\ 1)\,.\,(1 - i,\ -1,\ 1 - i)$$

$$= i\,.\left(\overline{1-i}\right) + 2i\left(\overline{-1}\right) + 1\,.\left(\overline{1-i}\right)$$

$$= i.\,(1 + i) - 2i + 1\,(1 + i) = i + i^2 - 2i + 1 + i = 0$$

and $$(x,\ u) = (1 - i)\,\bar{i} - 1\,.\,2\bar{i} + (1 - i)\,\bar{1}$$

$$= (1 - i)\,(-i) + 2i + 1 - i = -i + i^2 + 2i + 1 - i = 0$$

Similarly, we can see that

$$(v,\ x) = (x,\ v) = 0,\ (w,\ x) = (x,\ w) = 0.$$

Theorem 9.5.1: *If $u, v \in V$ and $a, b \in F$, then*

(1) $$(au + bv,\ au + bv) = a\,\bar{a}\,(u,\ u) + a\bar{b}\,(u,\ v) + \bar{a}\,b\,(v,\ u) + b\bar{b}\,(v,\ v).$$

(2) $$\|au\| = |a|\ \|u\|$$

Proof: **(1)** By property (3) of the definition of inner product,

$$(au + bv,\ au + bv) = a\,(u,\ au + bv) + b\,(v,\ au + bv)$$

$$= a\,\bar{a}\,(u, u) + a\bar{b}\,(u, v) + b\bar{a}\,(v, u) + b\,\bar{b}\,(v, v)$$

Since $(u, au + bv) = \bar{a}\,(u, u) + \bar{b}\,(u, v)$

and $(v, au + bv) = \bar{a}\,(v, u) + \bar{b}\,(v, v).$

(2) $\|au\|^2 = (au, au) = a\,(u, au) = a\bar{a}\,(u, u) = |a|^2\,\|u\|^2$

Since $a\bar{a} = |a|^2$ and $(u, u) = \|u\|^2$

$\Rightarrow$ $\|au\| = |a|\,\|u\|.$

Theorem 9.5.2: *(Schwarz Inequality)*
For all $u, v \in V$,

$$|(u, v)| < \|u\|\,\|v\|$$

Proof: If $u = 0$, then L.H.S. = $(0, v) = 0$ and R.H.S = $\|0\|\,\|v\| = 0$.

So $|(u, v)| = \|u\|\,\|v\|.$

Assume now that $u \neq 0$, $\|u\| \neq 0$. If $\|u\| = 0$, then $(u, u) = 0 \Rightarrow u = 0$ contradiction to $u \neq 0$.

Put $$w = v - \frac{(v, u)}{\|u\|^2}\,u.$$

Since $\|w\|^2 \geq 0$, then

$$\left(v - \frac{(v, u)}{\|u\|^2}\,u,\; v - \frac{(v, u)}{\|u\|^2}\,u,\right) \geq 0.$$

$$\Rightarrow \quad (v, v) - \frac{\overline{(v, u)}}{\|u\|^2}(v, u) - \frac{(v, u)}{\|u\|^2}(u, v) + \frac{(v, u)}{\|u\|^2}\,\frac{\overline{(v, u)}}{\|u\|^2}(u, u) \geq 0$$

$$\Rightarrow \quad \|v\|^2 - \frac{|(v, u)|^2}{\|u\|^2} - \frac{\overline{(u, v)}(u, v)}{\|u\|^2} + \frac{|(v, u)|^2}{\|u\|^4}\,\|v\|^2 \geq 0$$

$$\Rightarrow \quad \|v\|^2 - \frac{|(u, v)|^2}{\|u\|^2} \geq 0$$

$$\Rightarrow \quad \|v\|^2\,\|u\|^2 \geq |(u, v)|^2$$

$$\Rightarrow \quad |(u, v)|^2 \leq \|u|^2\,\|v\|^2$$

$$\Rightarrow \quad |(u, v)| \leq \|u\|\,\|v\|.$$

Remark: This inequality is called Cauchy Schwarz inequality.

Now, we define the distance between two vectors.

Definition 9.5.3: For $u, v \in V$, $\|u - v\|$ is called the distance between u and v. The distance between u and v will be denoted by $d\,(u, v)$. That is, $d\,(u, v) = \|u - v\|$.

***Theorem 9.5.3:** Every inner product space is a metric space.*

Proof: Let V be the inner product space.

For any $u, v \in V$, we have $d(u, v) = \|u - v\|$

1. Since the norm is always positive, $d(u, v) \geq 0$.

2. $d(u, v) = 0 \Rightarrow \|u - v\| = 0$.

$$\Rightarrow (u - v, u - v) = 0.$$

$$\Rightarrow u - v = 0 \Rightarrow u = v$$

$$d(u, v) = \|u - v\| = \sqrt{(u - v, u - v)}$$

$$= \sqrt{(-1)\overline{(-1)}\,(v - u, v - u)}$$

$$= \sqrt{(v - u, v - u)} = \|v - u\| = d(v, u).$$

Now, we claim that

$$\|u + v\| \leq \|u| + \|v\| \text{ for all } u, v \in V.$$

Now,

$$\|u + v\|^2 = (u + v, u + v)$$

$$= (u, u) + (u, v) + (v, u) + (v, v)$$

$$= \|u\|^2 + (u, v) + \overline{(u, v)} + \|v\|^2$$

Since $\quad (u, v) + \overline{(u, v)} = 2$ Real part of (u, v),

$$\|u + v\|^2 = \|u\|^2 + 2R(u, v) + \|v\|^2$$

$$\leq \|u\|^2 + 2|(u, v)| + \|v\|^2 \text{ since } R(u, v) \leq |(u, v)|$$

$$\leq \|u\|^2 + 2\|u\|\|v\| + \|v\|^2 \text{ by Schwarz inequality}$$

$$\leq (\|u\| + \|v\|)^2$$

$$\Rightarrow \quad \|u + v\| \leq \|u\| + \|v\| \qquad \ldots(1)$$

Finally, for any $u, v, w \in V$

$$d(u, v) + d(v, w) = \|u - v\| + \|v - w\|$$

$$\geq \|u - v + v - w\| \text{ by (1)}$$

$$\geq \|u - w| = d(u, w).$$

Hence (V, d) is a metric-space.

Orthogonality plays a very important role in inner product space.

Definition 9.5.4: If $W(F)$ is a sub space of $V(F)$, the orthogonal complement of W, denoted by $W^{\perp}$, is defined by

CHAPTER 9

$$W^{\perp} = \{x \in V \mid (x, w) = 0, \text{ for all } w \in W\}.$$

Lemma 9.5.1: *For any sub-space W of V, $W^{\perp}$ is a sub-space of V (F) such that $W \cap W^{\perp} = (0)$*

Proof: If $u, v \in W^{\perp}$, then for all $a, b \in F$ and for all $w \in W$

$$\begin{aligned}(au + bv, w) &= a(u, w) + b(v, w) \\ &= a \cdot 0 + b \cdot 0 \text{ since } u, v \in W^{\perp} \\ &= 0\end{aligned}$$

So $\quad au + bv \in W^{\perp}$

Hence, $W^{\perp}$ is a subspace of V.

Finally
$$\begin{aligned} x \in W \cap W^{\perp} &\Rightarrow x \in W \text{ and } x \in W^{\perp} \\ &\Rightarrow (x, x) = 0 \\ &\Rightarrow x = 0. \\ &\Rightarrow W \cap W^{\perp} = (0).\end{aligned}$$

Hence our goal is to show that $V = W + W^{\perp}$. If this is done then we can say that V is the direct sum of W and $W^{\perp}$.

Definition 9.5.5: The set of vectors $\{V_i\}$ in V is an orthonormal set if

(1) each v_i is of length 1, *i.e.*, $(v_i, v_i) = 1$.

(2) For $i \neq j$, $(V_i, V_j) = 0$.

Example 9.5.5: Let V be the inner product space of all real valued functions defined on [0, 1] to R with inner product

$$(f, g) = \int_0^1 f(x)g(x)\,dx, \text{ for all } f, g \in V$$

Define $f_n(x) = \sqrt{2}\cos 2\pi nx$,

and $g_n(x) = \sqrt{2}\sin 2\pi nx$, for all $n \in N$.

We consider $X = \{1, f_1, g_1, f_2, g_2, \ldots\}$. Then prove that X is orthonormal subset

Solution: Clearly for all $n \in N$

$$\begin{aligned}(f_n, f_n) &= \int_0^1 2\cos^2 2\pi nx\,dx \\ &= \int_0^1 (1 + \cos 4\pi nx)\,dx \\ &= \left[x + \frac{1}{4\pi n}\sin 4\pi nx\right]_0^1 = 1 + 0 = 1.\end{aligned}$$

$\Rightarrow \qquad \|f_n\| = 1.$

Similarly, for all $n \in N$

$$(g_n, g_n) = \int_0^1 2\sin^2 2\pi nx\,dx$$

$$= \int_0^1 (1 - 4 \cos 4\pi\, n x)\, dx$$

$$= \left[x - \frac{1}{4\pi n} \sin 4\pi\, n x \right]_0^1 = 1.$$

$$\Rightarrow \qquad \|g_n\| = 1.$$

Further for $n \neq m$

$$(f_n, f_m) = \int_0^1 2 \cos 2\pi\, n\, x \cos 2\pi\, mx\, dx$$

$$= \int_0^1 \left[\cos 2\pi\,(n+m)x + \cos 2\pi\,(n-m)x \right] dx$$

$$= \left[\frac{\sin 2\pi(n+m)x}{2\pi(x+m)} + \frac{\sin 2\pi\,(n-m)x}{2\pi\,(n-m)} \right]_0^1$$

$$= 0$$

Similarly,

$$(g_n, g_m) = \int_0^1 2 \sin 2\pi\, nx \sin 2\pi\, mx\, dx.$$

$$= \int_0^1 \left[\cos 2\pi\,(n-m)n - \cos 2\pi\,(n+m)x \right] dx = 0.$$

and

$$(f_n, g_m) = \int_0^1 2 \cos 2\pi\, nx \sin 2\pi\, mx\, dx$$

$$= \int_0^1 \left[\sin 2\pi\,(n+m)x - \sin 2\pi\,(n-m)x \right] dx$$

$$= 0.$$

Also for $n \in N,\ (1, f_n) = \int \sqrt{2} \cos 2\pi nx\, dx = 0$

and $(1, g_m) = \int \sqrt{2} \sin 2\pi mn\, dx = 0$

Hence X is an orthonormal subset of V.

Example 9.5.6: In the vector space C^2 with the standard inner product $(u, v) = a_1\bar{a}_2 + b_1\bar{b}_2$ for all $u = (a_1, b_1)$, $v = (a_2, b_2) \in C^2$, the set $\{e_1, e_2\}$ is an orthonormal set, where $e_1 = (1, 0)$ and $e_2 = (0, 1)$, since $(e_1, e_1) = 1.\bar{1} + 0.\bar{0} = 1$

and $(e_2, e_2) = 0.\bar{0} + 1.\bar{1} = 1 \Rightarrow \|e_1\| = 1,\ \|e_2\| = 1$

and $(e_1, e_2) = 1.\bar{0} + 0.\bar{1} = 0.$

Lemma 9.5.2: *Any orthonormal set $\{v_i\}$ of non-zero vectors is linearly independent. If $w = \alpha_1 V_1 + \alpha_2 v_2 + ... + \alpha_n v_n$, then*

$$\alpha_i = (w, v_i), \text{ for all } i = 1, 2, ..., n.$$

CHAPTER 9

Proof: Let $\{v_i\}$ be an orthonormal subset of an inner product space V with the property $0 \notin \{v_i\}$.

Suppose that

$$\alpha_1 v_1 + \alpha_2 v_2 + \ldots + \alpha_n v_n = 0.$$

Therefore

$$(\alpha_1 v_1 + \alpha_2 v_2 + \ldots + \alpha_n v_n, v_i) = (0, v_i)$$

$\Rightarrow$ $\alpha_1 (v_1, v_i) + \alpha_2 (v_2, v_i) + \ldots + \alpha_n (v_n, v_i) = 0$

$\Rightarrow$ $\alpha_i (v_i, v_i) = 0$ since $(v_i, v_j) = 0$ for $i \neq j$ and $(V_i, V_i) = 1$.

$\Rightarrow$ $\alpha_i = 0$.

Thus the set $\{V_i\}$ is linearly independent.

Moreover, if $w = \alpha_1 v_1 + \alpha_2 v_2 + \ldots + \alpha_n v_n$, then

$$\begin{aligned}(w, v_i) &= ((\alpha_1 v_1 + \alpha_2 v_2 + \ldots + \alpha_n v_n), v_i)\\ &= \alpha_i (v_1, v_i) + \alpha_2 (v_2, v_i) + \ldots + \alpha_n (v_n, v_i)\\ &= \alpha_i (v_i, v_i) \text{ since } (v_i, v_j) = 1, i = j = 0, i \neq j\\ &= \alpha_i \text{ since } (v_i, v_i) = 1.\end{aligned}$$

Lemma 9.5.3: *If $\{v_1, v_2, \ldots, v_n\}$ is orthonormal set in v and if $w \in V$, then $u = w - (w, v_1) v_1 - (w, v_2) v_2 - \ldots - (w, v_i) v_i - \ldots - (w, v_n) v_n$ is orthogonal to each of $v_1, v_2, \ldots, v_n$.*

Proof: $\{v_1, v_2, \ldots v_n\}$ is an orthonormal set of v, then for $i \neq j$ $(v_i, v_j) = 0$ and $(v_i, v_i) = 1$.

Now, $u = w - (w, v_1) v_1 - \ldots - (w, v_i) v_i - \ldots - (w, v_n) v_n$ for $i \leq n$,

$\Rightarrow$ $(u, v_i) = (w, v_i) - (w, v_1)(v_1, v_i) - \ldots - (w, v_i)(v_i, v_i) \ldots - (w, v_n)(v_n, v_i)$

$\Rightarrow$ $(u, v_i) = (w, v_i) - 0 \ldots - (w, v_i) .1 - \ldots\ldots 0$

$\Rightarrow$ $(u, v_i) = 0$

$\therefore$ u is orthogonal to each v_i.

From a given set of linearly independent vectors, it is possible to construct by their linear combination another set of linearly independent vectors which are orthogonal to each other. This can be converted to an orthonormal set of vectors. The process of obtaining such a new set is called the orthogonalization or orthonormalization of the given set of vectors. This process is called Gram-schmidt process.

Theorem 9.5.4: *Let V be a finite dimensional inner product space, then V has an orthonormal set as a basis.*

Proof: Let $\{v_1, v_2, v_3, \ldots v_n)$ be a basis of finite dimensional vector space of dimension n over F. From this basis we shall construct an orthonormal set of n vectors.

And by lemma 9.5.2. This set is linearly independent and forms a basis of V.

We seek n vectors $w_1, w_2, \ldots, w_n$ each of length one such that $(w_i, w_j) = 0$ for $i \neq j$. We shall produce them in the following form;

(i) w_1 will be multiple of v_1, w_2 will be in the linear span of w_1, v_2; w_3 is in the linear span of w_1, w_2, v_3 and so on, more generally, w_i in the linear span of w_1 w_2,..., $w_{(i-1)}$, v_i

Let $$w_1 = \frac{v_1}{\|v_1\|};$$

Then $$(w_1, w_1) = \left(\frac{v_1}{\|v_1\|}, \frac{v_1}{\|v_1\|}\right) = \frac{1}{\|v_1\|^2}(v_1, v_1) = \frac{\|v_1\|^2}{\|v_1\|^2} = 1$$

$$\Rightarrow \qquad \|w_1\| = 1.$$

Now, we define a vector u_1 by the linear combination

$$u_2 = v_2 + \alpha\, w_1$$

where α is scalar. If u is orthogonal to w_1, then their inner product is zero, so that

$$(u_2, w_1) = 0$$

$$\Rightarrow \qquad (v_2 + \alpha w_1, w_1) = 0$$

$$\Rightarrow \qquad (v_2, w_1) + \alpha\,(w_1, w_1) = 0$$

$$\Rightarrow \qquad \alpha = -(v_2, w_1) \text{ since } (w_1, w_1) = 1$$

$\therefore$ $u_2 = v_2 - (v_2, w_1)\, w_1$; u_2 is orthogonal to w_1

Since v_1 and v_2 are linearly independent, so is w_1 and v_2 and $u_2 \neq 0$

Let $w_2 = \dfrac{u_2}{\|u_2\|}$; $\{w_1, w_2\}$ is an orthonormal set.

We continue this process.

Let $u_3 = -(v_3, w_1)\, w_1 - (v_3, w_2)\, w_2 + v_3$.

We see that

$$(u_3, w_1) = -(v_3, w_1)(w_1, w_1) - (v_3, w_2)(w_2, w_1) + (v_3, w_1) = 0$$

Similarly,

$$(u_3, w_2) = 0$$

Since w_1, w_2 and v_3 are linearly independent (for w_1, w_2 are in the linear span of v_1 and v_2); $u_3 \neq 0$.

Let $w_3 = \dfrac{u_3}{\|u_3\|}$; then (w_1, w_2, w_3) is an orthonormal set.

Repetition of this procedure gives the vector w_i, where

$w_i = \dfrac{u_i}{\|u_i\|}$ and for $i \leq n$.

$u_i = v_i - (v_i, w_1)\, w_1 - (v_1, w_2)\, w_2 - \ldots\ldots - (v_i, w_{i-1})\, w_{i-1}$,

$u_i \neq 0$ and it is orthogonal to each of $w_1, \ldots\ldots w_{i-1}$.

In this way, given n linearly independent vectors in v, we can construct an orthonormal set having n elements. This is the required basis.

Example 9.5.7: Construct a set of three mutually orthonormal vectors which are the Linear combination of vectors $v_1 = (1, 0, 2, 2)$, $v_2 = (1, 1, 0, 1)$ and $v_3 = (1, 1, 0, 0)$.

Solution: We see that v_1, v_2, v_3 are linearly independent. For this, let $a, b, c \in F$ such that

$$a_1 v_1 + b v_2 + c v_3 = 0$$

$$\Rightarrow \quad (a, 0, 2a, 2a) + (b, b, 0, b) + (c, c, 0, 0) = (0, 0, 0, 0)$$

$$\Rightarrow \quad a + b + c = 0,\ b + c = 0,\ 2a = 0,\ 2a + b = 0$$

$$\Rightarrow \quad a = b = c = 0.$$

Hence v_1, v_2, v_3 are linearly independent.

Now let

$$w_1 = \frac{v_1}{\|v_1\|} = \frac{(1, 0, 2, 2)}{\sqrt{1+0+4+4}} = \left(\frac{1}{3}, 0, \frac{2}{3}, \frac{2}{3}\right)$$

$$u_2 = v_2 - (v_2, w_1)\, w_1$$

$$= (1, 1, 0, 1) - \left(1 \cdot \frac{1}{3}, 1 \cdot 0 + 0 \cdot \frac{2}{3} + 1 \cdot \frac{2}{3}\right)\left(\frac{1}{3}, 0, \frac{2}{3}, \frac{2}{3}\right)$$

$$= (1, 1, 0, 1) - 1\left(\frac{1}{3}, 0, \frac{2}{3}, \frac{2}{3}\right) = \left(\frac{2}{3}, 1, \frac{-2}{3}, \frac{1}{3}\right)$$

and

$$w_2 = \frac{u_2}{\|u_2\|} = \left(\frac{\sqrt{2}}{3}, \frac{1}{\sqrt{2}}, \frac{-\sqrt{2}}{3}, \frac{\sqrt{2}}{6}\right) \text{ and } \|u_2\| = \sqrt{2}$$

$$u_3 = v_3 - (v_3, w_1)\, w_1 - (w_3, w_2)\, w_2$$

Here

$$(v_3, w_1) = (1, 1, 0, 0), \left(\frac{1}{3}, 0, \frac{2}{3}, \frac{2}{3}\right) = \left(\frac{1}{3} + 0 + 0 + 0\right) = \frac{1}{3}$$

and

$$(v_3, w_2) = (1, 1, 0, 0), \left(\frac{\sqrt{2}}{3}, \frac{1}{\sqrt{2}}, \frac{-\sqrt{2}}{3}, \frac{\sqrt{2}}{6}\right)$$

$$= \frac{\sqrt{2}}{3}\ \frac{1}{\sqrt{2}} + 0 + 0 = \frac{5}{3\sqrt{2}}$$

$\therefore$

$$u_3 = (1, 1, 0, 0) - \frac{1}{3}\left(\frac{1}{3}, 0, \frac{2}{3}, \frac{2}{3}\right) - \frac{5}{3\sqrt{2}}\left(\frac{\sqrt{2}}{3}, \frac{1}{\sqrt{2}}, \frac{-\sqrt{2}}{3}, \frac{\sqrt{2}}{6}\right)$$

$$= \left(\frac{1}{3}, \frac{1}{6}, \frac{1}{3}, -\frac{1}{2}\right)$$

and
$$w_3 = \frac{u_3}{\| u_3 \|} = \frac{\left(\frac{1}{3}, \frac{1}{6}, \frac{1}{3}, -\frac{1}{2}\right)}{\sqrt{\frac{1}{9}+\frac{1}{36}+\frac{1}{9}+\frac{1}{4}}} = \left(\frac{\sqrt{2}}{3}, \frac{\sqrt{2}}{6}, \frac{\sqrt{2}}{3}, \frac{-\sqrt{2}}{2}\right)$$

Thus, the required orthonormal vectors are w_1, w_2, w_3 whose values are given above.

Example 9.5.8: Let V be the vector space of all polynomials in x over the real field F of degree 2 or less. In V the inner product is defined by

$$(f(x), q(x)) = \int_{-1}^{1} p(x) g(x)\, dx \text{ for all } p(x), g(x) \in V.$$

Construct the orthonormal set from the basis $\{1, x, x^2\}$.

Solution: We use the following construction.

Let
$$w_1 = \frac{v_1}{\| v_1 \|} = \frac{1}{\sqrt{\int_{-1}^{1} 1\, dx}} = \frac{1}{\sqrt{2}};$$

$$u_2 = v_2 - (v_2, w_1)\, w_1$$

$$= x - \left(x, \frac{1}{\sqrt{2}}\right) \cdot \frac{1}{\sqrt{2}} = x - \int_{-1}^{1} x \frac{1}{\sqrt{2}}\, dx \cdot \frac{1}{\sqrt{2}}$$

$$= x - \frac{1}{2}\left[\frac{x^2}{2}\right]_{-1}^{1} = x - \frac{1}{2}\left[\frac{1}{2} - \frac{1}{2}\right] = x$$

$$w_2 = \frac{u_2}{\| u_2 \|} = \frac{x}{\sqrt{\int_{-1}^{1} x^2\, dx}} = \frac{\sqrt{3}}{\sqrt{2}}\, x;$$

Finally,

$$u_3 = v_3 - (v_3, w_1)\, w_1 - (v_3, w_2)\, w_2$$

$$(v_3, w_1) = \int_{-1}^{1} x^2 \frac{1}{\sqrt{2}}\, dx = \frac{1}{\sqrt{2}} \frac{2}{3} = \frac{\sqrt{2}}{3}$$

$$(v_3, w_2) = \int_{-1}^{1} x^2 \frac{\sqrt{3}}{\sqrt{2}}\, x\, dx = \frac{\sqrt{3}}{\sqrt{2}} \cdot \int_{-1}^{1} x^3\, dx = 0$$

$$u_3 = x^2 - \frac{\sqrt{2}}{3} \frac{1}{\sqrt{2}} - \frac{\sqrt{3}}{\sqrt{2}}\, x = x^2 - \frac{1}{3}$$

CHAPTER 9

$$w_3 = \frac{u_3}{\| u_3 \|} = \frac{x^2 - \frac{1}{3}}{\sqrt{\int_{-1}^{1}\left(x^2 - \frac{1}{3}\right)^2 dx}} = \frac{\sqrt{10}}{4}(3x^2 - 1)$$

Thus $\{w_1, w_2, w_3\}$ is an orthonormal set.

Theorem 9.5.5: *If V is a finite-dimensional inner product space and if W is a subspace of V, then $V = W + W^\perp$, Particularly, V is the direct sum of W and $W^\perp$.*

Proof: Since W is a subspace of the inner product space V, W is itself an inner product space (inner product of V is restricted to W). Thus, we can find an orthonormal set $w_1, \ldots, w_r$ in W which is a basis of W.

If $v \in V$, $v_0 = v - (v, w_1) w_1 - (v, w_2) w_2 - \ldots - (v, w_r) w_r$ is orthogonal to each of $w_1, w_2, \ldots, w_r$. So v_0 is orthogonal to W. Thus $v_0 \in W^\perp$.

Since

$$v = v_0 + (v_1\ w_1)\ w_1 + (v_1\ w_2)\ w_2 + \ldots + (v_1\ w_r)\ w_r,$$

$$v \in W + W^\perp$$

Therefore $V = W + W^\perp$.

Since $W \cap W^\perp = (0)$, so V is the direct sum of W and $W^\perp$. Hence $V = W \oplus W^\perp$.

Corollary: If V is a finite-dimensional inner product space and W is a subspace of V, then $(W^\perp)^\perp = W$.

Proof: If $w \in W$, then for any $u \in W^\perp$, $(w, u) = 0$

$\Rightarrow \quad w \in (W^\perp)^\perp \Rightarrow W \subseteq (W^\perp)^\perp$.

Now $\quad V = W + W^\perp$ and $V = W^\perp + (W^\perp)^\perp$

Since V is the direct sum, then $\dim (W) = \dim (W^\perp)^\perp$

Since $W \subseteq (W^\perp)^\perp$ and $\dim (W) = \dim (W^\perp)^\perp$, then

$$W = (W^\perp)^\perp.$$

Exercise 9.5.1: Prove that two vectors u, v in a Euclidean space are orthogonal if and only if

$\|u + v\|^2 = \|u\|^2 + \|v\|^2$. Further show that this result is not necessarily true for Unitary spaces.

Solution: Let u, v be orthogonal to each other, then

$$(u, v) = 0 \text{ and } (v, u) = 0. \qquad \ldots(1)$$

Now

$$\|u + v\|^2 = (u + v, u + v)$$

$$= (u, u) + (u, v) + (v, u) + (v, v)$$

$$= \|u\|^2 + \|v\|^2 \text{ by } (1)$$

Conversely, let $\|u + v\| = \|u\|^2 + \|v\|^2$ ....(2)

Again
$$\|u + v\|^2 = (u + v, u + v)$$
$$= (u, u) + (u, v) + (v, u) + (v, v)$$
$$= \|u\|^2 + (u, v) + (u, v) + \|v\|^2$$
$$\Rightarrow \quad 2(u, v) = 0 \text{ by } 2$$
$$\Rightarrow \quad (u, v) = 0$$
$\Rightarrow$ u is orthogonal to v.

Consider the complex vector space C^2 with standard inner product. Take $u = (0, i)$ and $v = (0, 1)$, then

$$(u, v) = 0 . \bar{0} + i . \bar{i} = i \neq 0$$

$\Rightarrow$ u is not orthogonal to v.

But
$$\|u+v\| = \|0, 1 + i\|^2 = ((0, 1 + i), (0, 1 + i))$$
$$= (0, 0) + (0, 1 + i) + ((1 + i) 0) + ((1 + i), (1 + i))$$
$$= \|1 + i\|^2 = 2$$

and
$$\|u\|^2 = ((0, i), (0, i)) = 0 . \bar{0} + i . \bar{i} = 1$$

and
$$\|v\|^2 = ((0, 1), (0, 1)) = 0 . \bar{0} + 1 . \bar{1} = 1$$

Hence $\|u + v\|^2 = \|u\|^2 + \|v\|^2$.

Exercise 9.5.2: Let V be the set of real functions $y = f(x)$ satisfying

$$\frac{d^3y}{dx^3} - 6\frac{d^2y}{dx^2} + 11\frac{dy}{dx} - 6y = 0.$$

(*a*) Prove that V is 3-dimensional real vector space

(*b*) In V define

$$(u, v) = \int_{-\infty}^{0} u\,v\,dx.$$

Show that this defines an inner product on V and find an orthogonal basis for V.

Solution: (*a*) Here $y = 0(x) = 0$, that is, zero function,

0 satisfies the given differential equation. So $0 \in V$, $V \neq \phi$. Let $\lambda_1, \lambda_2, \lambda_3 \in R$ and y_1, y_2, $\in V$.

Then
$$D^3 y_1 - 6D^2 y_1 + 11D y_1 - 6y_1 = 0$$

and $D^3 y_2 - 6D^2 y_2 + 11D y_2 - 6y_2 = 0$, where $\frac{d}{dx} \equiv D$.

Now $D^3 (\lambda_1 y_1 + \lambda_2 y_2) - 6D^2 (\lambda_1 y_1 + \lambda_2 y_2) + 11D (\lambda_1 y_1 + \lambda_2 y_2) - 6 (\lambda_1 y_1 + \lambda_2 y_2) = 0$

$\Rightarrow \quad \lambda_1 (D^3 y_1 - 6D^2 y_1 + 11D - 6y_1) + \lambda_2 (D^3 y_2 - 6D^2 y_2 + 11D y_2 - 6 y_2) = 0$

$\Rightarrow \quad \lambda_1 0 + \lambda_2 . 0 = 0.$

$\Rightarrow \quad \lambda_1 y_1 + \lambda_2 y_2 \in V.$

Hence V is a vector space over reals.

By theory of differential equation the given differential has the solution of the form

$$y = c_1 e^x + c_2 e^{2x} + c_3 e^{3x}, \text{ for some } c_1, c_2, c_3 \in R$$

thus $\{e^x, e^{2x}, e^{3x}\}$ spans the vector space V.

Further for some $a, b, c \in R$, we assume that

$$a e^x + be^{2x} + c e^{3x} = 0. \quad ...(1)$$

$$\frac{d}{dx}\left(ae^x + be^{2x} + ce^{3x}\right) = \frac{d}{dx}(0)$$

$$\Rightarrow \quad a e^x + 2be^{2x} + 3ce^{3x} = 0. \quad ...(2)$$

and $$\frac{d^2}{dx^2}\left(ae^x + be^{2x} + ce^{3x}\right) = \frac{d^2}{dx^2}(0)$$

$$\Rightarrow \quad ae^x + 4be^{2x} + 9ce^{3x} = 0. \quad ...(3)$$

the coefficient determinant of a, b, c in (1), (2), (3) is

$$\begin{vmatrix} e^x & e^{2x} & e^{3x} \\ e^x & 2e^{2x} & 3c^{3x} \\ e^x & 4e^{2x} & 9c^{3x} \end{vmatrix} = e^x e^{2x} e^{3x} \begin{vmatrix} 1 & 1 & 1 \\ 1 & 2 & 3 \\ 1 & 4 & 9 \end{vmatrix}$$

$$= e^x e^{2x} e^{3x} \begin{vmatrix} 1 & 0 & 0 \\ 0 & 1 & 2 \\ 0 & 3 & 8 \end{vmatrix} = e^{6x}(8-6) = 2e^{6x} \neq 0$$

$$\therefore \quad a = b = c = 0.$$

Hence $\{e^x, e^{2x}, e^{3x}\}$ forms the basis for V.

(*b*) Now inner product defined in V is

$$(u, v) = \int_{-\infty}^{0} uv\, dx$$

If $$u = e^x, (u, u) = \int_{-\infty}^{0} uv\, dx = \int_{-\infty}^{0} e^x . e^x\, dx = \left|\frac{e^{2x}}{2}\right|_{-\infty}^{0} = \frac{1}{2}$$

$$\Rightarrow \quad \|u\| = \frac{1}{\sqrt{2}}$$

We define

$$w_1 = \frac{u_1}{\|u_1\|} = \frac{e^x}{\frac{1}{\sqrt{2}}} = \sqrt{2}\, e^x$$

$$u_2 = v_2 - (v_2\ w_1)\ w_1$$

$$= e^{2x} - \left(\int_{-\infty}^{0} e^{2x}\sqrt{2}\ e^{x}\,dx\right)\sqrt{2}\ e^{x}$$

$$= e^{2x} - \sqrt{2}\left(\frac{e^{3x}}{3}\int_{-\infty}^{0}\sqrt{2}e^{x}\right)$$

$$= e^{2x} - \frac{\sqrt{2}}{3}\times\sqrt{2}\ e^{x} = \frac{3e^{2x}-2e^{x}}{3}$$

$$w_2 = \frac{u_2}{\|u_2\|} = \frac{\left(3e^{2x}-2e^{x}\right)/3}{\left[\int_{-\infty}^{0}\frac{\left(3e^{2x}-2e^{x}\right)^2}{3}dx\right]^{1/2}}$$

$$= \frac{\left(3e^{2x}-2e^{x}\right)/3}{\left[\frac{1}{9}\left[\frac{9e^{4x}}{4}+\frac{4e^{2x}}{2}-\frac{12e^{3x}}{3}\right]_{-\infty}^{0}\right]^{1/2}}$$

$$= \frac{\left(3e^{2x}-2e^{x}\right)/3}{\left[\frac{1}{9}\left[\frac{9}{4}+2-4\right]\right]^{1/2}} = \frac{3\times\left(3e^{2x}-2e^{x}\right)}{3\times\left(\frac{1}{2}\right)} = 2\left(3e^{2x}-2e^{x}\right)$$

Further

$$u_3 = v_3 - (v_3,\ w_1)\ w_1 - (v_3,\ w_2)\ w_2$$

$$= e^{3x} - \left(\int_{-\infty}^{0} e^{3x}\sqrt{2}e^{x}\,dx\right)\sqrt{2}\ e^{x} - \left(\int_{-\infty}^{0} e^{3x}\,2\left(3e^{2x}-2e^{x}\right)dx\right)$$

$$2\left(3c^{2x}-2e^{x}\right)$$

$$= e^{3x} - \frac{\sqrt{2}}{4}\times\sqrt{2}\ c^{x} - \left[4\left(\frac{3}{5}-\frac{2}{4}\right)\left(3e^{2x}-2c^{x}\right)\right]$$

$$= e^{3x} - \frac{1}{2}\times e^{x} - \frac{2}{5}\times 3e^{2x} + \frac{2}{5} + 2e^{x}$$

$$= \frac{10e^{3x}-12e^{2x}+3e^{x}}{10}$$

We can compute

$$(u_3,\ u_3) = \int_{-\infty}^{0}\left(10\ e^{3x}-12e^{2x}+3e^{x}\right)^2 dx = \frac{1}{600}$$

$$\|u_3\| = \frac{1}{10\sqrt{6}}$$

CHAPTER 9

$$w_3 = \frac{u_3}{\| u_3 \|} = \frac{10e^{3x} - 12e^{2x} + 3e^x}{10 \times \frac{1}{10\sqrt{6}}}$$

$$= \left(10e^{3x} - 12e^{2x} + 3e^x\right)\sqrt{6}.$$

Exercise 9.5.3: (Bissel's Inequality) If $\{w_1, \ldots, w_k)$ is an orthonormal subset of V, then prove that

$(w_1, v)^2 + (w_2, v)^2 + \ldots + (w_k, v)^2 \le \|v\|^2$ for all $v \in V$.

Proof: Let $v \in V$, we can write $v = \sum_{i=1}^{k} (v, w_i)w_i + x$, where $x = v - \sum_{i=1}^{k} (v, w_i)w_i$. By lemma 9.5.3 $(x, w_i) = 0$ for all $i = 1, 2, \ldots, k$. Putting $w = \sum_{i=1}^{k} (v, w_i)w_i$, we get $v = w + x$, with $(w, x) = \left(\sum_{i=1}^{k} (v, w_i)w_i, x\right)$

$$= \sum_{i=1}^{k} (v, w_i)(w_i, x) = 0.$$

Thus $\|v\|^2 = (w + x, w + x) = (w, w) + (w, x) + (x, w) + (x, x)$

$$= \|w\|^2 + \|x\|^2$$

$\Rightarrow$ $\|w\|^2 \le \|v\|^2$

But $\|w\|^2 = \left(\sum_{i=1}^{k} (v, w_i)w_i, \sum_{i=1}^{k} (v, w_i)w_i\right)$

$$= \sum_{j=1}^{k} \sum_{i=1}^{k} (v, w_i)(v, w_j)(w_i, w_j)$$

$$= \sum_{i=1}^{k} (v, w_i)^2 \text{ since } (w_i, w_j) = 0,\ i \neq j = 1,\ i = j$$

$$= \sum_{i=1}^{k} \left|\overline{(v, w_i)}\right|^2 = \sum_{i=1}^{k} \left|(w_i, v)\right|^2$$

Hence $\sum_{i=1}^{k} \left|(w_i, v)\right| \le \| v \|^2$

Exercise 9.5.4: Prove that

(*i*) $\|u + v\|^2 + \|u - v\|^2 = 2\|u\|^2 + 2\|V\|^2$

(*ii*) $4\,(u, v) = \|u + v\|^2 - \|u - v\|^2 + i\,\|u + V\|^2 - i\,\|u - V\|^2$.

Proof: (*i*) L.H.S $= (u + v, u + v) + (u - v, u - v)$

$= (u, u) + (u, v) + (v, u) + (v, v) + (u, u) - (v, u) - (u, v) + (v, v)$

$$= 2\|u\|^2 + 2\|v\|^2 = \text{R.H.S.}$$

(*ii*) $\|u + v\|^2 = (u + v, u + v) = (u, u) + (u, v) + (v, u) + (v, v)$

Replacing v by $-v$, iv by $-iv$, we get

$$\|u - v\|^2 = (u, u) - (u, v) - (v, u) + (v, v)$$

$$\|u + iv\|^2 = (u, u) - i\,(u, v) + i\,(v, u) + (v, v)$$

$$\|u - iv\|^2 = (u, u) + i\,(u, v) - i\,(v, u) + (v, v)$$

Now $\|u + v\|^2 - \|u - v\|^2 + i\,\|u + iv\|2 - i\,\|u - v\|^2$

$$= (u, u) + (u, v) + (v, u) + (v, v)$$
$$- (u, u) + (u, v) + (v, u) - (v, v)$$
$$+ i\,(u, u) + (u, v) - (v, u) + i\,(v, v)$$
$$- i\,(u, u) + (u, v) - (v, u) - i\,(v, v) = 4\,(u, v)$$

CHAPTER 9

Exercise 9.5.5: Prove these following inequalities in complex numbers vector space If $a, b, a_1, a_2, ..., a_n$ are complex numbers, then

(1) $|a + b| \le |a| + |b|$

(2) and $|a_1 + a_2 + ... + a_n| \le |a_1| + |a_2| + ... + |a_n|$

(3) $\dfrac{|a+b|}{1+|a+b|} \le \dfrac{|a|}{1+|a|} + \dfrac{|b|}{1+|b|}$

Solution: (1) and (2) are simple and left as exercise (3) Let there be continuous differentiable function $f(t) = t\,(1 + t)^{-1}$, where $t > -1$. Then $f'(t) = (1 + t)^{-2}$ which is positive. Thus, the rate of change of $f(t)$ is positive, so that $f(t)$ increases as t increases.

It means that $t_1 < t_2 \Rightarrow f(t_1) < f(t_2)$.

Now by (1) $\quad |a + b| \le |a| + |b|$

$\Rightarrow \quad f(|a + b|) \le f(|a| + |b|)$

$$\Rightarrow \quad \frac{|a+b|}{1+|a+b|} \le \frac{|a|+|b|}{1+|a|+|b|}$$

$$\Rightarrow \quad \frac{|a+b|}{1+|a+b|} \le \frac{|a|}{1+|a|+|b|} + \frac{|b|}{1+|a|+|b|}$$

$$\Rightarrow \quad \frac{|a+b|}{1+|a+b|} \le \frac{|a|}{1+|a|} + \frac{|b|}{1+|b|},$$

Since $|a| > 0$, $|b| > 0$.

Exercise 9.5.6: If $(a_1, a_2, ...a_n)$ and $(b_1, b_2, ..., b_n)$ are complex, then prove that

$$\sum_{r=1}^{n} |a_r b_r| \le \left[\sum_{r=1}^{n} |a_r|^2\right]^{1/2} \left[\sum_{r=1}^{n} |b_r|^2\right]^{1/2}$$

Proof: If x_1 and y_1 are reals, then

$(x_1 + y_1)^2 = 4\, x_1\, y_1 + (x_1 - y_1)^2$. Since $(x_1 - y_1)^2 \geq 0$

then $4x_1y_1 \leq (x_1 + y_1)^2$

If x_1, y_1 are assumed positive, it gives

$$\sqrt{x_1 y_1} \leq \frac{1}{2}(x_1 + y_1) \qquad ...(1)$$

Let us put $x_1 = \left[\frac{|a_1|}{\lambda}\right]^2$, $y_1 = \left[\frac{|b_1|}{\mu}\right]^2$, where ...(2)

$$\lambda = \left[\sum_{r=1}^{n} |a_r|^2\right]^{1/2}, \quad \mu = \left[\sum_{r=1}^{n} |b_r|^2\right]^{1/2} \qquad ...(3)$$

Since λ and μ are real and positive, by (2) x_1 and y_1 are real and positive.
Now putting (2) is (1)

$$\left[\frac{|a_1|}{\lambda}\right]\left[\frac{|b_1|}{\mu}\right] \leq \frac{1}{2}\left[\frac{|a_1|^2}{\lambda^2} + \frac{|b_1|^2}{\mu^2}\right]$$

Similarly,

$$\left[\frac{|a_2|}{\lambda}\right]\left[\frac{|b_2|}{\mu}\right] \leq \frac{1}{2}\left[\frac{|a_2|^2}{\lambda^2} + \frac{|b_2|^2}{\mu^2}\right]$$

....

$$\left[\frac{|a_n|}{\lambda}\right]\left[\frac{|b_n|}{\mu}\right] \leq \frac{1}{2}\left[\frac{|a_n|^2}{\lambda^2} + \frac{|b_n|^2}{\mu^2}\right]$$

Adding these, we get

$$\sum_{r=1}^{n} |a_r b_n| \leq \frac{\lambda\mu}{2}\left[\frac{1}{\lambda^2}\sum_{r=1}^{n} |a_r|^2 + \frac{1}{\mu^2}\sum_{r=1}^{n} |b_r|^2\right]$$

$$\leq \frac{\lambda\mu}{2}[1+1] \leq \lambda\,\mu$$

$$\leq \left[\sum_{r=1}^{n} |a_r|^2\right]^{1/2} \left[\sum_{r=1}^{n} |b_r|^2\right]^{1/2}$$

PROBLEMS

In all the problems V is an inner product space are F.

1. If F is a field of real numbers, then show that the Schwarz Inequality in F^3. Holds and the cosine of an angle is of absolute value at must 1.

2. If V in finite-dimensional and if $\{w_1,...,w_n\}$ is an orthonormal set in F such that

$$\sum_{i=1}^{n} |(w_i, v)|^2 = \|v\|^2,$$

for every $v \in V$, prove that $\{w_1,...., w_m\}$ must be basis of V.

3. If dim $V = n$ and if $\{w_1,..., w_m\}$ is an orthonormal set in V, prove that there exist vectors $w_{m+1},...., w_n$ such that $\{w_1,...., w_m, w_{m+1},...., w_n\}$ is an orthonormal set and basis of V.

4. Let V be the set of real valued functions $y = f(x)$ satisfying $\frac{a^2y}{dx^2} + ay = 0$.

   (a) Prove that V is a two-dimensional real vector space.

   (b) In V define $(y, z) = \int_0^\pi y\,z\,dx$. Find an orthonormal basis in V.

5. Let V be the set of all real functions $y = f(x)$ satisfying

$$\frac{a^2y}{ax^2} + 4y = 0.$$

   (a) Prove that V is the two dimensional real vector space.

   (b) In V define $(y, z) = \int_0^\pi yz\,dx$ for all $y, z \in V$. Find an orthonormal basis.

6. Find if possible all 4-tuples of real numbers a, b, c, d such that $u = (\alpha_1, \alpha_2)$, $v = (\beta_1, \beta_2) \in R^2$, and $(u, v) = a\alpha_1^2 + b\beta_2^2 + c\alpha_1\beta_2 + d\alpha_2\beta_1$, define an inner product on R^2.

7. Let $u = (a, b)$, $v = (c, d) \in R^2$, define $(u, v) = ac - bc - ad + 4\,bd$. Prove that R^2 is an inner product space.

8. If W is a subspace of V and if $v \in V$ satisfies $(v, w) + (w, v) \le (w, w)$, for every $w \in W$, prove that $(v, w) = 0$ for every $w \in W$.

9. If V is a finite-dimensional inner product space and if f is linear functional on V. (*i.e.*, $f: V \to F$), prove that there is $u_0 \in V$ such that $f(v) = (v, u_0)$ for all $v \in V$.

CHAPTER 9

9.6 MODULES

Now we generalise the concept of a vector space to a left R– module where R is a ring with identity.

Definition 9.6.1: An abelian group $(M, +)$ together with a map $f: R \times M \to M$ given by $(a, x) = a\,.\,x$, is called a left R-module satisfying the following axioms:

$$a\,.\,(x + y) = a\,.\,x + a\,.\,y,\ a \in R,\ x, y \in M.$$

$$(a + b)\,.\,x = a\,.\,x + b\,.\,x,\ a, b \in R,\ x \in M$$

$$a\,.\,(b\,.\,x) = (ab)\,.\,x,\ a, b \in R,\ x \in M.$$

If R has a unit element 1, $1\,.\,x = x$, for all $x \in M$, M is called Unital R-module.

Hence by an R-module we mean a left R-module throughout this section.

Example 9.6.1: (1) Every abelian group $(G, 0)$ is a module over the ring of integers. It is a unital Z-module.

(2) Let $(M, +, .)$ be a left ideal of the *ring* $(R, +, .)$ then M is a R-module.

(3) Any ring $(R, +, .)$ is an R-module over itself.

(4) Any vector space $V(F)$ over a field F is an F-module.

(5) $M = R[x]$, the ring of polynomials is an R-module for the usual addition and multiplication of a polynomial by a scalar.

Definition 9.6.2: A subset $N \subseteq M$ is called sub-module of M if $(N, +)$ is a sub-group of the abelian group $(M, +)$ and $a\,x \in N$ for all $a \in R$ and $x \in N$.

Example 9.6.2: (1) Any subspace W of a vector space over F is a sub-module of V.

(2) The set of all polynomials of degree atmost n is a sub-module of the R-module $R[x]$.

(3) Every R-module M contains 0 and M sub-modules. They are called improper sub-modules.

Definition 9.6.3: Let N be a sub-module of R-module M.

Then the quotient abelian group M/N with addition and scalar multiplication given by

$$(x + M) + (y + M) = x + y + M.$$

and

$$a \,.\, (x + M) = ax + M, \text{ for all } a \in R,\ x, y \in M$$

is well-defined and $(M/N +,)$ acquires the structure of an R-module. This is called the Quotient R-module M/N.

Definition 9.6.4: Let M_1 and M_2 be two R-modules. The mapping $f : M_1 \to M_2$ is called a R-homomorphism if for all $x, y \in M, a \in R$

$$f(x + y) = f(x) + f(y)$$

and

$$f(a \,.\, x) = a\,f(x).$$

Definition 9.6.5: The homomorphism f is said to be an isomorphism of R-modules if it is both one-to-one and onto.

Example 9.6.3: Show that the map $f : M \to M/N$ defined by

$$f(x) = x + N, \text{ for all } x \in M \text{ is a homomorphism of } R\text{-modules.}$$

Solution: We have the mapping $f: M \to M/N$ defined by

$$f(x) = x + N, \text{ for all } x \in M$$

For $x, y \in M$

$$\begin{aligned} f(x + y) &= x + y + N \\ &= (x + N) + (y + N) \\ &= f(x) + f(y) \end{aligned}$$

and $$f(a \, . \, x) = a \, . \, x + N$$
$$= a\,(x + N) = a\, f(x).$$

Hence f is a R-homomorphism.

Theorem 9.6.1: *Let f: $M \to N$ be a homomorphism of M onto N, where M and N are* R*-modules. Then the Ker $f = \{x \in M \mid f(x) = 0\}$ is a sub-module of M and the quotient module $M/Kerf \cong N$.*

Proof: Here $Kerf = \{x \in M \mid f(x) = 0\}$.

For $x, y \in Kerf$, $f(x) = 0$, $f(y) = 0$.

$$f(x - y) = f(x) - f(y) = 0 - 0 = 0.$$

$\Rightarrow$ $$x - y \in Ker\, f$$

and for $a \in R$

$$f(a \, . \, x) = a\, f(x) = a \, . \, 0 = 0$$

$\Rightarrow$ $$a \, . \, x \in Kerf.$$

Hence $Kerf$ is a R-module and M/K is meaningful.

Now, we define a map $T : M/K \to N$ by

$$T\,(x+K) = f(x), \ \forall \ x \in M.$$

For $x + K$, $y + K \in M/K$, x, $y \in M$.

$$T_{((x+K) \, + \, (y+K))} = T_{(x+y+K)}$$
$$= f_{(x+y)} = f(x) + f(y) = T_{(x+K)} + T_{(y+x)}$$

and $$T_{(x+K)} = T_{(y+K)}$$

$\Rightarrow$ $$f(x) = f(y)$$

$\Rightarrow$ $$f(x) - f(y) = 0$$

$\Rightarrow$ $$f(x - y) = 0$$

$\Rightarrow$ $$x - y \in K$$

$\Rightarrow$ $$x + K = y + K.$$

It is clearly onto, hence T is an isomorphism of R-modules.

$\therefore$ $$M/K \cong N.$$

Corollary 1: Let M be an R-module, N a sub-module of M and K a sub-module of N, then

$$M/N \cong \frac{M/K}{N/K}$$

Proof: We define a mapping $\phi : M/K \to M/N$ by $\phi\,(x + K) = x + N$, for all $x \in M$.

For $x + K$, $y + K \in M/K$

$$\phi\,(x + K + y + K) = \phi\,(x + y + K) = (x + y + N)$$

CHAPTER 9

$$= (x + N) + (y + N)$$
$$= \phi\,(x + K) + \phi\,(y + K).$$

This show ϕ is a homomorphism of R-modules.

Now we determine the $Ker\ \phi$,

$$Ker\ \phi = \left\{x+K \mid \phi(x+K) = N,\ x+K \in \frac{M}{K},\ x \in M\right\}$$

$$= \{x+K \mid x+N = N,\ x \in M\}$$

$$= \{x+K \mid x \in N\} = N/K.$$

Since K is sub-module of N.

Hence by fundamental theorem of isomorphism.

$$\frac{M/K}{N/K} \cong M/N$$

Corollary 2: Let N and K be sub-module of M, then

$$\frac{N+K}{K} \cong \frac{N}{N \cap K}$$

Proof: Since N and K are sub-modules of M, $N + M$ is also a sub-module of M. We have

$$N + K = \{x + y \mid x \in N,\ y \in K\}$$

Now $\qquad y \in K \Rightarrow 0 + y \in N + K \Rightarrow K \subset N + K.$

Hence $\dfrac{N+K}{K}$ is meaningful.

Again, since N and K and R-sub-modules $N \cap K$ is also a sub-module of N and K. So $\dfrac{N}{N \cap K}$ is meaningful.

Now we define a mapping $T : N \rightarrow \dfrac{N+K}{K}$ by

$$T\,(x) = x + K, \text{ for all } x \in N.$$

For $x, y \in N$, $f\,(x + y) = x + y + K = (x + K) + (y + K) = T\,(x) + T\,(y)$.

Moreover, T is onto because any $x + K \in \dfrac{N+K}{K}$ can be written as $x + K = h + k + K$, $h \in N$, $k \in K$

$$x + K = h + (k + K) = h + K.$$

Since $\qquad k + K = K$

Hence $\qquad x + K = h + K = T\,(h),\ h \in N.$

The kernel of T

$$Ker\ (T) = \{x \in N \mid T(x) = K\}$$
$$= \{x \in N \mid x + K = K\}$$
$$= \{x \in N \mid x \in K\} = N \cap K.$$

By fundamental theorem of isomorphism

$$\frac{N}{N \cap K} \cong \frac{N+K}{K}.$$

Now we consider some operations on modules. If N and K are sub-modules of M, $N \cap K$ is a sub-module of M but $N \cup K$ many not be a sub-module of M. The smaller sub-module of M containing $N \cup K$ is called the sub-module generated by N and K.

Theorem 9.6.2: *The sub-module S generated by N and K is the sub-module*

$$N + K = \{x + y \mid x \in N,\ y \in k\}$$

Proof: Let $u, v \in N + K$, then for some $x_1, x_2 \in N$ and $y_1, y_2 \in K$, $u = x_1 + y_1$, $v = x_2 + y_2$, then

$$u - v = (x_1 + y_1) - (x_2 + y_2) = (x_1 - x_2) + (y_1 - y_2) \in N + K$$

and
$$a\ (u) = a\ (x_1 + y_1) = ax_1 + ay_1 \in N + K.$$

Hence $N + K$ is a sub-module of M.

Since $N \subset N + K$ and $K \subset N + K$, then $S \subset N + K$.

Conversely, for any $x \in N$, $y \in K$, we have $x, y \in S$.

So that $x + y \in S$. Thus $N + K \subset S$ and $N + K = S$.

Corollary 1: If $N_1, N_2, \dots N_R$ are sub-modules and M is sub-module generated by $N_1, N_2, \dots . N_k$, then

$$M = \left\{\sum_{i=1}^{k} x_i \mid x_i \in N_i\right\}$$

Let A be a subset of M and U a left ideal of R. Then the set

$$A = \{x = \Sigma a_i x_i \mid a_i \in U \text{ and } x \in N_i\}$$

is an R-sub-module of M and is denoted by $\cup A$. In particular if $U = R$ and $A = \{x\}$, $\cup A$ is denoted by $R\ x$.

Definition 9.6.6: An R-module M is called cyclic if $M \in Rx$ for some $x \in M$.

Theorem 9.6.3: *An R-module M is cyclic if and only if $M \cong R/U$ for some left ideal U in R.*

Proof: Let M be cyclic, then for some $x \in M$,

$M = Rx$. The map ϕ: $R \to M$ defined by $\phi\ (a) = ax$, is an R-homomorphism which is on to

$$Ker\ \phi = \{a \in R \mid \phi\ (a) = 0\}$$

If $U = Ker\ \phi$, U is a left ideal of R and by theorem 9.6.1 $\dfrac{R}{U} \cong M$

CHAPTER 9

Conversely, if $M \cong R/U$ for some left ideal u of R, then M is cyclic as R/U is cyclic being generated by

$$T = 1 + U.$$

Definition 9.6.7: The annihilator of an R-module M is defined by $Ann\ (M) = \{a \in R \mid aM = 0\}$.

Clearly $Ann\ (M)$ is a left ideal of R. If M is cyclic and generated by x, then $Ann\ (M)$ is denoted by $Ann\ (x)$.

Definition 9.6.8: An R-module M is called a faithful R-module if $Ann\ (M) = (0)$.

Definition 9.6.9: An R-module M is said to be finitely generated if there exist elements $a_1, a_2, \ldots, a_n \in M$ such that every $m \in M$ can be expressed as

$$m = r_1 a_1 + v_2 a_2 + \ldots + r_n a_n, r_i \in R.$$

Since if $(a_1) = M_1, (a_2) = M_2, \ldots, (a_n) = M_n$, then

$M = M_1 + M_2 + \ldots + N_n$, where each M_i is a cyclic, the set $\{a_1, a_2, \ldots, a_n\}$ is called the generating set.

Definition 9.6.10: An R-module M is called a direct sum of sub-modules $M_1, M_2, \ldots, M_n$ if every $m \in M$ can be expressed uniquely as

$$m = m_1 + m_2 + \ldots + m_n, m_i \in M_i, 1 \le i \le n.$$

The direct sum is denoted by the symbol

$$M = M_1 \oplus M_2 \oplus \ldots \oplus M_n.$$

Theorem 9.6.4: *An R-module $M = M_1 \oplus M_2 \oplus \ldots \oplus M_n$ if and only if*

(1) $M = M_1 + M_2 + \ldots + M_n$ *and*

(2) $M_i \cap (M_1 + M_2 + \ldots + M_{i-1} + M_{i+1} + \ldots + M_n) = \{0\}$ *for all i, $1 \le i \le n$.*

Proof: Suppose $M = M_1 \oplus M_2 \oplus \ldots \oplus M_n$. Then every element $m \in M$ can be expressed uniquely as

$$m + m_1 + m_2 + \ldots + m_n, m_i \in M_i, 1 \le i \le n.$$

$$\Rightarrow \qquad M = M_1 + M_2 + \ldots + M_n.$$

To prove (*ii*) Let $x \in M_i \cap (M_1 + M_2 + \ldots + M_{i-1} + M_{i+1} + \ldots + M_n)$

Then $x \in M_i$ and $x \in M_1 + M_2 + \ldots + M_{i-1} + M_{i-1} + \ldots + M_n$.

So $x = y_1 + y_2 + \ldots + y_{i-1} + y_{i+1} + \ldots + y_n + y_j \in M_j, i \ne j$

and $\quad x = 0 + 0 + \ldots + 0 + x + 0 + \ldots + 0.$

Comparing two expressions of x and by uniqueness we get $x = 0$.

This proves that $M_i \cap (M_1 + M_2 + \ldots + M_{i-1} + M_{i+1} + \ldots + M_n) = \{0\}$

Conversely, Assume conditions (1) and (2).

By (1) $m \in M$ can be expressed as

$$m = m_1 + m_2 + \ldots + m_n, m_i \in M_i, 1 \le i \le n.$$

To prove the uniquesness of expression, suppose

$$m = n_1 + n_2 + \ldots + n_n, n_i \in M_i$$

Then $\qquad m_1 + m_2 + \ldots + m_n = n_1 + n_2 + \ldots + n_n$

$\Rightarrow \qquad (m_1 - n_1) + (m_2 - n_2) + \ldots + (m_n - n_n) = 0.$

So that $m_i - n_i \in M_i$ and

$$m_i - n_i = [(m_1 - n_1) + (m_2 - n_2) + \ldots + (m_{i-1} - n_{i-1}) + (m_{i+1} - n_{i+1}) + \ldots + (m_n - n_i)]$$

But condition (II) $m_i - n_i = 0 \Rightarrow m_i = n_i$, $1 \le i \le n$.

Thus $$M = M_1 \oplus M_2 \oplus \ldots \oplus M_n.$$

This proves the theorem.

Again we consider $M = M_1 \oplus M_2$, then $x \in M$ can be expressed uniquely as

$$m = m_1 + m_2,\ m_1 \in M,\ m_2 \in M_2,$$

The mappings $p_1 : M \to M_1$ and $p_2 : M \to M_2$ defined by $p_1(x) = x_1$ and $p_2(x) = x_2$ are called projection mappings.

Definition 9.6.11: A cyclic R-module $M = Rx$ is called free of $Ann(x) = 0$.

In this case, we shall see that every element $m \in M$ can be expressed uniquely as $m = ax$, for $a \in R$. If $m = bx$, then

$$ax = bx \Rightarrow (a - b)x = 0 \Rightarrow a - b \in Ann(x) = 0$$
$$\Rightarrow a = b.$$

Definition 9.6.12: An R-module M is called free on a finite basis if it can expressed as a direct sum $M = M_1 \oplus M_2 \oplus \ldots \oplus M_n$, where each M_i is a free cyclic R-module. If $M_i = Rm_i$, then the set $\{m_1, m_2, \ldots, m_n\}$ is called a basis of the free module M.

In this case, every $m \in M$ can be expressed uniquely

as $$m = a_1 m_1 + a_2 m_2 + \ldots + a_n m_n,\ a_i \in R. \text{ For if}$$
$$m = b_1 m_1 + b_2 m_2 + \ldots + b_n m_n,\ b_i \in R, \text{ then}$$

$a_i m_i = b_i m_i$ since $M = M_1 \oplus M_2 \oplus \ldots \oplus M_n$.

$$\Rightarrow \quad (a_i - b_i) m_i = 0$$
$$\Rightarrow \quad a_i - b_i \in Ann(x_i) = 0$$
$$\Rightarrow \quad a_i = b_i \text{ for all } i.$$

Theorem 9.6.5: *Let $(R, +, .)$ be a commutative ring. Any two basis of a free R-module (with finite base) have the same number of elements.*

Proof: Let M be a free R-module with basis $\{x_1, x_2, \ldots x_n\}$ choose a maximal ideal $(U, +, .)$ then $(R/U, +\ .)$ is a field, say, F. Then $V = \dfrac{M}{mM}$ is annihilated by m and hence it is vector space over F. If $\bar{x}_i = x_i + mM$ $1 \le i \le n$, then we see that for $a_1, a_2, \ldots, a_n \in F$,

$$\sum_{i=1}^{n} a_i \bar{x}_i = mM$$

$$\Rightarrow \quad a_1\bar{x}_1 + a_2\bar{x}_2 + \ldots\ldots + a_n\bar{x}_n = m\,M$$
$$\Rightarrow \quad a_1(x_1 + mM) + \ldots + a_n(x_n + mM) = mM.$$
$$\Rightarrow \quad (a_1x_1 + \ldots + a_n x_n) + mM = mM = 0 \text{ since it is annihilated by } m.$$
$$\Rightarrow \quad a_1 x_1 + a_2 x_2 + \ldots + a_n x_n = 0$$
$$\Rightarrow \quad a_1 = a_2 = \ldots = a_n = 0 \text{ since } \{x_1, \ldots, x_n\} \text{ is a basis.}$$

Hence $\{\bar{x}_1, \ldots, \bar{x}_n\}$ is a basis of V over F. Since any two basis of a vector space have the same number of elements. The result follows.

CHAPTER 9

Definition 9.6.13: If a free R-module F has a basis with n elements, then any other basis of F also has n elements. Then n is called rank of F.

PROBLEMS

1. Verify that every abelian group is a module over the ring of integers.
2. Suppose that $(R, +, .)$ is a ring with identity and M is a module over R but is not unital. Prove that there exists an $m \neq 0$ in M such that $rm = 0$ for all $r \in R$.
3. If T is a homomerphism of M into N, let $Ker\ (T) = \{x \in M \mid T\ (x) = 0\}$. Prove that $Ker\ (T)$ is a sub-module of M and $I\ (T) = \{T\ (x) | x \in M\}$ is a sub module of N.
4. Prove the homomorphism $T : M \to N$, where M and N are R-modules is an isomorphism if and only if $Ker\ (T) = (O)$.
5. Let M, N, Q be three R-modules and let $T: M \to N$ be a homomorphism and $S : N \to Q$ be homomorphism of N into Q. Define $S.oT.\ M \to Q$ by
$$(S\ o\ T)\ (m) = S\ (T\ (m) \text{ for ever } m \in M.$$
Prove that SoT is an R-homomorphism of M into Q and determine its Kernel, $Ker\ (SoT)$.
6. An R-module M is said to be irreduable if, its only sub-modules are (0) and M. Prove that any unital, irreduable R-module is cyclic.
7. If M is an irreduable R-module, prove that either M is cyclic or that every $m \in M$ and $r \in R$, $rm = 0$.
8. If M is an irreduable R-module such that $rm \neq 0$ for some $r \in R$ and $m \in M$, prove that any R-homomorphism T of M into M is either an isomorphism of M onto M or that $T\ (m) = 0$ for every $m \in M$.
9. Let M be an R-module, if $m \in M$ and $\lambda (m) = \{x \in R | xm = 0\}$. Show that $\lambda\ (m)$ is a left ideal of R. It is called the order of m.
10. If λ is a left ideal of R and if M is an R-module, show that for $m \in M$, $\lambda\ m = \{xm | x \in \lambda\}$ is a sub-module of M.
11. Let M be an irreducible R-module in which $r\ m \neq 0$ for some $r \in R$ and $m \in M$. Let $m_0 \neq 0 \in M$ and let $\lambda\ (m_0) = \{x \in R \mid x\ m_0 = 0\}$.
    (a) Prove that $\lambda\ (m_0)$ is maximal left ideal of R (that is, if λ is a left ideal of R such that $\lambda\ (m_0) \subset \lambda \subset R$, then $\lambda = R$ or $\lambda = \lambda\ (m_0)$.
    (b) As R-modules, prove that $M \cong R - \lambda\ (m_0)$.

CHAPTER 10

LINEAR TRANSFORMATION

10.1 DEFINITION AND EXAMPLES OF LINEAR TRANSFORMATIONS

We have studied homomorphisms from one algebraic system to another algebraic system, namely, group homomorphism, ring homomorphism. On parallel lines we shall study vector space homomorphism, Since the vector space $U(F)$ is comprised of two algebraic system, group $(V, +)$ and a field $(F, +,.)$, there may be some confusion as to what operations are to be preserved by such functions. Generally vector space homomorphism are called *Linear mappings* or *linear transformation.*

Definition 10.1.1: Let $U(F)$ and $V(F)$ be two vector spaces over the same field $(F, +,.)$. A function f from U to V, $f: U \to V$, is said to be *linear transformation* from $U(F)$ into $V(F)$ if

(1) $f(x + y) = f(x) + f(y)$,

(2) $f(c\,x) = cf(x)$,

$\forall\ x, y \in U$ and $c \in F$.

Thus $f: U \to V$ is linear if it preserves the vector addition and scalar multiplication. Since $U(F)$ and $V(F)$ are vector spaces over the same field F, then some times we call the linear transformation f an F – linear transformation. The set of all linear mappings from $U(F)$ into $V(F)$ will be denoted by $L(U, V)$.

From (2), if $c = 0$, we get $f(0\,.\,x) = f(0) = 0\,f(x) = 0$, that is, every linear transformation maps *zero vector* of U onto *zero vector* of V.

Theorem 10.1.1: *Let $U(F)$ and $V(F)$ be two vector spaces over the same field F. Then $f: U \to V$ is a linear mapping if and only if $f(ax + by) = a\,f(x) + b\,f(y)$, $\forall\ x, y \in U$ and $a, b \in F$.*

Proof: Let $f : U \to V$ be a linear mapping. Then

$\forall\ x, y \in U, a, b \in F, ax + by \in U$, since $U(F)$ is a vector space over F.

Then
$$f(ax + by) = f(ax) + f(by) \text{ by 1 of Def.}$$
$$= af(x) + bf(y) \text{ by 2 of Def.}$$

Conversely, let $f(ax + by) = af(x) + bf(y)$.

Taking $a = 1 = b \in F$, the identity of F, then
$$f(x + y) = f(1.x + 1.y) = 1.\,f(x) + 1.\,f(y)$$
$$= f(x) + f(y)$$

Again taking $a \neq 0$, $b = 0$, then

$f(ax) = f(ax + 0y) = af(x) + 0\, f(y) = af(x)$.

Hence f is a linear transformation.

Example 10.1.1: *Let $M_{m\times n}$ (F) be a vector space over the field F, and let $[a_{ij}]$ be* $m \times n$ a fixed matrix over F.

Define a function $f : M_{mn} \to M_{mn}$ by

$f[b_{ij}] = [a_{ij}] \,.\, [b_{ij}],\ \forall\, [b_{ij}], \in M_{mn},$

we see that for $[b_{ij}], [c_{ij}] \in M_{mn}$ and for $r, s \in F$,

$$\begin{aligned} f(r[b_{ij}] + s[c_{ij}]) &= [a_{ij}] \,.\, [r[b_{ij}] + s[a_{ij}]] \\ &= r[a_{ij}] \,.\, [b_{ij}] + s[a_{ij}][c_{ij}] \\ &= rf[b_{ij}] + sf[c_{ij}]. \end{aligned}$$

Hence f is a linear transformation from M_{mn} *(F)* into M_{mn} *(F)*.

Example 10.1.2: Let $F[x]$ be a vector space of polynomials in the indeterminant x with coefficients from F. A Function $f : F[x] \to F[x]$ is defined by the differentiation, that is, *if* $p(x) = a_0 + a_1x + a_2x^2 + ... + a_n\, x^n \in F[x]$, then $f(p(x)) = a_1 + 2a_2\, x + ... + n\, a_n\, x^{n-1} \in F[x]$ *i.e.*, $f(p(x)) = p'(x)$. We see that $\forall\, p(x), g(x) \in F(x)$

$$\begin{aligned} f(ap(x) + bg(x)) &= [a(p(x) + bg(x)]' \\ &= (a(p(x))' + (bg(x))' \\ &= ap'(x) + bg'(x) \\ &= af(p(x)) + bf(g(x)). \end{aligned}$$

which shows f, that is, the differentiation is a L.T. (Linear transformation).

Example 10.1.3: *Let F^3 and F^2 be two vector spaces over the same field F.* Define a mapping $f : F^3 \to F^2$

by $(a, b, c) = (a, b),\ \forall\, (a, b, c) \in F^3$

For any $(a, b, c), (a', b', c') \in F^3$,

$$\begin{aligned} f[(a, b, c) + (a', b', c')] &= f(a + a, b + b', c + c') \\ &= (a + a', b + b') \\ &= (a, b) + (a', b') \\ &= f(a, b, c) + f(a, b', c'). \end{aligned}$$

And
$$\begin{aligned} f(r(a, b, c) &= f(ra.\, rb, rc) = (ra, rb) \\ &= r(a.b) = rf(a, b, c). \end{aligned}$$

Hence f is *a* L.T.

Example 10.1.4: *Let W (F) be a subspace of the vector space V (F).* Then the natural mapping $f : V \to V/W$ defined *by* $f(x) = x + W, x \in V$,

is a linear transformation. We can check that

$$\begin{aligned} f(ax + by) &= ax + by + W \\ &= (ax + W) + (by + W) \\ &= a(x + W) + b(y + W) \\ &= af(x) + bf(y). \end{aligned}$$

Example 10.1.5: *Let $f : R^2 \to R^2$* be the translation mapping defined by $f(x, y) = (x + 1, y + 2)\ \forall\, (x, y) \in R^2$. We observes that $f(0, 0) = (1, 2) \neq (0, 0)$. That is, f does not map $(0, 0) \in R^2$ onto $(0, 0) \in R^2$. This shows that f is not a linear transformation.

Example 10.1.6: *Let $f : U \to V$* be the mapping which assigns $0 \in V$ to every $u \in U$, *i.e.*, $f(u) = 0,\ \forall\, u \in U$. Than for any $u, v \in U$ and for any $k \in F$, we have

$f(u + v) = 0 = 0 + 0 = f(u) + f(v)$ and

$f(ku) = 0 = k0 = kf(u)$

Hence f is L.T. which is called the *zero mapping* usually denoted by O.

Example 10.1.7: We consider the identity mapping $I : V \to V$ defined by $I(v) = v, v \in V$. Then for any $u, v \in V$, $I(u + v) = u + v = I(u) + I(v)$ and $I(ku) = kI(u)$. Hence I is L.T.

Example 10.1.8: In R^2 let f_1, f_2, and f_3 be defined by

$$f_1(x, y) = (x, 0),$$
$$f_2(x, y) = (0, y),$$
$$f_3(x, y) = (y, x).$$

Hence f_1, f_2, and f_3 are all L.T. f_1 is a *projection* of each point of the plane onto the x-axis, f_2 is a *projection* of each point of a plane onto *y-axis*, and f_3 is *reflexion* across the *line* $y = x$.

We observe that $f_1(x, y) = (x, 0)$

and $$f_2(x, 0) = (0,0),$$

that is, $$f_2 \, o \, f_1(x, y) = f_2(f_1(x, y)) = f_2(x, 0) = (0, 0).$$

So $f_2 \, o \, f_1 = 0$ but $f_1 \neq 0$ and $f_2 \neq 0$. Hence the product of non-zero linear transformation can be the zero transformation. Also $f_3 \, o \, f_2((x, y)) = f_3((0, y) = (y, 0)$; however $f_2 \, o \, f_3(x, y) = f_2(y, x) = (0, x)$. Hence $f_3 \, o \, f_2 \neq f_2 \, o \, f_3$, so the multiplication of transformation is not commutative. Finally, we observe that $f_1(f_1(x, y)) = f_1(x, 0) = (x, 0) = f_1(x, y)$, So $f_1^2 = f_1$. Thus there exists *idempotent transformation* other than I and 0.

Example 10.1.9: In the vector space $F[x]$ of polynomials $p(x)$ of degree not exceeding n, Let

$D(p(x)) = \dfrac{d}{dx}(p(x)) = p'(x)$. we have seen that D is linear. We also note that $D^{n+1}(P(x)) = 0$ for every $p(x) \in F[x]$, so $D^{n+1} = 0$. Thus there exists a non-zero transformation f such that a finite power of f is 0. A transformation f is called *nilpotent of index n* if $f^n = 0$ but $f^{n-1} \neq 0$.

Definition 10.1.2: *Two* linear transformations f_1 and f_2 from $U(F)$ to $V(F)$ are said to be equal if and only if $f_1(x) = f_2(x), \forall x \in U$.

Definition 10.1.3: A linear transormation $f : U \to V$ is called an *isomorphism* if f is one-to-one. The vector spaces $U(F)$ and $V(F)$ are said to be *isomorphic* if there is an isomorphism of U onto V. Symbolically, we write $U(F) \cong V(F)$.

Example 10.1.10: Let $V(F)$ be a vector space over the field F of dimension n and let $(e_1, e_2, \ldots, a_n,\}$ be the basis of V we define $f : V \to F^n$ by $f(v) = f(a_1 e_1 + a_2 e_2 + \ldots a_n e_n) = (a_1, a_2, \ldots, a_n)$, since $v \in V$ can be written as $v = a_1 e_1 + a_2 e_2 + \ldots + a_n e_n$. It can be easily seen that f is one to one mapping. Hence it is isomorphism from V onto F^n So V (F) $\cong F^n$.

Theorem 10.1.2: *Let $f : U \to V$ be a linear transformation from $U(F)$ into $V(F)$. If $\{x_1, x_2, \ldots, x_n\}$ is L.D. in U, then $f(x_1), f(x_2), \ldots, f(x_n)$ are also L.D. in V.*

Proof: Since $x_1, x_2 \ldots x_n$ are L.D., then there exists $a_1, a_2, \ldots, a_n \in F$ not all zero such that

$$a_1 x_1 + a_2 x_2 + \ldots + a_n x_n = 0 \in U.$$

Therefore $$0 = f(0) = f(a_1 x_1 + a_2 x_2 + \ldots + a_n x_n) = a_1 f(x_1) + a_2 f(x_2) + \ldots + a_n f(x_n),$$

which *shows* $f(x_1), f(x_2), \ldots, f(x_n)$ are L.D. in V.

CHAPTER 10

Theorem 10.1.3: *Let $f : U \to V$ be linear, and suppose $x_1, x_2, ..., x_n \in U$ have the property that their images $f(x_1), f(x_2), f(x_3), ... f(x_n)$ are linearly independent (L.I.). Then the vectors $x_1, x_2, ..., x_n$ are also L.I.*

Proof: Since $f(x_1), f(x_2), ..., f(x_n)$ are L.I., than these exists scalars $a_1, a_2, ..., a_n \in F$ with $a_1 = a_2 = ... = a_n = 0$ such that

$$a_1 f(x_1) + a_2 f(x_2) + ... + a_n f(x_n) = 0 \in V$$

$$\Rightarrow \ f(a_1 x_1 + a_2 x_2 + ... + a_n x_n) = 0 = f(0)$$

$$\Rightarrow \ a_1 x_1 + a_2 x_2 + ... + a_n x_n = 0 \in U, \text{ where}$$

$a_1 = a_2 = ... = a_n = 0$. This shows that $x_1, x_2, ..., x_n$ are linearly independent.

But the converse is not true, that is, if $x_1, x_2, ..., x_n$ are *L.I*, then their images $f(x_1), f(x_2), ..., f(x_n)$ may not be linearly independent. This can be shown by an example. In R^2 let f be a linear transformation defined by

$$f(x, y) = (0, y).$$

We know $\{(1, 1), (0, 1)\}$ is a basis of R^2, that is, (1, 1) and (0, 1) are linearly independent. Their images $f(1, 1) = (0, 1)$ and $f(0, 1) = (0, 1)$ are not L.I. For this, let $k_1, k_2 \in F$, then

$$k_1 f(1, 1) + k_2 f(0, 1) = (0, 0)$$

$$\Rightarrow k_1 (0, 1) + k_2 (0, 1) = (0, 0) \Rightarrow (0, k_1) + (0, k_2) = (0, 0)$$

$$\Rightarrow (0 + 0, k_1 + k_2) = (0, 0)$$

$$\Rightarrow k_1 + k_2 = 0$$

$$\Rightarrow k_1 = -k_2$$

Thus, $f(1, 1)$ and $f(0, 1)$ are not L.I., since k_1 and k_2 are not zero.

Theorem 10.1.4: *Let $\{x_1, x_2, ..., x_n\}$ be a basis for the finite dimensional vector space V (F) and $\{y_1, y_2, ... y_n\}$ be an arbitrary set of n vectors from W (F). Then there exists a unique linear mapping $f : V \to W$ such that*

$$f(x_1) = y_1, f(x_2) = y_2, ..., f(x_n) = y_n.$$

Proof: Let $x \in V$. Since $\{x_1, x_2, ..., x_n]$ is a basis for V, then there exist scalars $a_1, a_2, ..., a_n \in F$ for which

$$x = a_1 x_1 + a_2 x_2 + ... + a_n x_n$$

Let us define $f : V \to W$ at the vector x by taking

$$f(x) = a_1 y_1 + a_2 y_2 + ... + a_n y_n.$$

Since $a_1, a_2, ..., a_n$ are unique, the mapping f is well defined. For $i = 1, 2, ..., n$,

$$x_i = o\, x_1 + o\, x_2 + ... + 1\, x_i + ... + o\, x_n.$$

$$\Rightarrow \qquad f(x_i) = o\, y_1 + o\, y_2 + ... + 1\, y_i + ... + o\, y_n.$$

Now we see that f is linear mapping. Let

$x, y \in V$, where $x = a_1 x_1 + a_2 x_2, + ... + a_n x_n$ and

$$y = b_1 x_1 + b_2 x_2 + ... + b_n x_n.$$

Then
$$\begin{aligned} f(x + y) &= f\{(a_1 + b_1) x_1 + (a_2 + b_2) x_2 + ... + (a_n + b_n) x_n\} \\ &= (a_1 + b_1) y_1 + (a_2 + b_2) y_2 + ... + (a_n + b_n) y_n \\ &= (a_1 y_1 + a_2 y_2 + ... + a_n y_n) + (b_1 y_1 + b_2 y_2 + ... + b_n y_n) \\ &= f(x) + f(y). \end{aligned}$$

Further
$$cx = ca_1 x_1 + ca_2 x_2 + ... + ca_n x_n,$$

$$f(c\, x) = ca_1 y_1 + ca_2 y_2 + ... + ca_n y_n.$$

$$= c\,(a_1 y_1 + a_2 y_2 + \ldots + a_n y_n)$$
$$= cf\,(x).$$

Thus f is linear.

Finally we shall show that f is unique linear mapping. Now suppose $g : V \to W$ is linear with the property that $g\,(x_i) = y_i$, $i = 1, 2, \ldots, n$.

If $x = a_1 x_1 + a_2 x_2 + \ldots + a_n x_n$, then

$$g\,(x) = a_1 g\,(x_1) + a_2 g\,(x_2) + \ldots + a_n g\,(x_n)$$
$$= a_1 y_1 + a_2 y_2 + \ldots + a_n y_n = f\,(x).$$

Since $g\,(x) = f\,(x)$, $\forall\, x \in V$, $g = f$. Thus f is unique. Hence the theorem is proved.

Example 10.1.11: *Let $f : R^2 \to R$ be* the linear mapping for which $f\,(1, 1) = 3$ and $f\,(0, 1) = -2$. Since $\{(1, 1), (0, 1)\}$ is a basis for R^2, such mapping exists and is unique by above theorem. Find $f\,(a, b)$.

Solution: Since $\{(1, 1), (0, 1)\}$ is a basis of R^2, then $(a, b) \in R^2$ can be expressed uniquely as

$$(a, b) = x\,(1, 1) + y\,(0, 1), \text{ where } x, y \text{ are scalars.}$$
$$= (x, x) + (0, y)$$
$$= (x, x + y) \Rightarrow x = a \text{ and } x + y = b$$

Thus $\quad x = a,\; y = b - a.$

Now $\quad f\,(a, b) = f\,(x\,(1, 1) + y\,(0, 1)) = x\,f\,(1, 1) + y\,f\,(0, 1)$

$$= x\,(3) + y\,(-2)$$
$$= a \,.\, 3 + (b - a)\,(-2)$$
$$= 3a - 2b + 2a = 5a - 3b.$$

10.2 KERNEL AND IMAGE OF LINEAR MAPPING

Definition 10.2.1: Let $f : U \to W$ be a linear mapping from a vector space $U\,(F)$ into a vector space $W\,(F)$. Then $Ker\,(f)$ is the set of all $x \in U$ which are mapped on 0, the additive identity of W, by f. That is,

$\text{Ker}\,(f) = \{x \in U \,|\, f\,(x) = 0 \in W\}.$

Definition 10.2.2: Let $f : V \to W$ be a linear mapping. Then the set of all images of all element $x \in U$ is called the *Range of f*, written as R_f, or the image of V under f, written as $f\,(V)$. Thus

$R_f = f\,(V) = \{y \in W \,|\, y = f\,(x) \text{ for some } x \in U\}.$

Theorem 10.2.1: *Let $f : V \to W$ be a Linear mapping from a vector space $V\,(F)$ into $W\,(F)$ over the same field F. Then*

(1) *$Ker\,(f)\,(F)$ is a subspace of $U\,(F)$;*

(2) *$f\,(V)$ is a subspace of $W\,(F)$;*

(3) *$dim\ V = dim\ Ker\,(f) + dim\ f\,(V)$* (Sylvester's law)

Proof: **(1)** Let $x, y \in \text{Ker}\,(f)$, then $x, y \in V$ and

$f\,(x) = 0$ and $f\,(y) = 0$.

Now for $a, b \in F$, we have

$f\,(ax + by) = af\,(x) + bf\,(y) = a.0 + b.0 = 0 + 0 = 0$

So $ax + by \in \text{Ker}\,(f)$.

Hence $\text{Ker}\,(f)\,(F)$ is a subspace of $V\,(F)$.

Definition 10.2.3: (1) Ker *(f (F)* is called a null space and the nullity $\nu(f)$ of a linear transformations f is the dimension of its null space, $\nu(f) = \dim \text{Ker}(f)$.

(2) Let $y_1, y_2 \in f(V)$. Then there exist $x_1, x_2 \in V$ such that $f(x_1) = y_1, f(x_2) = y_2$, Now for $a, b \in F$, we have

$$\begin{aligned} f(a_1 x_1 + a_2 x_2) &= a_1 f(x_1) + a_2 f(x_2) \\ &= a_1 y_1 + a_2 y_2 \in f(V). \end{aligned}$$

Thus $f(V)(F)$, is a subspace of $W(F)$.

Definition 10.2.4: $f(V)(F)$ is called *range space* and $\dim f(V)$ is called the *rank* $r(f)$ of a linear transformation f.

(3) Let $\dim V = n$. Suppose $\text{Ker}(f) \neq \{0\}$, so that $\text{Ker}(f)$ is a subspace of $V(F)$ of finite dimension, say, $\dim \text{Ker}(f) = r \leq n$.

Let $\{x_1, x_2, \ldots x_r\}$ be a bases of Ker (f). Then we can extend $\{x_1, x_2, \ldots, x_r)$ to a basis $\{x_1, x_2, \ldots, x_r, y_1, y_2, \ldots, y_{n-r}\}$ of V.

Now for any $y \in f(V)$, there exists $x \in V$ such that $y = f(x)$. In terms of basis for $V(F)$, the vector x can be written as

$$x = a_1 x_1 + a_2 x_2 + \ldots + a_r x_r + b_1 y_1 + b_2 y_2 + \ldots + b_{n-r} y_{n-r}$$

But $$\begin{aligned} f(a_1 x_1 + a_2 x_2 + \ldots + a_r x_r) &= a_1 f(x_1) + \ldots + a_r f(x_2) \\ &= a_1 0 + \ldots + a_r . 0 = 0. \end{aligned}$$

which implies

$$\begin{aligned} y &= f(x) = a_1 f(x_1) + a_2 f(x_2) + \ldots + a_r f(x_r) + b_1 f(y_1) \\ &\quad + \ldots + b_{n-r} f y_{n-r} \\ &= b_1 f(y_1), + b_2 f(y_2) + \ldots + b_{n-r} f(y_{n-r}). \end{aligned}$$

Thus from this, we get that every element $y \in f(V)$ can be expressed as a Linear combination of vectors *of* $f(y_1), f(y_2), \ldots, f(y_{n-r})$. *Let* $B = \{f(y_1), f(y_2), \ldots f(y_{n-r})$

If we show that the set B is L.I., then theorem is proved, for this,

Let

$$a_1 f(y_1) + a_2 f(y_2) + \ldots + a_{n-r} f(y_{n-r}) = 0, \text{ then}$$
$$f(a_1 y_1 + a_2 y_2 + \ldots + a_{n-r} y_{n-r}) = 0$$
$$\Rightarrow a_1 y_1 + a_2 y_2 + \ldots + a_{n-r} y_{n-r} \in \text{Ker}(f).$$

Since $\{x_1, x_2, \ldots, x_r\}$ is basis for Ker (f), then there must exist scalars $d_1, d_2, \ldots, d_r$ such that

$$d_1 x_1 + d_2 x_2 + \ldots d_r d_r = a_1 y_1 + a_2 y_2 + \ldots + a_{n-r} y_{n-r}$$

or $d_1 x_1 + d_2 x_2 + \ldots + d_r x_3 - a_1 y_1 - a_2 y_2 \ldots - a_{n-r} y_{n-r} = 0$

Since $\{x_1, x_2, \ldots x_r, y_1, y_2, \ldots y_{n-r}\}$ is a basis for $V(F)$, it is L.I. consequently $d_1, = d_2, = \ldots = d_r = a_1 = a_2 = \ldots = a_{n-r} = 0$

Hence $\{f(y_1), f(y_2), \ldots f(y_{n-r})\}$ is a basis of $f(V)(F)$.

Thus

$\dim V = r + (n - r) = \dim \text{Ker}(f) + \dim f(V)$.

Hence the theorem is proved.

If $\text{Ker}(f) = 0$, then f maps any basis of $V(F)$ onto a basis of $f(V)(F)$. Thus $\dim V = \dim f(V)$.

Corollary: *Let $V(F)$ and $W(F)$ be finite-dimensional vector spaces with $\dim V = \dim W$ and let $f: V \to W$ be a L.T. Then f is one-to-one mapping if and only if f maps V onto W.*

Proof: First of all suppose f is one-to-one, so that Ker $(f) = \{0\}$ which means dim Ker $(f) = 0$. By Sylvester's law we have dim V = dim f (V). We know that if dim V = dim f (V) and f is one-to-one, then f is an onto mapping.

Conversely, if f maps V onto W, $f(V) = W$.

Now from $f(V) = W$ and dim V = dim W = dim $f(V)$, the equation dim V = dim Ker (f) + dim $f(V)$ yields to dim Ker $(f) = 0$ which implies Ker $(f) = \{0\}$. Hence f is one-to-one mapping.

Theorem 10.2.2: *If V (F) is a finite dimensional vector space of dimension n, then V (F) $\cong V_n$ (F) or F^n.*

Proof: Let dim $V = n$, then there exists a basis of n elements $x_1, x_2, \ldots x_n$ in V. So every $x \in V$ can be expressed as $x = a_1 x_1 + a_2 x_2 + \ldots + a_n x_n$, $a_i \in F$, $1 \le i \le n$.

Now we define a mapping $f : V \to V_n$ by

$f(x) = f(a_1 x_1 + a_2 x_2 + \ldots + a_n x_n) = (a_1, a_2, \ldots a_n)$.

Now we prove that f is linear. For any $x, y \in V$, $a_i \in F$, $b_i \in F$, $1 \le i \le n$, we have

$$x = a_1 x_1 + a_2 x_2 + \ldots + a_n x_n,$$
$$y = b_1 x_1 + b_2 x_2 + \ldots + b_n x_n.$$

For $k_1, k_2 \in F$, we have

$$\begin{aligned} f(k_1 x + k_2 y) &= (k_1 a_1 + k_2 b_1, k_1 a_2 + k_2 b_2, \ldots, k_1 a_n + k_2 b_n) \\ &= (k_1 a_1, k_1 a_2, \ldots, k_1 a_n) + (k_2 b_1, k_2 b_2, \ldots, k_2 b_n) \\ &= k_1 (a_1, a_2, \ldots, a_n) + k_2 (b_1, b_2, \ldots, b_n) \\ &= k_1 f(x) + k_2 f(y). \end{aligned}$$

Again we show that f is one-to-one mapping.

For any $x, y \in V$, assume

$$\begin{aligned} f(x) = f(y) &\Rightarrow (a_1, a_2, \ldots a_n) = (b_1, b_2, \ldots, b_n) \\ &\Rightarrow a_1 = b_1, a_2 = b_2, \ldots, a_n = b_n \\ &\Rightarrow x = y. \end{aligned}$$

Further for $(a_1, a_2, \ldots, a_n) \in V_n$, $x \in V$ such that $f(x) = (a_1, a_2, \ldots, a_n)$. Which shows f is an onto mapping. Hence f is an isomorphism. Hence $V(F) \cong V_n(F)$.

Corollary 1: *Two finite-dimensional vector spaces V (F) and W (F) are isomorphic if and only if dim V = dim W.*

Proof: Let dim V = dim $W = n$, then we shall prove $V(F) \cong W(F)$. Since dim V = dim $W = n$, then by the above theorem, we have

$V(F) \cong V_n(F)$ and $W(F) \in V_n(F)$ which implies

$V(F) \cong W(F)$.

Conversely, if $V(F) \cong W(F)$, then there exists one-to-one and onto mapping $f : V \to W$, so that Ker $(f) = \{0\}$. By Sylvester's Law, we have dim V = dim Ker (f) + dim $f(V)$ = dim $\{0\}$ + dim W = dim W.

Corrollary 2: *Let U (F) and W (F) be complementary subspace relative to finite-dimensional vector space V (F), that is V = U ⊕ W.*

$(V/W)(F) \cong U(F)$.

Proof: Let dim $V = n$. We have seen in chapter 9, dim V/W = dim V–dim W = dim U, since dim (V) = dim $(U \oplus W)$ = dim U + dim W. Since dim V/W = dim U, $(V/W)(F) \cong U(F)$.

Theorem 10.2.3: *Let $f \in L(V, W)$. A mapping $\bar{f}$ is defined from $(V/Ker(f))(F)$ into $W(F)$ as follows:*
$f(x + Ker(f)) = f(x)$.
Then the mapping $\bar{f}$ is well defined, linear, one-to-one, and onto R_f. In short $\bar{f}$ is an isomorphism of $\left(\dfrac{V}{\text{Ker } f}\right)(F)$ onto $R_f(F)$.

(1) To prove $\bar{f}$ is well defined we have to show that if $x + \text{Ker}(f) = y + \text{Ker}(f)$, then $\bar{f}(x + \text{Ker}(f)) = \bar{f}(y + \text{Ker}\, ff)$ That is, $f(x) = f(y)$.

Now $x + \text{Ker}(f) = y + \text{Ker}(f) \Rightarrow x - y \in \text{Ker}(f)$
$\Rightarrow f(x - y) = 0 \in W$, since $f : V \to W$
$\Rightarrow f(x) - f(y) = 0$, since f is linear
$\Rightarrow f(x) = f(y)$
$\Rightarrow \bar{f}(x + \text{Ker}(f)) = \bar{f}(y + \text{Ker}(f))$, by definition of $\bar{f}$.

Hence $\bar{f}$ is well defined.

(2) For any $x + \text{Ker}(f)$, $y + \text{Ker}(f) \in V/\text{Ker}(f)$, $a, b \in F$,

$$\begin{aligned}\bar{f}(a(x + \text{Ker}(f)) + b(y + \text{Ker}(f))) &= \bar{f}(ax + \text{Ker}(f) + (by + \text{Ker}(f))\\ &= \bar{f}(ax + by + \text{Ker}(f))\\ &= f(ax + by)\\ &= af(x) + bf(y), \text{ since } f \text{ is linear}\\ &= a\bar{f}(x + \text{Ker}(f)) + b\bar{f}(y + \text{Ker}(f)).\end{aligned}$$

Hence $\bar{f}$ is a linear transformation from $(V/\text{Ker}(f))(F)$ into $W(F)$.

(3) To show that $\bar{f}$ is one-to-one. We have, for $x + \text{Ker}(f)$, $y + \text{Ker}(f) \in V/\text{Ker}(f)$,

$$\begin{aligned}\bar{f}(x + \text{Ker}(f)) = \bar{f}(y + \text{Ker}(f)) &\Rightarrow f(x) = f(y)\\ &\Rightarrow f(x) - f(y) = 0 \in W\\ &\Rightarrow f(x - y) = 0 \text{ Since } f \text{ is linear}\\ &\Rightarrow x - y \in \text{Ker}(f)\\ &\Rightarrow x + \text{Ker}(f) = y + \text{Ker}(f).\end{aligned}$$

This shows that $\bar{f}$ is one-to-one linear function.

(4) For any $w \in R_f$, there exists some $x \in V$ for which $w = f(x)$ and $w = f(x) = f(x + \text{Ker}(f))$ which implies that every element $w \in R_f$ is an image element of some $x + \text{Ker}(f)$ under $\bar{f}$. Hence $\bar{f}$ is an onto linear mapping.

It follows

$(V/\text{Ker}(f))(F) \cong R_f(F)$.

If $R_f = W$, then $(V/\text{Ker}(f))(F) \cong W(F)$.

Theorem 10.2.4: If *$W_1(F)$ and $W_2(F)$ are subspaces of a vector space $V(F)$, then $(W_1 + W_2)/W_2)(F) \cong \left(\dfrac{W_1}{W_1 \cap W_2}\right)(F)$.*

Proof: We know that if $W_1(F)$ and $W_2(F)$ are subspaces of $V(F)$, then $(W_1 + W_2)(F)$ is a subspace of $V(F)$ and $W_1 \subseteq W_1 + W_2$ and $W_2 \subseteq W_1 + W_2$. Thus $W_2 \subseteq W_1 + W_2 \subseteq V$ which means $W_2(F)$ is a subspace of the space $(W_1 + W_2)(F)$.

So we have the quotient space $(W_1 + W_2) / W_2 (F)$.

Now we define a mapping $f : W_1 \to \dfrac{W_1 + W_2}{W_2}$ by

$f(x) = x + W_2, \forall x \in W_1$.

Now we observe the following things about f.

(1) f is well defined, since for: $x, y \in W_1$,

$x = y \Rightarrow x + W_2 = y + W_2 \Rightarrow f(x) = f(y)$.

(2) For any $x, y \in W_1$ and $a, b \in F$, we have

$$\begin{aligned} f(ax + by) &= ax + by + W_2 \\ &= (ax + W_2) + (by + W_2) \\ &= a(x + W_2) + b(y + W_2) \\ &= af(x) + bf(y). \end{aligned}$$

Thus f is linear.

(3) For any $x + W_2 \in \dfrac{W_1 + W_2}{W_2}$, $x \in W_1 + W_2 \Rightarrow x = x_1 + x_2$ for $x_1 \in W_1, x_2 \in W_2$

$$\begin{aligned} &\Rightarrow W_2 + x = W_2 + (x_1 + x_2) \\ &\Rightarrow W_2 + x = W_2 + x_1 \\ &\Rightarrow x - x_1 + W_2 = f(x'), \; x - x_1 = x' \in W_2, \end{aligned}$$

which proves f is onto linear function. Hence f is a homomorphism from $W_1 (F)$ onto $\dfrac{W_1 + W_2}{W_2} (F)$. Since f is not one-to-one as more than one $x \in W_1$ are mapped on the same $x + W_2$ by f.

4. Since f is not one-to-one, Ker (f) exists.

So

$$\begin{aligned} \text{Ker } (f) &= [x \in W_1 \mid f(x) = x + W_2 = W_2\}. \\ &= \{x \mid x \in W_1, x + W_2 = W_2 \Rightarrow x \in W_2\} \\ &= [x \mid x \in W_1 \text{ and } x \in W_2] \\ &= W_1 \cap W_2. \end{aligned}$$

5. Now $f : W_1 \to \dfrac{W_1 + W_2}{W_2}$ is linear and

Ker $(f) = W_1 \cap W_2$. Then by theorem 10.2.3

$$\left(\frac{W_1}{W_1 \cap W_2}\right)(F) \cong \left(\frac{W_1 + W_2}{W_2}\right)(F).$$

This completes the theorem.

Corollary: If $V = W_1 \oplus W_2$, then $\left(\dfrac{V}{W_2}\right) (F) \cong W_1 (F)$.

Proof: Since $V = W_1 \oplus W_2$, then $W_1 \cap W_2 = \{0\}$ and $V = W_1 + W_2$.

So Ker $(f) = W_1 \cap W_2 = \{0\}$ of the mapping

$f : W_1 \to \dfrac{W_1 \oplus W_2}{W_2}$ Thus by the above theorem 10.2.4.

$W_1 (F) \cong \dfrac{W_1 + W_2}{W_2}(F)$. or $W_1 (F) \cong (V/W_2) (F)$

CHAPTER 10

10.3 NON-SINGULAR TRANSFORMATIONS

Definition 10.3.1: A linear transformation f from $V(F)$ into $W(F)$ is said to be *non-singular* if and only if there exists mapping f^* from $R_f(F)$ onto $V(F)$ such that

$f^* \circ f = I$, where I is *identity mapping* on $V(F)$.

Now we have that f is linear and $f^* \circ f = I$ and we prove with these conditions that f^* is also linear. If $x, y \in R_f, v_1, v_2 \in V$ such that $f(v_1) = x$ and $f(v_2) = y$.

Then $(f^* \circ f)(v_1) = f^*(f(v_1)) = f^*(x) = v_1$ and

$f^*(y) = v_2$. *For any* $a, b \in F$,

$$\begin{aligned} f^*(ax + by) &= f^*(a f(v_1) + b f(v_2)) \\ &= f^*(f(av_1) + f(bv_2)) \\ &= f^*(f(av_1 + bv_2)) \\ &= av_1 + bv_2 = af^*(x) + bf^*(y). \end{aligned}$$

Hence f^* is linear when f is linear and $f^* of = I$.

Theorem 10.3.1: *Let $f \in L(V, W)$ be linear mapping; the following statements are equivalent.*

(a) f is non-singular.

(b) For all $x, y \in V$, if $f(x) = f(y)$, then $x = y$.

(c) Ker $(f) = \{0\}$.

(d) $\gamma(f) = 0$

(e) $r(f) = \dim V$.

(f) f maps any basis of $V(F)$ onto a basis of $W(F)$.

Proof: **(a)** implies (*b*). Assume f is non-singular and also assume $f(x) = f(y)$. Then

$$\begin{aligned} (f^*f(x)) = f^*(f(y)) &\Rightarrow (f^* \circ f)\, x = (f^*(x) \circ f)(y) \\ &\Rightarrow I(x) = I(y) \\ &\Rightarrow x = y. \end{aligned}$$

(b) implies (*c*). If $x \in$ Ker (f), then

$$\begin{aligned} f(x) = 0 = f(0) &\Rightarrow f^*(f(x)) = f^*(f(0)) \\ &\Rightarrow I(x) = I(0) \\ &\Rightarrow x = 0. \end{aligned}$$

(c) implies (*d*). $\nu(f) = \dim \text{Ker}(f) = \dim \{0\} = 0$.

(d) implies (*e*). By Sylvester's law

$\dim V = \nu(f) + r(f) = r(f)$ as $\nu(f) = 0$

(e) implies (f). By (*e*) $\dim V = \dim f$

$\Rightarrow V(F) \cong R_f(F)$.

Let $\{x_1, x_2, \ldots, x_n\}$ be a basis of V, then there exist $a_1, a_2, \ldots a_n \in F$ such that

$a_1 x_1 + a_2 x_2 + \ldots + a_n x_n = 0$. where $a_1 = a_2 = \ldots = a_n = 0$.

$\Rightarrow f(a_1 x_1 + a_2 x_2 + \ldots + a_n x_n) = f(0) = 0$

$\Rightarrow a_1 \cdot f(x_1) + a_2 f(x_2) + \ldots a_n f(x_n) = 0$

$\Rightarrow f(x_1), f(x_2), \ldots f(x_n)$ are L.I. since $a_1 = a_2 = \ldots a_n = 0$

(f) implies (*a*). Let $\{x_1, x_2, \ldots x_n\}$ be a basis of V *and* $f(x_1), f(x_2), \ldots, f(x_n)$ bases of R_f. Hence each $y \in R_f$ has unique expression of the form $y = b_1 f(x_1) + b_2 f(x_2) + \ldots + b_n f(x_n)$, $b_k \in F$,

$1 \le k \le n$. Let f^* be a mapping from R_f onto V defined by $f^*(y) = b_1 x_1 + b_2 x_2 + \ldots + b_n x_n = \sum_{k=1}^{n} b_k x_k$. We must show that $f^* \circ f = I$ on $V(F)$. For each $x \in V$, $x = \sum_{k=1}^{n} a_k x_k$,

$$f(x) = f\left(\sum_{k=1}^{n} a_k x_k\right) = \sum_{k=1}^{n} a_k f(x_k) \in R_f,$$

$$f^*(f(x)) = \sum_{k=1}^{n} a_k f^*(f(x_k)) = \sum_{k=1}^{n} a_k x_k = x. \text{ Hence } f^* \circ f = I.$$

Example 10.3.1: Prove that a linear mapping $f : V \to W$ is one-to-one if and only if the image of an independent set is independent.

Proof: Let f be one-to-one, and let $\{x_1, x_2, \ldots, x_n\}$ be a linearly independent subset of V.

Now For, $a_1, a_2, \ldots, a_n \in F$, we assume

$a_1 f(x_1) + a_2 f(x_2) + \ldots + a_n f(x_n) = 0$

$\Rightarrow f(a_1 x_1 + a_2 x_2 + \ldots + a_n x_n) = 0$, since f is linear

$\Rightarrow a_1 x_1 + a_2 x_2 + \ldots + a_n x_n \in \text{Ker } f$.

Since f is one-to-one, Ker $f = \{0\}$.

So $a_1 x_1 + a_2 x_2 + \ldots + a_n x_n = 0$. Since $x_1, x_2, \ldots, x_n$ are linearly independent, $a_1 = a_2 = \ldots = a_n, = 0$ which implies $a_1 f(x_1) + a_2 (x_2) + \ldots + a_n f(x_n) = 0$, where all a's are zero.

Hence $f(x_1), f(x_2), \ldots, f(x_n)$ are L.I.

Conversely, Let the image of any independent set be independent. If $v \in V$ is non-zero, $\{v\}$ is L.I. By assumption $\{f(v)\}$ is also L.I. which means $f(v) \neq 0$. Thus non-zero elements of V have non-zero images which implies that Ker (f) does not contain non-zero elements of V. So Ker (f) contains only $0 \in V$ *i.e.*, Ker $(f) = \{0\}$ which implies f is one-to-one.

Remark: If f *is* one-to-one, then f is non-singular. Therefore *if* $f : V \to W$ *is* a linear mapping, then f is non-singular if and only if the image of L.I. set is L.I. set.

Theorem 10.3.2: *A linear mapping $f : V \to W$ is an isomorphism of a vector space $V(F)$ to a vectors space $W(F)$ if and only if f is non-singular.*

Proof: Let $f : V \to W$ be linear mapping from $V(F)$ onto $U(F)$. Let f be non-singular. Assume $f(x) = f(y) \Rightarrow f(x) - f(y) = 0 \Rightarrow f(x - y) = 0 \Rightarrow (x - y) \in \text{Ker } f \Rightarrow x - y = 0 \Rightarrow x = y$, Since f is non-singular, Ker $(f) = \{0\}$ which implies f is one-to-one. Hence f is an isomorphism.

Conversely, if f is an isomorphism, then $f : V \to W$ is one-to-one. So Ker $(f) = \{0\} \Rightarrow f$ *is* non-singular.

Theorem 10.3.3: *If f is a linear mapping from $V(F)$ to $W(F)$ and if f^* is a mapping from R_f to $V(F)$ such that $f^* \circ f = I$, then $f \circ f^* = I$ on R_f.*

Proof: By hypothesis, f is non-singular, so any $y \in R_f$ can be written as $y = f(x)$ for some $x \in V$.

Then

$$(f \circ f^*)(y) = (f \circ f^*)(f(x)) = f(f^*(f(x))) = f(x) = y$$

Hence $f \circ f^* = I$ on R_f

This shows that f is invertible and its inverse is f^* which is written $f^{-1} = f^*$.

CHAPTER 10

10.4 ALGEBRA OF LINEAR TRANSFORMATIONS

Let $V(F)$ and $W(F)$ be two vector spaces over the same field F. The set of all Linear transformation is denoted by $L(V, W)$. Now we have to see that $L(V, W)$ under defined operations over the field F is a vector space. Some authors write $L(V, W)$ as, $\text{Hom}_F(V, W)$.

Definition 10.4.1: The sum $f+g$ of two Linear mappings $f, g \in L(V, W)$ is defined by the rule

$$(f+g)(x) = f(x) + g(x), x \in V, \text{ and}$$

the scalar multiplication is given by

$$(cf)(x) = cf(x), \text{ where } c \in F, x \in V.$$

***Lemma: 10.4.1:** If $f, g \in L(V, W)$, then sum $f+g \in L(V, W)$.*

Proof: Let $x, y \in V$, $a, b \in F$. Since f and g are L.T., then $f(ax+by) = af(x) + bf(y)$, and $g(ax+by) = ag(x) + bg(y)$

Now
$$\begin{aligned}(f+g)(ax+by) &= f(ax+by) + g(ax+by)\\ &= af(x) + bf(y) + ag(x) + bg(y)\\ &= a(f(x)+g(x)) + b(f(y)+g(y))\\ &= a(f+g)(x) + b(f+g)(y)\end{aligned}$$

Which shows $f+g$ is linear.

Lemma 10.4.2: For any $f \in L(V, W)$, $c \in F$, then $cf \in L(V, W)$.

Proof: Let $x, y \in V$, $a, b \in F$. Since f is linear, $f(ax+by) = af(x) + bf(y)$.

$$\begin{aligned}\text{Now } (cf)(ax+by) &= cf(ax+by)\\ &= (ca)f(x) + (cb)f(y)\\ &= (ac)f(x) + (bc)f(y)\\ &= a(cf(x)) + b(cf(y)).\end{aligned}$$

Hence $cf \in L(V, W)$.

***Theorem 10.4.1:** Let $V(F)$ and $W(F)$ be vector spaces over the field F. With addition and scalar multiplication defined as in Definition 10.4.1 for mappings, $L(V, W)(F)$ is itself a vector space.*

Proof: To prove the theorem we shall first prove that $(L(V, W), +)$ is a commutative group under the addition defined by $(f+g)(x) = f(x) + g(x)$, $\forall f, g \in L(V, W)$ where $x \in V$.

Now we verify group axioms.

1. Closure property: For any $f, g \in L(V, W)$,
$f+g \in L(V, W)$ by Lemma 10.4.1
which shows $L(V, W)$ is closed under the operation.

2. Associativity: For any $f, g, h \in L(V, W)$, and $x \in V$,

$$\begin{aligned}[(f+g)+h](x) &= [(f+g)(x)] + h(x)\\ &= [f(x) + g(x)] + h(x)\\ &= f(x) + [g(x) + h(x)] \text{ since } f(x),\end{aligned}$$

$g(x), h(x) \in W$ and addition in W is associative.

$= [f + (g+h)](x)$.

Thus $(f+g)+h = f+(g+h)$. Hence addition in $L(V, W)$ is associative.

3. Existence of identity: We define a zero function $Z ! V \rightarrow W$ which maps every element of V on 0 of W. that is $Z(x) = 0, \forall x \in V$. Now we see that Z is linear. For this, if $x_1, x_2, \in V, a, b \in F$, then

$Z(a x_1 + b x_2) = 0 = 0 + 0 = aZ(x_1) + bZ(x_2)$.

which show $Z \in L(V, W)$.

Now for any $f \in L(V, W)$, $x \in V$, we have

$(f + Z)(x) = f(x) + Z(x)) = f(x) + 0 = f(x)$.

$(f + Z) = f \Rightarrow Z$ is the additive identity in $L(V, W)$.

4. Existence of inverse: For any $f \in L(V, W)$. Let $-f$ be defined by $(-f)(x) = -f(x)$. Now for any $x, y \in V, a, b \in F, (-f)(ax + by) = -f(ax + by)$

$$= [-(a f(x) + b f(y)]$$
$$= -af(x) - bf(y)$$
$$= a(-f(x)) + b(-f(y)).$$

which shows $-f \in L(V, W)$.

Now $(f + (-f))(x) = f(x) + (-f)(x)$

$$= f(x) - f(x) = 0 \text{ additive identity of } W$$
$$= Z(x).$$

Thus $f + (-f) = Z$. Which shows $-f$ is the additive inverse of f. Hence every $f \in L(V, W)$ has additive inverse $-f \in L(V, W)$,

5. Commutativity: For any $f, g \in L(V, W)$, $x \in V$,

We have $(f + g)(x) = f(x) + g(x)$

$= g(x) + f(x)$ since $f(x)$ and $g(x) \in W$ and addition in W is commutative.

$= (g + f)(x)$.

Therefore $f + g = g + f$. Hence addition in $L(V, W)$ is commutative.

Hence $(L(V, W), +)$ is an abelian group under the addition defined above.

For showing $(L(V, W))(F)$ is a vector space we have to verify the following properties with respect to scalar multiplication. By Lemma 10.4.2 for any $f \in L(V, W)$ and $c \in F$, $(cf) \in L(V, W)$. Hence $L(V, W)$ is closed under scalar multiplication. For any, $f, g \in L(V, W)$, $a, b \in F$, we have

6. (a) : $c(f + g)(x) = c[(f + g)(x)]$

$$= c[f(x) + g(x)] = cf(x) + cg(x)$$
$$= (cf)(x) + cg)(x)$$
$$= (cf + cg)(x).$$

Hence $c(f + g) = cf + cg$.

(b) $((a + b)f)(x) = (a + b)f(x) = af(x) + bf(x)$

$$= (af)(x) + (bf)(x)$$
$$= (af + bf)(x).$$

Therefore $(a + b)f = af + bf$.

(c) $((ab)f)(x) = (ab)f(x) = a(bf(x))$

$$= a(bf)(x)$$

Hence $(ab)f = a(bf)$.

(d) (If) (x) = I (f (x)) = f (x) which implies

$If = f.$

Hence L (V, W) (F) is a vector space over the field F.

Theorem 10.4.2: *If V (F) and W (F) are vector spaces of dimension m and n, respectively, over F, then the space L (V, W) (F) is of dimension m.n over F.*

Proof: To prove this theorem we shall show that there exists a basis of L (V, W) (F) of $m.n$ elements. Let $(x_1, x_2, \ldots, x_m)$ be a basis of V (F) and let $\{y_1, y_2, \ldots, y_n\}$ be a basis of W (F). By theorem. 10.17 there exists a linear mapping in L (V, W) which maps elements of the basis of V (F) onto arbitrary elements of W (F), so we define f_{ij} $V \to W$ by, $1 \leq i \leq m$, $1 \leq j \leq n$,

$$f_{ij}(x_k) = 0 \text{ if } i \neq k$$

$$= y_j \text{ if } i = k.$$

That is, f_{ij} maps x_i onto y_j and other x's onto 0.

We observe that the set $\{f_{ij}\}$ contains $m.n$ elements. Hence the theorem is proved if the set (f_{ij}) forms a basis of L (V, W) (F).

First we shall see that $\{f_{ij}\}$ generates L (V, W). Let $f \in L$ (V, W). For each index i ($1 \leq i \leq m$), $f(x_i) \in W$. So $f(x_i)$ can be expressed as a linear combination of vectors $y_1, y_2, \ldots, y_n$ of basis vectors of W. Thus

$$f(x_i) = a_{i1} y_1 + a_{i2} y_2 + \ldots + a_{in} y_n,\ a_{ij} \in F.\ 1 \leq i \leq m$$

$$= \sum_{j=1}^{n} a_{ij} y_j$$

Since $f_{ij}(x_i) = y_j$ and $f_{ij}(x_k) = 0$ if $i \neq k$.,

$$f(x_k) = \sum_{j=1}^{n} a_{kj} y_j$$

$$f(x_k) = \sum_{j=1}^{n} a_{kj} f_{kj}(x_k) = \sum_{i=1}^{m}\left(\sum_{j=1}^{n} a_{ij} f_{ij}(x_k)\right)$$

$$= \sum_{i=1}^{m}\left(\sum_{j=1}^{n} a_{ij} f_{ij}(x_k)\right)$$

which implies $f = \sum_{j=1}^{n}\left(\sum_{i=1}^{m} a_{ij} f_{ij}\right)$. Thus every $f \in L$ (V, W) can be written as the linear combination of the elements of the set $\{f_{ij}\}$. Thus we conclude the set $\{f_{ij}\}$ spans L (V, W) (F).

In order to prove that $\{f_{ij}\}$ forms a basis of L (V, W). There remains to show that $\{f_{ij}\}$ is linearly independent. Suppose that there is some linear combination

$$\sum_{i=1}^{m}\sum_{j=1}^{n} c_{ij} f_{ij} = 0, \text{ where } c_{ij} \in F.$$

If we evaluate this expression at the vector x_k,

$$0 = \sum_{i=1}^{m}\sum_{j=1}^{n} c_{ij} f_{ij}(x_k) = \sum_{i=1}^{m}\sum_{j=1}^{n}\left(c_{ij} f_{ij}(x_k)\right)$$

$$= \sum_{j=1}^{n} c_{kj}\, y_j \text{ , since } f_{ij}(x_k) = y_j$$

if $i = k$ and $f_{ij}(x_k) = 0$, if $i \neq k$. $1 \leq j \leq n$.

Thus $\sum_{j=1}^{n} c_{kj}\, y_j = 0$. Since $\{y_1, y_2,\ldots, y_n\}$ is a basis for $W(F)$, then $\sum_{j=1}^{n} c_{kj}\, y_j = 0$ gives $c_{k1} = c_{k2} = \ldots = c_{kn} = 0$.

By varying k, we conclude that all the coefficient $c_{ij} = 0$.

Hence the theorem is proved.

Corollary 1: *If dim $V = n$, then dim $L(V, V) = n^2$.*

Corollary 2: *If dim $V = n$, then dim $L(V, F) = n$.*

Proof: Since the field F is a vector space over itself and the field F is generated by its multiplicative identity, then dim $F = 1$. So by the theorem dim (V, F) = dim V. dim $F = n$. $1 = n$.

Note: Sometimes we denote the set $L(V, V)$ of all linear transformation from the vector space $V(F)$ into itself by $A(V)$ or by Hom (V, V).

Theorem 10.4.3: *Let $V(F)$, $U(F)$, and $W(F)$ be vector spaces over F. Let $f : V \to U$ be a linear mapping from $V(F)$ into $U(F)$ and let $g ! U \to W$ be a linear mapping from $U(F)$ into $W(F)$. Then $g \circ f$ is a linear mapping from $V(F)$ into $W(F)$.*

Proof: Since $f : V \to U$ be a linear mapping, then

For $x, y \in V$, $a, b \in F$ we have, $f(ax + by) = af(x) + bf(y) \in U$.

Again, since $g : U \to W$ be a linear mapping,

Then for $f(x), f(y) \in U$, we have

$$(g \circ f)(ax + by) = g(f(ax + by)$$
$$= g(a f(x)) + b f(y) = a\, g(f(x)) + b\, g(f(y))$$
$$= a (g \circ f)(x) + b (g \circ f)(y).$$

This shows g *of is* a linear mapping from $V(F)$ into $W(F)$.

Corollary 1 *: $k(g \circ f) = (kg) \circ f = g \circ (kf)$.*

Corollary 2: Let $V(F)$ be a vector space. Then $L(V, V)(F)$ is also a vector space over F. *If $f, g \in L(V, V)$*, then $g \circ f$ and $f \circ g \in L(V, V)$.

Theorem 10.4.4: *For each vector space $V(F)$, the triple $(L(V, V), +, .)$ is ring with identity under addition defined in $L(V, V)$ and composition of functions.*

Proof: The proof is left as an exercise for the students. But from this discussion we have a very important another fact.

Corollary: If $GL(V)$ denotes the set of all invertible mappings in $L(V, V)$, then $(GL(V), o)$ forms a group, called the *general linear group*.

Proof: Let $f, g \in GL(V)$. That is, f and g are invertible mappings from V into V . f is invertible, if and only if it is an one-to-one and onto mapping. Since f and g are invertible, that is, they are one-to-one mappings, their compositions $f \circ g$ and $g \circ f$ are also one-to-one. Thus $f \circ g$ and $g \circ f$ both are invertible. Since f and g are linear, $f \circ g$ and $g \circ f$ are also linear by the theorem 10.4.3.

Hence $f \circ g$, $g \circ f \in GL(V)$. Hence $GL(V)$ is closed under the composition 'o'.

Since the composition o of functions in $L(V, V)$ is associative, it is also associative in $GL(V)$ as $GL(V) \subset L(V, V)$.

Identity mapping e_v from V onto V is always one-to-one, so it is invertible. Since $e_v(ax + by) = a\,x + by = ae_v(x) + be_v(y)$, and $e_v \in L(V, V)$.

Since f is invertible, f^{-1} exists and f^{-1} is one to one. So $(f^{-1})^{-1} = f \in GL(V) \Rightarrow f^{-1}$ is invertible. Thus every $f \in V$ has its inverse $f^{-1} \in GL(V)$. To see that f^{-1} is linear, Let $x_1, x_2, \in V$. Since f is one-to-one and onto, there exists $y_1, y_2 \in V$ such that $f(x_1) = y_1, f(x_2) = y_2$.

Since f is linear $f(x_1 + x_2) = f(x_1) + f(x_2) = y_1 + y_2$ and $f(ax) = af(x) = ay_1$

By definition of f^{-1}, we have $f^{-1}(y_1) = x_1, f^{-1}(y_2) = x_2$ and

$$f^{-1}(y_1 + y_2) = x_1 + x_2 = f^{-1}(y_1) + f^{-1}(y_2),$$
$$f^{-1}(ay_1) = ax_1 = a\,f^{-1}(y_1).$$

Hence f^{-1} is Linear.

The structure of $L(V, V)$ can be approached from one more direction. For this we define what is meant by *algebra* over a field.

Definition 10.4.2: A vector space $V(F)$ is said to be an *algebra* over the field F if its elements can be multiplied in such a way that $V(F)$ becomes a ring in which scalar multiplication is related to ring multiplication (denoted by.,) by the following mixed associative law:

$c\,(xy) = (cx)\,.\,y = x\,.\,(cy), (\forall\, x, y \in V, c \in F)$.

Theorem 10.4.5: *The triple (L (V, V), +, o) is an algebra over the field F.*

Proof: We observe the following things about the triple $(L(V, V), +, o)$.

(i) $L(V, V)$ is a vector space over F by theorem 10.4.1.

(ii) $L(V, V)$ is a vector space of dimension n^2 over F by corollary of theorem 10.4.2.

(iii) $(L(V, V), +, 0)$ is ring with identity under the addition of functions and composition of functions by theorem 10.4.4.

To show that $(L(V, V), +, o)$ is an algebra we have to prove that mixed associative law holds. To get this, Let $c \in F$, $g, f \in L(V, V)$, and $x \in V$,

Then

$$(f\,o\,(cg))\,(x) = f\,((cg)\,(x)) = f\,(c\,g\,(x)) = cf\,(g\,(x))$$
$$= c\,(f\,o\,g)\,(x).$$

Since this equality holds for $x \in V$, we conclude that

$$f\,o\,(c\,g) = c\,(f\,o\,g).$$

Similarly,

$$((cf)\,o\,g\,(x) = (cf)\,(g\,(x)) = c\,(f\,g\,(x)$$
$$= c\,(f\,o\,g)\,(x).$$

Hence $(cf)\,o\,g = c\,(f\,o\,g)$.

Thus $c\,(f\,o\,g) = (c\,f)\,o\,g = f\,o\,(c\,g)$.

Therefore the triple $(L(V, V), +, o)$ is an algebra over F of dimension n^2. Hence the theorem.

Example 10.4.1: We consider the set M_n, (F) of all square matrices of order n over the field F. We know that $(M_n(F), +,.)$ is a ring and vector space over F. The mixed associative Law also holds in $M_n(F)$. For this Let $A = [a_{ij}]_{n \times n}$, $B = [b_{ij}]_{n \times n} \in M_n(F)$

Then $A \cdot B = [c_{ij}]_{n \times n}$
For $k \in F$,
$(kA)B = k\,(A)\,B = k\,(AB)$
$A(kB) = k\,(AB)$. So
$k\,(AB) = (kA)B = A\,(k\,B)$

(2) The rings $(R, +,.)$ and $(C, +,.)$ of real and complex numbers are also algebras over R and C respectively.

Theorem 10.4.6: *Let $f: V \to U$ and $g: U \to W$ be linear. Hence $(g \circ f): V \to W$ is Linear. Then*

(i) rank $(g \circ f) \le$ rank g,
(ii) rank $(g \circ f) \le$ rank f.

Proof: (i) Since $f(V) \subseteq U$, $g(f(V)) \subseteq g(U)$ and so $\dim g(f(V)) \le \dim g(U)$. Then rank $(g \circ f) = \dim (g \circ f)(V) = \dim g(f(V) \le \dim g(U) =$ rank g. Hence rank $g \circ f \le$ rank g.

(ii) Since $f(V) \subseteq U$ and $g: f(V) \to W$, then by Sylvester's law $\dim f(V) = \dim \text{Ker}(g) + \dim g(f(V))$
which implies $\dim g(f(V)) \le \dim f(V)$

Now rank $(g \circ f) = \dim (g \circ f)(V) = \dim g(f(V)) \le \dim f(V) =$ rank f.
Hence, rank $(g \circ f) \le$ rank f.

Theorem 10.4.7: *For any Linear transformation $f: V \to W$, $r(f) \le \min(\dim V, \dim W)$.*

Proof: Let $\dim V = n$ and $\dim W = m$. Since $f(V) \subseteq W$, then $\dim f(V) \le \dim W$. That is, $r(f) \le \dim W$.

Again, since $\dim V = n$, then there exists $n + 1$ vectors $x_1, x_2, \ldots, x_n, x_{n+1}$ which are L.D. Therefore for some $a_1, a_2, \ldots, a_n, a_{n+1}, \in F$, not all zero, we have $a_1 x_1 + a_2 x_2 + \ldots + a_n x_n + a_{n+1} x_{n+1} = 0$. Since f is linear $f(0) = 0$. Thus

$0 = f(0) = f(a_1 x_1 + a_2 x_2 + \ldots + x_n + a_{n+1} x_{n+1})$.
$= a_1 f(x_1) + a_2 f(x_2) + \ldots + a_n f(x_n) + a_{n+1} f(x_{n+1})$. *which shows* $f(x_1), f(x_2), \ldots f(x_n), f(x_n), f(x_{n+1})$ are L.D., since all a's are not zero. Hence $f(V)$ cannot contain $(n + 1)$ L.I. vectors, which implies $\dim f(V) \le n = \dim V$, Thus $r(f)$ is less then $\dim V$ and $\dim W$ which means $r(f) \le \min(\dim V, \dim W)$.

Theorem 10.4.8: *Let $f: V \to W$ be a linear transformation. For any subspace $H(F)$ of $V(F)$, $\dim f(H) \ge \dim H - \nu(f)$.*

Proof: Let f_1 be a *restriction* of f to H, that is, $f(H) = f_1(H)$. Since f is linear, f_1 is linear of H into W. For this, if $x, y \in H$, $a, b \in F$, we have

$f_1(ax + by) = f(ax + by) = af(x) + bf(y) = af_1(x) + bf_1(y)$. By Sylvester's law $\dim H = \dim \text{Ker}(f_1) + \dim f_1(H)$
or $\dim f_1(H) = \dim H - \nu(f_1)$ ... (1)

Now since $f(H) = f_1(H)$, $\dim f(H) = \dim f_1(H)$ ... (2)

$\text{Ker } f_1 = \{x \in H \mid f_1(x) = 0\}$
$= \{x \in H \mid f(x) = 0\} \subseteq \{x \in V \mid f(x) = 0\}$
$\Rightarrow \text{Ker}(f_1) \subseteq \text{Ker}(f) \Rightarrow \nu(f_1) \le \nu(f)$ ... (3)

CHAPTER 10

From (1), (2) and (3), we have

$\dim f(H) = \dim f_1(H) = \dim H - \nu(f_1)$

$\geq \dim H - \nu(f)$.

Hence the result is proved.

Theorem 10.4.9: *If $f, g \in L(V, W)$, then*

(*i*) $r(af) = r(f), \forall\, a \neq 0 \in F$,

(*ii*) $|r(f) - r(g)| \leq r(f + g) \leq r(f) + r(g)$

(*i*) Since $f(V)(F)$ is a subspace of $W(F)$, then $(af)(V) = a\, f(V) \subseteq f(V)$ and $a^{-1} f(V) \subseteq f(V)$.

Therefore $a\,[a^{-1} f(V)] \subseteq a\, f(V)$.

or $(a\, a^{-1}) f(V) \subseteq a\, f(V) \Rightarrow f(V) \subseteq a\, f(V)$

Hence $f(V) = a\, f(V)$ or $\dim f(V) = \dim af(V)$

Hence $r(f) = r(a\, f)$.

(*ii*) We know, for any $x \in V$,

$$(f + g)(x) = f(x) + g(x).$$

Therefore $(f + g)(V) \subseteq f(V) + g(V)$

which implies $\dim (f + g)(V) \leq \dim (f(V) + g(V))$

$\leq \dim f(V) + \dim (g)(V)$

$\Rightarrow r(f + g) \leq r(f) + r(g)$ ...(1)

Now $f = f + g - g$. By (1), we have

$r(f) = r[(f + g) + (-g)] \leq r(f + g) + r(-g)$

By (*i*) we have $r(ag) = r(g)$, if $a = -1$. then

$$r(-g) = r(g).$$

So $r(f) \leq r(f + g) + r(-g) = r(f + g) + r(g)$

or $r(f) - r(g) \leq r(f + g) \leq r(f) + r(g)$ ...(2)

Similary $r(g) - r(f) \leq r(f + g) \leq r(f) + r(g)$ ...(3)

From (2) and (3) we have

$|r(f) - r(g)| \leq r(f) + r(g)$.

Theorem 10.4.10: *Let $U(F)$, $V(F)$ and $W(F)$ be vector spaces over F, Let $f : U \to V$ and $g : V \to W$ be linear mappings.*

then

(*i*) $r(g \circ f) \leq \min \{r(f), r(g)\}$,

(*ii*) $r(f) + r(g) - n \leq r(g \circ f)$, *where* $n = \dim V$.

Proof (i): By theorem 10.4.7,

$r(g \circ f) \leq \min(\dim U, \dim W)$

$r(g \circ f) = \dim (g \circ f)(U) = \dim (g(f(U))$

$\leq \min(\dim f(U), \dim W)$

Since, by applying theorem 10.4.7 to, $g : f(U) \to W$, we have $r(g) = \dim g(fU)) \leq \min(\dim f(U), \dim W,)$

$\leq \dim f(U) = r(f)$.

Thus $r(g \circ f) \leq r(f)$ ...(1)

Since $(g \circ f)(x) = g(f(x))$, it follows that range *of* $(g \circ f)$ is contained in range of g, *so* that

$(g \circ f)(U) \subseteq g(V) \Rightarrow r(g \circ f) \leq r(g)$ ...(2)

From (1) and (2), we have

$r\ (g\ o\ f) \leq \min\ (r\ (f),\ r\ (g)).$ ...(3)

(ii) Since dim $V = n$, then by Sylvester's law, $n = \dim V = \dim \text{Ker}\ (g) + \dim\ (g\ (V)).$

or $n = \nu\ (g) + r\ (g)$

or $\nu\ (g) = n - r\ (g)$ ... (4)

Since $f\ (U) \subseteq V$, then we can choose g_1 to be the restriction of g *to* $f(U)$.

Then

Ker $g_1 = \{x \in f\ (U) \mid g_1\ (x) = 0\} = \{x \in f\ (U) \Rightarrow g\ (x) = 0\}$

$\subseteq \{x \in V \mid g\ (x) = 0\} =$ Ker g

Thus Ker $g_1 \subseteq$ Ker $g \Rightarrow \nu\ (g_1) \leq \nu\ (g).$ ...(5)

From (4) and (5) we get

$\nu\ (g_1) \leq \nu\ (g) = n - r\ (g)$

or $\nu\ (g_1) \leq n - r\ (g)$ ...(6)

Now $g_1\ (f(U)) = g\ (f\ (U)) = (g\ o\ f)\ (U) \Rightarrow r\ (g_1) = r\ (g\ o\ f).$

But by Sylvester's law

$\dim f\ (U) = \nu\ (g_1) + r\ (g_1)$

$= \nu\ (g_1) + r\ (g\ o\ f),$

Thus, we get

$r\ (f) = r\ (g_1) + r\ (g\ o\ f)$

$\leq r\ (g\ o\ f) + n - r\ (g)$ by (6)

i.e., $r\ (f) + r\ (g) \leq r\ (g\ o\ f) + n$

or $r\ (f) + r\ (g) - n \leq r\ (g\ o\ f).$

Thus the.theorem is proved.

Corollary: If one of f and g is an isomorphism, the $r\ (g\ o\ f) = \min\ (rf),\ r\ (g)).$

Let g be an isomorphism, then, $f\ (U) \subseteq V$

and $(g\ o\ f)\ (U) = g\ (f\ (U) \subseteq W$, say $W_1 = g\ (f\ (U)).$

Since g is an isomorphism, $f\ (U)\ (F) \cong W_1\ (F).$

or $\dim f\ (U) = \dim W_1 = \dim g\ (f\ (U)$

$\gamma\ (f) = r\ (g\ o\ f)$ ...(1)

Again, If f is an isomorphism, then

$f\ (U) = V \Rightarrow r\ (f) = \dim V.$

and

$r\ (g\ o\ f) = \dim\ (g\ o\ f)\ (U) = \dim g\ (f(U)) = \dim g\ (V) \Rightarrow r\ (g\ o\ f) = r\ (g)$ ...(2)

From (1), and (2)

We have $r\ (g\ o\ f) = \min\ (r\ (f),\ r\ (g)),$

Example 10.4.2: *Let* $V\ (F)$ be a vector space over F of dimension 5. Let $\{x_1, x_2, x_3, x_4, x_5\}$ be a basis of $V\ (F)$. Let $f : V \to V$, and $g : V \to V$ be defined by

$$f\ (x) = a_1\ x_1 + a_2\ x_2 + a_3\ x_3,$$

and $g\ (x) = a_2\ x_2 + a_3\ x_3 + a_4\ x_4$. For any $x = a_1\ x_1 + a_2 x_2 + a_3 x_3 + a_4 x_4 + a_5 x_5 \in V$, $a_i \in F$, $1 \leq i \leq 5$.

It is clear that $f\ (V)$ has a basis $\{x_1, x_2, x_3\}$ and $g\ (V)$ has a bases $\{x_2, x_3, x_4\}$.

Thus $r\ (f) = r\ (g) = 3$.

Now $(f o\ g)\ (x) = f\ (g\ (x)) = f\ (a_2\ x_2 + a_3\ x_3 + a_4\ x_4)$

$= a_2\ x_2 + a_3\ x_3.$

This gives $(f\ o\ g)\ (V)$ has a bases $[x_2, x_3\}$ and so $r\ (f\ o\ g) = 2$.

CHAPTER 10

Thus from (1) and (2) it is clear that

$r(f) + r(g) - 5 < r(f \circ g) < \min(r(f), r(g))$

$\Rightarrow 3 + 3 - 5 < 2 < \min\{3, 3\}. = 3.$

PROBLEMS

1. If we regard the complex numbers as a vector space over the real field, is the conjugate mapping, $f(a + ib) = a - ib$, a linear transformation?
2. Determine which of the following functions are linear mappings of $V_3(R)$ over the real field R into itself:
   (a) $f(a_1, a_2, a_3) = (a_1 + 1, a_2 + 1, 0)$,
   (b) $f(a_1, a_2, a_3) = (a_2, a_1, a_3)$,
   (c) $f(a_1, a_2, a_3) = (a_1, a_2, I)$,
   (d) $f(a_1, a_2, a_3) = (a_1, -a_2 -a_3)$,
   (e) $f(a_1; a_2, a_3) = (a_1, a_1 + a_2, a_1 + a_2 + a_3)$,
   (f) $f(a_1, a_2, a_3) = (a_1 + 2a_2 - a_3, -a_1 + a_2, 2a_1 + a_2)$.
3. Let $f \in L(V, W)$. then
   (1) Show that any subspace $V_1(F)$ of $V(F)$ is mapped by f into a subspace $W_1(F)$ of $W(F)$.
   (2) Conversely, show that if $W_1(F)$ is a subspace of $W(F)$, the set of all vectors of $V(F)$ which are mapped into $W_1(F)$ is a subspace of $V(F)$.
4. Show that a linear transformation f from $V_n(F)$ to $W(F)$ is determined by the effect of f on any basis of $V_n(F)$.
5. Let $p(x) = a_0 + a_1 x + a_2 x^2 + ... + a_n x^n \in R[x]$, the vector space of polynomials in x over the real field R. Let the mapping D and M be defined by
   $D(p(x)) = \frac{d}{dx}(p(x))$,
   $Mp(x) = xp(x)$.
   (I) Show that both D and M are linear transformations.
   (II) Is D nilpotent on this space?
   (III) Prove that $MD - DM = I$
   (IV) Deduce that $(DM)^2 = D^2 M^2 + DM$.
6. Let the mapping $f \in L(V, V)$ and S denote the set of vectors of V which are left fixed by f:
   $$S = \{x \in V \mid f(x) = x\}.$$
   Verify that $S(F)$ forms a subspace of the vector space $V(F)$
7. Show that by example that the ϕ conclusion of the corollary to theorem 10.2.2 is false if $V(F)$ is infinite dimensional.
8. Suppose the mapping $f \in L(V, W)$ with dim V > dim W. Show that there exists a non zero vector $x_0 \in V$ for which $f(x_0) = 0$.
9. Let $V(F)$ be finite-dimensional with basis $\{x_1, x_2, ..., x_n\}$, and let $\{y_1, y_2, ..., y_n\}$ be any n elements of V. If the function $f : V \to V$ is defined by taking
   $f(a_1 x_1 + ... + a_n x_n) = a_1 y_1 + ... + a_n y_n, a_k \in F$,
   Prove that f is Linear; determine when f will be an isomorphism.
10. Prove that if the mapping $f \in L(V, V)$ is such that Ker (f) = Ker (f^2), then V = Ker $(f) \oplus f(V)$.
11. For a fixed element $a \in F$, define the Scalar multiplication $f_a : V \to V$ by $f_a(x) = ax, x \in V$. Given $R = (f_a \mid a \in F\}$ and $R' = R - \{0\}$, show that

(a) the triple $(R, +, o)$ forms a subring of $(L(V, V), +, o)$ isomorphic to $(F, +,.)$.

(b) The pair $(R', 0)$ is a normal subgroup of the linear group $(GL(V), o)$.

12. A linear mapping $f \in L(V, V)$ is said to be *nilpotent* if $f^n = 0$ for some $n \in N$. If f is nilpotent and if $f^{n-1}(x_0) \neq 0$
Prove that
$\{x_0, f(x_0), f^2(x_0), \ldots, f^{n-1}(x_0)\}$
is a linearly independent set of vectors.

13. If the linear mappings $f, g \in L(V, W)$
(a) Ker $(g \, o \, f) \supseteq$ Ker (f) and $\nu(g \, o \, f) \geq \nu(f)$
(b) if $f, g \in L(V_n, V_n)$, then
$\nu(f + g) \geq \nu(f) + \nu(g) - n.$
and $\nu(f) + \nu(g) \geq \nu(g \, o \, f) \geq max \{\nu(f), \nu(g)\}$

14. Let $V(F)$, $U(F)$, and $W(F)$ be vector spaces over F. Let f, f' be linear mappings from V into U and Let g, g' be linear mappings from U into W; then prove that
(i) $g \, o \, (f + f') = g \, o \, f + g \, o \, f'$
(ii) $(g + g') \, o \, f = g \, o \, f + g' \, o \, f$.

10.5 DUAL SPACES

We have studied in the last section the structure of $L(V, W)$. Now if we take vector space $V(F)$ in place of $W(F)$, then we get the set of linear transformations $L(V, V)$ which we have also done in brief. Now we take up a special case when $W(F) = F$. The field F is considered a vector space over itself, so our aim in this section is to confine our attention to the set $L(V, F)$ which is called the set of *linear functionals*. So *scalar valued functions* are *Linear functionals* which are linear mappings form a vector space $V(F)$ to F.

Definition 10.5.1: Let $V(F)$ be a vector space over F. A linear transformation from $V(F)$ into F is called a *linear functional*. That is, if $\phi : V \to F$ is a mapping satisfying that $\phi(ax + by) = a \, \phi(x) + b \, \phi(y)$ for every $x, y \in V$ and $a, b \in F$ is a linear functional. The set of all linear functionals is denoted by V^*.

Example 10.5.1: Let $V(F)$ is a vector space over F. Then $\phi_i : V_n(F) \to F$, defined by $\phi_i(a_1, a_2, \ldots, a_n), = a_i$, is a linear functional on $V_n(F)$.

Example 10.5.2: Let $V(F)$ be a vector space of all polynomials in x over F. Let $f : V \to R$ be the integral operator defined by $f(p(x)) = \int_a^b p(x) \, d \, x$,
where $a \leq x \leq b$. Now for $k_1, k_2 \in F$, $p(x), q(x) \in V(F)$,

$$\begin{aligned} I(k_1 p(x) + k_2 q(x)) &= \int_a^b (k_1 p(x) + k_2 q(x)) \, dx \\ &= \int_a^b k_1 p(x) \, dx + \int_a^b k_2 q(x) \, dx \\ &= k_1 \int_a^b (x) \, dx + k_2 \int_a^b q.(x) \, dx \\ &= k_1 I(p(x)) + k_2 I(q(x). \end{aligned}$$

Hence I is linear, and hence it is a linear functional on $V_n(F)$.

Example 10.5.3: Let $V(F)$ be the vector space of n-square matrices over F. Let $\phi : V \to F$ be the trace mapping defined by $\phi(A) = a_{11} + a_{22} + \ldots + a_{nn}$, where $A = [a_{ij}]$.

That is, ϕ assigns to the matrix A the sum of the diagonal elements. To see that ϕ is linear, let

$A = [a_{ij}]$, $B = [b_{ij}] \in V(F)$ and $k_1, k_2 \in F$, then

$$\begin{aligned} \phi(k_1 A + k_2 B) &= \phi(k_1 [a_{ij}] + k_2 [b_{ij}] \\ &= \phi([k_1 a_{ij}] + [k_2 b_{ij}]) \end{aligned}$$

CHAPTER 10

$$= \phi([k_1 a_{ij} + k_2 b_{ij}])$$
$$= k_1 a_{11} + k_1 a_{22} + \ldots + k_1 a_{nn} + k_2 b_{11} + k_2 b_{22} + \ldots + k_2 b_{nn}$$
$$= k_1 \phi(A) + k_2 \phi(B).$$

Hence ϕ is linear and so it is linear functional on $V(F)$.

In the theorem 10.4.1 we have seen that $L(V, W)$ is a vector space. So the set V^* of all linear functionals on $V(F)$ also forms a vector space under addition and scalar multiplication defined by

$(\phi + \sigma)(x) = \phi(x) + \sigma(x)$, and $(k\phi)(x) = k\phi(x)$. where $\phi, \sigma \in V^*$ and $k \in F$.

Definition 10.5.2: The vector space $L(V, F)(F)$, consisting of all linear functionals is called *dual space* of $V(F)$ and is denoted by $V^*(F)$. A dual space is also called a *Conjugate space.*

Since F is vector space over itself of dimension one, then for any finite-dimensional vector space $V(F)$, $\dim V^* = \dim(L(V, F)) = \dim V. \dim F = \dim V.$ by theorem 10.4.2. Now we establish an important relation between a vector space and its dual.

Theorem 10.5.1: *Let $\{x_1, x_2, \ldots, x_n\}$ be a basis of $V(F)$ over F. Let $\phi_1, \ldots \phi_n, \in V^*$ be the linear functionals defined by*

$$\phi_i(x_j) = \delta_{ij} = 1 \text{ if } i = j$$
$$= 0, \text{ if } i \neq j$$

Then $\{\phi_1, \phi_2, \ldots \phi_n\}$ is a bases of $V^*(F)$, where δ_{ij} is Kronecker delta.

Proof: Kronecker delta δ_{ij} is a short form of writing

$\phi_1(x_1) = 1, \phi_1(x_2) = 0, \ldots, \phi_1(x_n) = 0$

$\phi_2(x_1) = 0, \phi_2(x_2) = 1, \ldots, \phi_2(x_n) = 0$

....

$\phi_n(x_1) = 0, \phi_n(x_2) = 0, \ldots, \phi_n(x_n) = 1.$

We first show that $\{\phi_1, \phi_2, \ldots, \phi_n\}$ spans V^*, Let ϕ be an arbitrary element of V^* and let

$\phi(x_1) = k_1, \phi(x_2) = k_2, \ldots, \phi(x_n) = k_n$

now

$k_1\phi_1 + k_2\phi_2 + \ldots + k_n\phi_n = \sigma \in V^*.$

So, for $1 \leq i \leq n$, we have

$\sigma(x_i) = (k_1\phi_1 + k_2\phi_2 + \ldots + k_i\phi_i + \ldots + k_n\phi_n)(x_i)$

$= k_1\phi_1(x_i) + \ldots + k_i\phi_i(x_i) + \ldots + k_n\phi_n(x_i)$

$= k_1 0 + \ldots + k_i . I \ldots + k_n 0 = k_i$

Thus $\phi(x_i) = \sigma(x_i) = k_i$ for $i = 1, 2, \ldots n$. Since ϕ and σ agree on the basis vectors, $\phi = \sigma = k_1\phi_1 + k_2\phi_2 + \ldots + k_n\phi_n$. This shows $\{\phi_1 \phi_2, \ldots, \phi_n\}$ spans V^*.

To prove the theorem there remains to show that $\phi_1, \phi_2, \ldots, \phi_n$ is L.I. Assume that $a_1\phi_1 + a_2\phi_2 + \ldots + a_n\phi_n = 0 \in V^*, a_i \in F.$

So $(a_1\phi_1 + a_2\phi_2 + \ldots + a_n\phi_n) x_i = 0(x_i) = 0 \in F$

$a_1\phi_1(x_1) + a_2\phi_2(x_i) + a_i\phi_i(x_i) + \ldots + a_n\phi_n(x_i) = 0$

$a_1 . 0 + a_2 0 + \ldots + a_i + \ldots a_n 0 = 0 \Rightarrow a_i = 0.$

That is, $a_1 = a_2 \ldots = a_n = 0$. Hence $\{\phi_1, \phi_2 \ldots, \phi_n\}$ is L.I. and so it is a basis of V^*.

The set $\{\phi_1, \phi_2, \ldots \phi_n\}$ is termed *dual basis*. It also follows that $\dim V = n = \dim V^*$.

Example 10.5.4: Consider the following basis of R^2: $\{x_1 = (2, 1), x_2 = (3, 1)\}$. Find the dual bases $\{\phi_1, \phi_2\}$.

Solution: We seek linear functionals

$\phi_1 (x, y) = ax + by$ and $\phi_2 (x, y) = cx + dy$ such that

$\phi_1 (x_1) = I, \phi_1 (x_2) = 0, \phi_2 (x_1) = 0, \phi_2 (x_2) = 1.$

Thus $\phi_1 (x_1) = \phi_1 (2, 1) = a . 2 + b . 1 = I$

$\phi_1 (x_2) = \phi_1 (3, 1) = a . 3 + b . I = 0$

both expressions give $a = -1, b = 3$.

$\phi_2 (x_1) = \phi_2 (2, 1) = c . 2 + d . 1 = 0$

$\phi_2 (x_2) = \phi_2 (3, 1) = c . 3 + d . 1 = 1$

these both expressions give $c = 1$ or $d = -2$.

Hence the dual basis is $\{\phi_1 (x, y) = -x + 3y, \phi_2 (x, y) = x - 2y\}$.

Theorem 10.5.2: *Let $\{x_1, x_2,....x_n\}$ be a basis of V and let $(\phi_1, \phi_2,.... \phi_n\}$ be the dual basis of V* then any vector $x \in V$,*

$$x = \phi_1 (x) x_1 + \phi_2 (x) x_2 +...+ \phi_n (x) x_n. \quad ...(1)$$

and, for any linear functional $\sigma \in V^*$,

$$\sigma = \sigma (x_1) \phi_1 + \sigma (x_2) \phi_2 +...+ \sigma (x_n) \phi_n \quad ...(2)$$

Proof: Since $\{x_1, x_2,..., x_n\}$ is a basis, then any $x \in V$ can be written as $x = a_1 x_1 + a_2 x_2 +...+ a_n x_n$ for some $a_i \in F$.

Then $\phi_i (x) = \phi_i (a_1 x_1 +...+ a_i x_i +...+ a_n x_n)$

$= a_1 \phi_i (x_1) +... a_i \phi_i (x_i) +...+ a_n \phi_i (x_n)$

$= a_1 0 +...+ a_i I +...a_n 0 = a_i$

that is, $\phi_1 (x) = a_1, \phi_2 (x) = a_2,... \phi_n (x) = a_n$.

Hence $x = \phi_1 (x) x_1 + \phi_2 (x) x_2 +...+ \phi_n (x) x_n$.

Again let $\sigma \in V^*$ then

$\sigma (x) = \sigma (a_1 x_1 + a_2 x_2 +...+ a_n x_n)$

$= a_1 \sigma (x_1) + a_2 \sigma (x_2) +...+ a_n \sigma (x_n)$

$= \phi_1 (x) \sigma (x_1) + \phi_2 (x) \sigma (x_2) +...+ \phi_n (x) \sigma (x_n)$.

$= \sigma (x_1) \phi_1 (x) + \sigma (x_2) \phi_2 (x) +...+ \sigma (x_n) \phi_n (x)$. Since $\sigma (x_i) \in F$ and $\phi_i (x) \in F$, and multiplication in F is commutative $= \sigma (x_1) \phi_1 + \sigma (x_2) \phi_2 +....+ \sigma (x_n) \phi_n) (x)$

which implies

$\sigma = (\sigma (x_1)) \phi_1 + (\sigma (x_2)) \phi_2 +...+ (\sigma x_n)) \phi_n$.

This completes the proof of the theorem.

Theorem 10.5.3: *Let V (F) be a finite-dimensional vector space over F. Then there exists $\phi \in V^*$ such that $\phi (x) \neq 0$, for any $x \in V$.*

Proof: Since $x \neq 0$, $\{x\}$ is L.I subset of V. So it can be extended to a basis $\{x = x_1 x_2...,$ $x_n\}$ of V. Let $\{\phi_1, \phi_2,...., \phi_n\}$ be a basis of V^* dual to the basis $\{x_1, x_2,...., x_n\}$. Then by definition $\phi_1 (x_1) = 1$. Thus taking $\phi = \phi_1$, we get $\phi (x) \neq 0$.

Since

$\phi (x) = \phi (a_1 x_1 + a_2 x_2 +....+ a_n x_n) = \phi_1 (a_1 x_1 + a_2 x_2 +...+ a_n x_n)$

$= a_1 \phi_1 (x_1) + a_2 \phi_1 (x_2)....+ a_n \phi_1 (x_n)$

CHAPTER 10

$= a_1 + a_2\, 0 + ... + a_n \, . \, 0$
$= a_1 \neq 0.$

We have seen that every space $V(F)$ has its dual space $V^*(F)$. So consequently the dual space will have its dual space. We shall denote the dual space of $V^*(F)$ by $V^{**}(F)$ and refer to it as the *second dual* of $V(F)$ or the *bidual space of* $V(F)$. That is, V^{**} is the set of linear function as on $V^*(F)$ which map $f \in V^*$ onto some $k \in F$. Thus any mapping $f \in V^*$ is a linear functional on V and at the same time f is a vector for finding a linear functional on V^*, More over we know that dim V = dim V^*. Similarly, dim V^* = dim V^{**}. Which results that dim V = dim V^{**} and by the corollary of the theorem 10.2.2 $V(F) \cong V^{**}(F)$.

For any vector $x \in V$ and $f \in V^*$, $f(x)$ is scalar (here it is not a function of x). In this case x ranges over V and f is fixed. Now we keep x fixed and allow f to range over V^*.

Now we specially define the function

$$T_x : V^* \to F \text{ by}$$
$$T_x(f) = f(x),\ f \in V^*.$$

To see T_x is linear, we have, for $f, g \in V^*$, $a, b \in F$

$$\begin{aligned} T_x(af + bg) &= (af + bg)(x) \\ &= af(x) + bg(x) \\ &= a\,T_x(f) + b\,T_x(g). \end{aligned}$$

which means T_x is linear functional on V^* and so $T_x \in V^{**}(F)$. T_x, defined in this way, is called the *evalution functional induced by the vector x.*

Theorem 10.5.4: *If V (F) is a finite-dimensional vector space, then V (F) ≅ V** (F) by the mapping*
$\phi : V \to V^{**}$ defined by $\phi(x) = T_x$.

Proof: First we show that ϕ is linear. For this, let $x, y \in V$ and $a, b \in F$, then $\phi(a\,x + by) = T_{ax+by}$.

Now for any $f \in V^*$, we have

$$\begin{aligned} T_{(ax+by)}(f) &= f(ax + by) \\ &= af(x) + bf(y) \\ &= aT_x(f) + bT_y(f) \\ &= (aT_x + bT_y)(f) \end{aligned}$$

which implies $\quad T_{(ax+by)} = aT_x + bT_y$

Thus $\phi(ax + by) = T_{(ax+by)} = aT_x + bT_y = a\,\phi(x) + b\,\phi(y)$

This shows that ϕ is linear.

Now we shall see that ϕ is one-to-one function. For this,

$$\begin{aligned} x \in \text{Ker}\,\phi &\Rightarrow T_x(f) = 0,\ f \in V^* \\ &\Rightarrow f(x) = 0,\ f \in V^*. \end{aligned}$$

But by theorem 10.5.3 there exists $f \in V^*$ for which $f(x) \neq 0$ when $x \neq 0$, which implies that $x \in \text{Ker}(\phi) \Rightarrow f(x) = 0,\ f \in V^* \Rightarrow x = 0 \Rightarrow \text{Ker}\,\phi = \{0\}$. Hence ϕ is one-to-one linear function. Since dim V = dim V^{**}, ϕ is also onto function which establishes that ϕ is an isomorphism from $V(F)$ onto $V^{**}(F)$. ϕ is called *canonical isomorphism*.

Hence the theorem

Corollary: If dim V = finite, then each linear functional in V^{**} is of the form T_x for some unique $x \in V$.

Theorem 10.5.5: *If V (F) is a finite-dimensional vector space. If* $\{\phi_i\}$ *is the basis of* V^* *(F) dual to a basis* $\{x_i\}$ *of V (F), then* $\{x_i\}$ *is the basis of V (f) =* V^{**} *(F) which is dual to* $\{\phi_i\}$.

Proof: Let $\{\phi_1, \phi_2, ...\phi_n\}$ be a basis of V^* (F). Then we can find a basis $[T_1, T_2, ..., T_n]$ which is dual to the basis $\{\phi_1, \phi_2, ...\phi_n\}$. This implies

$$T_i (\phi_j) = \delta_{ij} = 1 \text{ if } i = j$$
$$= 0 \text{ if } i \neq j$$

But according to the preceding corollary, there exist vectors $x_1, x_2, ..., x_n$ for which $Tx_1, Tx_2, ..., Tx_n \in V^{**}$

So we can have

$T_i (\phi) = T_{xi} (\phi) = \phi (x_1), \phi \in V^*$. In particular $\phi = \phi_j \in \{\phi_1, \phi_2, ..., \phi_n\} \subset V^*$, we have

$T_i (\phi_j) = \delta_{ij} = T_{xi} (\phi_j) = \phi_j (x_i)$

Hence we have proved that

$\phi_j (x_i) = \delta_{ij}$ $(i, j = 1, 2, ...n)$

where $\{x_1, x_2, ... x_n\}$ is a basis for $V (F)$ having $\{\phi_1, \phi_2, ...\phi_n)$ as its dual.

Definition 10.5.3: *Let V (F) be a* vector space over *F.* and V^* *(F)* dual space of *V (F)*. For any subset *W* of *V*, if there exists $f \in V^*$ such that $f(x) = 0$ for every $x \in W$, *i.e.*, $f(W) = 0$, then *f* is called an *annihilator of W*, the set $W^\perp = \{f \in V^* \mid f(x) = 0$, for all $x \in W)$ is called the *annihilator* or *orthogonal of W in* V^*. For any set $T \subset V^*$, the set $T^\perp = \{x \in V \mid f(x) = 0, \forall f \in T)$ is called the *orthogonal* or *annihilator* of *T* in *V*.

Lemma 10.5.1: *Let* V^* *(F) be a dual space* of *vector space* V (F) *over F. For any subsets W and T of V and* V^* *respectively* $W^\perp$ *(F) and* $T^\perp$ *(F) are subspace of* V^* *(F) and V (F) respectively.*

Proof: Since V^* (F) is a vector space, there is zero function Z which maps every element of V onto 0, zero of F. i.e.,

$Z(x) = 0, x \in V$. *So* $Z \in W^\perp$ and $W^\perp \neq \phi$.

Now for $f, g \in W^\perp$ and $a, b \in F$, we have

$(af + bg)(x) = af(x) + bg(x) = a.0 + b.0 = 0$. Since $f(x) = g(x) = 0, \forall x \in W$. Thus $af + bg \in W^\perp$. Hence $W^\perp$ (F) is a subspace of V^* (F).

Again $f(0) = 0, f \in T \Rightarrow 0 \in T^\perp \Rightarrow T^\perp \neq \phi$.

Let $x, y \in T^\perp$, $a, b \in F$, and $f \in T$. Then $f(x) = f(y) = 0$

by definition of $T^\perp$ and $f(ax + by) = af(x) + bf(y)$

$= a.0 + b.0 = 0 \Rightarrow ax + by \in T^\perp$.

This shows that $T^\perp$ (F) is a subspace of a vector space V (F). This completes the proof.

Lemma 10.5.2: *Let* V^* *(F) be the dual space of a vector space V (F) over F. Then*

(i) *For any subsets* $W_1, W_2 \subseteq V, W_1 \subseteq W_2 \Rightarrow W_2^\perp \subseteq W_1^\perp$

(ii) *For any subsets* $T_1, T_2 \subseteq V^*, T_1, \subseteq T_2 \Rightarrow T_2^\perp \subset T_1^\perp$

(iii) *For any subset* $W \subseteq V, W \subseteq W^{\perp\perp}$,

(iv) *For any subset* $T \subseteq V^*. T \subseteq T^{\perp\perp}$

we leave the proof of (II) and IV as exercise. However, we shall prove (I) and (III).

Proof: **(i)** Let $\phi \in W_2^\perp$ Then $\phi(x) = 0, x \in W_2$. But $W_1 \subseteq W_2$ so $x \in W_1$, and $\phi(x) = 0, \forall x \in W_1$. Hence $\phi \in W_1^\perp$. Therefore $W_2^\perp \subseteq W_1^\perp$.

CHAPTER 10

(ii) Let $x \in W$. Then for every linear functional $f \in W^{\perp}$, $f(x) = 0 \Rightarrow T_x(f) = f(x) = 0$, $\forall f \in W^{\perp} \Rightarrow T_x \in W^{\perp\perp}$. Therefore, under the identification of $V(F)$ and $V^{**}(F)$, $x \in W^{\perp\perp}$ which implies $W \subseteq W^{\perp\perp}$.

Theorem 10.5.6: *If W (F) is a subspace of the finite dimensional vector space V (f), the annihilator of W is the set $W^{\perp}$. Then*

(i) dim W + dim $W^{\perp}$ = dim V,

(ii) $W^{\perp\perp}$ = W.

Proof: **(*i*)** Let dim $V = n$ and dim $W = r$. Then there are r independent vectors $x_1, x_2, \ldots, x_r$ which form the basis of $W(F)$. This basis $\{x_1, x_2, \ldots, x_r]$ of W can be extended to a basis $\{x_1, x_2, \ldots, x_r, y_1, \ldots, y_{n-r})$ of $V(F)$.

Let $\{\phi_1, \phi_2, .. \phi_r. f_1. f_2, . f_{n-r}\}$ be a dual basis of $\{x_1, x_2, \ldots, x_r, y_1, y_2, \ldots y_{n-r}\}$,

the bases of $V(F)$. Then $\phi_i(x_j) = \delta_{ij} = \begin{cases} 1 \text{ if } i = j, \\ 0 \text{ if } i \neq j, \end{cases}$

and $f_r(x_s) = \delta_{rs} = 1$ if $r = s$

0 if $r \neq s$

Thus it is clear that each $f_r(x_i) = 0$, $x_i \in \{x_1, x_2, \ldots, x_r\}$. Hence $f_1, f_2, \ldots, f_{n-r} \in W^{\perp}$. Now we claim that $\{f_j\}$ is a basis of $W^{\perp}(F)$. It is obvious that $\{f_j)$ is a part of the basis of $V^*(F)$ and so it is *L.I.* To show that $\{f_j\}$ is a basis of $W^{\perp}(F)$, there remains to show that $\{f_j\}$ spans $W^{\perp}(F)$. For this, Let $f \in W^{\perp}$. By theorem 10.5.2, we have,

$$\begin{aligned} f &= f(x_1)\,\phi_1 + f(x_2)\,\phi_2 + \ldots + f(x_r)\,\phi_r \\ &\quad + f(y_1)\,f_1 + f(y_2)\,f_2 + \ldots + f(y_{n-r})\,f_{n-r} \\ &= 0\,\phi_1 + 0\,\phi_2 + \ldots + 0.\,\phi_r + f(y_1)\,f_1 + f(y_2)\,f_2 \\ &\quad + \ldots + f(y_{n-r})\,f_{n-r} \\ &= f(y_1)\,f_1 + f(y_2)\,f_2 + \ldots + f(y_{n-r})\,f_{n-r}. \end{aligned}$$

Thus $\{f_1, f_2, \ldots, f_{n-r}\}$ spans $W^{\perp}$, which follows that dim $W^{\perp} = n - r =$ dim V – dim W or dim W+ dim $W^{\perp}$ = dim V.

This completes the first part of the theorem.

(*ii*) Let dim $V = n$ and dim $W = r$, then dim V = dim $V^* = n$. by

(1) dim $W^{\perp} = n - r$. Thus again by (1) dim $W^{\perp\perp} = n - (n - r) = r$ which implies dim W = dim $W^{\perp\perp}$

By the Lemma 10.5.2: $W \subseteq W^{\perp\perp}$, which shows $W = W^{\perp\perp}$. Hence the theorem.

Theorem 10.5.7: *Let W_1 (F) and W_2 (F) be two subspaces of a finite dimensional vector space V (F), then*

(a) $(W_1 + W_2)^{\perp} = W_1^{\perp} \cap W_2^{\perp}$

(b) $(W_1^{\perp} + W_2^{\perp}) = (W_1 \cap W_2)^{\perp}$

Proof: $f \in (W_1^{\perp} + W_2^{\perp}) \Rightarrow f(x) = 0,\ x \in W_1 + W_2$

$\Rightarrow f(w_1 + w_2) = 0$ since $x = w_1 + w_2,\ w_1 \in W_1,\ w_2 \in W_2$

$\Rightarrow f(w_1) + f(w_2) = 0,\ w_1 \in W_1,\ w_2 \in W_2$

$\Rightarrow f(w_1) = 0$ and $f(w_2) = 0$ since $W_1 \subseteq W_1 + W_2$

and $W_2 \subseteq W_1 + W_2$

$\Rightarrow f \in W_1^{\perp}$ and $f \in W_2^{\perp}$

$\Rightarrow f \in W_1^{\perp} \cap W_2^{\perp}$

Thus $\left(W_1^{\perp} + W_2^{\perp}\right) \subseteq W_1^{\perp} \cap W_2^{\perp}$ ....(1)

Conversely, $g \in W_1^{\perp} \cap W_2^{\perp} \Rightarrow g \in W_1^{\perp}$ and $g \in W_2^{\perp}$

$\Rightarrow g(w_1) = 0$ and $g(w_2) = 0$,

$w_1 \in W_1$ and $w_2 \in W_2$

$\Rightarrow g(w_1) + g(w_2) = 0,\ w_1 \in W_1,\ w_2 \in W_2$

$\Rightarrow g(w_1 + w_2) = 0,\ \forall\ w_1 \in W_1,\ w_2 \in W_2$

$\Rightarrow g \in (W_1 + W_2)^{\perp}$ since $w_1 + w_2 \in W_1 + W_2$

Thus $W_1^{\perp} \cap W_2^{\perp} \subseteq \left(W_1 + W_2\right)^{\perp}$. ....(2)

From (1) and (2), we have

$(W_1 + W_2)^{\perp} = W_1^{\perp} \cap W_2^{\perp}$.

(*b*) We have that $W_1^{\perp}(F)$ and $W_2^{\perp}(F)$ are subspaces of $V^*(F)$. Now by (*a*) we obtain,

$\left(W_1^{\perp} + W_2^{\perp}\right) = \left(W_1^{\perp}\right)^{\perp} \cap \left(W_2^{\perp}\right)^{\perp} = W_1 \cap W_2$

Thus $((W_1^{\perp} + W_2^{\perp})^{\perp})^{\perp} = \left(W_1 \cap W_2\right)^{\perp} \Rightarrow W_1^{\perp} + W_2^{\perp} = \left(W_1 \cap W_2\right)^{\perp}$.

Hence the theorem.

10.6 TRANSPOSE OF A LINEAR MAPPING

Definition 10.6.1: Let $f : V \to U$ be a linear mapping from a vector space $V(F)$ into a vector space $U(F)$. Now for any linear functional $\phi \in U^*$, the composition $(\phi \circ f)$ is a linear mapping from $V(F)$ to F,

That is, $\phi \circ f \in V^*$. Thus the correspondence $\phi \to (\phi \circ f)$ is a mapping from $U^*(F)$ into $V^*(F)$, we denote it by f^T and call it *transpose of f*. Thus $f^T : U^* \to V^*$ defined by

$f^T(\phi) = \phi \circ f,\ \phi \in U^*$.

i.e., $(f^T(\phi))(x) = \phi(f(x)),\ \forall\ x \in V$

Theorem 10.6.1: *Let $f : V \to U$ be linear mapping and let $f^T : U^* \to V^*$ be its transpose. Then*

(i) f^T is a linear mapping from $U^*(F)$, into $V^*(F)$

(ii) $Ker(f^T) = f(V)^{\perp}$

Proof: ***(i)*** For any scalars $a, b \in F$ and for any linear functional $\phi, \sigma \in U^*$.

$f^T(a\phi + b\sigma) = (a\phi + b\sigma) \circ f$

$= a(\phi \circ f) + b(\sigma \circ f)$

$= af^T(\phi) + bf^T(\sigma)$.

Thus f^T is linear function from $U^*(F)$ into $V^*(F)$.

(ii) Let $\phi \in \ker(f^T)$. That is, $f^T(\phi) = \phi \circ f = 0 \in V^*$

If $u \in f(V)$, then $u = f(x)$ for some $x \in V$.

CHAPTER 10

Thus we have $\phi(u) = 0$ for every $u \in f(V)$ which means $\phi \in f(V)^{\perp}$. Hence $\ker(f^T) \subseteq f(V)^{\perp}$.

Conversely, let $\sigma \in f(V)^{\perp}$. Then $\sigma(y) = 0$, $\forall\, y \in f(V)^{\perp}$ which implies $\sigma(f(V)) = \{0\}$. Then for every $x \in V$, $(f^T(\sigma))(x) = (\sigma \circ f)(x) = \sigma(f(x)) = 0 = 0(x)$

$$\text{which implies } (f^T(\sigma))(x) = 0 \Rightarrow f^T(\sigma) = 0$$
$$\Rightarrow \sigma \in \ker(f^T)$$

Therefore $f(V)^{\perp} \subseteq \ker(f^T)$

From (1) and (2) we conclude that

$\text{Ker}(f^T) = f(V)^{\perp}$

Theorem 10.6.2: *Let $f : V \to W$ be linear and let $V(F)$ and $W(F)$ be finite-dimensional vector spaces. Then rank (f) = rank (f^T).*

Proof: Let $\dim V = n$ and $\dim W = m$ and also let rank $(f) = \dim f(V) = r$. Now by theorem 10.5.6, we have

$\dim W = \dim(f(V)) + \dim(f(V)^{\perp}$.

or $\dim f(V)^{\perp} = \dim W - \dim f(V)$

$= m - r$

By the theorem 10.6.2, we have

$\text{Ker}(f^T) = f(V)^{\perp} \Rightarrow$ mulity $(f^T) = m - r$.

By Sylvester's Law, we obtain

$\dim W^* = \dim(\text{Ker}(f^T) + \dim(f(W^*)$

or $\dim(f^T(W)^*) = \dim W^* - \dim \text{Ker}(f^T)$.

Since $\dim W = m = \dim W^*$, then

rank $(f^T) = m - (m - r) = r =$ rank (f). Hence the theorem.

Theorem 10.6.3: *It $V = U \oplus W$, then the function $f_w : V \to W$ which assigns to each vector $x \in V$ its uniquely determined component x_2, in the representation $x = x_1 + x_2$, $(x_1 \in U, x_2 \in W)$ is an idempotent linear mapping with Ker $(f_w) = U$, $f_w(V) = W$. Conversely, every idempotent linear mapping $f \in L(V, V)$ defines a direct sum decomposition $V = \text{Ker}(f) \oplus f(V)$.*

Proof: First we show that $f_w \in L(V, V)$. For this, suppose $x, y \in V$, then $x = u_1 + w_1$ and $y = u_2 + w_2$ where $u_1, u_2 \in U$, $w_1, w_2 \in W$. For $a, b \in F$, we have

$ax + by = a(u_1 + w_1) + b(u_2 + w_2)$

$= (au_1 + bu_2) + (aw_1 + bw_2) \in V$, *where*

$au_1 + bu_2 \in U$ *and* $aw_1 + bw_2 \in W$.

So $f_w(ax + by) = f_w((au_1 + bu_2) + (aw_1 + bw_2))$

$= aw_1 + bw_2 = af_w(x) + bf_w(y)$.

which shows $f_w \in L(V. V)$.

We observe something more:

Since $u = u + 0$ and $w = 0 + w$ as every element of V is uniquely expressed as sum of elements of U and W respectively. Then we have

(i) $f_w(u) = f_w(u + 0) = 0$ if and only if $u \in U$,

(ii) $f_w(w) = f_w(0 + w) = w$ if and only if $w \in W$.

Now suppose $x = u + w$. with $u \in U$, $w \in W$, then

$f_w^2(x) = f_w(f_w(x)) = f_w(f_w(u + w)) = f_w(w) = w = f_w(x)$

$f_w^2 = f_w$. Hence f_w is idempotent.

For the converse. Let $f \in L(V, V)$ with $f^2 = f$. Taking $u = x - f(x)$ and $w = f(x)$, we may write $x \in V$ as $x = u + w$. Since f is idempotent, then

$$f^2(x) = f(x) \Rightarrow f(x) - f^2(x) = 0 = f(u)$$

while $f(w) = f^2(x) = f(x) = w$.

This shows that $V = \text{Ker}(f) + W$, where the set $W = \{x \in V \mid f(x) = x\}$.

Now we have to show that $V = \text{Ker}(f) + W$ is direct sum, that is, $\text{Ker}(f) \cap W = \{0\}$.

if $x \in \text{Ker}(f)$, then $f(x) = 0$, where as if $x \in W$, $f(x) = x$. if $x \in \text{Ker}(f)$ and $x \in W$, $f(x) = 0$. By def. of W, it is clear that $W \subseteq f(V)$. For the reverse inclusion, we have $f^2 = f \Rightarrow f(x) \in W, \forall x \in V \Rightarrow f(V) \subseteq W$. Hence $W = f(V)$. This completes the proof of the theorem

Example 10.6.1: *Let* $f : V_4 \to V_4$ be defined by

$$f(a_1, a_2, a_3, a_4) = (a_1, a_2, 0, 0).$$

Then $\text{Ker}(f) = \{(0, 0, a_3, a_4) \mid a_3, a_4 \in F\}$

and $f(V) = \{(a_1, a_2, 0, 0) \mid a_1, a_2 \in F\}$.

Now $\text{Ker}\, f \cap f(V_4) = \{(0, 0, 0, 0)\}$

and any $(a_1, a_2, a_3, a_4) \in V$ can be written as $(a_1, a_2, a_3, a_4) = (0, 0, a_3\, a_4) + (a_1, a_2, 0, 0) \in \ker(f) + f(V_4)$.

$$\begin{aligned} \text{and } f^2(a_1, a_2, a_3, a_4) &= f(f(a_1, a_2, a_3, a_4) \\ &= f(a_1, a_2, 0, 0) \\ &= (a_1, a_2, 0, 0) = f(a_1, a_2, a_3, a_4) \end{aligned}$$

Hence f is idempotent.

Theorem 10.6.4: *Let $U(F)$ and $W(F)$ be complementary subspaces of a finite dimensional space $V(F)$, i.e., $V = U \oplus W$. Then*

(i) $U^*(F) \cong W^\perp(F)$, $U^\perp(F) \cong W^*(F)$.

(ii) $V^* = U^\perp + W^\perp$

Proof: *(i)* Since $V(F)$ is finite dimensional, and $V = U \oplus W$, then $\dim V = \dim U + \dim W$.

Let $\dim V = n$, $\dim U = m$. So $\dim W = \dim V - \dim U$

$$= n - m.$$

$$\dim W + \dim W^\perp = \dim V \Rightarrow \dim W^\perp = \dim V - \dim W$$
$$= n - (n - m)$$
$$= m = \dim U.$$

Since $\dim W^\perp = \dim U$, then $W^\perp(F) \cong U(F)$

Since $\dim U = \dim U^*$, so $W^\perp(F) \cong U^*(F)$.

Again $\dim U + \dim U^\perp = \dim V \Rightarrow \dim U^\perp = \dim V - \dim U$
$$= n - m$$
$$= \dim W.$$

But $\dim W = \dim W^*$, so $\dim U^\perp = \dim W^*$.

Which implies $U^\perp(F) \cong W^*(F)$.

(ii) For proving $V^* = U^\perp + W^\perp$, we shall show that $U^\perp \cap W^\perp = \{0\}$, where 0 is a zero function.

$$\begin{aligned} \text{Now } f \in U^\perp \cap W^\perp &\Rightarrow f \in U^\perp \text{ and } f \in W^\perp \\ &\Rightarrow f(u) = 0,\ u \in U \text{ and } f(w) = 0,\ w \in W. \\ &\Rightarrow f(u) + f(w) = 0,\ u \in U \text{ and } w \in W \\ &\Rightarrow f(u + w) = 0,\ u \in U \text{ and } w \in W, \text{ since } f \text{ is linear} \end{aligned}$$

$\Rightarrow f(V) = 0$ since $v = u + w$ as $V = U \oplus W$

$\Rightarrow f$ is a zero function

$\Rightarrow U^{\perp} \cap w^{\perp} = \{0\}$, where 0 is the zero function.

Now there remains to show that every element of $f \in V^*$ is uniquely expressible as $f = f_u + f_w$, where $f_u \in W^{\perp}$ and $f_w \in U^{\perp}$.

Since $V = U \oplus W$, and $x \in V$ can be written as

$x = u + w,\ u \in U,\ w \in W.$

Now define $f_u : V \to F$ and $f_w : V \to F$ by

$f_u(x) = f(u)$ and $f_w(x) = f(w)$

So $f_w, f_w \in V^*$

For $x, y \in V$, $x = u_1 + w_1$, $y = u_2 + w_2$ and for $a, b \in F$,

$$\begin{aligned} f_u(ax + by) &= f_u(au_1 + bu_2 + aw_1 + bw_2) \text{ where } a\,u_1 + bu_2 \in U,\ aw_1 + bw_2 \in W. \\ &= f(au_1 + bu_2) \\ &= af(u_1) + bf(u_2) \\ &= af_u(x) + bf_u(y) \end{aligned}$$

which implies f_u is linear. Similarly, f_w can be shown linear.

Since $u = u + 0,\ w = 0 + w$

Now for $w \in W$,

$$f_u(w) = f_u(0 + w) = f(0) = 0 \Rightarrow f_u \in W^{\perp}$$

and for $u \in U$

$$f_w(u) = f_w(u + 0) = f(0) = 0 \Rightarrow f_w \in U^{\perp}$$

And
$$\begin{aligned} (f_u + f_w)(x) &= f_u(x) + f_w(x) \\ &= f_u(u + w) + f_w(u + w)) = f(u) + f(w) \\ &= f(u + w) = f(x) \end{aligned}$$

which implies $f_u + f_w = f$.

This completes the proof of the theorem.

PROBLEMS

1. Let $V(F)$ be a finite dimensional vector space over F. Prove that
   (a) if $x_1, x_2 \in V$ with $x_1 \neq x_2$, then there exists a linear functional $f \in V^*$ for which $f(x_1) \neq f(x_2)$.
   (b) If $W(F)$ is proper non-zero subspace of $V(F)$ and the vector $x_0 \notin W$, then there exists some $f \in V^*$ such that $f(x_0) = 1$, $f(x) = 0$, $\forall\, x \in W$.
2. If $W(F)$ is subspace of the vector space $V(F)$, and $W^{\perp}$ is the annihilator of W, then prove that $(V/W)^*(F) \cong W^{\perp}(F)$ and $(V/W^{\perp})(F) \cong W^*(F)$.
3. For each linear mapping $f \in L(V, W)$ there is a transpose f^{T} of f. Then prove that $L(V, W) \cong L(W^*, V^*)$ under the mapping that sends each function $f \in L(V, W)$ to its transpose f^{T}.
4. Let $\{x_1, x_2, ..., x_n\}$ be a basis for finite dimensional vector space $V(F)$, and let $\{f_1, f_2, ... f_n\}$ be the corresponding dual basis for $V^*(F)$. Suppose that $[a_{ij}]$ is the representing matrix, relative to $\{x_1, x_2, ..., x_n\}$, of linear mapping $f \in L(V, V)$. Prove that the transpose f^{T} of f is represented by the matrix $[a_{ij}]^T$ relative to $\{f_1, f_2, ..., f_n\}$.
5. If the functionals $f, g \in V^*$ are such that $\text{Ker}(f) \subseteq \text{Ker}(g)$, prove that there exists a scalar a for which $f = ag$.

6. Suppose $f: U \rightarrow V$ and $g : V \rightarrow W$ are linear. Prove that $(g\ o\ f^{T} = f^{T}\ o\ g^{T}$.
7. Let $V(F)$ be a vector space over F. Let $\phi_1, \phi_2, \in V^*$ and suppose $f : V \rightarrow R$ is defined by $f(v) = \phi_1(v).\ \phi_2(v)$ also belongs to V^*, show that either $\phi_1 = 0$ or $\phi_2 = 0$.

10.7 MATRICES AND LINEAR TRANSFORMATIONS

Let $U(F)$ and $V(F)$ be two vector spaces over the field $(F, +,.)$ and let $T: U \rightarrow V$ be a linear transformation. Again let $\{u_1, u_2,..., u_m]$ and $\{v_1, v_2,..., v_n)$ be the bases of $U(F)$ and $V(F)$ respectively. We know that T is determined on any vector as soon as we know its action on a basis of U. Since T maps U into V, $T(u_1), T(u_2),..., T(u_m)$ will belong to V. Since every element of V can be expressed as a linear combination of $v_1, v_2,..., v_n$ in a unique way. Thus:

$$T(u_1) = a_{11} v_1 + a_{12} v_2 +....+ a_{1n} v_n$$
$$T(u_2) = a_{21} v_1 + a_{22} v_2 +...,+ a_{2n} v_n$$
$$T(u_m) = e_{m1} v_1 + a_{m2} v_2 +....+ a_{m2} v_n,$$

where each $a_{ij} \in F$. This system of equation can be written as

$$T(u_i) = \sum_{j=1}^{n} a_{ij} v_j, \text{ for } i = 1, 2,...., m.$$

The ordered set of $m.n.$ numbers of F completely describes T. They will serve the purpose of representing T.

Definition 10.7.1: Let $U(F)$ and $V(F)$ be two vector spaces of dimension m and n over field $(F, +,.)$ respectively and Let $\{u_1, u_2,.., u_m\}$ and $\{v_1, v_2,..., v_n\}$ be bases of $U(F)$ and $V(F)$ respectively. If $T \in Hom(U, V)$, then the matrix of T or $\{T(u_1), T(u_2),... T(u_m)\}$ relative to the basis $\{v_1, v_2,..., v_n\}$, written as $m(T)$, is

$$m(T) = \begin{bmatrix} a_{11}\, a_{12} & ....a_{1n} \\ a_{21}\, a_{12} & ....a_{2n} \\ . & \\ . & \\ . & \\ a_{m1}\, a_{m2} & ....a_{mn} \end{bmatrix} = (a_{ij})$$

where $T(u_i) = \sum_{j=1}^{n} a_{ij} v_j, \quad i = 1, 2, 3,......, m.$

If $U = V$, $u_i, = v_i$, $I \leq i \leq n$, then the square matrix $m(T)$ of order n is called the matrix of T relative to the basis $(v_1, v_2,..., v_n)$.

We observe that $m(T)$ depends not only on T but also on the bases $\{u_1, u_2,...., u_m\}$ and $\{v_1, v_2,....., v_n\}$.

Example 10.7.1: Let F be a field and let V be the set of all polynomials in x of degree $n - 1$ or less over F.

Define $T : V \rightarrow V$ by $T(f) = f'$. Find the matrix of T with respect to the bases $\{1, x, x^2,...., x^{n-1}\}$.

Solution: Let us compute the matrix of T in the basis

$v_1 = 1,\ v_2 = x,\ v_3 = x^2,....., v_i = x^{i-1},...., v_n = x^{n-1}.$

CHAPTER 10

Now

$$\begin{aligned} T(v_1) &= T(1) = 0 = 0\ v_1 + 0\ v_2 + ... + 0\ v_n \\ T(v_2) &= T(x) = 1 = 1\ v_1 + 0\ v_2 + ... + 0\ v_n \\ T(v_3) &= T(x^2) = 2x = 2v_2 = 0\ v_1 + 2\ v_2 + ... + 0\ v_n \\ T(v_4) &= T(x^3) = 3x^2 = 3v_3 = 0\ v_1 + 0\ v_2 + 3\ v_3 + ... + 0\ v_n \end{aligned}$$

$$T(v_n) = T(x^{n-1}) = (n-1)\ x^{n-2} = (n-1)\ v_{n-1} = 0\ v_1 + 0\ v_2 + ... + (n-1\ v_{n-1} + 0\ v_n.$$

By the definition of the matrix of a linear transformation in a given basis, we see the matrix $m \times n$ of T in the basis $1, x, ..., x^{n-1}$ is

$$m(T) = \begin{vmatrix} 0 & 0 & 0 & ...0 & 0 \\ 1 & 0 & 0 & ...0 & 0 \\ 0 & 2 & 0 & ...0 & 0 \\ 0 & 0 & 3 & ...0 & 0 \\ - & - & - & - & - \\ 0 & 0 & 0 & (n-1) & 0 \end{vmatrix}$$

Example 10.7.2: Compute the matrix of T with respect to another basis of V over F. Here $u_1 = 1$, $u_2 = 1 + x$, $u_3 = 1 + x^2, ..., u_n = 1 + x^{n-1}$.

It is easy to verify that $\{u_1, u_2, u_3, ..., u_n\}$ forms the basis of V over F. Since.

$$\begin{aligned} T(u_1) &= T(1) = 0 = 0\ u_1 + 0\ u_2 + ... + 0\ u_n \\ T(u_2) &= T(1 + x) = 1 = 1\ u_1 + 0\ u_2 + ... + 0\ u_n \\ T(u_3) &= T(1 + x^2) = 2x = 2u_2 - 2u_1 \\ &= -2u_1 + 2u_2 + 0\ u_3 + ... + 0\ u_n \\ T(u_4) &= T(1 + x^3) = 3x^2 = 3(u_3 - u_1) \\ &= -3u_1 + 0\ u_2 + 3u_3 + ... + 0\ u_n \\ T(u_n) &= T(1 + x^{n-1}) = (n-1)\ x^{x-2} = (x-1)(u_{n-1} - u_n) \\ &= -(n-1)\ u_1 + 0\ u_2 + 0\ u_3 + ... + (n-1)\ u_{n-1} + 0\ u_n. \end{aligned}$$

The matrix $m(T)$ of D in this basis is

$$m(T) = \begin{bmatrix} 0 & 0 & 0 & .....0 & 0 \\ 1 & 0 & 0 & .....0 & 0 \\ -2 & 2 & 0 & .....0 & 0 \\ -3 & 0 & 3 & .....0 & 0 \\ — & — & — & — & — \\ -(n-1) & 0 & 0 & (n-1) & 0 \end{bmatrix}$$

Remark 1: By these examples we see that matrices $m(T)$ of transformation T, for two bases used, depended completely on the basis. Although different from each other, they represent the same transformation T and we could reconstruct T from any of them if we know the basis used in their determination.

Remark 2: Let $m(T) = [a_{ij}]_{m \times n}$. If we have to find out the image of a vector $u \in U$, then u can be written as $u = (x_1, x_2, ..., x_m)$ and the image of (u) is obtained by writing u on the left of $m(T)$.

$$uT = (x_1, x_2, ..., x_m)\ [a_{ij}]_{m \times n}.$$

If we write the vector u on the right of $m(T)$, then we have to take the transpose of $[m(T)]$

Thus

$$T(u) = \left[a_{ji}\right]_{n \times m} \begin{bmatrix} x_1 \\ x_2 \\ \cdot \\ \cdot \\ \cdot \\ x_m \end{bmatrix}$$

Example 10.7.3: Find the matrix representation, relative to the basis $E = \{e_1 = (1, 0, 0), e_2 = (0, 1, 0), e_3 = (0, 0, 1)\}$ of each of the following transformations.

(*a*) $T(x, y, z) = (2x - y, y + z)$

(*b*) $T(x, y, z) = (x, y - z, y + 2z)$

(*c*) $(x, y, z)\, T = (x + y, y + 2z, x + 3z)$

Solution: (*a*) Hence

$$\begin{aligned} T(1, 0, 0) &= (2, 0) = 2(1, 0) + 0(0, 1) \\ T(0, 1, 0) &= (-1, 0) = -1(1, 0) + 0(0, 1) \\ T(0, 0, 1) &= (0, 1) = 0(1, 0) + 1(0, 1) \end{aligned}$$

Therefore

$$\begin{matrix} & T(e_1) & T(e_2) & T(e_3) \end{matrix}$$

$$m(T) = \begin{bmatrix} 2 & -1 & 0 \\ 0 & 0 & 1 \end{bmatrix}$$

Remark: In the associated matrix of transformation T, $T(e_1)$ constitute the first column, $T(e_2)$ second column, and $T(e_n)$ constitute nth column.

(*b*) $T(1, 0, 0) = (1, 0, 0) = 1(1, 0, 0) + 0(0, 1, 0) + 0(0, 0, 1)$

$T(0, 1, 0) = (0, 1, 1) = 0(1, 0, 0) + 1(0, 1, 0) + 1(0, 0, 1)$

$T(0, 0, 1) = (0, -1, 2) = 0(1, 0, 0) - 1(0, 1, 0) + 2(0, 0, 1)$

Therefore

$$m(T) = \begin{bmatrix} 1 & 0 & 0 \\ 0 & 1 & -1 \\ 0 & 1 & 2 \end{bmatrix}$$

(c) $(x, y, z)\, T = (x + y, y + 2z, x + 3z)$

$(1, 0, 0)\, T = (1, 0, 1) = 1(1, 0, 0) + 0(0, 1, 0) + 1(0, 0, 1)$

$(0, 1, 0)\, T = (1, 1, 0) = 1(1, 0, 0) + 1(0, 1, 0) + 0(0, 0, 1)$

$(0, 0, 0)\, T = (0, 2, 3) = 0(1, 0, 0) + 2(0, 1, 0) + 3(0, 0, 1)$

Therefore

$$m(T) = \begin{bmatrix} 1 & 0 & 1 \\ 1 & 1 & 0 \\ 0 & 2 & 3 \end{bmatrix}$$

Example 10.7.4: Let $f : R^3 \to R^2$ be the linear map defined by

$$f(x, y, z) = (3x + 2y - 4z, x - 5y + 3z)$$

Find the matrix of f if the base vectors in R^3 and R^3 are respectively $u_1 = (1, 1, 1)$, $u_2 = (1, 1, 0)$, $u_3 = (1, 0, 0)$; $v_1 = (1, 3)$, $v_2 = (2, 5)$.

Solution: Here

$f(u_1) = f(1, 1, 1) = (1, -1)$ then $(1, -1)$ is a linear combination of v_1 and v_2. So for some $a, b \in R$

$$a\, v_1 + b\, v_2 = (1, -1)$$
$$\Rightarrow a\,(1, 3) + b\,(2, 5) = (1, -1)$$
$$\Rightarrow a + 2b = 1 \text{ and } 3a + 5b = -1$$
$$\Rightarrow a = -7,\ b = 4.$$

So $f(u_1) = f(1, 1, 1) = (1, -1) = -7\,(1, 3) + 4\,(2, 5) = -7\,v_1 + 4\,v_2$

Similarly, $f(u_2) = f(1, 1, 0) = (5, -4) = -33\,(1, 3) + 19\,(2, 5) = -33\,v_1 + 19v_2$

and $f(u_3) = f(1, 0, 0) = (3, 1) = -13\,(1, 3) + 8\,(2, 5) = -13\,v_1 + 8\,v_2$.

Hence the matrix of f

$$[f] = m(f) = \begin{bmatrix} -7 & -33 & -13 \\ 4 & 19 & 8 \end{bmatrix}$$

Example 10.7.5: Let $f(v) = Av$, where v in Av is written as column vector, while A is the matrix $A = \begin{bmatrix} 2 & 5 & -3 \\ 1 & -4 & 7 \end{bmatrix}$.

(*i*) Show that the matrix $m(f)$ of f is A itself relative to the usual bases of R^3 and R^2.

(*ii*) Find the matrix of f relative to the bases of R^3 and R^2 respectively as $u_1 = (1, 1, 1)$, $u_2 = (1, 1, 0)$, $u_3 = (1, 0, 0)$ and $v_1 = (1, 3)$, $v_2 = (2, 5)$.

Solution: Taking $u = (x, y, z)$, then $f(u) = f(x, y, z) = \begin{bmatrix} 2 & 5 & -3 \\ 1 & -4 & 7 \end{bmatrix}\begin{bmatrix} x \\ y \\ z \end{bmatrix}$.

$\Rightarrow f(x, y, z) = (2x + 5y - 3z,\ x - 4y + 7z)$.

Here $e_1 = (1, 0, 0)$, $e_2 = (0, 1, 0)$, $e_3 = (0, 0, 1)$, and $e_1^1 = (1, 0)$, $e_2^1 = (0, 1)$.

(*i*) Now $f(e_1) = f(1, 0, 0) = (2, 1)$

$f(e_2) = f(0, 1, 0) = (5, -4)$

$f(e_3) = f(0, 0, 1) = (-3, 7)$

Therefore the matrix of f is $m(f) = \left[T(e_1)\,T(e_2)\,T(e_3)\right]$

$$= \begin{bmatrix} 2 & 5 & -3 \\ 1 & -4 & 7 \end{bmatrix} = \text{A}.$$

(*ii*) $f(u_1) = f(1, 1, 1) = (4, 4)$ and $(4, 4)$ can be written as linear combination of v_1 and v_2.

So $(4, 4) = a\,v_1 + b\,v_2 = a\,(1, 3) + b\,(25)$

$$\Rightarrow 4 = a + 2b \text{ and } 4 = 3a + 5b$$
$$\Rightarrow a = -12,\ b = +8$$

$\therefore f(u_1) = (4, 4) = -12\,v_1 + 8\,v_2$

Similarly $f(u_2) = f(1, 1, 0) = (7, -3) = -41\, v_1 + 24\, v_2$
and $f(u_3) = f(1, 0, 0) = (2, 1) = -8\, v_1 + 5\, v_2$.
Therefore matrix of f

$$m(f) = \begin{bmatrix} T(e_1) & T(e_2) & T(e_3) \\ -12 & -41 & -8 \\ 8 & 24 & 5 \end{bmatrix}.$$

Example 10.7.6: Determine the mapping $T : R^2 \to R^2$ with reference to the basis (1, 0), (1, 1) if the matrix of T is $\begin{bmatrix} \frac{1}{2} & 1 \\ \frac{2}{3} & 4 \end{bmatrix}$.

Solution: Here $u_1 = (1, 0)$, $u_2\ (1, 1)$. By definition of $m(T)$

$$a_{11} = \frac{1}{2},\ a_{12} = \frac{2}{3},\ a_{21} = 1,\ a_{22} = 4.$$

$$T(u_1) = a_{11}\, u_1 + a_{12}\, u_2 = \frac{1}{2}(1, 0) + \frac{2}{3}(1, 1) = \left(\frac{7}{6}, \frac{2}{3}\right)$$

$$T(u_2) = a_{21}\, u_1 + a_{22}\, u_2 = 1\,(1, 0) + 4\,(1, 1) = (5, 4).$$

Now $(x, y) \in R^2$ can be written as

$$(x, y) = \alpha_1 u_1 + \alpha_2 u_2 = \alpha_1 (1, 0) + \alpha_2 (1, 1)$$

$$\Rightarrow \quad x = \alpha_1 + \alpha_2,\ y = \alpha_2$$

$$\alpha_1 = x - \alpha_2 = x - y$$

Hence $T(x, y) = T(\alpha_1 u_1 + \alpha_2 u_2) = \alpha_1 T(u_1) + \alpha_2 T(u_2)$

$$= \left\{(x-y)\left(\frac{7}{6}, \frac{2}{3}\right) + y\,(5, 4)\right\}$$

$$= \left\{\frac{1}{6}(7x + 23y), \frac{1}{3}(2x + 10y)\right\}.$$

Theorem 10.7.1: *Let $U(F)$ and $V(F)$ be two vector spaces of dimension m and n over the field $(F, +, .)$ respectively. Then Hom $(U, V) \cong M_{m\times n}(F)$, where $M_{m\times n}$ is a vector space of all $m \times n$ matrices over $(F, +, .)$.*

Proof: Let $\{u_1, u_2, \ldots, u_m\}$ and $\{v_1, v_2, \ldots, v_n\}$ be bases of the vector spaces $U(F)$ and $V(F)$ respectively. Let $T \in$ Hom (U, V), then

$$T(u_i) = \sum_{j=1}^{n} a_{ij} v_j, \text{ for } i = 1, 2, \ldots, m.$$

When a vector space is defined on a field $(F, +, .)$ then there is no distinction between left and right scalar multiplication of a vector. So

$$T(u_i) = \sum_{j=1}^{n} v_i\, a_{ij}, \text{ for } i = 1, 2, \ldots, m.$$

CHAPTER 10

and $[a_{ij}]$ is an $m \times n$ matrix of T.

Let $T, T' \in \text{Hom}(U, V)$. Then

$$T(u_i) = \sum_{j=1}^{n} a_{ij} v_j, \text{ and } T^1(u_i) = \sum_{j=1}^{n} b_{ij} v_j$$

for $i = 1, 2, \ldots, m$, and a_{ij} and $b_{ij} \in F$.

Since T and T' are Linear transformation, then

$$(T + T')(u_i) = T(u_i) + T'(u_i) = \sum_{j=1}^{n} a_{ij} v_j + \sum_{j=1}^{n} b_{ij} v_j$$

$$= \sum_{j=1}^{n} (a_{ij} + b_{ij}) v_i$$

Thus if $[a_{ij}]$ and $[b_{ij}]$ are $m \times n$ matrices of T and T' respectively, $[a_{ij} + b_{ij}]$ is an $m \times n$ matrix of $T + T'$.

For any $c \in F$, we have

$$(cT)(u_i) = cT(u_i) = c \sum_{j=1}^{n} a_{ij} v_j$$

$$= \sum_{j=1}^{n} (ca_{ij}) v_j$$

which implies $[ca_{ij}]$ is an $m \times n$ matrix of the linear transformation (cT).

Let $M_{m \times n}(F)$ be a vector space of all $m \times n$ matrices over the field $(F, +, .)$. We define ϕ: Hom $(U, V) \to M(F)$ by $\phi(T) = [a_{ij}]$. Then we see that

$$\phi(T + T') = [a_{ij} + b_{ij}] = [a_{ij}] + [b_{ij}] = \phi(T) + \phi(T')$$

and $\phi(cT) = [ca_{ij}] = c(a_{ij}) = c\,\phi(T)$.

It shows that ϕ is a linear transformation.

Now, for any T and T', we have

$$\phi(T) = \phi(T') \Rightarrow [a_{ij}] = [b_{ij}]$$

$$\Rightarrow a_{ij} = b_{ij}, \text{ for all } i, j$$

$$\Rightarrow a_{ij} v_j = \Sigma\, b_{ij} v_j \Rightarrow T(u_i) = T'(u_i) \Rightarrow T = T'.$$

Again for any matrix $[a_{ij}] \in M_{m \times n}(F)$ there exists a linear transformation T such that $\phi(T) = [a_{ij}]$. Hence ϕ is onto. This proves the theorem.

Corollary: Hom $(V, V) \cong M_n(F)$, the algebra of all matrices over the field $(F, +, .)$.

Proof: In the above theorem 10.7.1 we have seen that there is an isomorphism ϕ: Hom $(V, V) \to M_n(F)$ defined by $\phi(T) = [a_{ij}]$, $[a_{ij}]$ is a square matrix of order n and such that

$$T(v_i) = \sum_{j=1}^{n} a_{ij} v_j, \text{ for any } T \in \text{Hom}(V, V).$$

Let $T' \in$ Hom (V, V) and $T'(v_i) = \sum_{j=1}^{n} a_{ij} v_j$

Then $(TT')(V_i) = T(T'(V_i)) = T\left(\sum_{j=1}^{n} b_{ij} v_j\right)$

$$= T\left(\sum_{j=1}^{n} v_j v_{ij}\right)$$

$$= \sum_{j=1}^{n} T(v_j) b_{ij}$$

$$= \sum_{j=1}^{n}\left(\sum_{i=1}^{n} v_i a_{ki}\right) b_{ij}$$

$$= \sum_{j=1}^{n}\sum_{i=1}^{n} a_{ki} b_{ij} v_i = \sum_{i=j}^{n} C_{kj} v_j$$

Where $$c_{kj} = \sum_{i=1}^{n} a_{ki} b_{ij}$$

Hence $$\phi (T. T') = \phi (T) \phi (T')$$

This proves the corollary.

Theorem 10.7.2: *Let V be a vector space over F with basis $v_1, v_2, \ldots, v_n$ and let $T \in L(V, V)$. If m (T) is the matrix of T with respect to the basis $v_1, v_2, \ldots v_n$ and $v_1, v_2, \ldots, v_n$, then*

$$m (T_1 o T_2) = m (T_2)\, m (T_1) \text{ for all } T_1, T_2 \in L (V_1 V).$$

Proof: Since $T_1: V \to V$ is a linear map, then

$$T_1 (V_i) = \sum_{j=1}^{n} a_{ij} v_j,$$

So that $m (T_1) = [a_{ij}]_{n \times n}$

Again, since $T_2 : V \to V$ is a linear map, then

$$T_2 (V_i) = \sum_{j=1}^{n} b_{ij} v_j,$$

So that $$m (T_2) = [b_{ij}]_{n \times n}.$$

Now $$(T_1 o T_2) (v_i) = T_1 (T_2 (v_i)) = T_1\left(\sum_{j=1}^{n} b_{ij} v_j\right)$$

$$= \sum_{j=1}^{n} b_{ij} T_1(v_j)$$

$$= \sum_{j=1}^{n} b_{ij} \sum_{k=1}^{n} a_{jk} v_k$$

$$= \sum_{j,k=1}^{n} b_{ij} a_{jk} v_k$$

$$= \sum_{k=1}^{n} c_{ik} v_k,$$

where $$c_{ik} = \sum_{j=1}^{n} b_{ij} a_{jk}$$

CHAPTER 10

Hence $m(T_1 \, o \, T_2) = [b_{jk}] \, o \, [a_{jk}] = [b_{ij}] \, [a_{ij}] = m(T_2)$ and (T_1).

Corollary: T is an isomorphism if and only if $m(T)$ is non-singular.

Proof: If T is an isomorphism, then there exists another isomorphism T' such that $T \, o \, T' = I = T' \, o \, T$.

Hence $m(T') \, m(T) = I_n = m(T) \, m(T')$

$\Rightarrow m(T)$ is non-singular with $m(T)^{-1} = m(T')$

or
$$m(T)^{-1} = m(T^{-1}).$$

Conversely, If $m(T)$ is non-singular, then there exists an $n \times n$ matrix A such that
$$A \, . \, m(T) = m(T) \, . \, A = I_n.$$

If T_A is the linear map associated to A, then
$$m(T_A) = A.$$

So that
$$m(ToT_A) = m(T_A) \, . \, m(T) = A \, . \, m(T) = I_n$$

Since $T \to m(T)$ is a one-to-one, $T \, o \, T_A = I$.

Similarly, $T_A \, o \, T = I$ which show that T is an isomorphism.

The following theorem shows how the matrices of T with respect to different bases are related to each other.

Theorem 10.7.3: *Let V be a vector space over the field F with basis $v_1, v_2, \ldots, v_n$ and let $T : V \to V$ be a linear mapping. Let A be the matrix of T with respect to the basis $v_1, v_2, \ldots, v_n$ and let B be the matrix of T with respect to another basis $u_1, u_2, \ldots, u_n$. Then there exists a non-singular matrix C such that $C \, A \, C^{-1} = B$.*

Proof: Let $T(v_i) = \sum_{j=1}^{n} a_{ij} v_j$, Then $A = [a_{ij}]$

and let $T(u_i) = \sum_{j=1}^{n} b_{ij} u_j$, Then $B = [b_{ij}]$.

Let $u_i = \sum_{j=1}^{n} c_{ij} v_j$ and Let $C = [c_{ij}]$.

Then C is non-singular matrix, for let $v_i = \sum_{j=1}^{n} d_{ij} u_j$ and $D = [d_{ij}]$.

Then
$$v_i = \Sigma \, d_{ij} \, u_j = \sum_{j=1}^{n} d_{ij} \sum_{k=1}^{n} c_{jk} v_k$$
$$= \sum_k \left(\sum_j d_{ij} c_{jk} \right) v_k$$

Since $v_1, \ldots\ldots, v_n$ are linearly independent, we have
$$\sum_{j=1}^{n} d_{ij} \, c_{jk} = 1 \text{ if } i = k$$
$$= 0 \text{ if } i \neq k, \text{ i.e, } DC = I_n.$$

Similarly, substituting for v_j in the expression for u_j and using the fact that $u_1, u_2 \ldots.,$ u_n are L.I, we have

$CD = I_n$. Thus C is non-singular.

Now $$T(u_i) = \sum_{j=1}^{n} b_{ij}\, u_j = \sum_{j=1}^{n} b_{ij} \sum_{k=1}^{n} c_{jk}\, v_k$$

$$= \sum_{k=1}^{n}\left(\sum_{j=1}^{n} b_{ij}\, c_{jk}\right) v_k$$

But $$T(u_i) = T\left(\sum_{j=1}^{n} c_{ij}\, v_j\right)$$

$$= \sum_{j=1}^{n} c_{ij}\, T(v_j)$$

$$= \sum_{j=1}^{n} c_{ij} \sum_{k} a_{jk} v_k$$

$$= \sum_{k=1}^{n}\left(\sum_{j=1}^{n} c_{ij}\, a_{jk}\right) v_k$$

Since $v_1, v_2, \ldots, v_n$ are *L.I.*, we have by comparing the coefficient of v_k,

$$\sum_{j=1}^{n} b_{ij}\, c_{jk} = \sum_{j=1}^{n} c_{ij}\, a_{jk} \text{ for all } i, k. \text{ Thus}$$

$$BC = CA \Rightarrow B = CAC^{-1}.$$

Example 10.7.7: Let V be the vector space of all polynomials over F of degree 3 or less. Let $T : V \to V$ be defined by $T(f) = f'$.

Let $\{1, x, x^2, x^3\}$ be a bases of V.

In this case $v_1 = 1$, $v_2 = x$, $v_3 = x^2$, $v_4 = x^4$. Then by Example 10.7.1 the matrix of T is

$$m_1(T) = \begin{bmatrix} 0 & 0 & 0 & 0 \\ 1 & 0 & 0 & 0 \\ 0 & 2 & 0 & 0 \\ 0 & 0 & 3 & 0 \end{bmatrix}.$$

In the basis $u_1 = 1$, $u_2 = 1 + x$, $u_3 = 1 + x^2$, $u_4 = 1 + x^3$, the matrix of T is

$$m_2(T) = \begin{bmatrix} 0 & 0 & 0 & 0 \\ 1 & 0 & 0 & 0 \\ -2 & 2 & 0 & 0 \\ -3 & 0 & 3 & 0 \end{bmatrix}.$$

Let S be the linear transformation of V defined by

$S(v_1) = w_1 (= v_1)$, $S(v_2) = w_2 = 1 + x = v_1 + v_2$, $S(v_3) = w_3 = 1 + x^2$ $= v_1 + v_3$ and $S(v_4) = w_4 = 1 + x^3 = v_1 + v_4$

The matrix of S in the basis v_1, v_2, v_3, v_4 is

CHAPTER 10

$$C = \begin{bmatrix} 1 & 0 & 0 & 0 \\ 1 & 1 & 0 & 0 \\ 1 & 0 & 1 & 0 \\ 1 & 0 & 0 & 1 \end{bmatrix}$$

By simple computation we can see that

$$C^{-1} = \begin{bmatrix} 1 & 0 & 0 & 0 \\ -1 & 1 & 0 & 0 \\ -1 & 0 & 1 & 0 \\ -1 & 0 & 0 & 1 \end{bmatrix}, \text{ Then}$$

$$C\, m_1\,(T)\, C^{-1} = \begin{bmatrix} 1 & 0 & 0 & 0 \\ 1 & 1 & 0 & 0 \\ 1 & 0 & 1 & 0 \\ 0 & 0 & 0 & 1 \end{bmatrix} \begin{bmatrix} 0 & 0 & 0 & 0 \\ 1 & 0 & 0 & 0 \\ 0 & 2 & 0 & 0 \\ 0 & 0 & 3 & 0 \end{bmatrix} \begin{bmatrix} 1 & 0 & 0 & 0 \\ -1 & 1 & 0 & 0 \\ -1 & 0 & 1 & 0 \\ -1 & 0 & 0 & 1 \end{bmatrix}$$

$$= \begin{bmatrix} 0 & 0 & 0 & 0 \\ 1 & 0 & 0 & 0 \\ -2 & 2 & 0 & 0 \\ -3 & 0 & 3 & 0 \end{bmatrix} = m_2\,(T).$$

Example 10.7.8: Verify $m\,(T)\,[v] = [T\,(v)]$ in the problem of example 10.7.4 if $v = (2, 1, 0)$.

Solution: In the example 10.7.4 the matrix of T is

$$m\,(T) = \begin{bmatrix} -7 & -33 & -13 \\ 4 & 19 & 8 \end{bmatrix}$$

Given vector (2, 1, 0) can be written as linear combination of basis elements;
$v = (2, 1, 0) = a\,(1, 1, 1) + b\,(1, 1, 0) + c\,(1, 0, 0).$
$\Rightarrow 2 = a + b + c,\ 1 = a + b,\ 0 = a$
$\Rightarrow a = 0,\ b = 1,\ c = 1.$
$\therefore$ (2, 1, 0) = 0 (1, 1, 1) + 1 (1, 1, 0) + (1, 0, 0).

$$\therefore\ [v] = \begin{bmatrix} 0 \\ 1 \\ 1 \end{bmatrix} \text{ and } m\,(T)\,[v] = \begin{bmatrix} -7 & -33 & -13 \\ 4 & 19 & 8 \end{bmatrix} \begin{bmatrix} 0 \\ 1 \\ 1 \end{bmatrix} = \begin{bmatrix} -46 \\ 27 \end{bmatrix}$$

and $T\,(v) = T\,(2, 1, 0) = (8, -3) = l\,(1, 3) + m\,(2, 5)$
$\Rightarrow 8 = l + 2\,m$ and $3\,l + 5\,m = -3.$
$\Rightarrow m = 27,\ l = -46.$

$$\therefore \qquad [T\,(v)] = \begin{bmatrix} -46 \\ 27 \end{bmatrix}.$$

Hence $\qquad m\,(T)\,[v] = [T]\,[v] = [T\,(v)].$

PROBLEMS

1. Find the matrix representation of each of the following linear maps relative to usual bases $(e_1, e_2, \ldots, e_n)$ of R^n.
   (i) $T : R^2 \to R^3$ defined by $T(x, y) = (3x - y, 2x + 4y, 5x - 6y)$.
   (ii) $T : R^4 \to R^2$ defined by $T(x, y, z, t) = (3x - 4y + 2z - 5t, 5x + 7y - z - 2t)$.
   (iii) $T : R^3 \to R^4$ defined by $T(x, y, z) = (2x + 3y - 8z, x + y + z, 4x - 5z, 6y)$.

$$\left[\textbf{Ans. } (i) \begin{bmatrix} 3 & -1 \\ 2 & 4 \\ 5 & 6 \end{bmatrix}, (ii) \begin{bmatrix} 3 & -4 & 2 & -5 \\ 5 & 7 & -1 & -2 \end{bmatrix}, (iii) \begin{bmatrix} 2 & 3 & -8 \\ 1 & 1 & 1 \\ 4 & 0 & -5 \\ 0 & 6 & 0 \end{bmatrix}\right]$$

2. If $T : R^2 \to R^2$ defined by $T(x, y) = (2x - 3y, x + 4y)$, then find the matrix of T in the bases $e_1 = (1, 0)$, $e_2 = (0, 1)$, and $v_1 = (1, 3)$, $v_2 = (2, 5)$ respectively.

$$\left[\textbf{Ans. } \begin{bmatrix} -8 & 2 & 3 \\ 5 & -1 & 3 \end{bmatrix}\right]$$

3. If $f(v) = Av$, where v in Av is written as a column vector white A is the matrix $A = \begin{bmatrix} 1 & 2 \\ 3 & 4 \end{bmatrix}$,
   (i) Show that the matrix $m(f)$ of f is A itself relative to the usual bases of R^2.
   (ii) Find the matrix $m(f)$ of relative to the bases $v_1 = (1, 3)$, $v_2 = (2, 5)$ in R^2.

$$\left[\textbf{Ans. } m(f) = \begin{bmatrix} -5 & -8 \\ 6 & 10 \end{bmatrix}\right]$$

4. Find the matrix representation of each of the following linear mapping T relative to basis $v_1 = (1, 3)$, $v_2 = (2, 5)$,
   (i) $T(x, y) = (2y, 3x - y)$
   (ii) $T(x, y) = (3x - 4y, x + 5y)$.

$$\left[\textbf{Ans. } (i) \begin{bmatrix} -30 & -48 \\ 18 & 29 \end{bmatrix}, (ii) \begin{bmatrix} 77 & 124 \\ -43 & -69 \end{bmatrix}\right]$$

5. Let there be a linear mapping $T : R^2 \to R^2$ whose matrix relative to the standard basis $(1, 0), (0, 1)$ is (i) $\begin{bmatrix} 1 & 1 \\ 1 & 1 \end{bmatrix}$, (ii) $\begin{bmatrix} 2 & -3 \\ 1 & 1 \end{bmatrix}$.

   Find T and the matrix $m(T)$ of T relative to the basis $B = \{(1, 1), (1, -1)\}$ in each case.

$$\left[\textbf{Ans. } (i)\ T(x, y) = (x + y, x + y), (ii)\ T(x, y) = (2x - 3y, x + y), \begin{bmatrix} \frac{1}{2} & \frac{5}{2} \\ -\frac{1}{2} & \frac{5}{2} \end{bmatrix} (i) \begin{bmatrix} 2 & 0 \\ 0 & 0 \end{bmatrix}\right]$$

6. Let there be a linear mapping $T : R^3 \to R^3$ relative to the standard basis $e_1 = (1, 0, 0)$, $e_2 = (0, 1, 0)$, $e_3 = (0, 0, 1)$ such that

$$m(T) = \begin{bmatrix} 0 & 1 & 1 \\ 1 & 0 & -1 \\ -1 & -1 & 0 \end{bmatrix} \text{ with respect to given basis.}$$

CHAPTER 10

Find T and the matrix $m\ (T)$ of T with respect to the basis $v_1 = (0, 1, -1)$, $v_2 = (1, -1, 1)$, $v_3 = (-1, 1, 0)$.

$$\left[\textbf{Ans. } T\ (x, y, z) = (y + z, x - z, -x - y),\ m\ (T) = \begin{bmatrix} 1 & 0 & 0 \\ 0 & 0 & 0 \\ 0 & 0 & -1 \end{bmatrix} \right].$$

7. Let there be two mappings $f : U \to V'$ and $g: V \to V'$ defined by $f\ (x, y, z) = (3x + 2y - 4z, x - 5y + 3z)$; and $g\ (a, b, c) = (2a + 5b - 3c, a - 4b + 7c)$ where basis of V and V' are respectively

   $v_1 = (1, 1, 1)$, $v_2 = (1, 1, 0)$, $v_3 = (1, 0, 0)$ and $v'_1 = (1, 3)$, $v'_2 = (2, 5)$.

   Verify $m\ (f) + m\ (g) = m\ (f + g)$ and $m\ (xf) = km\ (f)$.

8. Let there be two linear mappings $f : V \to V'$ and $g: V'' \to V'$ defined by $f\ (x, y, z) = (2x - 4y + 9z, 5x + 3y - 2z)$ and $g\ (a, b) = (3a + 4b, 5a - 2b, a + 7b, 4a)$, So that V, V′, V″ are of dimensions 3, 2, 4 respectively. Let there bases be $v_1 = (1, 0, 0)$, $v_2 = (1, 1, 0)$, $v_3 = (1, 1, 1)$; $v'_1 = (1, -1)$, $v'_2 = (0, 1)$;

   $v''_1 = (0, 0, 1, 0)$, $v''_2 = (-1, 0, 0, 1)$, $v''_3 = (1, 1, 1, -1)$, $v''_4 = (0, 1, 2, 1)$

   verify $m\ (g\ o\ f) = m\ (f)\ m\ (g)$.

CHAPTER 11 FIELD THEORY

Field as a system with two binary operations defined on a non-empty set having at least two elements has been done in the chapter of Ring Theory. That is, Field is a commutative ring with identity whose non-zero elements have their multiplicative inverse. In this chapter we shall make use of it in theory of equations. Generally the aspects of field theory has impact on the theory of equations. In the present section we shall see the relation between two fields.

11.1 EXTENSION FIELDS

Definition 11.1.1: If $(F, +, .)$ is a subfield of the field $(K, +, .)$, then the field K is said to be an extension of F That is, K is an extension of F if $(F, +, .)$ is a subfield of the field $(K, +, .)$.

Throughout this chapter $(F, +, .)$ will denote a given field and $(K, +, .)$ an extension of F.

Example 11.1.1: The field $(R, +, .)$ of real numbers is a subfield of the field $(C, +, .)$ of complex numbers. So $(C, +, .)$ is an extension of R.

Example 11.1.2: The field $(Q, +, .)$ of rational numbers is a subfield of the field $(R, +, .)$ of real numbers. So $(R, +, .)$ is an extension of Q.

Example 11.1.3: The field $(\{a + b\sqrt{2} \mid a, b \in Q\}, +, .\}$ contains the field $(Q, +, .)$ of rational numbers. So $(\{a + b)\sqrt{2} \mid a, b \in Q|, +, .\}$ is an extension of Q.

Let $(K, +, .)$ be an extension of F. That is, $(F, +, .)$ is a subfield of the field $(K, +, .)$. Now we observe that $K(F)$ is a vector space over the field $(F, +, .)$. If we consider $K(F)$ as a vector space over the field $(F, +, .)$. then we can determine the subset of *linearly independent vectors* (elements) of K which will form the *basis* of the vector space $K(F)$. So we can also find out the *dimension* of the vector space $K(F)$.

Definition 11.1.2: The dimension of the vector space $K(F)$ over the field $(F, +, .)$ is called the *degree* of K over F and is denoted by $[K : F]$.

Definition 11.1.3: Let K be an extension of F, then $K(F)$ is the vector space over the field $(F, +, .)$. If the dimension of the vector space $K(F)$ is finite, then K is a *finite extension* of F.

Example 11.1.4: (*i*) In example 11.1.1 C is an extension of R. So $C(R)$ is the vector space over the field $(R, +, .)$ of real numbers. The vector space is generated by the set $\{1, i\}$ which is linearly independent. Hence $\{1, i\}$ is a basis of $C(R)$ of two elements. Therefore $[C : R] = 2$.

(ii) In example *11.1.2. R* is an extension of *Q*. So *R (Q)* is the vector space over the field (*Q*, +, .) of rational numbers. The set {1} generates the vector space *R (Q)* which is obviously *a L.I.* Set. Thus the set {1} is a basis of *R (Q)*. Hence [*R* : *Q*] = 1.

(iii) In example 11.1.3 $\{a + b\sqrt{2} \mid a, b \in Q\}$ is an extension of *Q*. So $\{a + b\sqrt{2} \mid a, b \in Q\}$ *(Q)* is a vector space over the field (*Q*, +, .) with the basis $\{1, \sqrt{2}\}$ Hence $\{a + b\sqrt{2}\}$: *Q* = 2.

Theorem 11.1.1: If *L is a finite extension of K and* if *K is a finite extension of F, then L is a finite extension of F. Moreover,* [*L* : *F*] = [*L* : *K*] [*K* : *F*].

Proof: Since *L* is a finite extension of *K* and *K* is a finite extension of *F*, then [*L* : *K*] and [*K* : *F*] are finite numbers. Let [*L* : *K*] = *m* and [*K* : *F*] = *n*. Then the vector space *L (K)* has a basis of *m* elements and the vector space *K (F)* has a basis of *n* element.

Let $\{v_1, v_2,..., v_m\}$ be a basis of *L* (*K*) and let $\{w_1, w_2,..., w_n]$ be a basis of *K* (*F*). Then we have to prove that the basis of *L* (*F*) has *m*. *n* elements. For this, we must show that the elements $v_i\ w_j$, where *i* = 1, 2,..., *m*, *j* = 1, 2,..., *n*, form a basis of *L (F)* over the field (*F*, +, .). That is, every element of *L* is a linear combination of the elements $v_i\ w_j$, where *i* = 1, 2,..., *m*, *j* = 1, 2,...,*n*, with the coefficients in *F* and these *m* . *n* elements $v_i\ w_j$ are linearly independent over *F*.

Let *t* be any element of *L*. Since *L* (*K*) is the vector space over the field (*K*, +, .) with the basis $\{v_1, v_2,..., v_m\}$ then *t* can be written as the linear combination of $v_1, v_2,..., v_m$. Thus $t = k_1\ v_1 + k_2\ v_2 +....+ k_m\ v_m$, where $k_i \in K$,

i = 1, 2,.., *m*. that is, $t = \sum_{i=1}^{m} k_i v_i$

Again, since *K (F)* is the vector space over the field (*F*, +, .) with the basis $(w_1, w_2,..., w_n\}$, then k_i, *i* = 1, 2,..., *m*, can be expressed as the linear combination of $w_1, w_2,..., w_n$.

Thus
$$k_i = a_{ii}\ w_1 + a_{i2}\ w_2 +....+ a_{in}\ w_n$$

or
$$k_i = \sum_{j=1}^{n} a_{ij} w_j, \text{ where } a_{ij} \in F \text{ for all } i \text{ and } j.$$

Hence
$$t = \sum_{i=1}^{m}\left(\sum_{j=1}^{n} a_{ij}\ w_j\right) v_i$$

$$= \sum_{i=1}^{m}\sum_{j=1}^{n} a_{ij}\, w_j\ v_i., \forall a_{ij} \in F.$$

Thus we have that $t \in L$ is a linear combination of $w_j\ v_i$ for all *i*, *j* over the field *F*.

Now to prove the theorem it remains to show that the *mn* elements $v_i\ w_j$ are linearly independent over *F*. For this, suppose that

$$a_{11}\ v_1\ w_1 + a_{12}\ v_1\ w_2 +....+ a_{1n}\ v_1\ w_n +....+$$
$$a_{m1}\ v_m\ w_1 + a_{m2}\ v_m\ w_2 +..... + a_{mn}\ v_m\ w_n = 0$$

or
$$(a_{11}\ w_1 + a_{12}\ w_2 +.....+ a_{1n}\ w_n)\ v_1 +...+$$
$$(a_{m1}\ w_1 + a_{m2}\ w_2 +....+ a_{mn}\ w_n)\ v_n = 0$$

Since $v_1, v_2,... v_n$ are linearly independent over K, it follows

$a_{11} w_1 + a_{12} w_2 +....+ a_{1n} w_n = 0, ...$

$$\begin{matrix} . & . & . \\ . & . & . \\ . & . & . \end{matrix}$$

$a_{m1} w_1 + a_{m2} w_2 +....+ a_{mn} w_n = 0$

That is,

$a_{i1} w_1 + a_{i2} w_2 +....+ a_{in} w_n = 0, i = 1, 2,....m.$

Since $w_1, w_2,..... w_n$ are linearly independent over F, it follows

$a_{i1} = a_{i2} =....= a_{in} = 0, i = 1, 2,...m.$

That is, all $a_{ij} = 0, i = 1, 2,...m, j = 1, 2,....n.$

Hence $v_i\ w_j$ are linearly independent.

This shows that these mn elements $v_i\ w_j$ form a basis for the vector space $L\ (F)$ over the field $(F, +, .)$.

Thus $(L : F) = mn$; since $m = [L : K]$ and $n = [K : F]$, hence $[L : F] = [L : K] . [K : F]$.

Corollary: If L is a finite extension of F and $(K, +, .)$ is a subfield of $(L, +, .)$ which contains F, then $[K : F] \mid [L : F]$.

Proof: Let $(F, +, .)$ be a subfield of the field $(K, +,.)$ and let $(K, +, .)$ be a subfield of the field $(L, +, .)$. Let $[L : F]$ be finite. It is clear that any elements of L which are linearly independent over K are also linearly independent over F. Therefore assumption $[L : F]$ is finite implies $[L : K]$ is finite. Also $K\ (F)$ is a sub-space of $L\ (F)$. So $[K : F]$ is finite. By the theorem 11.1.1 we have $[L : F] = [L : K]\ [K : F]$ which implies $[K : F] \mid [L : F]$.

Let S be a subset of the field $(K, +, .)$. Then the inter section of all subfields of the field $(K, +, .)$ which contain S is known as the subfield generated by S. It is denoted by $[S]$.

Definition 11.1.4: Now Let K be a field extension of F and let S be any subset of K, then the intersection of all subfields of the field $(K, +, .)$ which contain both F and S is called the subfield of $(K, +, .)$ *generated by S over F* and this field is denoted by $F\ (S)$. However if S is a finite set having n elements $a_1, a_2..... a_n$, then we write $F\ (S) = F\ (a_1, a_2,...., a_n)$, or the field obtained by adjoining the elements $a_1, a_2,...., a_n$ to F.

Definition 11.1.5: A field $(K, +, .)$ is said to be *finitely generated over F* if there exists a finite number of elements $a_1, a_2..... a_n$ in K such that

$K = F\ (a_1, a_2,...., a_n)$.

Definition 11.1.6: A field $(K, +, .)$ is said to be *simple extension* of F if $K = F\ (a), a \in K$. That is, the field $(K, +, .)$ is obtained from F by the adjunction of a single element a.

Let us remark on $F\ (a)$. Let $a \in K$ and let $(F, +, .)$ be a subfield of the field $(K, +, .)$. Then the expression of the form $b_0 + b_1, a +.....+ b_m a^m$ belongs to K, where the $b_0, b_1,..... b_m$ are in F and m is a any non negative integer. Since $(K, +, .)$ is a field, one such element can be divided by another, provided that another such elements which divides the former is not zero. If U is the collection of all such quotients, $U \subset K$ and $(U, +, .)$ is a subfield of the field $(K, +, .)$

If $(F, +, .)$ is a field, then we have the ring $(F\ [x], +, .)$ of polynomials over the field $(F, +, .)$. If $f(x) \in F\ [x]$, then $f\ (x) = a_0 + a_1 x + a_2 x^2 +....+ a_n x^n$ and for any $d \in K, f(d)$ denotes the expression $a_0 + a_1 d + a_2 d^2 +....+ a_n d^n$ which is the element of K . $f\ (d)$ is the value of the polynomial $a_0 + a_1 x + a_2 x^2 +......+ a_n x^n$. It can also be shown that the mapping $f\ (x) \rightarrow f(a)$ is a ring homomorphism of $(F\ [x], +,.)$ into $(K, +, .)$.

We know that an element $a \in K$ is a root of the polynomial $f(x) \in F[x]$ if $f(a) = 0$. In this case we also say that $f(x)$ is satisfied by a. If in the expression of the polynomial $f(x) = a_0 + a_1 x + ... + a_n x^n$, $a_n \neq 0$, $f(x)$ *is* said to be of *degree* n and if $a_n = 1$, the identity of the field $(F, +, .)$, $f(x)$ is said to be *monic polynomial*.

Definition 11.1.7: If $K \supset F$, then an element $c \in K$ is said to be *algebraic over F* if $f(c) = 0$, for some non-zero polynomial $f(x) \in F[x]$. That is, c is the root *of* $f(x) \in F[x]$. An element $c \in K$ is *transcedental over F* it is not algebraic over F. K is said to be an *algebraic extension of F* if every element of K is algebraic over F. That is, every element of K is a root of some non-zero polynomial $f(x) \in F[x]$.

The extensions which are not *algebraic are* called *transcendental extensions.*

Theorem 11.1.2: *An element u which is algebraic over a field F is a root of unique monic irreducible polynomial* $f(x) \in F(x)$. *If* $g(x) \in F[x]$ *is another polynomial with* $g(u) = 0$, *then* $p(x) | g(x)$.

Proof: Since u *is* algebraic over the field F, then there exists a non-zero polynomial $f(x) \in F[x]$ such that $f(u) = 0$.

If $f(x)$ *is* not irreducible, then by unique factorization theorem $f(x) = c\, p_1(x) \,.\, p_2(x) ...... p_m(x)$, where $c \in F$ and $p_1(x), p_2(x), ...p_m(x)$ are all monic irreducible polynomials in $F[x]$. Since $f(u) = 0$, then there exists at least one factor $p_i(x)$ for which $p_i(u) = 0$. thus we have found a monic irreducible polynomial $p(x)$ in $F[x]$ with $p(u) = 0$. Let deg $p(x) = n$.

Now we shall show that u is not a root of any polynomial $g(x)$ of degree less than n. That is, there does not exist any polynomial $g(x) \in F[x]$ with deg $g(x) < n$ such that $g(u) = 0$. If there exists $g(x)$ with deg $g(x) < n$ and $g(u) = 0$, then $p(x)$ and $g(x)$ have the greatest common diviser 1 since $p(x)$ is irreducible. So $1 = t(x)\, p(x) + s(x) \,.\, g(x)$, where $t(x), s(x) \in F[x]$. On substituting $x = u$, we get

$1 = t(u)\, p(u) + s(u) \,.\, g(u). = 0$, since $p(u) = 0 = g(u)$. which is never possible. Hence $g(u) = 0$ holds only for a polynomial $g(x) \neq 0$ if the degree of $g(x)$ is at least $n = \deg p(x)$. Thus the monic irreducible polynomial $p(x)$ with $p(u) = 0$ is of the lowest degree.

Furthermore, we shall show that the monic irreducible polymial $p(x)$ is unique.

Let $p(x)$ and $p_1(x)$ be two monic irreducible polynomials with $p(u) = 0$, $p_1, (u) = 0$, and deg $p(x) = n =$ deg $p_1(x)$.

Then

$p(x) - p_1(x) = r(x) \in F[x]$, where the degree of $r(x)$ will be less than n, since $p(x)$ and $p_1(x)$ are monic polynomials. For $x = u$, we get

$p(u) - p_1(u) = r(u) = 0$, as $p(u) = 0 = p_1(u)$. Thus we have a polynomial $r(x)$ of degree less than n with $r(u) = 0$ which is not possible because $p(x)$ is of lowest degree n with $p(u) = 0$. Hence $r(x) = 0$ which implies $p(x) = p_1(x)$.

Now we consider $g(x) \in F[x]$ for which $g(u) = 0$. On dividing $g(x)$ by $p(x)$ have

$g(x) = q(x) \,.\, p(x) + r(x)$, where $r(x) = 0$ or deg $r(x) <$ deg $p(x)$

For $x = u$, we obtain

$g(u) = q(u) \,.\, p(u) + r(u)$ which means

$r(u) = 0$, as $g(u) = 0 = p(u)$.

But we cannot have a polynomial $r(x)$ of degree less than n with $r(u) = 0$, since $p(x)$ is of lowest degree with $p(u) = 0$. Hence $r(x) = 0$ which implies

$g(x) = q(x) \cdot p(x)$

or $p(x) \mid g(x)$.

This completes the proof of the theorem.

Definition 11.1.8: The unique monic irreducible polynomial $p(x)$ of which u is a root will be called the *minimal polynomial* of u over F, while u is said to be of *degree n over F*, $n = \deg p(x)$, denoted by $n = [u : F]$.

Definition 11.1.9. The element $a \in K$ is said to be the *algebraic of degree n* over F if it satisfies a non-zero monic *irreducible polynomial p (x) of lowest degree n.*

Theorem 11.1.3: *An element $a \in K$ is algebriac over F if and only if $[F(a) : F]$ is finite.*

Proof: Let $[F(a) : F]$ be finite, say, n, a positive integer. Then $(n + 1)$ elements of $F(a)$ are linearly dependent over F. If we consider $(n + 1)$ elements $1, a, a^2, \ldots, a^{n-1}, a^n$ of $F(a)$ which are linearly dependent, then there are $d_0, d_1, \ldots d_{n-1}, d_n \in F$, not all zero, such that

$d_0 . 1 + d_1 . a + d_2 . a^2 + \ldots + d_{n-1} a^{n-1} + d_n a^n = 0.$

This shows a satisfies a polynomial

$f(x) = d_0 + d_1 x + \ldots + d_{n-1} x^{n-1} + d_n x^n \in F[x]$. That is, a is algebraic over F.

Conversely, let a be algebraic over F and let $p(x)$ be the minimal polynomial of a in $F[x]$. Let $n = \deg p(x)$. Now we shall consider the integral domain $F[a]$. Let

$f(x) = d_0 + d_1 x + \ldots + d_n x^n \in F[x]$ or

$0 \neq f(a) = d_0 + d_1 a + \ldots + d_n a^n \in F[a]$, then $p(x)$ does not divide $f(x)$.

Therefore the greatest common divisor of $f(x)$ and $p(x)$ is I and there exist two polynomials $t(x)$ and $s(x)$ in $F[x]$ such that

$1 = t(x) . f(x) + s(x) . p(x).$

For $x = a$,

$1 = t(a) . f(a) + s(a) . p(a) = t(a) f(a)$ as $p(a) = 0$.

Hence

$(f(a))^{-1} = t(a) \in F[a].$

Since $F(a)$ is a field of quotients of elements of $F[a]$, then $F(a) = F[a]$.

Now we shall show that $F(a) = F[a]$ is a vector space over F. Let us consider the elements $1, a, a^2, \ldots, a^{n-1} \in F(a)$. If for some $d_0, d_1, \ldots, d_{n-1} \in F$, not all zero, there holds a relation $d_0 . 1 + d_1 a + \ldots + d_{n-1} a^{n-1} = 0$, then a satisfies a polynomials of degree less than $n = \deg p(x)$ which is not possible since $p(x)$ is the minimal polynomial of lowest degree satisfied by a, Hence the relation

$d_0 . 1 + d_1 + \ldots + d_{n-1} a^{n-1} = 0$ holds if and only if

$d_0 = d_1 = \ldots = d_{n-1} = 0$. Hence the set $\{1, a, a^2, \ldots a^{n-1}\}$ is linearly independent over F.

Now there remains to show that every element of $F(a)$ can be expressed as a linear combination of $1, a, a^2, \ldots a^{n-1}$. If $f(a) \in F(a)$, then we can find the remainder $r(x)$ on dividing $f(x)$ by $p(x)$, as

$f(x) = q(x) p(x) + r(x)$, $\deg r(x) < \deg p(x)$.

Since $p(a) = 0$, the substitution $x = a$ gives

$f(a) = r(a)$, we may have

$r(x) = b_0 + b_1 x + \ldots + b_{n-1} x^{n-1} \in F[x]$, as $\deg r(x) < \deg p(x)$. Thus every $f(a) \in F(a)$ can be expressed as a linear combination of $1, a, a^2, \ldots, a^{n-1}$. Hence the set $\{1, a, a^2, \ldots, a^{n-1}\}$ is a basis of the vecter space $F(a)$ over the field F. Which contain n elements.

Hence $[F(a): F] = n$. That is, if $a \in K$ is algebraic over F then $[F(a) : F] = n$.

Theorem 11.1.4: *Every finite extension of a field F is algebraic extension.*

Proof: Let K be a finite extension of the field F and Let $[K : F] = n$. If $a \in K$, then $F(a)$ is a subfield of the field K. Thus $F \subset F(a) \subset K$. So by the theorem 11.1.1 $n = [K : F] = [K : F(a)] \, . \, [F(a): F]$ which implies $[F(a): F] \mid n$, that is, $[F(a): F]$ is finite. Therefore by the theorem 11.1.3 a is algebraic over F which proves K is algebraic extension of F.

Let K be a finite extension of the field F. that is, K is an algebraic extension of F. Now we consider a field F_1 such that $F \subseteq F_1 \subseteq K$. Since K is algebraic over F, then $a \in K$ will satisfy the minimal polynomial $p(x) \in F[x]$. Since $F \subset F_1$, then the coefficients of $p(x)$ also belong to F_1, that is, $p(x) \in F_1[x]$ such that $p(a) = 0$. If $p_1(x)$ is the minimal polynomial of a over F_1, then $p_1(a) = 0$ and $p_1(x)$ divided $p(x)$. Thus we get $p(x) = q(x). p_1(x)$, $q(x) \in F_1[x]$, which implies $\deg p_1(x) \leq \deg p(x)$. But we know $[F_1(a): F] = \deg p_1(x)$ and $[F(a): F] = \deg p(x)$. Therefore $[F_1(a): F] \leq [F(a) : F]$.

If a is algebraic over F, then there exists a finite extension $F(a)$ of F such that $[F(a): F] = m$ = degree of the minimal polynomial of a over F. Again if b is algebraic over F, it is also algebraic over $F(a)$ since $F \subseteq F(a)$. So there exists a finite extension $(F(a))(b)$ of $F(a)$ since $F(a) \subseteq (F(a))(b)$ which is denoted by $F(a, b)$. Similarly, we can establish the result that if $a_1, a_2, \ldots, a_n$ are in K which are algebraic over F, then there exists a finite extension $F(a_1, a_2, \ldots, a_n)$ in K such that

$$F \subseteq F(a_1) \subseteq F(a_1, a_2) \subseteq \ldots \subseteq F(a_1, a_2, \ldots, a_n).$$

Now we shall prove one more important theorem.

Theorem 11.1.5: *If L is an algebraic extension of K and K is algebriac extension of F, then L is an algebraic extension of F.*

Proof: $u \in L$ is algebraic over K if and only if $[K(u) : K] = n$, a finite positive integer. Thus there exists a minimal polynomial $f(x) = x^n + a_1 x^{n-1} + \ldots + a_n \in K[x]$. But K is algebraic over F, and $a_1, a_2, \ldots, a_n$ are algebraic over F. Thus we get finite extension $F(a_1, a_2, \ldots, a_n) = M$ of F. Since u satisfies the polynomial $x^n + a_1 x^{n-1} + \ldots + a_n$, where $a_1, a_2, \ldots a_n \in$ M, u is algebraic over M. That is, there exists a finite extension $M(u)$ of M. But we know that if $F \subseteq M(u)$ then

$[M(u) : F] = [M(u): M] \, [M : F]$ = finite, because $M(u)$ and M are finite extensions of M and F respectively. Which implies $M(u)$ is finite extension of F and consequently u is algebraic over F. This completes the proof of the theorem.

Theorem 11.1.6: If *$a, b \in K$ are algebraic over F. That $a \pm b$, $a b$, and a/b (if $b \neq 0$) are all algebraic over F. That is, the elements of K which are algebraic over F form a subfield of K.*

Proof: Let $a \in K$ be algebraic of degree m over F and let $b \in K$ be algebraic of degree n over F. Then there exists a finite extension $F(a)$ of F such that $[F(a): F] = m$. Since $F \subseteq F(a) = T$, then if b is algebraic over F it is also algebraic over $T = F(a)$ of degree n. Thus the field $W = T(b)$ is a finite extension of T containing F and $W \subseteq K$ with $[T(b): T] = n$.

But we know that if $F \subseteq T \subseteq W$, then

$[W : F] = [W: T] \, [T : F]$ which implies

$[W : F] \leq n \,.\, m$. Consequently W is a finite extension of F which means W is an algebraic extension of F. We have also seen that $W = T(b)$ is a subfield of K. Therefore, if $a, b \in W$, then $a \pm b$, $a \,.\, b$, a/b, $b \neq 0$ all belong to W which are algebraic over F. Hence the theorem

Corollary: If a and b are algebraic over F of degrees m and n respectively, then $a \pm b$, $a\,b$, and a/b (if $b \neq 0$) are algebraic over F of degree at most $m \,.\, n$.

Proof: In the theorem 11.1.6 we have proved that if $a, b \in K$ are algebraic over F of degree m and n then $a \pm b$, $a\,b$, a/b are algebraic over F and the set of all elements which are algebraic over F form a subfield W of K such that $[W: F] \leq m \,.\, n$. That is, the elements of W satisfy the minimal polynomial $f(x) \in F(x)$ of degree at most mn.

Theorem 11.1.7: *If a is algebraic over F, then the field $F(a)$ coincides with the ring $F[a]$.*

Proof: If a is algebraic over F, then there exists a minimal polynomial $p(x)$ of a over F. Let $\dfrac{h(a)}{g(a)}$ be any element of $F(a)$. Since $g(a) \neq 0$, then $p(x)$ does not divide $g(x)$. Therefore $p(x)$ and $g(x)$ are relatively prime since $p(x)$ is irreducible. Hence 1 is the greatest common divisor of $p(x)$ and $g(x)$ and can be expressed as

$1 = A(x)\,p(x) + B(x)\,.\,g(x)$, where $A(x), B(x) \in F[x]$. Substituting $x = a$, we have

$1 = A(a)\,p(a) + B(a)\,.g(a) = B(a)\,.\,g(a)$, since $p(a) = 0$. which shows $g(a)$ is a unit in $F[a]$. This implies that $\dfrac{h(a)}{g(a)} \in F[a]$, which completes the proof of the theorem.

Theorem 11.1.8: *If an element $a \in K$ is algebraic over F and if $p(x)$ is the minimal polynomial of a over F, then $\dfrac{F(x)}{(p(x))} \cong F(a)$.*

Proof: Since a is algebraic over F and there is the minimal polynomial $p(x)$ of a over F, i.e., $p(a) = 0$, then any $f(x) \in F[x]$ with $f(a) = 0$ is a multiple of $p(x)$. Thus $f(x)$ belongs to the principal ideal $(p(x))$ generated by $p(x)$. That is, $f(x) \in (p(x))$. Now we define a homomorphism $\phi: F[x] \to F[a]$ by $\phi(g(x)) = g(a)$, for any $g(x) \in F[x]$. It is clearly an onto homomorphism. Then kernel of the homomorphism consists of those polynomials $g(x) \in F[x]$ for which $g(a) = 0$. That is, $g(x) \in \text{Ker}\,\phi$ if and only if $g(a) = 0$. Thus $\text{Ker}\,\phi = (p(x))$. By the fundamental theorem of homomorphism we have $\dfrac{F[x]}{(p(x))} \cong F[a]$. But we know $F[a] = F(a)$.

Hence $\dfrac{F[x]}{(p(x))} \cong F(a)$

Theorem 11.1.9: *If any two elements $a, b \in K$ are algebraic over F and satisfy the same minimal polynomial $p(x)$ over F, then there exists an isomorphism ϕ of $F(a)$ onto $F(b)$ such that $\phi(c) = c$, $\forall\, c \in F$ and $\phi(a) = b$, that is, ϕ is the identity mapping of F which carries a onto b.*

Proof: By theorem 11.1.8 there are isomorphisms

CHAPTER 11

h_1 of $\dfrac{F[x]}{(p(x))}$ onto $F(a)$ and h_2 of $\dfrac{F[x]}{(p(x))}$ onto $F(b)$ such that for any $f(x) \in F[x]$, h_1 $[f(x) + (p(x))] = f(a)$ and

$h_2\,[f(x) + (p(x))] = f(b)$.

Since $h_1 : \dfrac{F[x]}{(p(x))} \to F(a)$ and $h_2 : \dfrac{F[x]}{(p(x))} \to F(b)$, and $h_1^{-1} : F(a) \to \dfrac{F[x]}{(p(x))}$ are isomorphisms, then

$\phi = h_2\, h_1^{-1} : F(a) \to F(b)$ is also an isomorphism of $F(a)$ onto $F(b)$ such that

$$\phi(c) = h_2\, h_1^{-1}(c) = h_2(c + (p(x))) = c, \ \forall\ c \in F$$

and
$$\phi(a) = h_2\, h_1^{-1}(a) = h_2(a + (p(x))) = h_2(a) = b$$

This completes the theorem.

Definition 11.1.10: Two elements a and b of one and the same extension field K of F are conjugate over F if they are *algebraic* over F and have the same *minimal polynomial over F.*

So far we have discussed the elements of a given extension K of F which were algebraic over F. That is, the elements which are roots of some polynomials $f(x) \in F[x]$. Here from a given polynomial $p(x) \in F[x]$ we shall find a field K such that $F \subseteq K$ and $p(x)$ has a root in K.

That is, here it is prime object to construct a field K which is an extension of F.

Definition 11.1.11: If $p(x) \in F[x]$, then an element a belonging to some extension field K of F is called a root of $p(x)$ if $p(a) = 0$.

Lemma 11.1.1: *If $p(x) \in F[x]$ and if K is an extension of F, then for any element $b \in K$, $p(x) = (x - b)\, q(x) + p(b)$, where $q(x) \in K[x]$ and $\deg q(x) = \deg p(x) - 1$.*

Proof: Since $F \subset K$, then $F[x] \subset K[x]$. Thus $p(x) \in F[x] \Rightarrow p(x) \in K[x]$, $x - b \in K[x]$. Now dividing $p(x)$ by $(x - b)$ there exists $q(x)$, $r \in K[x]$, the quotient and remainder respectively such that.

$p(x) = (x - b)\, q(x) + r$, where $r = 0$ or $\deg r < \deg (x - b) = 1$. That is, either $r = 0$ or $\deg r = 0$. The substitution $x = b$ gives.

$p(b) = (b - b)\, q(b) + r = r$. Therefore,

$p(x) = (x - b)\, q(x) + p(b)$ where $\deg q(x) = \deg p(x) - 1$ because in all cases r is constant element of K and $\deg (x - b) = 1$.

Corollary: If $a \in K$ is a root of $p(x) \in F[x]$, where $F \subset K$, then $(x - a)$ divides $p(x)$ in $K[x]$.

Proof: Since $F \subset K$, $F[x] \subset K[x]$.

$p(x) \in F[x] \Rightarrow p(x) \in K[x]$ and $(x - a) \in K[x]$. Then by division algorithm we have

$p(x) = (x - a)\, q(x) + r$, where $q(x) \in K[x]$.

For $x = a$, we get

$0 = p(a) = (a - a)\, q(a) + r = r$, since $p(a) = 0$ because a is the root of $p(x) \in F[x]$. which follows $p(x) = (x - a)\, q(x) \Rightarrow (x - a) \mid p(x)$.

Definition 11.1.12: The element $a \in K$ is a root of $p(x) \in F[x]$ is called of *multiplicity* m if $(x - a)^m \mid p(x)$, where as $(x - a)^{m+1} \nmid (x)$ in $K[x]$.

Thus if a is of multiplicity m in K, then $(x - a)^m \mid f(x)$ that is, $f(x) = (x - a)^m q(x)$, where $q(x) \in K[x]$, If a is a root of $q(x)$, then $(x - a) \mid q(x)$, that is, $q(x) = (x - a) q_1(x)$, where $q_1(x) \in K[x]$ which follows

$f(x) = (x - a)^{m+1} q_1(x) \Rightarrow (x - a)^{m+1} \mid f(x)$ which is a contradiction to the definition of multiplicity m of a. Hence $(x - a) \nmid q(x)$ or $q(a) \neq 0$.

Now we shall answer to two questions about roots.

(1) How many roots can a polynomial $f(x)$ of positive degree in $F[x]$ have in any field extension of F?

(2) What is the largest number of roots which lie in an extension field of F?

Theorem 11.1.10: *A polynomial of degree n over a field can have at most n roots in any extension field.*

Proof: We shall prove the theorem by induction on n, the degree of the polynomial $p(x)$. If deg $p(x) = 1$, then $p(x) = a x + b$, $a, b \in F$, $a \neq 0$ has one root, say, α. That is,

$p(\alpha) = 0 = a\alpha + b \Rightarrow \alpha = a^{-1} b$. Thus $\left(-\dfrac{b}{a}\right)$ is the unique root of $p(x)$ lying in an extension field of F. Hence for $n = 1$ The theorem holds good.

Now we assume that the theorem is true for all polynomials of degree less than $n =$ deg $p(x)$ over the field F. Let K be an extension field of F. If $p(x)$ has no root in K. The theorem is trivially true. Suppose that $p(x)$ has at least one root α in K and that α is a root of multiplicity m. Then $(x - \alpha)^m \mid p(x)$ or.

$p(x) = (x - \alpha)^m . q(x)$, where $q(x) \in K[x]$ is of degree $n - m < n$.

By induction $q(x)$ cannot have more than $n - m$ roots in K. It follows that $p(x)$ cannot have more than $m + (n - m) = n$ roots in K. This completes the proof of the theorem.

Now we shall construct an extension field of F in which a given polynomial has roots.

Theorem 11.1.11: (*Kronecker*) *If p (x) is an irreducible polynomial in F (x) of degree n ≥ 1, then there is an extension E of F, such that [E : F] = n, in which p (x) has a root.*

Proof: Since $p(x)$ is an irreducible polynomial in $F[x]$, the principal ideal $(p(x)) = V$ generated by $p(x)$ is a maximal ideal in $F[x]$. So $F[x] / V$ is a field. If we set $E = \dfrac{F[x]}{V}$, then we show that E is the required extension field of F.

Now we define a homomorphism $\phi ! F \to E$ by $\phi(d) = d + V$, $\forall d \in F$. For $d, b \in F$ we assume $\phi(d) = \phi(b) \Rightarrow d + V = b + V \Rightarrow d - b \in V$.

Since $p(x)$ is irreducible of degree at least one, then we have $d = b$. Hence ϕ is an isomorphism of F onto $\phi(F)$ which is contained in E. Thus we can consider E is an extension field of F. We identity F with its isomorphic image $\phi(F)$ in E and d is in F with $d + V = d$ in E.

Let $p(x) = a_0 + a_1 x + ... + a_n x^n$, $a_n \neq 0$, $a_i \in F$.

The degree $p(x) = n$. If $x \in F$, then $x + V \in E$. For $0 \in F$, in particular

$$p(x) \in V \Rightarrow p(x) + V = \overline{p(x)} = V = \bar{0}$$

$$\begin{aligned} \bar{0} &= 0 + V = \overline{p(x)} = p(x) + V \\ &= (a_0 + a_1 x + ... + a_n x^n) + V \end{aligned}$$

CHAPTER 11

$$= (a_0 + V) + (a_1 x + V) + ... + (a_n x^n + V)$$
$$= a_0 (1 + V) + a_1 (x + V) + ... + (a_n (x^n + V))$$
$$= a_0 (1 + V) + a_1 (x + V) + ... + a_n (x + V)^n.$$
$$= a_0 + a_1 \bar{x} + ... + a_n (\bar{x})^n.$$

Hence $\bar{x}$ is a roots of $\overline{p(x)}$, consequently root of p (x). Thus E contains a root of p *(x)*.

Since $\dfrac{F[x]}{V} = E$ is an extension field of F, $\dfrac{F[x]}{V}$ (F) can be considered a vector space over F. Now we show that the elements $1 + V$, $x + V$, $(x + V)^2 = x^2 + V$,..., $(x + V)^i = x^i + V$,.... $(x + V)^{n-1} + V$ form a basis of E over F. For any $f(x) \in F[x]$, $f(x) + V \in E$. On dividing $f(x)$ by $p(x)$ we have $f(x) = p(x)\, g(x) + r(x)$ where $q(x)$ and $r(x) \in F[x]$ and either $r(x) = 0$ or $\deg r(x) < \deg p(x)$.

Now

$$f(x) + V = (p(x).\, q(x) + r(x)) + V$$
$$= r(x) + V \text{ since } p(x)\, q(x) \in V$$

or $\overline{f(x)} = \overline{r(x)}$, we can write

$$r(x) = b_0 + b_1 x + b_2 x^2 + ... + b_{n-1} x^{n-1}, \text{ then}$$
$$\overline{f(x)} = b_0 + b_1 x + b_2 x^2 + ... + b_{n-1} x^{n-1} + V$$
$$= (b_0 + V) + (b_1 x + V) + ... + (b_{n-1} x^{n-1} + V)$$
$$= b_0 (1 + V) + b_1 (x + V) + ... + b_{n-1} (x^{n-1} + V)$$
$$= b_0 (1 + V) + b_1 (x + V) + ... + b_{n-1} (x + V)^{n-1}$$

Hence every $\overline{f(x)} \in E$ can be expressed as Linear combination of $1 + V$, $x + V$, $x^2 + V$,..., $x^{n-1} + V$. For $a_0, a_1, a_2,...., a_{n-1}$, we have $a_0 (1 + V) + a_1 (x + V) + ... + a_{n-1} (x^{n-1}) + V) = V = \bar{0}$.

or $\quad a_0 + a_1 x + ... + a_{n-1} x^{n-1} + V = V$

which implies $a_0 + a_1 x + ... + a_{n-1} x^{n-1} \in V$ which is a contradiction. Because V contains p *(x)* of degree n and all multiples of p *(x)*. This forces $a_0 = a_1 = a_2 = ... = a_{n-1} = 0$ Hence $\{1 + V, x + V,..., x^{n-1} + V\}$ is a basis of E over F containing n elements. Thus $[E : F] = n$.

Corollary: If $f(x) \in F[x]$, then there is a finite extension E of F in which $f(x)$ has a root. Moreover, $[E : F] \leq \deg f(x)$.

Proof: Let $p(x)$ be an irreducible factor of $f(x)$ in $F[x]$. Then any root of $p(x)$ is a root of $f(x)$ and $\deg p(x) \leq \deg f(x)$. But by the theorem there exists an extension E of F such that $[E : F] = \deg p(x) \leq \deg f(x)$ in which $p(x)$ has a root, consequently $f(x)$ has a root.

Theorem 11.1.12: *Let $f(x) \in F[x]$ be of degree $n \geq 1$. Then there is an extension E of F of degree at most n ! in which $f(x)$ has n roots (and so, a full complement of roots).*

Proof: We prove the theorem by induction on $n = \deg f(x)$, Let $\deg f(x) = 1$, then $f(x)$ is of the form $ax + b$, $a \neq 0$, $a, b, \in F$. Thus $f(x) = a x + b$ has only a root $\left(-\dfrac{b}{a}\right) \in F$.

So in this case we can consider $E = F$. Hence theorem is true for $n = 1$. To apply induction we assume that the theorem is true for all polynomials of degree less than n.

Now by the above corollary there is an extension E_0 *of* F such that $[E_0 : F] \leq n$ and $f(x)$ has a root α in E_0. Therefore $x - \alpha$ divides $f(x)$ in $E_0[x]$. Thus there is $f_1(x) \in E_0[x]$ of degree $n - 1$ such that $f(x) = (x - \alpha).\ f_1(x)$. By induction there exists an extension E of E_0 of degree at most $(n - 1)!$ in which $f_1(x)$ has $n - 1$ roots. Since n roots of $f(x)$ are α and $(n - 1)$ roots of $f_1(x)$ in E, we get all n roots of $f(x)$ in E.

Since $[E_0 : F] \leq n$, and $[E : F_0] \leq (n - 1)!$ with

$F \subset E_0 \subset E$, then

$[E : F] = [E : E_0] . [E_0 ! F] \leq (n - 1)! . n = n!$

This completes the proof of the theorem.

From the above theorem we conclude that if $f(x) \in F[x]$ over a field F has a root, say, α, that is $f(a) = 0$, then a is algebraic over F and there exists an extension $F(x)$ of F such that $[F(a) : F] = n = \deg f(x)$. Thus if $f(x)$ has n roots $a_1, a_2, \ldots, a_n$, then there exists a finite extension $F(a_1, a_2, \ldots, a_n)$ of F which contains F and $a_1, a_2, \ldots, a_n$. If $a_1, a_2, \ldots a_n$ are roots of $f(x)$, then $x - a_1$, $x - a_2, \ldots, x - a_n$ will divide $f(x)$. So $f(x) = a(x - a_1)(x - a_2)\ldots(x - a_n)$, the product of linear factors belonging to $E[x]$. The extension field $E = F(a_1, a_2, \ldots a_n)$ is a minimal extension of F which has n roots of $f(x)$. That is, there does not exist any extension $E^1 \subseteq E$ of F which has n roots of $f(x)$. Thus $f(x)$ splits up completely over E as a product of linear factors which is a *minimal extension* of F.

Then we arrive at the following

Definition 11.1.13: Let $f(x) \in F[x]$ be any polynomial of degree $n \geq 1$. Then a field extension E of F is called a *splitting field* (*root field* or *stenfield*) of $f(x)$ if

(*i*) $f(x)$ can be written as the product of n linear factors over E.

(*ii*) There does not exists any subfield E' of E which contain F and n roots of $f(x)$. That is, $f(x)$ cannot be factored into n linear factors over E'.

Thus, it is evident that E is spliting field if E has n roots of $f(x)$ and $E = F(a_1, a_2, \ldots, a_n)$, generated by F and n roots $a_1, a_2, \ldots, a_n$ of $f(x)$. Now we shall see if there exists any relation between two spliting fields E and E' of the same polynomial $f(x)$ over *a* field F.

Let F and F' be two fields and let σ be an isomorphism of F onto F' defined by $\sigma(a) = a'$, $\forall\ a \in F$ and $a' \in F'$. Then there exists an extension ρ of σ which is an isomorphism or $F(x)$ onto $F'[t]$ where $F(x)$ and $F'(t)$ are rings of polynomials over the fields F and F' in indeternates x and t respecively such that $\rho(x) = t$.

For an arbitrary polynomial $f(x) = a_0 x^n + a_1 x^{n-1} + \ldots + a_n \in F[x]$ we define ρ by

$$\begin{aligned} \rho(f(x)) &= \rho(a_0 x^n + a_1 x^{n-1} + \ldots + a_n) \\ &= \rho(a_0)\rho(x^n) + \rho(a_1)\rho(x^{n-1}) + \ldots + \rho(a_n) \\ &= \sigma(a_0)\rho(x^n) + \sigma(a_1)\rho(x^{n-1}) + \ldots + \sigma(a_n) \\ &= a_0' t^n + a_1' t^{n-1} + \ldots + a_n' \end{aligned}$$

since ρ is an extension of σ.

If $f(x) \in F(x)$ then we write $\rho(f(x))$ by $f'(t)$ in $F'[t]$.

Now we shall see that if $p(x)$ is irreducible, then so is $p(t)$ and if $p(x)$ is reducible, $p'(t)$ is also reducible. For this, let $p(x)$ be any irreducible polynomial in $F[x]$. Suppose $p'(t) = f'(t) . g'(t)$ for some polynomials $f'(t)$ and $g'(t)$ of positive degree. Since ρ is an isomorphism of the ring $F[x]$ onto the ring $F'[t]$, then there exist $f(x)$ and $g(x) \in F[x]$ such that $\rho(f(x)) = f'(t)$ and $\rho(g(x)) = g'(t)$. An isomorphism preserves the degrees of polynomials. So $f(x)$ and g (x) are of positive degree.

Now we have

$$\begin{aligned} \rho(p(x)) = p'(t) = f'(t) . g'(t) &= \rho(f(x)) . \rho(g(x)) \\ &= \rho(f(x) . g(x)), \end{aligned}$$

CHAPTER 11

which implies $p(x) = f(x) \cdot g(x)$, that is, $p(x)$ is reducible which is a contradiction that $p(x)$ is irreducible. Hence if $p(x)$ is irreducible, then $p'(t)$ is also irreducible.

If $\rho : F[x] \to F'[t]$ is an isomorphism and if A is an ideal in $F[x]$, then $\rho(A)$ is an ideal in $F'[t]$. Thus we have two quotient Rings $\dfrac{F[x]}{A}$ and $\dfrac{F'[t]}{\rho(A)}$.

Now we can define an isomorphism $\eta : \dfrac{F[x]}{A} \to \dfrac{F'[t]}{\rho(A)}$ by $\eta(f(x) + A) = \rho(f(x)) + \rho(A))$ $= f'(t) + \rho(A)$.

If $p(x)$ is irreducible, then $p'(t)$ is also irreducible under the isomorphism ρ of $F[x]$ onto $F'[t]$. Then $(p(x))$ is an ideal generated by $p(x)$ and $(p'(t))$ is the corresponding ideal generated by $p'(t)$ in $F'[t]$. So we can define an isomorphism.

$\eta : \dfrac{F[x]}{(p(x))} \to \dfrac{F'[t]}{(p'(t))}$ by

$\eta(f(x) + (p(x))) = f'(t) \neq p'(t_1) \forall f(x) \in F[x]$

Now we have the following

Theorem 11.1.13: *If $p(x)$ is irreducible in $F[x]$ and if u is a root of $p(x)$, then $F(u)$ is idomorphic to $F'(w)$, where w is a root of $p'(t)$; moreover, this isomorphism σ can so be chosen that*

(i) $\sigma(u) = w$

(ii) $\sigma(\alpha) = \alpha', \forall \alpha \in F.$

Proof: Since $p(x)$ is irreducible, $p'(t)$ is also irreducible by the above remarks. Then there exist isomorphisms

$$\sigma_1 : \frac{F[x]}{(p(x))} \to F(u),$$

and
$$\sigma_2 : \frac{F'[t]}{(p'(t))} \to F'(w),$$

such that $\sigma_1 [f(x) + (p(x))] = f(u)$ and $\sigma_2 [f'(t) + (p'(t))] = f'(w)$. which leave the elements of F and F' fixed.

Here $\dfrac{F[x]}{(p(x))}$ and $\dfrac{F'[t]}{(p'(t))}$ are extension fields of F and F' in which $p(x)$ and $p'(t)$ have roots respectively. By the above remark there is an isomorphism η of $\dfrac{F[x]}{(p(x))}$ onto $\dfrac{F^1[t]}{(p^1(t))}$,

i.e.,
$$\eta : \frac{F[x]}{(p(x))} \to \frac{F'[t]}{(p'(t))} \text{ defined by}$$

$\eta[f(x) + (p(x))] = f'(t) + (p'(t))$.

we notice that

$\sigma_1 [x + (p(x))] = u$

and $\sigma_2 [t + (p'(t))] = w$.

Now $\sigma_2 \eta \sigma_1^{-1}$ is an isomorphism of $F(u)$ onto $F'(w)$.

$$F(u) \xrightarrow{\sigma_1^{-1}} \frac{F[x]}{(p(x))} \xrightarrow{\eta} \frac{F^1[t]}{(p^1(t))} \xrightarrow{\sigma_2} F'(w)$$

Now for any $\alpha \in F$,

$$\begin{aligned} \sigma(\alpha) &= (\sigma_2 \eta \sigma_1^{-1})(\alpha) \\ &= (\sigma_2 \eta)(\sigma_1^{-1})(\alpha) \\ &= \sigma_2 \eta (\alpha + (p(x)) \\ &= \sigma_2 [\alpha' + (p'(t))] \\ &= \alpha'. \end{aligned}$$

$\sigma(u) = (\alpha_2 \eta \sigma_1^{-1})(u) = [\sigma_2 \eta (x + (p(x))] = \sigma_2 [t + (p'(t))] = w$.

This completes the proof of the theorem.

Corollary: If $p(x) \in F[x]$ is irreducible and if a, b are roots of $p(x)$, then $F(a)$ is isomorphic to $F(b)$ by an isomorphism, which takes a onto b and which leaves every element of F fixed.

Proof: We have

$$\sigma_1 : \frac{F[x]}{(p(x))} \to F(a)$$

and

$$\sigma_2 : \frac{F[x]}{(p(x))} \to F(b)$$

such that

$$\sigma_1 [f(x) + (p(x))] = f(a), \ \sigma_1 [x + (p(x))] = a,$$
$$\sigma_1 (\alpha + (p(x))] = \alpha.$$

and

$$\sigma_2 [f(x) + (p(x))] = f(b), \ \sigma_2 [x + (p(x)] = b,$$
$$\sigma_2 (\alpha + (p(x))] = \alpha$$

$\sigma = \sigma_2 \sigma_1^{-1}$ is an isomorphism of $F(a)$ onto $F(b)$. We also have

$$\begin{aligned} \sigma(a) &= \sigma_2 \sigma_1^{-1}(a) \\ &= \sigma_2 (x + (p(x))) = b, \end{aligned}$$

and $\forall \alpha \in F$

$$\begin{aligned} \sigma(\alpha) &= \sigma_2 \sigma_1^{-1}(\alpha) \\ &= \sigma_2 [\alpha + (p(x))] = \alpha. \end{aligned}$$

Hence the corollary.

Theorem 11.1.14: *Let σ be a isomorphism of the field F onto the field F' and let $f(x) \in F[x]$ of positive degree and $f'(t)$ be the corresponding polynomial in $F'[t]$. If E and E' are splitting fields of $f(x)$ and $f'(t)$ over F and F' respectively, then there exists an isomorphism ϕ of E onto E' such that $\phi(\alpha) = \alpha'$, $\forall \alpha \in F$.*

Proof: Let $\deg f(x) = n$, then the splitting field E over F is finite, that is, $[E : F]$ is finite. We shall prove the theorem by applying induction on the degree $[E : F]$. So now we suppose $[E : F] = 1$, then $E = F$. So all the n roots $a_1, a_2, \ldots, a_n$ of $f(x)$ are in F and

CHAPTER 11

$f(x) = \alpha (x - a_1)(x - a_2) \ ...(x - a_n)$ for some $\alpha \neq 0 \in F$. The corresponding polynomial $f'(t) = \sigma(f(x))$

$$= \sigma [\alpha (x - a_1)(x - a_2)... (x - a_n)]$$
$$= \sigma(\alpha) \sigma(x - a_1) \sigma(x - a_2) ... \sigma(x - a_n)$$
$$= \alpha' (t - a_1')(t - a_2') .. (t - a_n') \in F'[t].$$ That is,

$f'[t]$ splits over F'.

This means $\alpha', a_1', a_2', ...a_n' \in F'$. Thus $E' = F'$.

We can take $\phi = \sigma$ which provides us with an isomorphism of E onto E' coinciding with σ on F. Hence the theorem is true for $[E : F] = 1$.

Now we suppose $[E : F] \geq n > 1$, where E is the splitting field of $f(x)$ over F. Since $[E : F] > 1$, $f(x)$ has irreducible factor $p(x)$ of degree $r > 1$ and at least one root u of $f(x)$ is not in F. Since $p(x)) \mid f(x)$, $f(x) = p(x) \ q(x)$ for some $q(x) \in F[x]$. Then $f'(t) = \sigma(f(x)) = \sigma(p(x) \ . \ q(x)) = \sigma(p(x)). \ \sigma(q(x)) = p'(t) \ . \ q'(t)$. Since $p'(t)$ is the corresponding polynomial of $p(x)$, $p'(t)$ is also irreducible and $f'(t)$ has irreducible factor $p'(t)$.

Since E splites $f(x)$, all roots of $f(x)$ are in E and consequently all roots of $p(x)$ are in E. Thus for some v $\in E$, $p(v) = 0$ and by theorem 11.1.11 $[F(v): F] = r$. Similarly, there is $w \in E'$ such that $p'(w) = 0$ and $[F'(w): F'] = r = \deg p'(t) - \deg p(x)$. By theorem. 11.1.13. there is an isomorphism of $F(v)$ onto $F'(\omega)$ such that $\sigma(\alpha) = \alpha'$, $\forall \ \alpha \in F$.

Since $[E : F(v)] = r > 1$, we have

$\{E : F(v)] \ . \ [F(v) : F] = [E : F]$

or $\quad [E : F(v)] = \dfrac{[E : F]}{[F(v) : F]} = \dfrac{n}{r} < n.$

To apply the induction we claim that E is the spilitting field of $f(x)$ taken a polynomial over $F_0 = F_1(v)$, for no subfield of E which contain F_0 and F can split $f(x)$. Since E is the splitting field of $f(x)$ over F, E' is the splitting field of $f'(t)$. over $F_0' = F'(w)$ by our induction hypothesis there exists an isomorphism ϕ of E onto E' such that $\phi(a) = \sigma(a)$, $\forall \ a \in F_0$. But we have $\sigma(a) = a'$, $\forall \ a \in F_0$ where $a \in F \subset F_0$, thus $\phi(a) = a'$, $\forall \ a \in F$. Therefore if $[E : F] = n$, by induction hypothesis the theorem is true. That completes the proof of the theorem.

PROBLEMS

Note: In the following problems F stands for the field $(F, +, .)$ and K stands for an extension field $(K, +, .)$ of F.

1. Prove that the mapping $\phi : F[x] \to F(a)$ defined by $\phi(h(x)) = h(a)$ is a homomorphism.
2. Let $(F, +, .)$ be a field and let $(F[x] +, .)$ be the ring of polynomials in x over $(F, +, .)$. Let $g(x)$ be a polynomial in $F[x]$ of degree n and let $V = (g(x))$ be the ideal generated by $g(x)$ in $F[x]$. Prove that $F[x] / V$ is an n-dimensional vector space over $(F, +, .)$.
3. If $[K : F] = 1$, then prove that $K = F$.
4. For any two elements $a, b \in K$, prove that $F(a, b) = F(b, a)$.
5. If $F \subset K$ such that $[K: F]$ is finite and $V(K)$ is a finite dimensional vector space over the field $(K, +, .)$, show that $V(F)$ is a finite dimensional vector space over F and that more over $\dim F(V) = (\dim K(V)). \ [K : F]$.
6. Let $(R, +, .)$ be the field of real numbers and $(Q, +, .)$ be the field of rational numbers. In R, $\sqrt{2}$ and $\sqrt{3}$ are both algebraic over Q. Exhibit a polynomial of degree 4 over Q satisfied by $\sqrt{2} + \sqrt{3}$.

7. If a, $b \in K$ are algebraic over F of degrees m and n, respectively, and if m and n are relatively prime, prove that $F(a, b)$ is of degree $m \cdot n$ over F.
8. Suppose that F is a field having a finite number of elements, and q.
   (a) Prove that there is a prime number p such that

$$\underbrace{a+a+.....+a}_{p\,.\,times} = 0 \text{ for all } a \in F.$$

   (b) Prove that $q = p^n$ for some integer n.
   (c) If $a \in F$ prove that $a^q = a$.
   (d) If $b \in K$ is algebraic over F prove that $b^{q^m} = b$ for some $m > 0$.

Definition 11.1.14: An algebraic number a is said to be an *algebraic integer* if it satisfies a monic polynomial in $Z[x]$. Since i satisfies $x^2 + 1 \in Z[x]$, i is an *algebraic integer*.

Prove the following.

(i) If a is any algebraic number, prove that there is a positive integer n such that na is an algebraic integer.
(ii) If the rational number r is also algebraic number, prove that r must be an ordinary integer.
(iii) Prove that the sum of two algebraic integer is an algebraic integer,
(iv) Prove that the product of two algebraic integers is an algebraic integer.

11.2 SEPARABLE AND INSEPARABLE EXTENSION

Let $(F, +, .)$ be a field and let $(F[x], +, .)$ be the polynomial ring in x over the field $(F, +, .)$. We recall the definition of a root of multiplicity m of a polynomial $f(x)$ in any extension field that any element a of any extension field is said to be *multiple root* if a is a root of $f(x)$ of multiplicity $m \geq 2$. Here we shall study the existence of multiple roots of a polynomial $f(x)$. So for this we give.

Definition 11.2.1: If $f(x) = a_0 + a_1, x + a_2 x^2 + ... + a_n x^n \in F[x]$, the devivative of $f(x)$, denoted by $f'(x)$, is defined by

$f'(x) = a_1 + 2\, a_2\, x + ... + n\, a_n\, x^{n-1}$

We see that if $f(x) = a_0$, then $f'(x) = 0$, that is the derivative of a constant polynomial is always zero.

In the polynomial $f(x) \in F[x]$, the coefficients of $f(x)$ belong to the field $(F, +, .)$ The derivative $f'(x)$ is again a polynomial in $F[x]$. We observe that if the characteristic of the field $(F, +, .)$ is zero, then the coefficients of $f'(x)$, $2a_2$, $3\, a_3$,......, $(n-1)\, a_{n-1}\, na_n$ can be zero if and only if $a_2, a_3, ... a_{n-1}, a_n$ are zero. This shows that if the characteristic of $(F, +, .)$ is zero, then

$f'(x) \neq 0$, when $n > 0$.

We again suppose that the characteristic of the field $(F, +, .)$ is a prime number $p \neq 0$, then the coefficient $(n - i)\, a_{n-i}$, where $i = 0, 1,....., n - 2$ of $f'(x)$ can be zero if either $a_{n-i} = 0$ or p divides $n - i$, $i = 0, 1,...., n - 2$. In particular $na_n = 0$, if and only if $p \mid n$ since $a_n \neq 0$. This implies that $f'(x) = 0$, if and only if n is divisible by p and all those coefficients a_{n-i}, $i = 1, 2,...., n - 2$, of $f'(x)$ are zero for which $p \nmid (n - i)$. When that is so, the terms $a_{n-i}\, x^{n-i}$ occured in $f(x)$ are such that $p \mid n - i$, $a_{n-i} \neq 0$ That implies that $f'(x)$ is a polynomial $f(x)$ in x^p because $n - i = pt$, for some integer t and $x^{n-i} = x^{pt} = (x^p)^t$. Thus $f(x) \in F[x^p]$. Thus we conclude that $f(x) \in F[x^p]$ is a necessary and sufficient condition for vanishing of $f'(x)$. Therefore the degree of $f'(x)$ is zero or $n - 1$ according to $f'(x) = 0$ or $na_n \neq 0$ *i.e.*, $f'(x) \neq 0$.

CHAPTER 11

***Lemma 11.2.1:** For any $f(x), g(x) \in F[x]$ and $a \in F$,*

(1) $(f(x) + g(x))' = f'(x) + g'(x)$;

(2) $(a f(x))' = af'(x)$;

(3) $(f(x). g(x))' = f'(x). g(x) + f(x) . g'(x)$.

Proof: parts (1) and (2) are easy and are left to the reader to prove. However we prove (3). Let $f(x) = x^i$, $g(x) = x^j$ where $i > 0$, $j > 0$. Then

$f(x) . g(x) = x^{i+j}$

and $\qquad (f(x) . g(x))' = (i + j) x^{i+j-1}$

$$f'(x) = i x^{i-1}$$

and $\qquad g'(x) = j x^{j-1}$

So
$$\begin{aligned} f'(x) . g(x) + f(x) . g'(x) &= i x^{i-1} . x^j + j x^i x^{j-1} \\ &= i x^{i+j-1} + j x^{i+j-1} \\ &= (i + j) x^{i+j-1} \\ &= (f(x). g(x))'. \end{aligned}$$

***Lemma 11.2.2:** An element α of an extension field $(K, +, .)$ of the field $(F, +, .)$ is a multiple root of non-constant polynomial $f(x) \in F[x]$ if and only if α is a common root of $f(x)$ and $f'(x)$.*

Proof: Let α be a root of $f(x) \in F(x)$ of multiplicity m. Then $(x - \alpha)^m \mid f(x)$ and $(x - \alpha)^{m+1} \nmid f(x)$. So for some $g(x) \in F[x]$, $f(x) = (x - \alpha)^m . g(x)$, and $g(\alpha) \neq 0$. We have

$f'(x) = m (x - \alpha)^{m-1} g(x) + (x - \alpha)^m . g'(x)$.

Now putting $x = \alpha$, we obtain

$f'(\alpha) = m (\alpha - \alpha)^{m-1} g(\alpha) + (\alpha - \alpha)^m g(\alpha) = 0$

$\Rightarrow \alpha$ is a root of $f'(x)$.

Hence α is a common root of $f(x)$ and $f'(x)$.

Conversely, if α is not a root of $f(x)$ of multiplicity $m \geq 2$, then $m = 1$, and $f(x) = (x - \alpha) . g(x)$

$f'(x) = g(x) + (x - \alpha) g'(x) \Rightarrow f'(\alpha) = g(\alpha) + (\alpha - \alpha) g'(\alpha) \Rightarrow f'(\alpha) = g(\alpha) \neq 0$. Hence α is not a common root of $f(x)$ and $f'(x)$ if α is not a multiple root.

***Lemma 11.2.3:** Let $f(x)$ be an irreducible polynomial in $F[x]$. Then $f(x)$ has multiple root in some extension field $(K, +, .)$ of $(F, +, .)$ (definitely in the splitting field of $f(x)$ over the field $(F, +, .)$) if and only if $f'(x) = 0$.*

Proof: Let α be an element of some extension field $(K, +, .)$ of $(F, +, .)$ which is multiple root of an irreducible polynomial $f(x) \in F[x]$. Then by the lemma 11.2.1 α is also a root of $f'(x)$, *i.e.*, $f'(\alpha) = 0$. This implies that $f(x) \mid f'(x)$ because if $f'(x)$ is an irreducible polynomial in $F[x]$ with $f(\alpha) = 0$ and $g(x)$ is another polynomial in $F[x]$ such that $g(\alpha) = 0$, then $f(x) \mid g(x)$.

If $f'(x) \neq 0$, then $\deg f'(x) < \deg f(x)$, and hence $f(x)$ cannot divides $f'(x)$. This forces $f'(x) = 0$

Conversely, let $f'(x) = 0$ and let $\deg f(x) = n$. Since $(K, +, .)$ is an extension field (splitting field) of $(F, +, .)$, $f(x)$ can be factored into n linear factors in $K[x]$. Hence $f(x) = c(x - \alpha_1)(x - \alpha_2) .... (x - \alpha_n)$, where α_i, s are all distinct $f'(x) = c \sum_{i=1}^{n} (x - \alpha_1) \;...\; (x - \alpha_i)$, where .... $(x - \alpha_n)$ denotes the omitted term.

Now $f'(\alpha_1) = c(\alpha_i - \alpha_1) \ldots (\alpha_i - \alpha_i) \ldots (\alpha_i - \alpha_n)$
$= c(\alpha_i - \alpha_1)(\alpha_i - \alpha_2)\ldots.(\alpha_i - \alpha_{i-1})(\alpha_i - \alpha_{i+1})\ldots.(\alpha_i - \alpha_n)$
$= c\,\Pi(\alpha_i - \alpha_i) \neq 0$, since the all roots are distinct. This implies that

α_i is not a root of $f'(x) \Rightarrow f'(x)$ and $f(x)$ have no common root. Thus $f'(x) = 0$ cannot hold unless $f'(x)$ and $f(x)$ have some common roots, that is, unless one of the root α_i is a multiple root of $f(x)$.

Hence the Lemma is proved.

Corollary 1: An irreducible polynomial $f(x)$ over $(F, +,.)$ of *characteristic zero* has no multiple root in any extension field.

Proof: Let $f(x)$ be an irreducible polynomial over the field $(F, +,.)$ of characteristic zero of degree $n \geq 1$.

Then $f'(x) \neq 0$, where $\deg f'(x) = n - 1 \geq 0$.

This implies for any root α for $f(x)$, $f'(\alpha) \neq 0$. Hence $f(x)$ and $f'(x)$ have no common root. Hence $f(x)$ has no multiple root.

Corollary 2: If $f(x) \in F[x]$ is irreducible, then if the characteristic of $(F, +,.)$ is $p > 0$, $f(x)$ has a multiple root only if it is of the form $f(x) = g(x^p)$.

Proof: If the irreducible polynomial $f(x)$ has a multiple roots, then by lemma 11.2.3 $f'(x) = 0$. We have seen that $f'(x) = 0$ holds over the field of characteristic $p > 0$ if $p \mid n = \deg f(x)$ and in all coefficient of $f(x)$, $p \mid n - i$ for which $a_{n-i} \neq 0$ and $a_{n-i} = 0$ for which $p \nmid n - i$. So every power $n - i$ and n are multiple of p so x^{n-1} can be replaced by (x^p) where $n - i = p \,.\, t$. Hence $f(x)$ is polynomial of the form $g(x^p) \in F[x^p]$.

In the following example we shall see that an irreducible polynomial over a field of characteristic $p \neq 0$ might have multiple roots.

Example 11.2.1: Let $(Z_2, +_2, \odot_2)$ be a field of integers modulo 2. That is, the characteristic of the field $(Z_2, +_2, \odot_2)$ is 2 and let $(Z_2(x), +,.)$ be the field of rational functions in x of the form $\dfrac{f(x)}{g(x)}$, $f(x), g(x) \in Z_2[x]$ over the field $(Z_2, +_2, \odot_2)$. Now we shall consider a polynomial $t^2 - x$ over the field $(Z_2[x], +,.)$. $t^2 - x$ is irreducible over $(Z_2(x), +,.)$ since $t^2 - x$ has no root in $Z_2(x)$. That is, there exists no rational function $\dfrac{f(x)}{g(x)}$ such that $\left(\dfrac{f(x)}{g(x)}\right)^2 = x$, $f(x), g(x) \in Z_2[x]$, $g(x) \neq 0$.

Let $f(x), g(x) \in Z_2[x]$ such that $(f(x) \mid g(x))^2 = x$ and let $\deg f(x) = m$ and $\deg g(x) = n$. Then we have

$(f(x))^2 = x(g(x))^2 \Rightarrow \deg(f(x))^2 = \deg x(g(x))^2$ If $m = n$, $\deg(f(x))^2 \neq \deg x(g(x))^2 \Rightarrow f(x)^2 \neq x(g)(x)^2$. Now if $m \neq n$, then $\deg(f(x))^2 >$ or $< \deg(g(x))^2$ by at least two. So $\deg(f(x))^2 >$ or $< \deg x(g(x))^2$ by at least one. Therefore $\deg(f(x))^2 \neq \deg x(g(x))^2 \Rightarrow (f(x))^2 \neq x(g(x))^2$. Hence $t^2 - x$ does not have solution of the form $\dfrac{f(x)}{g(x)}$ where $g(x), f(x) \in Z_2[x]$. Hence $t^2 - x$ is irreducible in $Z_2[x]$.

Let a_1, a_2 be two roots of $t^2 - x$. Then $t^2 - x = (t - a_1)(t - a_2)$ comparing the coefficients we get $a_1 + a_2 = 0$, and $a_1 a_2 = x$. but $a_1 + a_2 = 0 \Rightarrow a_1 = -a_2$ we have $2a_2 = 0 \Rightarrow a_2 = -a_2$ since $(F, +,.)$ is of characteristic 2. Thus $a_1 = a_2$. Hence $t^2 - x$ has two equal roots. That is, it has a multiple root.

Definition 11.2.2: An irreducible polynomical $f(x) \in F[x]$ of positive degree n is said to be *separable* if $f'(x) \neq 0$. That is, $f(x)$ and $f'(x)$ have no common roots which implies $f(x)$ has no multiple root, *i.e.*, $f(x)$ has all distinct roots which lie in some splitting field of $(F, +,.)$.

Definition 11.2.3: An irreducible polynomial $f(x) \in F[x]$ of positive degree n is said to be inseparable if $f'(x) = 0$. An arbitrary polynomial $f(x) \in F[x]$ is *separable* if all its irreducible factors are *separable* otherwise $f(x)$ is *inseparable.*

If the field $(F, +,.)$ is of characteristic zero, every polynomial of $F[x]$ of positive degree is separable.

Example 11.2.2: We consider the polynomial $f(x) = (x - 2)^2 (x^2 + x + 1)$ over the field of rational numbers (Q, +,.). The factors of $f(x)$ are $x - 2$ and $x^2 + x + 1$ which are irreducible. Since $x - 2$ and $x^2 + x + 1$ are separable because if $h(x) = x - 2$, $g(x) = x^2 + x + 1$, then $h'(x) = 1 \neq 0$, $g'(x) = 2x + 1 \neq 0$ for any root of $h(x)\, g(x)$, $f(x)$ is separable by definition.

Definition 11.2.4: Let $(K, +,.)$ be an extension field of a field $(F, +,.)$. If $a \in K$ is algebraic over F, then the *element* a is said to be *separable* or *inseparable over F* if and only if the minimal polynomial of a over F is *separable* or *inseparable*.

Definition 11.2.5: An algebraic extension $(K, +,.)$ of $(F, +,.)$ is a *separable extension of* $(F, +,.)$ if every element of $(K, +,.)$ is separable over $(F, +,.)$. In the contrary case, $(K, +,.)$ is called an *inseparable extension* of $(F, +,.)$

We know that every polynomial of positive degree over a field of characteristic zero is separable. So the roots of a polynomial over the field of characteristic zero are separable over $(F, +,.)$. that is every algebraic extension of a field of characteristic zero is a *separable extension*.

Example 11.2.3: We have seen that there exists an infinite field of characteristic $p \neq 0$. For example, the field $(Z_2(x), +,.)$ of all rational function of the form $\frac{f(x)}{g(x)}$, $f(x), g(x) \in Z_2[x]$, $g(x) \neq 0$.

Let $F = Z_2[x]$ and let $(K, +,.)$ be the splitting field of $t^2 - x$. The $(K, +,.)$ is a finite extension of degree 2 = degree of $(t^2 - x)$. That is, $[K : F] = 2$. But the field $(K, +,.)$ is inseparable because $t^2 - x$ has a multiple root since $\frac{d}{dt}(t^2 - x) = 2t = 0$.

Now we shall show that any algebraic extension of a finite field is separable.

Lemma 11.2.4: *Let $(F, +,.)$ be a finite field of characteristic $p \neq 0$. Then the mapping $a \to a^p$, $\forall\, a \in F$ is an automorphism.*

Proof: Let f be a mapping from $(F, +,.)$ into itself defined by $f(a) = a^p$, $\forall\, a \in F$.

For any $a, b \in F$

$f(a + b) = (a + b)^p$

$= a^p + pc_1\, a^{p-1}\, b + pc_2\, a^{p-2}\, b_2 + ... pc_{p-1}\, a\, b^{p-1} + b^p.$

Since $(F, +,.)$ is of characteristic p, then $p \mid pc_1\, pc_2 \,......\, pc_{p-1}$ and consequently $p_{c_r}\, a^{p-r}\, b^r = 0$, $r = 1, 2,..., p - 1$.

So $f(a + b) = a^p + b^p = f(a) + f(b)$

And

$f(a\,.\,b) = (a\,.\,b)^p = a^p\,.\,b^p = f(a)\,.\,f(b).$

Hence f is a homomorphism.

For every $a^p \in F$, there exists an element $a \in F$ such that $f(a) = a^p$ which shows f is onto. Now we find out

$$\text{Ker}(f) = \{a \in F \mid a^p = 0\}$$
$$= \{a \in F \mid a = 0\} = \{0\}.$$

Hence f is one to one which establishes that f is an automorphism.

Theorem 11.2.1: *Any algebraic extension of a finite field (F, +,.) is a separable.*

Proof: Let $f(x)$ be an irreducible polynomial over a finite field $(F, +,.)$ of characteristic $p \neq 0$. Let $f(x)$ be an inseparable. Then $f(x)$ has a multiple root which implies $f(x)$ is of the form $g(x^p) \in F[x^p]$.

That is, if $f(x) = a_0 + a_1 x + a_2 x^2 + ... + a_n x^n$, $a_n \neq 0$.

Then $f'(x) = a_1 + 2a_2 x + ... + na_n x^{n-1}$

Since $f(x)$ is inseparable, $f'(x) = 0$

$a_1 = 2a_2 = ... = na_n = 0$.

For any r, $1 \leq r \leq n$, $ra_r = 0 \Rightarrow p \mid r$ or $a_r = 0$.

Thus if $a_r \neq 0$, then $p \mid r \Rightarrow r = r_1 p$ for some integer

$r_1 \geq 1 \Rightarrow a_1 x^r$ is of the form $a_r (x^p)^{r_1}$

$$= a_{r_1} (x^p)^{r_1}$$

Thus $f(x) = g(x) = b_0 + b_1 x^p + b_2 x^{2p} + ... + b_n x^{np}$.

Since $a \to a^p \in F$ is an automorphism of F. Thus we can find some a_i such that $b_i = a_i^p \ \forall i$.

$$f(x) = a_0 + a_1 x + ... + a_n x^n$$
$$= a_0^p x^p + ... + a_n^p x^{np}$$
$$= (a_0 + a_1 x + ... + a_n x^n)^p$$

$\Rightarrow f(x)$ *is* not irreducible. Hence this contradiction shows $f(x)$ is separable. Therefore every minimal polynomial of every element of $(K, +,.)$ over $(F, +,.)$ is separable, consequently the algebraic extension $(K, +,.)$ is separable over a finite field of characteristic $p > 0$.

Theorem 11.2.2: *If the field (F, +,.) is of characteristic zero and if a and b are algebraic over F, then there exists an element c ∈ F (a, b) such that F (a, b) = F (c).*

Proof: Let $f(x)$ and $g(x)$ be the irreducible polynomials, of degree m and n, satisfied by a and b respectively. Let K be an extension of F in which $f(x)$ and $g(x)$ split into linear factors completely. Since the characteristic of F is 0, all the roots of $f(x)$ are distinct and all the roots of $g(x)$ are distinct. Let the roots of $f(x)$ be $a = a_1, a_2, ..., a_m$ and let the roots of $g(x)$ be $b = b_1, b_2, ..., b_n$ in K. For $2 \leq i \leq m$, $2 \leq j \leq n$, we can define

$$\lambda_{ij} = \frac{a_i - a}{b - b_j}$$

Here λ_{ij} are finite in number. Since the field $(F, +,.)$ is of characteristic zero, F has infinite number of elements. So we can find an element $r \neq 0 \in F$ such that

$r \neq \lambda_{ij}$, for $i, j \geq 2$.

Then $r \neq \dfrac{a_i - a}{b - b_j} \Rightarrow r(b - b_j) \neq a_i - a$, $\forall\, i, j \geq 2$

$\Rightarrow a_i + rb_j \neq a + rb$, $\forall\, i, j \geq 2$

Now put $c = a + rb \in F(a, b)$, So $F(c) \subseteq F(a, b)$.

Now we shall prove that $F(a, b) \subseteq F(c)$.

CHAPTER 11

Since $F \subset F(c) = K$, $f(x) \in F[x] \Rightarrow f(x) \in K[x]$. So b satisfies a polynomial $g(x)$ over F, hence b also satisfies the polynomial $g(x)$ considered over $F(c) = K$. Moreover, if $h(x) = f(c - rx)$, then $h(x) \in K[x]$ and $h(b) = f(c - rb) = f(a) = 0$, since $a = c - rb$. Since b satisfies $h(x)$, $x - b$ is a factor of $h(x)$.

Thus in some extension $h(x)$ and $g(x)$ have $x - b$ as a common factor. We claim that $(x - b)$ is in fact a greatest common divisor of $h(x)$ and $g(x)$. For, if $b_j \neq b$ is another root of $g(x)$, then.

$h(b_j) = f(c - rb_j) \neq 0$, since r can be chosen such that for $j \neq 1$, $a_i \neq c - r\,b_j$, $i \geq 2$. Also, since $(x - b)^2 \mid g(x)$, $(x - b)^2$ cannot divide the greatest common divisor of $h(x)$ and $g(x)$. Thus $(x - b)$ is the greatest common divisor of $h(x)$ and $g(x)$. Thus $(x - b)$ is the greatest common divisor of $h(x)$ and $g(x)$ over some extension of K. Thus $x - b \in K[x] \Rightarrow b \in K$. We have $K = F(c)$, so $b \in F(c)$. Since $a = c - rb$, and $b, c \in K$, $r \in F \subset F(c)$, we get $a \in F(c)$ which implies $F(a, b) \subset F(c)$. Hence $F(a, b) - F(c)$.

Definition 11.2.6: The extension of a field $(F, +, .)$ is a *simple extension* of F if $K = F(\alpha)$ for some $\alpha \in K$.

Theorem 11.2.3: *Any finite field extension of a field of characteristic zero is a simple extension.*

Proof: Let K be a finite extension of F. Then there are finite number of elements $a_1, a_2, \ldots a_n$ in K such that $K = F(a_1, a_2, \ldots, a_n)$. We prove the result by induction. When $n = 1$, $K = F(a_1)$ is simple extension. Let $n > 1$, and let the theorem hold for all finite extension of F generated by less than n elements.

Let $K_1 = F(a_1, a_2, \ldots, a_{n-1})$, then by induction there exists $b \in K_1$ such that
$K_1 = F(a_1, a_2, \ldots, a_{n-1}) = F(b)$.

Thus $K = F(b, a_n)$. Now by the previous theorem 11.2.2 we have $K = F(c)$ for some $c \in F(b, a_n)$.

PROBLEMS

1. If F is of characteristic zero and $f(x) \in F[x]$ is such that $f'(x) = 0$, prove that $f(x) = \alpha \in F$.
2. If F is of characteristic $p \neq 0$, and if $f(x) \in F[x]$ is such that $f'(x) = 0$, prove that $f(x) = g(x^p)$ for some polynomial $g(x) \in F[x]$.

   Definition: A field $(F, +, .)$ is called *perfect* if all finite extension of F are separable.
3. Show that any field of characteristic zero is *perfect*.
4. (*a*) If F is of characteristic $p \neq 0$ show that for $a, b \in F$ $(a + b)^{p^m} = a^{p^m} + b^{p^m}$

   (*b*) If F is of characteristic $p \neq 0$, and if K is an extension of F, let $T = (a \in K \mid a^{p^m} \in F$ for some $n\}$. Prove that $(T, +, .)$ is a subfield of $(K, +, .)$.
5. Show that any finite field is *perfect*.
6. Show that a field F of characteristic $p \neq 0$ is perfect if and only if for every $a \in F$ we can find $a, b \in F$ such that $b^p = a$.
7. If K is an extension of F, prove that the set of elements in K which are separable over F form a subfield of K.
8. If K is a finite, separable extension of F prove that K is a simple extension of F.
9. If one of a or b is separable over F, prove that $F(a, b)$ is a simple extension of F.

CHAPTER 12

GALOIS THEORY

Let $(F[x], +, .)$ be a polynomial ring over the field. $(F, +, .)$. We shall associate with polynomial $p(x) \in F[x]$ a group, called the *Galois group of* $p(x)$. The association between the polynomial $p(x) \in F[x]$, its spliting field $(E, +, .)$ and the automorphism group $(A(E/F), 0)$ forms the basis for the Galois theory. There is a close relation between the roots of a polynomial and its Galois group. Infact, Galois group is a certain permutation group of the roots of the polynomial. Here we shall assume that all our fields are of characteristic 0. In this chapter we shall study the above ideas.

12.1 AUTOMORPHISMS AND GALOIS GROUPS

Definition 12.1.1: An *automorphism* f of the field $(K, +,.)$ is a one-to-one mapping f of K onto itself such that addition and multiplication are preserved, that is,

$f(a + b) = f(a) + f(b)$ and $f(a . b) = f(a) . f(b), \forall a, b \in K.$

Two automorphisms f and g are said to be distinct if $f(a) \neq g(a)$ for some element $a \in K$. The set of all automorphism is denoted by $A(K)$.

Definition 12.1.2: Let $(F, +, .)$ be a subfield of the field $(K, +, .)$. Then the set of all automorphisms of K which do not change the elements of' F is denoted by $A(K/F)$. The automorphism of K which leaves the elements of F fixed is called *F-automorphism*.

Theorem 12.1.1: *If $(K, +,.)$ is a field and if $\sigma_1, \sigma_2, ..., \sigma_n$ are distinict automorphisms of K, then $\sigma_1, \sigma_2, ... \sigma_n$ are linearly independent over K.*

Proof: We shall prove the result by induction. Suppose for $a_1, a_2, ..., a_n \in K$, not all zero.

$$a_1 \sigma_1 + a_2 \sigma_2 + ... + a_n \sigma_n = \bar{0}, \qquad ...(1)$$

Where $\bar{0}$ is a zero function.

For $n = 1$, $a_1 \sigma_1(x) = 0, \forall x \in K \Rightarrow a_1 = 0$ as $\sigma_1(x) \neq 0$ because σ_1 is an automorphism of K.

Hence the result holds good for $n = 1$.

Let us suppose that the result holds for any set of automorphisms which has less than n members.

In (1) Let $a_n \neq 0$, then (1) can be written as

$$a_1 a_n^{-1} \sigma_1 + a_2 a_n^{-1} \sigma_2 + ... + a_{n-1} - a_n^{-1} \sigma_{n-1} + \sigma_n = \bar{0}$$

or $b_1 \sigma_1 + b_2 \sigma_2 + ... + b_{n-1} \sigma_{n-1} + \sigma_n = \bar{0}$ ...(2)

Where $b_i = a_i \cdot a_n^{-1}$, $i = 1, 2, ..., n - 1$.

Since σ_1 and σ_2 are distinct, then there exists some $c \in K$ such that $\sigma_1(c) \neq \sigma_2(c)$ and $c \cdot x \in K$.

Thus $b_1 \sigma_1 (c. x) + b_2 \sigma_2 (c. x) + ... + b_{n-1} \sigma_{n-1} (cx) + \sigma_n (cx) = 0 (cx) = 0$

or $b_1 \sigma_1 (c) \sigma_1 (x) + b_2 \sigma_2 (c) \sigma_2 (x) + ... + b_{n-1} \sigma_{n-1} (c) . \sigma_{n-1} (x) + \sigma_n (c) \sigma_n (x) = 0$

or
$$b_1 \frac{\sigma_1(c)}{\sigma_n(c)} \sigma_1(x) + b_2 \frac{\sigma_2(c)}{\sigma_n(c)} \sigma_2(x) + ...$$

$$+ b_{n-1} \frac{\sigma_{n-1}(c)}{\sigma_n(c)} \sigma_{n-1}(x) + \sigma_n(x) = 0 \quad ...(3)$$

(2) Can be written as

$b_1 \sigma_1 (x) + b_2 \sigma_2 (x) + ... + b_{n-1} \sigma_{n-1} (x) + \sigma_n (x) = 0$ ...(4)

Subtracting (4) from (3) we get

$$b_1\left[\frac{\sigma_1(c)}{\sigma_n(c)} - 1\right]\sigma_1(x) + b_2\left[\frac{\sigma_2(c)}{\sigma_n(c)} - 1\right]\sigma_2(x) + b_{n-1}\left[\frac{\sigma_{n-1}(c)}{\sigma_n(c)} - 1\right]\sigma_{n-1}(x) = 0$$

Since $b_1\left[\frac{\sigma_1(c)}{\sigma_n(c)} - 1\right] \neq 0$, the last equation implies that $\sigma_1, \sigma_2, ..., \sigma_{n-1}$ are linearly dependent over K which contradicts the assumption of induction hypothesis.

Hence $\sigma_1, \sigma_2, ..., \sigma_n$ are linearly independent over K.

Theorem 12.1.2: *The collection A (K) of all automorphism of K is a group (A (K), 0) under functional composition. Moreover (A (K/F), 0) is a sub-group of the group (A (K), 0).*

Proof: We verify the group axioms to prove that $(A(K), 0)$ is group.

1. Closure: Let $f_1, f_2 \in A(K)$, then f_1 and f_2 are both one-to-one mapping. So $f_1 \, o \, f_2$ is also one-to-one mapping of K onto itself. For $a, b \in K$.

$$\begin{aligned}(f_1 \, o \, f_2)(a + b) &= f_1(f_2(a + b)) = f_1(f_2(a) + f_2(b)) \\ &= f_1(f_2(a)) + f_1(f_2(b)) \\ &= (f_1 \, o \, f_2)(a) + (f_1 \, o \, f_2)(b),\end{aligned}$$

and

$$\begin{aligned}(f_1 \, o \, f_2)(a \, . \, b) &= f_1(f_2(a \, . \, b)) = f_1(f_2(a) \, . \, f_2(b)) \\ &= f_1(f_2(a)) \, . \, f_1(f_2(b)) \\ &= (f_1 \, o \, f_2)(a) \, . \, (f_1 \, o \, f_2)(b),\end{aligned}$$

Thus $f_1 \, o \, f_2$ is an automorphism of K. Hence $f_1 \, o \, f_2 \in A(K)$.

2. Associative: We know that the functional composition is associative.

3. Existence of identity: Since identity mapping I of K is one-to-one mapping of K onto itself and preserves addition and multiplication, so identity mapping is an automorphism. Hence $I \in A(K)$ is the identity of $A(K)$.

4. Existence and Inverse: Let $f \in A(K)$. Then f is one-to-one mapping of K onto itself. So f^{-1} is also one-to-one mapping of K onto itself. For any $a \in K$, $f(a) = b \Rightarrow a = f^{-1}(b)$.

Now for any two $a_1, a_2 \in K$, $f(a_1) = b_1 \Rightarrow a_1 = f^{-1}(b_1)$ and $f(a_2) = b_2 \Rightarrow a_2 = f^{-1}(b_2)$.

So $f(a_1 + a_2) = f(a_1) + f(a_2) = b_1 + b_2 \Rightarrow a_1 + a_2 = f^{-1}(b_1 + b_2)$

$\Rightarrow f^{-1}(b_1) + f^{-1}(b_2) = f^{-1}(b_1 + b_2)$

and $f(a_1 \cdot a_2) = f(a_1) \cdot f(a_2) = b_1 \cdot b_2 \Rightarrow f^{-1}(b_1 \cdot b_2) = a_1 \cdot a_2$

$\Rightarrow f^{-1}(b_1 \cdot b_2) = f^{-1}(b_1) \cdot f^{-1}(b_2)$.

Thus f^{-1} is an automorphism. Hence $f^{-1} \in A(K)$.

This shows that (A (K), 0) is a group.

For the next part of theorem we consider two elements f and g of $A(K/F)$. Since f and $g \in$ A (K/F), then for $x \in F$,

$f(x) = x \Rightarrow x = f^{-1}(x)$ and $g(x) = x \Rightarrow x = g^{-1}(x)$.

So $(f \circ g^{-1})(x) = f(g^{-1}(x)) = f(x) = x$.

Hence $f \circ g^{-1} \in A(K/F)$. Thus (A (K/F), 0) is a sub-group of the group (A (K), 0).

Definition 12.1.3: The group $(A(K/F), 0)$ is the *automorphism group* of K relative to F.

Definition 12.1.4: If $(A(K), 0)$ is a group of automorphism of K, then the *fixed field* F_0 of $(A(K), 0)$ is the set of all elements $a \in K$ such that $\sigma(a) = a$ for all $\sigma \in A(K)$. That is, $F_0 = \{a \in K \mid \sigma(a) = a, \forall \sigma \in A(K)\}$

***Lemma 12.1.1:** The system $(F_0, +, .)$ is a subfield of the field $(K, +, .)$.*

Proof: Since for all $\sigma \in A(K)$, $\sigma(0) = 0$ and $\sigma(1) = 1$. So $0, 1 \in F_0$.

Let $a, b \in F_0$, then $\forall \sigma \in A(K)$, $\sigma(a) = a$ and $\sigma(b) = b$. But all $\sigma \in A(K)$ preserves addition and multiplication of the field $(K, +, .)$.

So $\sigma(a \pm b) = \sigma(a) \pm \sigma(b) = a \pm b$

and $\sigma(a \cdot b) = \sigma(a) \cdot \sigma(b) = a \cdot b$. Hence $a \pm b$, $ab \in F_0$, if $b \neq 0$, then $\sigma(b^{-1}) = [\sigma(b)]^{-1} = b^{-1}$, hence $b^{-1} \in F_0$. Thus the system $(F_0, +, .)$ is a subfield of the field $(K, +, .)$.

Remark 1: Let $(P, +, .)$ be a *prime subfield of the field* $(K, +, .)$. That is, there exists no subfield of $(K, +, .)$ which is also subfield of $(P, +, .)$. So $(P, +, .)$ is contained in every subfield of $(K, +, .)$. In particular, if $(F_0, +, .)$ is the fixed field of a group $(A(K), 0)$ of automorphisms of K, then $P \subseteq F_0$. Since $\sigma(x) = x$, $\forall x \in F_0$ and for every $\sigma \in A(K)$, $\sigma(x) = x$, $\forall x \in P$ and $\sigma \in A(K)$. Thus every automorphism of K leaves the elements of P fixed. So every automorphism of K is *a P–automorphism*.

Remark 2: Every automorphism of K is *a* F_0 *–automorphism* where F_0 is the fixed field of $A(K)$.

Remark 3: If $(K, +, .)$ is an extension field of the field $(F, +, .)$ and $(A(K/F), 0)$ is the automorphism group of K relative to F, Then $F \subset F_0$, the fixed field of $A(K)$.

Example 12.1.1: Let $(K, +, .)$ be the field of complex numbers and let $(F, +, .)$ be the field of real numbers. We compute $A(K/F)$. If $\sigma \in A(K)$, then since $i^2 = -1$, and $-1 = \sigma(-1) = \sigma(i^2) = \sigma(i \cdot i) = \sigma(i) \cdot \sigma(i) \Rightarrow \sigma(i) = \pm i$. If in addition σ leaves the element of F fixed

then
$$\begin{aligned} \sigma(a + ib) &= \sigma(a) + \sigma(ib) \\ &= \sigma(a) + \sigma(i)\,\sigma(b) \\ &= a \pm i\,b. \end{aligned}$$

each of the possibilities, namely the mapping

$\sigma_1(a + ib) = a + ib$ and $\sigma_2(a + ib) = a - ib$.

define automorphisms of K which leave the element of F fixed. So $A(K/F) = \{\sigma_1, \sigma_2\}$. It is clear that $(A(K/F), 0)$ is a group of order 2.

0	σ_1	σ_2
σ_1	σ_1	σ_2
σ_2	σ_2	σ_1

Now we compute the *fixed field* of $A(K/F)$. Since σ_1 and σ_2 leave the elements of F fixed, the fixed field F_0 of $A(K/F)$ contains F, *i.e.*, $F \subseteq F_0$. If $a + ib \in F_0$, then $\sigma_2(a + ib) = a - ib = a + ib$ which implies $b = 0$ and $a = a + ib \in F_0$. Hence F_0 does not contain any element of the form $a + ib$. Thus we see that the fixed field of $A(K/F)$ is the set F of all real numbers.

Example 12.1.2: The group $(A(F/R), 0)$, where $(F, +, .)$ is the field $(R[x] / x^2 - 2, +, .)$, is of order 2.

Here $F = \{a + b\sqrt{2} \mid a, b \in R\}$, where R is the set of all rational number. If $\sigma \in A(F/R)$, then since $\left(\sqrt{2}\right)^2 = 2$,

$$\sigma\left(\sqrt{2}\right)^2 = \sigma(2) \Rightarrow \sigma\left(\sqrt{2}\right)\sigma\left(\sqrt{2}\right) = 2$$

$$\Rightarrow \sigma\left(\sqrt{2}\right) = \pm\sqrt{2}$$

So for any $a + b\sqrt{2} \in F$,

$$\sigma\left(a + b\sqrt{2}\right) = \sigma(a) + \sigma\left(b\sqrt{2}\right)$$

$$= \sigma(a) + \sigma(b)\,\sigma\left(\sqrt{2}\right)$$

$$= a \pm b\sqrt{2}$$

So $\sigma_1\left(a + b\sqrt{2}\right) = a + b\sqrt{2}$ and

$\sigma_2\left(a + b\sqrt{2}\right) = a - b\sqrt{2}$ define R-automorphisms. Hence $A(F/R) = \{\sigma_1, \sigma_2\}$.

It is clear that $(A(F/R), 0)$ is a group of order 2.

0	σ_1	σ_2
σ_1	σ_1	σ_2
σ_2	σ_2	σ_1

It is clear that the fixed field F_0 contains R, *i.e.*, $R \subseteq F_0$. If $a + b\sqrt{2} \in F_0$, then

$\sigma_2(a + b\sqrt{2}) = a - b\sqrt{2} = a + b\sqrt{2} \Rightarrow b = 0.$

Hence $a = a + b\sqrt{2} \in F_0$. Thus the fixed field is the set R of all real numbers.

Example 12.1.3: Let *(Q, +, .)* be the field of rational numbers and let *(K, +, .)*, where $K = Q(\sqrt[3]{2})$ and $\sqrt[3]{2}$ is the real cube root of 2, be a field, every element in *K* is of the form $a_0 + a_1\sqrt[3]{2} + a_2\left(\sqrt[3]{2}\right)^2, a_0, a_1, a_2 \in Q$. If $\sigma \in A(K)$, then

$$\sigma\left(\sqrt[3]{2}\right)^3 = \sigma\left(\sqrt[3]{2}\right).\sigma\left(\sqrt[3]{2}\right).\sigma\left(\sqrt[3]{2}\right) = \sigma(2) = 2$$

$$\Rightarrow \sigma\left(\sqrt[3]{2}\right) = \sqrt[3]{2}.$$

But $\sigma\left[a_0 + a_1\left(\sqrt[3]{2}\right) + a_2\left(\sqrt[3]{2}\right)^2\right] = \sigma(a_0) + \sigma\left(a_1\sqrt[3]{2}\right) + \sigma\left(a_2\left(\sqrt[3]{2}\right)\right)^2 = a_0 + a_1\sqrt[3]{2} + a_2\left(\sqrt[3]{2}\right)^2$

Thus σ is the identity automorphism of *K*. We thus see that *A (K/F)* consists of only identity mapping and in this case the fixed field of A *(K/F)* is the field *(K, +,.)*.

Theorem 12.1.3: *If m (x) is an element of F_1 [x], where (F_1, +, .) is a subfield of the field (F, +, .) and if r ∈ F is a zero of m (x), then for each automorphism σ ∈ A (F/F_1) the image σ (r) of r is also zero of m (x).*

Proof: Let $m(x) = a_0 + a_1 x + ... + a_n x^n$ with $a_i \in F_1$, and $m(r) = 0 = a_0 + a_1 . r + ... + a_n r^n$.

Now we apply $\sigma \in A(F/F_1)$ to this equation which leaves the elements of F_1 fixed. Thus we have

$$\begin{aligned} 0 = \sigma(0) &= \sigma(a_0 + a_1 . r + ... + a_n . r^n) \\ &= \sigma(a_0) + \sigma(a_1)\sigma(r) + ... + \sigma(a_n) . \sigma(r^n) \\ &= a_0 + a_1 \sigma(r) + ... + a_n \sigma(r)^n, \end{aligned}$$

This equation implies that $m(\sigma(r)) = 0$.

So we see that the image σ *(r)* is a zero of *m (x)*. This theorem can be stated as **"The complex conjugate of a complex zero of a real polynomial is also a zero of the polynomial."**

Definition 12.1.5: The field (F_0, +, .) is the coefficient field of the polynomial *m (x)* if and only if *m (x)* is in the polynomial domain of (F_0, +,.) but not in the polynomial domain of any subfield of F_0. Then *(A (F/F_0)*, 0), where *F* is the splitting field of *m (x)*, is the *Galois group* of *m (x)*.

In example 12.1.1, we see that *(R, +, .)* is the coefficient field of $m(x) = x^2 + 1$, that $C = \{a + ib \mid a, b \in R\}$ is the splitting field of *m (x)* and that *(A (C/R)*, 0) is a group of order 2 consisting of two automorphisms

$$\sigma_1 : a + ib \to a + ib,$$
$$\sigma_2 : a + ib \to a - ib.$$

From the examples we conclude that if *K* is the splitting field of $f(x) \in F[x]$ where *(F, +, .)* is subfield of the field (*K*, +, .), then the *Galois group* is a *permulation group* of the roots of *f (x)*.

Theorem 12.1.4: *If m (x) is an irreducible polynomial of degree two in F [x], where (F, +, .) is a field of characteristic of not 2, then the group (A (F_1/F), 0), F_1 = F [x] / (m (x)), is of order two and the only elements of F_1 which are fixed by both automorphism in A (F_1/F) are the elements of F.*

Proof: Consider the polynomial $m(x) = a x^2 + b x + c$

CHAPTER 12

over F, we know that $m(x)$ has a zero, λ, in F_1, so that in F_1

$$a\lambda^2 + b\lambda + c = 0.$$

This can be written

$$\lambda^2 + a^{-1} . b\lambda = -a^{-1} . c$$

or $\lambda^2 + a^{-1} b . \lambda + (1/4)(a^{-1} . b)^2 = (1/4)(a^{-1} . b)^2 - a^{-1} . c,$

So that $(2a\lambda + b)^2 = b^2 - 4ac.$

Note how essential it is that the characteristic of F be different from two. From this formal manipulation we conclude that the polynomial

$$m_1(x) = x^2 - r.$$

Where $r = b^2 - 4ac,$

of $F[x]$ has a zero, namely $2a\lambda + b$, in F_1. Also the element $2a\lambda + b \notin F$ since $2a$ and $b \in F$ but $\lambda \notin F$. Thus $m_1(x)$ is an irreducible polynomial of degree two in $F[x]$, and certainly

$$F_1 \supseteq F[x] / (m_1(x)) = F_2$$

Since every element of F_2 is of the form

$h + k . \lambda_1, h, k \in F$, where λ_1 is a zero of $m_1(x_1)$,

$$\lambda_1 = 2a\lambda + b.$$

However, this last equation can be solved for λ,

$$\lambda = (2a)^{-1}(\lambda_1 - b),$$

So that each element of F_1,

$m + n\lambda$, where $m, n \in F$

is expressible as

$$m + n\lambda = (m - n(2a)^{-1} b) + n(2a)^{-1}\lambda_1.$$

Thus $F_2 \supseteq F_1$, and we conclude that

$$F_1 = F[x]/x^2 - r.$$

The element λ_1, a zero of $x^2 - r$, can be thought of a square root of r. Then we proceed exactly as in example 12.1.2 and show that the element of $A(F_1/F)$ are precisely

$\sigma_1 : h + k\lambda_1 \to h + k\lambda_1$

and $\sigma_2 : h + k\lambda_1 \to h - k\lambda_1$

Now suppose $m + n\lambda_1 \in F_1$ is left unchanged by all elements of $A(F_1/F)$. Then we have

$\sigma_2(m + n\lambda_1) = m - n\lambda_1 = m + n\lambda_1,$

So that $2n\lambda_1 = 0$

and $n = 0(2\lambda_1)^{-1} = 0.$

Thus $m + n\lambda_1 = m \in F$, and we conclude that the only elements of F_1 which are unchanged by both elements of $(A(F_1/F)$ are the elements of F.

Example 12.1.4: Determine the group $(A(F/R), o)$, where $F = R[x]/(x^2 + 4x + 2)$.

Solution: The polynomial $m(x) = x^2 + 4x + 2$ is irreducible over R since 1, 2, –1 and –2 are not roots of $m(x)$. In the field F it has a root λ and from

$$\lambda^2 + 4\lambda + 2 = 0$$

We get the equation

$$\lambda^2 + 4\lambda + 4 = 2$$

or

$$(\lambda + 2)^2 = 2$$

Thus $(\lambda + 2) = \sqrt{2}$ or $-\sqrt{2}$

and $F = R\,[x]/(x^2 - 2).$

Then the Galois group of

$$m\,(x) = x^2 + 4x + 2$$

is precisely the group of order 2 which was determined in Example 12.1.1. In general the Golois group of polynomial of degree n over the coefficient field F_0 is a finite group whose order is less than or equal to n, and the only elements of the splitting field which are left unchanged by all automorphisms in the Galois group are the elements of F_0 provided the characteristic of F_0 does not divide n.

Theorem 12.1.5: *If K is a finite extension of F then (A (K/F), o) is a finite group and its order,* 0 (A *(K/F)) satisfies*

$$0\,(A\,(K/F)) \leq [K : F].$$

Proof: Since K is a finite extension of F, $(F, +, .)$ is a subfield of the field $(K, +, .)$ and K is a finite vector space over the field $(F, +, .)$. Let the dimension of the vector space K (F) be finite, n *i.e.*, $[K : F] = n$. Then there is a basis $u_1, u_2, \ldots, u_n$ of K over F. Now we suppose $n + 1$ distinct automorphisms $\sigma_1\, \sigma_2, \ldots, \sigma_{n+1}$ in $A\,(K/F)$. We also consider the system of n homogeneous linear equations in the $n + 1$ unknowns $x_1, x_2, \ldots, x_{n+1}$.

$$\sigma_1\,(u_1)\,x_1 + \sigma_2\,(u_1)\,x_2 + \ldots + \sigma_{n+1}\,(u_1)\,x_{n+1} = 0$$
$$\sigma_1\,(u_2)\,x_1 + \sigma_2\,(u_2)\,x_2 + \ldots + \sigma_{n+1}\,(u_2)\,x_{n+1} = 0$$
$$\vdots$$
$$\sigma_1\,(u_n)\,x_1 + \sigma_2\,(u_n)\,x_2 + \ldots + \sigma_{n+1}\,(u_n)\,x_{n+1} = 0$$

Since the number of equations is less than the number of unknowns, the system of n linear equation has a non-trivial solution (not all zero)

$$x_1 = a_1,\ x_2 = a_2, \ldots, x_n = a_n,\ x_{n+1} = a_{n+1}.$$

Thus

$$a_1\,\sigma_1\,(u_i) + a_2\,\sigma_2\,(u_i) + a_3\,\sigma_3\,(u_i) + \ldots +_{n+1}\,\sigma_{n+1}\,(u_i) = 0$$
$$i = 1, 2, \ldots, n$$

Since every element of F is left fixed by all elements of σ_i and since any element $t \in K$ can be written as

$$t = b_1\,u_1 + b_2\,u_2 + \ldots + b_n\,u_n,\ b_i \in F,\ i = 1, 2, \ldots, n.$$
$$a_1\,\sigma_1\,(t) = a_1\,\sigma_1\,(b_1\,u_1 + b_2\,u_2 + \ldots + b_n\,u_n)$$
$$= a_1\,\sigma_1\,(b_1)\,.\,\sigma_1\,(u_1) + \ldots + \sigma_1\,(b_n)\,.\,\sigma_1\,(u_n)$$
$$= a_1\,[b_1\,\sigma_1\,(u_1) + \ldots + b_n\,.\,\sigma_1\,(u_n)]$$
$$a_i\,\sigma_i\,(t) = a_i\,[b_1\,\sigma_i\,(u_1) + \ldots + b_n\,\sigma_i\,((u_n)]$$
$$\vdots$$
$$a_{n+1}\,\sigma_{n+1}\,(t) = a_{n+1}\,[b_1\,\sigma_{n+1}\,(u_1) + \ldots + b_{n+1}\,(u_n)]$$

So

$$a_1\,\sigma_1\,(t) + \ldots + a_i\,\sigma_i\,(t) + \ldots + a_{n+1}\,\sigma_{n+1}\,(t)\ldots.$$
$$= b_1\,[a_1\,\sigma_1\,(u_1) + \ldots + a_{n+1}\,\sigma_{n+1}\,(u_1)] + \ldots +$$

$b_n [a_1 \sigma_1 (u_n) +. .+ a_i \sigma_i (u_n) +...+ a_{n+1} \sigma_{n+1} (u_n)]$
$= 0 +...+ 0 = 0$. From (1).

Since $a_1, a_2,...a_{n+1}$, are not all zero, the relation (2) shows that $\sigma_1 \sigma_2,..., \sigma_{n+1}$ are linearly dependent which is contradiction to the fact that any set of automorphisms of K is linearly independent by theorem 12.1.1. Hence in $A (K/F)$ There cannot be $n + 1$ distinct elements.

Thus 0 (A $(K/F)) \leq (K : F)$. In examples we discussed the groups of auto morphisms of fields and of fixed fields under such groups, we saw that the fixed field F_o of groups $(A (K/F), 0)$ contains F and some times $F_o = F$. Thus to impose the condition on an extension K of F that F is equal to the fixed field of $(A (K/F), 0)$ is an at most interesting.

Definition 12.1.6: K is a *normal extension* of F if K is a *finite extension* of F such that F is the fixed field of $(A (K/F), 0)$.

Since every finite extension is an algebraic extension, then an algebraic extension K is a normal extension of F if F is the fixed field of the group $(A (K/F), 0)$.

Theorem 12.1.6: *Let E be a splitting field of f (x) ∈ F [x]. Let p (x) be an irreducible polynomial in F [x]. If P (x) has one root in E, then p (x) can be factored into linear factors over E.*

Proof: Let α be a root of $p (x)$. Then we have to show that $p (x)$ has all its roots in E. We know that $F \subseteq F (\alpha) \subseteq E$. So $p (x)$ belong to $E [x]$. Let $\bar{E}$ be a splitting field of $p (x)$ over E and let $\bar{\alpha}$ be another root of $p (x)$. We shall prove that $\bar{E} = E$. We know that $F (\alpha) \cong F (\bar{\alpha})$ by a isomorphism ϕ by the corollary of theorem 11.1.13, Now from definition 11.1.13. It is clear that if $\alpha_1, \alpha_2,..., \alpha_n$ are roots of $f (x) \in F [x]$, then splitting field of $f (x)$ over F is $E = F (\alpha_1, \alpha_2,..., \alpha_n)$, generated by F and n roots $\alpha_1, \alpha_2,..., \alpha_n$. Since $\alpha \in E$ is a root of $p (x)$ in $F [x]$, then clearly $E = F (\alpha, \alpha_1, \alpha_2,..., \alpha_n)$. Now we consider $E (\bar{\alpha})$. We claim that $E (\bar{\alpha})$ is a splitting field for $f (x)$ over $F (\bar{\alpha})$. Certainly $f (x)$ splits in $E (\bar{\alpha})$. Suppose that $f (x)$ splits in K where $E (\bar{\alpha}) \supseteq K \supseteq F (\bar{\alpha})$. Then $\{\alpha_1, \alpha_2,..., \alpha_n\} \subseteq K$ and so $K \supseteq F (\bar{\alpha}, \alpha_1, \alpha_2,..., \alpha_n) = E (\bar{\alpha})$. Thus $K = E (\bar{\alpha})$. But by theorem 11.1.14 we may extend to an isomorphism ϕ of E onto $E (\bar{\alpha})$ and so, in particular, $[E : F (\alpha) = [E (\bar{\alpha}) : F (\bar{\alpha})]$; and since $[F (\alpha) : F] =[F (\bar{\alpha}) : F]$, if follows that $[E : F] = [E (\bar{\alpha}): F]$, and as $E \subseteq E (\bar{\alpha})$, that means that $E = E (\bar{\alpha})$. Thus E must contain all the roots of $p (x)$ and so $p (x)$ must split in E. (Thus $E = \bar{E}$),

Theorem 12.1.7: *Let K be a finite extension of F, a field of characteristic zero. If K is the splitting field of a polynomial f (x) ∈ F [x] then 0 (A (K/F)) = [K : F].*

Proof: We Lnow that a field which is a finite extension of a field of characteristic zero is simple extension. So K is a simple extension of F; $K = F (w)$, $w \in K$ is a root of an irreducible polynomial $f (x) \in F [x]$. Since $f (x) \in F [x]$ splits in K and $p (x)$ has a one root w in K, then by the theorem 12.1.6 $p (x)$ splits in K. Thus K is the splitting field of $p (x) \in F [x]$. Hence $0(A(K/F)) = [K:F] = \deg p(x)$

We know that if $f (x) \in F (x)$ is an irreducible polynomial and K is the splitting field of F, then the Galois group is $(A (K/F), 0)$ and the fixed field of $(A (K/F), 0)$ is precisely F. But in theorem 12.1.7 we have seen that if an irreducible polynomial $f (x) \in F [x]$ has a root in the field E, then $f (x)$ splits over E. Thus we can define normal extension of F as follows.

Definition 12.1.7: A finite extension $K \supseteq F$ is called a *normal extension* of F if, whenever an irreducible polynomial $p (x) \in F [x]$ having one root in K splits in $K [x]$.

Remark I. We can easily see that if F has characteristic zero then K is a normal extension of F if and only if K is the splitting field for a polynomial $f(x) \in F[x]$. And $o(A(K/F) = [K : F]$.

Remark II. If $F \subseteq E \subseteq K$ and if K is a normal extension of F, then K is a normal extension of E.

In Example 12.1.1, K is a normal extension of F,

in example 12.1.2, K is a normal extension of R, and

in example 12.1.3, K is not a normal extension of R.

In other words K is a normal extension of F if every element of K which is outside F is moked by some element in $A(K/F)$.

Theorem 12.1.8: *Let K be a finite extension of a field F of characteristic zero. Then K is a normal extension of F if and only if F is the fixed field of A (K/F).*

Proof: Let K be a normal extension of F. Then K is a finite extension and consequently K is a simple extension of F; say $K = F(w)$. So $o(A(K/F)) = [K : F] = n$, where n is a degree of an irreducible polynomial $p(x) \in F[x]$ of which w is a root.

Let us consider the fixed field F_0 of $(A(K/F), 0)$. It is clear that $F \subseteq F_0 \subseteq K$. It will suffice to show that $p(x)$ is irreducible over F_0, for then $[K : F] = [K : F_0]$ and so from $F \subseteq F_0$ we can conclude $F_0 = F$. Let $p(x)$ be not irreducible over F_0 and let $g(x)$ be the irreducible factor of $p(x)$ over F_0 of which w is a root. If $\bar{w}$ is another root of $p(x)$ then we know that there is an auto morphism $\phi \in A(K/F)$ such that $\phi(w) = \bar{w}$ and $F(w) \cong F(\bar{w})$. But since $\phi(a) = a$, $\forall a \in F_0$ it must show that $0 = \phi(0) = \phi(g(w)) = g(\phi(w)) = g(\bar{w})$. Thus all n distinct roots of $p(x)$ are roots of $g(x)$ and so $g(x)$ must have degree $\geq n$. But $g(x) \mid p(x)$ and thus $g(x)$ must be an associate of $p(x)$ in $F_0[x]$ and hence it follows that $p(x)$ too must be irreducible over F_0.

Conversely, suppose that $F = \{a \mid \sigma(a) = a, \forall \sigma \in A(K/F)\}$. Since $[K : F]$ is finite and F has characteristic zero we know that $K = F(w)$, for some $w \in K$. We shall exhibit a polynomial in $F[x]$ of which w is a root and for which K is a splitting field. From this it follows from theorem 12.1.6 that K is normal extension of F. Let $\sigma_1, \sigma_2, ..., \sigma_n$ be the automorphisms in $A(K/F)$. We consider

$g(x) = (x - \sigma_1(w))(x - \sigma_2(w))...(x - \sigma_n(w))$, clearly $g(x)$ splits in $K[x]$ and $g(w) = 0$, but we must show that $g(x)$ has coefficients in F.

We note that every automorphism $\sigma \in A(K/F)$ induces a natural automorphism σ of $K[x]$. We claim that if $f(x) \in K[x]$. Then $\sigma(1 f(x) = f(x)$, $\forall \sigma \in A(K/F)$ if and only if $f(x) \in F(x)$. Let

$f(x) = \Sigma a_i x^i$, then $\sigma(f(x)) = \Sigma (a_u) x^i = \Sigma a_i x^i$

Thus $f(x) = \sigma(f(x))$, $\forall \sigma \in A(K/F)$ if and only if $\sigma(a_j) = a_j$ $\forall \sigma$ if and only if $a_j \in F$. Now with this in mind we simply observe that in the automorphism σ of $K[x]$ we have

$\sigma(g(x)) = (x - \sigma\sigma_1(w)) .... (x - \sigma\sigma_n(w))$, $\forall \sigma \in A(K/F)$, and since $(A(K/F), o)$ is a group of automorphisms, we have that $\sigma(g(x)) = g(x)$, $\forall \sigma \in A(A/F)$. Hence $g(x) \in F[x\}$. Now $g(x)$ can split in no subfield $K \supseteq H \supseteq F$, since $K = F(w)$ and if $w \in H$, then $H \supseteq F(w) = K$. Thus K is the splitting field for $g(x) \in F[x]$.

Lemma 12.1.2: *Let K be a normal extension of F. Let $F \subseteq F_0 \subseteq K$. If a subgroup (H, o) of (A (K/F), 0) has F as its fixed field then $A(K/F_0) = H$.*

Proof: By definition we have $H \subseteq A(K/F_0)$. Since K is normal extension over F_0, $o(A(K/F_0)) = [K : F_0] = m$. Let $K = F_0(v)$. If $H = \{\sigma_1, \sigma_1, \sigma_2, ..., \sigma_s\}$ $(s \leq m$, then polynomial h

CHAPTER 12

$(x) = (x - \sigma_1 (v))....(x - \sigma_s (v))$ has coefficient in F_0, since F_0 is the fixed field of H, Thus v satisfies a polynomial in $F_0 [x]$ of degree $s \geq m$. Hence $m = [K : F_0] \leq s \leq m$. Thus $s = m$ and so $A (K/F_0) = H$.

Theorem 12.1.9: *(Fundamental Theorem of Galois Theory). Let E be a normal extension of F, a field of characteristic zero.*

1. *The mapping $K \to A (E/K)$ is a 1–1 mapping of the fields $\{K \mid E \supseteq K \supseteq F\}$ onto sub-groups of $A (E/F)$*
2. *If $E \supseteq H \supseteq F$, then $[E : K] = 0 (A (E/K))$ and $[K : F]$ = index of $A (E/A)$ in $A (E/F)$, in $A (E/F)$*
3. *$E \supseteq H \supseteq K \supseteq F$ if and only if $\{i\} \subseteq A (E/H) \subset A (E/K) \subseteq A (E/F)$.*
4. *K is a normal extension of F if and only if $A (E/K)$ is a normal sub-group of $A (E/F)$, in which case $(A (K/F), 0 \cong (A (E/F) /A (E/K), 0)$.*

Proof: The mapping T is defined by $T (K) = A (E/K)$. If E is the splitting field of some polynomial $f (x) \in F [x]$, then E is also splitting field of $f (x)$ overany subfield K containing F. So E is normal extension of K. Thus K is the fixed field of the group $(A (E/K), 0)$, the group of all automorphism which fix the elements of K. So $[E : A] = 0 (A (E/K))$. Moreover the mapping T is one-to-one because by theorem 12.1.8 K is the fixed field of $A (E/K)$. Thus if $A (E/K) = A (E/H) = S$, Then $H = K$ the fixed field of $S = \{a \mid \sigma (a) = a, \sigma \in S\}$. To show that T is onto let (S, o) be a sub-group $(A (E/K), 0)$. We define $K = \{a \mid \sigma (a) = \sigma \ \forall \ \sigma \in S\}$. The fixed field of S is K, by definition, hence by

Lemma 12.1.3: *We must have $S = A (E/K) = T (K)$. Thus the result 1 is proved.*

2. Since E is normal extension of K, then by theorem 12.1.7 we have $o (A (E/K)) = [E: K]$.

We also know that

$[E : F] = [E : K] \,.\, [K : F\} \Rightarrow o (A (E/F)) = o (A (E/K)).\ [K : F]$.

Hence $[K : F] = \dfrac{\underline{0}(A(E/F))}{\underline{0}(A(E/K))}$, the index of $A (E/K)$ in $A (E/F)$.

This proves 2.

3. It is clear that if $E \supseteq H \supset K \supseteq F$ then $A (E/H) \subseteq A (E/K)$. Since E is a normal extension of F, it is normal extension of H and K both, and consequently $o (A (E/H)) = [E : H]$ and $o (A . (E/K) = [E : K]$. So strict containment follows from $[E : H] > [E : K]$.

Conversely, if $A ((E/H) \subseteq A (E/K)$, then $E \supseteq H \supset K \supseteq F$ since H and K are fixed fields of $A (E/H)$ and $A (E/K)$ respectively.

4. Suppose K is a normal extension of F. Now let $E \in A (E/K)$ and $\sigma \in A (E/F)$. We have to show that $\sigma^{-1} o \rho o \sigma \in (E/K)$. to do this, in view of the $1 - 1$ correspondence T given above it suffices to show that for all $k \in K$, we have $\sigma^{-1} o E o \sigma (k) = k$, or that $E o \sigma (k) = \sigma (k)$, $\forall k \in K$ and $\forall \sigma \in A (E/F)$. To show that $(E o \sigma (k) = \sigma (k)$ it sufficis to show that $\sigma (k) \in K$ But this is the case, because in any event k and $\sigma (k)$ are conjugate over F and hence roots of the same irreducible polynomial $p (x) \in F [x]$. $p (x)$ has a root, namely k and as K is a normal extension of F it follows that K must contain all roots of $p (x)$, in particular $\sigma (k) \in K$.

Conversely, suppose that $(A (E/K), o)$ is a normal subgroup of $(A (E/F), o)$. We shall show that the fixed field of $A (K/F)$ is F. If $\sigma \in A (E/F)$, then for all $k \in K$, we have $\sigma (k) \in K$. To do this, let, $E \in A (E/K)$. Then $\sigma^{-1} o E o \sigma \in A (E/K)$ and so $\sigma^{-1} o E o \sigma (k) = k$

since k is fixed by $A(E/K)$. Thus $E(\sigma(k)) = \sigma(k)$, and so $\sigma(k)$ is fixed by all of $A(E/K)$ and so belongs to the fixed field of $A(E/K)$, which is K. Thus $\sigma(k) \in K$. This means that every automorphism $\sigma \in A(E/F)$ fixes K as a set and E the restriction of σ to K is a mapping σ^{-1} of K onto K, which is of course an automorphism. Now to complete the proof we prove that the fixed field of $A(E/K)$ is F. Let $f \in K$ and suppose $\sigma^*(f) = f$ for all $\sigma^* \in A(K/F)$. Then, in particular, $\sigma'(f) = f$ for all $\sigma' \in A(E/F)$ and so it follows that $\sigma(f) = f$ for all $\sigma \in A(E/F)$ since $\sigma(f) = \sigma'(f)$. Hence f belongs to the fixed field of $A(E/F)$, which is F.

This part of the proof suggests that if σ' denotes the restriction of σ to K, then the function $\phi : A(E/F) \to A(K/F)$ given by $\phi(\sigma) = \sigma'$ might be group homomorphism.

We prove this by verifying (*i*) $\phi(\sigma \, o \, \rho) = \phi(\sigma) \, o \, \phi(\rho)$, (*ii*) ϕ is an onto map, and (*iii*) the kernel of ϕ is $A(E/K)$. The proof of (*i*) is easy because the restriction of a function does not change the ordered pairs which constitute the function, thus since $(\sigma \, o \, \rho)(k) = \sigma(e(k)), (\sigma \, o \, e)' (\sigma \, o \, E) = (k) = \sigma(\rho(k) = \sigma(\rho'(k) = \sigma'(\rho'(k) = (\rho' \, o \, e')(k)$ (*iii*) If $\sigma \in A(E/F)$ is the identity automorphism when restricted to K this means that $\sigma \in A(E/K)$ by definition.

To prove (*ii*) we show that every automorphism in $A(K/F)$ arises as the restriction of an automorphism in $A(E/F)$ to K. To do this it suffices to show that every automorphism σ^* can be extended to an automorphism σ in $A(E/F)$. This is so since E is the splitting field of a polynomial in K, and every automorphism of K (fixing the elements of F) can be extended to an automorphism of E (fixing the elements of F). Thus our proof of the Fundamental theorem of Galois theory is complete.

PROBLEMS

1. Describe the Galois group of $m(x) = x^2 + x + 1$ over Q.
2. Describe the Galois group of $m(x) = x^2 + 2x + 1$ over Q.
3. Describe the Galois group of $m(x) = x^2 + x + 1$ over Z_3.
4. Let A be an automorphism of the field F. Show that the set $F_A = [x \mid x \in F \text{ and } A(x) = x\}$ is a subfield of F.
5. Show that the prime fields $(Q, +, .)$ and $(Z_p, +_p, o_p)$ have automorphism groups of order one.
6. Let $(F_1, +, .)$ be a subfield of $(F, +, .)$ containing F_0, then show that the set of all elements of $A(F/F_0)$ which leave the elements of F_1 unchanged is a sub-group of $A(F/F_0)$.
7. Show that the automorphism groups $(A(F), 0)$ and $(A(F,), o)$ are isomorphic if the fields $(F, +, .)$ and $(F_1, +, .)$ are isomorphic. By considering the fields $Q[x]/(x^2 - 2)$ and $Q[x] / (x^3 - 3)$, show that the converse is not true.
8. Show that the identity automorphism is the only automorphism of the field of real numbers. Is the same result true for the field of complex numbers? [**Ans.** No]
9. Show that Galois group of the splitting field K of $x^n - 1$ over Q is abelian.
10. Show that the Galois group of $x^6 - 2$ over $Q\left(\sqrt{-3}\right)$ is Z_6.
11. Let $K = F(a)$ be a simple algebraic extension of degree n of a field $(F, +, .)$ of characteristic zero. Show that the number of conjugates of a in K over F divides $[K : F]$.
12. Show that the Galois group of $(x^2 - 2)(x^3 - 3)$ over Q is isomorphic to $Z_3 \times Z_2$.
13. Let K be any algebraic extension of F. If $\{F_i\}$ $i \in Z$ be any non-void family of subfields containing F, such that each F_i is a normal extension of F_j show that the intersection $\cap F_i$ is also normal extension of F.
14. Let $F \subseteq K \subseteq L$ be three fields with the fields F of characteristic zero, K is a finite normal extension of F and L is a finite normal extension of K. Prove that if σ is any F- automorphism of K then there exists an F-automorphism η of L such that σ is restriction of η to K.

CHAPTER 12

15. Give an example of normal extension of degree 3 over Q and an example of an algebraic extension of Q of degree 3 which is not a normal extension.

16. Let K be a finite extension of a field F. Prove that K is a separable normal extension of F if and only if F is the fixed field under the group A (K/F).

17. Let K be a normal extension of a field F of characteristic zero such that $[K : F] = 4$. Prove that every subfield of K containing F is a normal extension of F.

18. The complex number w is a *Primitive nth root of unity* if $w^n = 1$ but $w^m \neq 1$ for $0 < m < n$. Prove that for each positive integer n.

(a) The number of primitive nth roots of unity is $\phi(n)$.

(b) $Q(w)$ is the splitting field of $x^n - 1$ over Q.

(c) If $w_1, w_2, \ldots, w_{\phi(n)}$ are the $\phi(n)$ primitive roots of unity, then any automorphism of $Q(w_1)$ takes w_1 into some w_i.

(d) $[Q(w_1) : Q] \leq \phi(n)$.

12.2 AN APPLICATION TO GEOMETRY (CONSTRUCTION WITH STRAIGHT EDGE AND COMPASS)

Now in this section we shall deal with two famous classical *Euclidean geometry problems.*

1. Construct a cube whose volume is twice the volume of a given cube.
2. Trisect a given plane triangle.

Here the tools permissible in completing those problems are the straight-edge (not a ruler with measurements on it) and the compass. We shall show that in general these problems are unsolvable with the permitted equipment. Naturally the problems must be translated from geometrical problems into algebraic problems, and so permissible constructions. This change over from geometry to algebra is familiar to the students from his study of analytic geometry we shall touch lightly upon it here. The permissible constructions correspond to the simple algebraic binary operation of addition, subtraction, multiplication, and division as well as the more complex unary operation of square root extraction. Thus from a given subfield $(F, +, .)$ of real numbers field we are permitted to 'construct' new numbers by using any finite sequence of the above five operations. So from the field $(Q, +, .)$ of rational numbers we build such numbers as $\sqrt{3}, \sqrt{1+\sqrt{3}}, \sqrt{2}+\frac{1}{\sqrt{7}}$, etc.

Now we replace the algebraic construction by quadratic field extensions. Suppose we start with the rational field $(Q, +, .)$.

If in attempting to construct a certain number we extract the square root of 2, then there after we are working in $Q[x]/(x^2 - 2) = F_1$. In constructing the number $-1 + \sqrt{2}$ we stay in $Q[x]/(x^2 - 2)$, but if we construct $\sqrt{-1+\sqrt{2}}$ then we have moved on into the field $F_2 = F_1[x]/(x^2 + 1 - \sqrt{2})$. For there is no number $a + b\sqrt{2}$ in F_1 such that

$$\left(a + b\sqrt{2}\right)^2 = -1 + \sqrt{2}$$

or
$$a^2 + 2b^2 + 2ab\sqrt{2} = -1 + \sqrt{2}$$

which implies

$$a^2 + 2b^2 = -1 \text{ and } 2ab = 1.$$

Since

$$a^2 + 2b^2 = -1 \text{ is impossible in } Q.$$

Thus we are faced with the problems of determining whether a root (zero) of an irreducible polynomial $m(x) \in Q[x]$ can lie in a field which can be reached through a finite sequence of quadratic extensions.

Let $m(x)$ be an irreducible polynomial of degree three in $Q[x]$, and suppose a field $(F, +, .)$ containing a root of $m(x)$ can reached through a finite sequence of quadratic extensions starting from the field $(Q, +, .)$. Then we consider a shortest such sequence,

$F_0 = Q$, $F_1 = Q[x]/(m_1(x)), \ldots, F_n = F_{n-1}[x]/(m_n(x))$, where $m_i(x)$ is an irreducible polynomial of degree two in $F_{i-1}[x]$, $i = 1, 2, 3, \ldots, n$. Thus $m(x)$ has a root in $F = F$ but not in F_i for $i < n$. We shall show that this is impossible.

Theorem 12.2.1: *A root of an irreducible cubic polynomial from $Q[x]$ lies in no field which is obtainable through a finite series of quadratic extensions from Q.*

Proof: We consider the automorphism group $(A(F_n/F_{n-1}), 0)$. We note here three important facts:

(1) $F_n = \{a + b\sqrt{c} \mid a, b, c \in F_{n-1}\}$ and $\lambda^2 - c$ is irreducibe in $F_{n-1}(x)$.

(2) $A(F_n/F_{n-1})$ contains two elements, the identity mapping E and the mapping σ, $\sigma(a + b\sqrt{c}) = a + b\sqrt{c}$ and $\sigma(a + b\sqrt{c}) = a - b\sqrt{c}$,

(3) The only elements of F_n which are left unchanged by σ are the elements of F_{n-1}.

Now let r be a root of $m(x)$ in F_n. Since $r \notin F_{n-1}$ because of the minimality of the sequence of extensions, we know that

$$r = a + b\sqrt{c} \text{ with } b \neq 0.$$

Thus $\sigma(a + b\sqrt{c}) = a - b\sqrt{c}$,

and since the coefficients of $m(x)$ are in Q, We know that $\sigma(r)$ is also a root of $m(x)$ by theorem 12.1.3 from this we conclude that the product

$$p(x) = (x - r)(x - \sigma(r))$$

is a factor of $m(x)$ in $F_n[x]$.

We examine $p(x)$ more closely:

$$p(x) = (x - a - b\sqrt{c})(x - a + b\sqrt{c})$$
$$= x^2 - 2ax + a^2 - b^2 c.$$

Since a, b, and c, are in F_{n-1}, we see that $p(x) \in F_{n-1}[x]$. But this means that the cubic polynomial $m(x)$ is reducible in $F_{n-1}[x]$ and hence has a root in F_{n-1}. As this contradicts the selection of F_n, we have proved the theorem.

Theorem 12.2.2: *It is impossible by straight-edge and compass alone, to trisect 60°.*

Proof: This geometrical problem of trisecting an angle of 60° changes to the algebrac problem of "constructing" a root of an irreducible cubic of $Q[x]$.

We know the trigonometric identity

$\cos 3\phi = 4\cos^3\phi - 3\cos\phi$, where ϕ is the measure of an angle.

Putting $\phi = 20°$ We have

$$4x^3 - 3x - 1/2 = 0$$

or

$$(2x)^3 - 3(2x) - 1 = 0,$$

where $2x = 2 \cos 20°$. *If the angle whose measure is 60° can be trisected by straight edge and compass constructions, then the number $2x = 2 \cos 20°$ is also constructable. This corresponds to the algebraic construction of a zero, of the polynomial $(2x)^3 - 3(2x) - 1 = 0$ or* $\lambda^3 - 3\lambda - 1 = 0$. which is obviously irreducible over Q since neither 1 nor −1 is a zero.

So by the above theorem 12.2.1 zeros of the polynomial $\lambda^3 - 3\lambda - 1$ are not obtainable through a finite sequence of quadratic extension from Q. Therefore $2 \cos 20°$ is not constructable, and consequently 60° cannot be trisected by straigth-edge and compass.

Theorem 12.2.3: *By straight-edge and compass it is impossible to duplicate the cube.*

Proof: Duplicating a cube means that constructing a cube whose volume is twice that of a given cube. If the original cube is a unit cube, the new cube is of length α such that $\alpha^3 = 2$. Since the polynomial $x^3 - 2$ is irreducible over Q. So by theorem 12.2.1 a root of $x^3 - 2$ does not lie in a field which can be reached through a finite sequence of quadratic extension. This α cannot be constructed. Hence the theorem is proved.

Theorem 12.2.4: *It is impossible to construct a regular septagon by straight-edge and compass.*

Proof: To carry out such a construction would require the constructibility of $\alpha = 2\cos\left(\frac{2\pi}{7}\right)$.

However, we claim that α satisfies $x^3 + x^2 - 2x - 1 = 0$ and this polynomial is irreducible over Q. But the constructibility of α corresponds to the root of the polynomial

$$x^3 + x^2 - 2x - 1$$

which does not lie in a field which is obtained through a finite sequence of quadratic extension from Q. Hence α is not constructible.

PROBLEMS

1. Show that the following polynomials are irreducible over Q :
(*a*) $8x^3 - 6x - 1$.
(*b*) $x^3 - 2$
(*c*) $x^3 + x^3 - 2x - 1$.

2. Prove that $2 \cos \frac{2\pi}{7}$ satisfies $x^3 + x^3 - 2x - 1$

Hint. use $2 \cos \frac{2\pi}{7} = e\frac{2\pi i}{7} + e\frac{-2\pi i}{7}$

3. Prove that the regular pentagon is constructible.

4. Prove that the regular hexagon is constructible.

5. Prove that it is possible to trisect 72°.

6. Prove that a regular 9–gon is not constructible.
Definition: A real number α is said to be a Constructible number if by use of straight-edge and compass alone we can construct a line segment of length α.

7. If α and β are constructible, then show that $\alpha \pm \beta$, $\alpha\beta$ and α/β, $\beta \# 0$ are constructible.
Definition: Let F be any subfield of real numbers. Consider $A = \{(x, y) \mid x, y \in F)$. Then A is called the *plane of F*. Any time joining two points of A is called a *line in F*. Moreover any circle having as centre point in the plane of F and having as radius an element of F is called a *circle in F*.

8. Show that the equations of a line and a circle in F are $ax + by + c = 0$ with $a, b, c \in F$ and $x^2 + y^2 + ax + by + c = 0$, with $a, b, c \in F$.

9. Prove that two lines in F, which intersect in the real plane, interesect at a point in the plane of F.
10. Prove that a line F and a circle in F which intersect in the real plane do so at a point either in the plane of $F\left(\sqrt{r}\right)$ or in the plane of F where r is a positive number in F.

12.3 GALOIS FIELDS

We know that if $p(x)$ is an irreducible polynomial in the domain $F[x]$ over the field F, then we have a field $F[x]/(p(x))$ which is the extension of the field F. Moreover, if F is finite so is $F[x]/p(x))$. Therefore if we could find all finite fields which contain F (or a subfield isomorphic with F), then we would have, in sense, found all of the irreducible polynomials of $F[x]$. Thus we change the problem of studying the irreducible polynomials of $Z_p[x]$ to the problem of determining all of the finite fields of characteristic p.

Theorem 12.3.1: *If F is a finite field, then F contains p^n elements, where p is the characteristic of F.*

Proof: Since a finite integral domain has characteristic a prime, F has characteristic p for some prime p. Now we consider $(F, +)$ as an additive group. It is finite and abelian and every element, except zero of course has additive order p. Using the theory of finite abelian groups we conclude that F contains p^n elements, for some positive integer n. For if the order of F is divisible by any other prime p_1, then by the Sylow's theorem F must contain some elements whose additive orders are p_1. Since this is not possible we know that F contains p^n elements.

Corollary: F is an extension of Z_p.

Proof: We note that F is an integral domain of characteristic p. Then we define a mapping $\sigma : F \to Z_p$ by $\sigma(em) = [m]$ for each of the p elements $e.\ 0, e.\ 1,..., e\ (p-1)$ of Z, where $[m]$ denotes the congruence class of integers mod p. Here $Z \subset F$, where e is the identity of F.

Since the p integers $0, 1,..., (p-1)$ are distributed on to each of the congrucence classes mod p, we know that σ is one-to-one mapping *of* $e\,Z$ onto Z_p. Now we consider $\sigma(en.em)$, clearly

$$en.em = enm = er$$

where $0 < r < p$ and $mn = qp + r$. Moreover, we also know that $[n] \odot p\ [m] = [mn] = [r]$.

Hence
$$\sigma(en.em) = \sigma(enm)$$
$$= \sigma(er) = [r] = [n] \odot_p [m]$$

Consider
$$en + em = e\,(n + m) = er^*,$$
where $o \le r^* < p$ and $n + m = pq^* + r^*$

$[n] +_p [m] = [n + m] = [r^*]$

So that $\sigma(en + em) = \sigma(er^*) = [r^*] = [n] +_p [m]$

Hence σ is an isomorphism from eZ onto Z_p. We say that an integral domain of characteristic p contains the finite field Z_p, identifying eZ with Z_p. Hence F is an extension of Z_p.

Example 12.3.1: Show that finite fields other than the fields Z_p exists.

Solution: This is done easily by showing an irreducible polynomial of degree two or higher from the polynomial domain of some Z_p. Surely $\lambda^2 + \lambda + 1$ is irreducible over Z_2 so that $Z_2[x]/(x^2 + x + 1)$ is a field of four elements which can be identified by the possible remainders 0, 1, x, $x + 1$. The operation tables of this field are given below:

Now we turn our attention to the multiplicative group F' of non-zero elements from F. Since the order of F is $q = p^n$ we know that multiplicative group $(F', .)$ contains $q - 1$ elements.

+	0	1	x	$x+1$
0	0	1	x	$x+1$
1	1	0	$x+1$	x
x	x	$x+1$	0	1
$x+1$	$x+1$	x	1	0

.	0	1	x	$x+1$
0	0	0	0	0
1	0	1	x	$x+1$
x	0	x	$x+1$	1
$x+1$	0	$x+1$	1	x

Theorem 12.3.2: *Every element of the field $(F, +, .)$ of $q = p^n$ elements is a root of the polynomial $f(x) = x^q - x$ from $F[x]$.*

Proof: Let λ be an element of F'. Then the order of λ, the least positive integer m is such that

$$\lambda^m = 1 \text{ in } F',$$

is a factor of $q - 1$, the order of F', since m is the order of the sub-group of F' generated by λ. Since m is a factor of $q - 1$ say

$$q - 1 = km,$$

then certainly

$$\lambda^q = \lambda^{1+km} = \lambda\,(\lambda^m)^k = \lambda,$$

So that λ is a root of $x^q - x$.

Corollary: The multiplicative group $(F', .)$ of the non-zero elements of the finite field F of order $q = p^n$ is a cyclic group.

Proof: We denote by r the largest of the multiplicative orders of the elements of F'. Then F' contains an element x such that $x^r = 1$, but no lower power of x equals 1. Now we claim that if y is an element of F', then

$$y^r = 1.$$

For otherwise F' contains an element t whose order k is a prime power which does not divide r. Then the order of t is greater than r, which contradicts the choice of r. Thus every element of F' is a zero of the polynomial.

$$z^r - 1$$

and since this polynomial has at most r roots in a field we know that

$$q - 1 \le r.$$

On the other hand we know from the above theorem that

$$r \le q - 1$$

Thus $r = q - 1$ and the elements of F' are precisely the powers.

$$x, x^2, \ldots, x^r = 1,$$

and F' is cyclic.

Example 12.3.2: Find a generator of the group *(F ′, .)* in Example 12.3.1.

Solution: The non-zero elements of example 12.3.1 form a group of order three. We see that both λ and $\lambda + 1$ serve as generators since

$$\lambda^2 = \lambda + 1,\ \lambda^3 = \lambda^2 + \lambda = 1,$$

and

$$(\lambda + 1)^2 = \lambda^2 + 2\lambda + 1 = \lambda^2 + 1 = \lambda,$$

$$(\lambda + 1)^3 = \lambda^3 + 1 + 3\lambda^2 + 3\lambda = 1 + 1 + \lambda^2 + \lambda = 1$$

Here we work with polynomials of the form $x^q - x$, $q = p^n$ over Z_p and investigate their splitting fields. It turns out that for each prime p and each positive integer n there is exactly one (in the sense of isomorphism) finite fields of order $q = p^n$, and it is the splitting field of the polynomial $x^q - x$.

Definition 12.3.1: The Galois field $GF(P^n)$, where p is a prime and n is a positive integer, is the splitting of the polynomials $f(x) = x^q - x$, $q = p^n$ of $Z_p[x]$.

When $n = 1$, we know that $GF(p)$ is the field $(Z_p', +_p, \odot_p)$ since each of the p element of Z_p is a zero of the polynomial.

$$x^q - x$$

For example if $p = 5$ and $Z_5 = \{0, 1, 2, 3, 4\}$, then it is clear that every element of Z_p is a root of the polynomial

$$f(x) = x^5 - x \text{ over } Z_5.$$

Theorem 12.3.3: *The Galois field $GF(P^n)$ contains $p^n = q$ elements.*

Proof: First we wish to show that $GF(p^n)$ has at least q elements, and this requires showing that $f(x) = x^q - x$ has q distinct roots in $GF(p^n)$. This is accomplished by using the formal derivative of the polynomial $f(x)$ is the polynomial domain $GF(p^n)$. For. If

$$f(x) = x^q - x = (x - t)\, g(x)$$

and t is a root of $g(x)$ also, then

$$D((x - t)\, g(x) = g(x) + (x - t)\, D(g(x)) = D(f(x))$$

has t as a root. But $D(-f(x)) = q\, x^{q-1} - 1 = -1$ since $q = p^n = 0 \pmod p$, and the polynomial $D(-f(x)) = -1$ has no roots. Hence the splitting field of $f(x) = x^q - x$ must contain at least $q = p^n$ elements since there are q different roots of $f(x)$.

Denote these q roots by

$$a_1 = 0,\ a_2 = 1, \ldots\ a_q$$

from $a_i^q = a_i$ and $a_m^q = a_m$ in $GF(p^n)$

We get

$$(a_i a_m)^q = a_i^q a_m^q = a_i a_m$$

and

$$(a_i + a_m)^q = a_i^q + a_m^q = a_i + a_m$$

Here we again use the fact that

$$q = p^n = 0 \bmod p$$

Furthermore, if $a_i \neq 0$, then $a_i^q = a_i \Rightarrow \left(a_i^{-1}\right)^q = a_i^{-q} = a_i^{-1}$. Then we see that the q roots of $f(x)$ form a field F, and since $f(x)$ factors into linear factors in $F[x]$,

$$f(x) = (x - a_1)(x - a_2) \ldots (x - a_q),$$

CHAPTER 12

we conclude that F is a roots field of $f(x)$. But the splitting field $GF(P^n)$ of $f(x)$ is the smallest root field of $f(x)$, so from

$$F \subseteq GF(p^n)$$

We conclude that $F = GF(p^n)$.

Thus $GF(p^n)$ contains p^n elements.

Corollary: For each prime p and each positive integer n there exists a field F of p^n elements, and F is isomorphic with $GF(p^n)$.

Proof: The existence of F is a consequence of the fact that the splitting field of $f(x) = x^q - x$ exists and contains p^n elements. And F is isomorphic with $GF(p^n)$ since the splitting field of a polynomial is unique up to an isomorphism.

Example 12.3.3: Find the Galois field $GF(9)$.

Solution: Here we have to find out the splitting field of the polynomial $f(x) = x^9 - x$ over $Z_3[x]$. $GF(9)$ will contain 9 elements

$0, 1, 2, \lambda, 2\lambda, \lambda + 1, 2\lambda + 1, \lambda + 2, 2\lambda + 2.$

which are roots of $f(x) = x^9 - x$. Hence $GF(9)$ is the splitting field of $f(x) = x^9 - x$.

Lemma 12.3.1: *If $a_1, \ldots, a_m$ are elements in a commutative ring of characteristic p, then for $q = p^n$*

$$(a_1 + \ldots + a_m)^q = a_1^q + \ldots + a_m^p$$

Proof: The proof involves a double extension of the binomial expansion

$$(a + b)^p = a^p + pa^{p-1} \cdot b + \frac{p(p-1)}{2!} a^{p-2} \cdot b^2 + \ldots + b^q.$$
$$= a^p + b^p \text{ since } pa = 0.$$

First

$$(a + b)^q = ((a + b)^r)^p, \text{ where } r = p^{n-1},$$

So, by the induction hypothesis,

$$(a + b)^r = a^r + b^r,$$

and

$$(a^r + b^r)^p = a^{rp} + b^{rp}.$$

Hence

$$(a + b)^q = a^q + b^q.$$

Now consider the expression

$$(a_1 + a_2 + \ldots + a_m)^q = (a_1 + (a_2 + \ldots + a_m))^q.$$

By the above

$$(a_1 + (a_2 + \ldots + a_m))^q = a_1^q + (a_2 + \ldots + a_m)^q,$$

and applying the second induction hypothesis to the second term on the right we get

$$(a_1 + a_2 + \ldots + a_m)^q = a_1^q + a_2^q + \ldots + a_m^q,$$

completing the second induction. This proves the Lemma.

Theorem 12.3.4: *If u is a generator of the multiplicative group of non-zero elements of $GF(p^n)$, $n > 1$, then the zeros of the monic irreducible polynomial $m(x)$ over $GF(p)$ which has u as a zero, are $u, u^p, \ldots, u^t$, where $t = p^{n-1}$.*

Proof: Consider a monic irreducible polynomial

$$m(x) = a_0 + a_1 x + \ldots + x^n,$$

which has u as a zero. We get the equation

$$0 = m(u) = a_0 + a_1 u + \ldots + u^n$$

Raising zero to the pth power yields

$$o^p = 0. = (a_0 + a_1 u + ... + u^n)^p$$
$$= a_0 + a_1^p u^p + ... + u^{np}.$$

Since the characteristic of the field is p.

By little theorem of Fermat we know that

$a_i^p = a_i \pmod p$. That is, we can replace a_i^p by a_i in $GF(p)$. So we get

$0 = a_0 + a_1 u^p + ... + (u^p)^n$.

This shows that u^p is a zero of $m(x)$. Hence the theorem.

Not only are the n elements w^k, $k = 1, p, p^2, ..., p^{n-1}$, the n zeros of $m(x)$, but they are all generators of the cyclic group of non-zero elements of $GF(p^n)$. (Although there are still other generators in general). This follows an integer relatively prime to powers of p.

Theorem 12.3.5: *If u generates the cyclic group of non-zero elements of $GF(p^n)$, then u^k is also a generator, when $k = 1, p, p^2, ..$ or p^{n-1}.*

Proof: We note that

$$1 = (k, p^{n-1})$$

If

$$(u^k)^m = 1, m > 0,$$

then km is a multiple of $p^n - 1$, and since k is relatively prime to p^{n-1} this means that m is a multiple of p^{n-1}. Thus u^k generates a sub-group $(H, .)$ of order at least p^{n-1}, and since this is the order of the whole group of non-zero elements, we conclude that u^k is a generator of that group.

Theorem 12.3.6: *The automorphism group A $(GF(p^n))$ of $GF(p^n)$ is a cyclicgroup of order n.*

Proof: Let σ be an automorphism of $GF(p^n)$. Since it fixes 1, it also fixes $2 = 1 + 1$, $3 = 2 + 1$, etc., so that the elements of $GF(p^n)$ are all fixed by σ. Using this we see that a generator u of the multiplicative group of non-zero elements of $GF(p^n)$ is mapped onto a zero of $m(x)$, where $m(x)$ is the irreducible polynomial of degree n over $GF(p)$,

$$m(x) = a_0 + a_1 x + ... + x^n,$$

Which has u as a zero. For

$$0 = a_0 + a_1 u + ... + u^n \text{ acted upon by } \sigma \text{ becomes}$$
$$\sigma(0) = 0 = \sigma(a_0) + \sigma(a_1 u) + ... + \sigma(u^n)$$
$$= a_0 + a_1 \sigma(u) + ... + \sigma(u^n)$$
$$= a_0 + a_1 \sigma(u) + ... + (\sigma(u))^n$$

Since σ preserves the arithmetic of $GF(p^n)$ and leaves the elements of $GF(p)$ fixed.

Thus $\sigma(u)$ is a zero of $m(x)$, so that by theorem 12.3.5

$\sigma(u) = u^k$, where k is some power of p.

This means that we can identify the automorphism of $GF(p^n)$ by looking at the n elements

$$u, u^p, ... u^{pn-1}$$

since each automorphism is identified completely by its effect on the generator u. Since the only possible images of u are the n elements u^k for $k = 1, p, ...p^{n-1}$, we know then that the automorphism group $AGF(p^n)$ contains at most n elements.

On the other hand, using the lemma 12.3.1 we see easily that $\underline{A}$ ($\underline{GF}$ (p^n)) contains at least n elements since mapping

$$\sigma_m : \lambda \to \lambda^r, \; r = p^m, \; \lambda \text{ in } GF(p^n),$$

is an automorphism of $\underline{GF}$ (p^n). If λ and μ are two elements of $\underline{GF}$ (p^n), then

$$(\lambda + \mu)^r = \lambda^r + \mu^r \text{ by lemma 12.3.1}$$

and certainly $(\lambda \mu)^r = \lambda^r \mu^r$.

This σ_m preserves the arithmatic of GF (p^n), and since $\lambda^r = 0$

if and only if $x = 0$, this means that σ_m is automorphism of GF (p^n). The n mappings $\sigma_1, \sigma_2, \ldots, \sigma_{n-1}$, and $\sigma_n = I$, the identity mapping, are all different since each of them sends u onto different zero of m (x). So we conclude that the group σ (1) GF (p^n) is of order n and contains exactly the elements $\sigma_1, \sigma_2, \ldots \sigma_n$. Furthermore we see that $\sigma_1^2 = \sigma_2$ since $\sigma_1^2 (\lambda) = \sigma_1 (\sigma_1 (\lambda)) = \sigma_1 (\lambda^p) = (\lambda^p)^p = \lambda^{p2} = \sigma_2 (\lambda)$, and that $\sigma_1^3 = \sigma_3$, $\sigma_1^4 = \sigma_4$,...etc. Hence the automorphism group of GF (p^n) is a cyclic group of order n.

12.4 RADICAL EXTENSIONS AND SOLVABILITY BY RADICALS

In this section we shall deal with the application of field theory in theory of equations. Before dealing with the solvability of a polynomial by radicals we give some more results of field theory which play an important role in the problem of solving equations by radicals.

Lemma 12.4.1: *$x^n - e$ has distinct roots in a splitting field over F.*

Proof: We have

$$f(x) = x^n - e \in F[x] \text{ and its derivative}$$
$$f'(x) = n\, x^{n-1} \in F[x].$$

Here we take the field F in such a way that the characteristic of F does not divide n. Then $n\, x^{n-1} = 0 \Rightarrow x = 0$. Thus $f'(x) = n\, x^{n-1} \neq 0$. Since for $x = 0$, $f'(x) = 0$, 0 is only the root of $f'(x)$ but 0 is not a root of $f(x)$. This implies $f(x)$ and $f'(x)$ do not have any common root in any extension field K of F. Thus in any extension field K of F, $f(x) \in F[x]$ does not have repeated root. So if K is the splitting field of $f(x)$ over F, then K contains n distinct roots of $f(x)$.

Definition 12.4.1: For any positive integer n, an element u of a field F is called an *nth* root of unity if $u^n = e$, e is the identity of F.

Theorem 12.4.1: *The roots of x^n-1 form a multiplicative cyclic group of order n.*

Proof: Let H be the set of all n roots of the polynomial $f(x) = x^n - 1$. $f(x)$ has n distinct roots in its splitting fields K. Let u and $v \in H$, then $u^n = 1$, $v^n = 1$.

We also have that

$(u.v^{-1})^n = u^n \,.\, (v^{-1})^n = u^n \,.\, (v^n)^{-1} = 1.1^{-1} = 1.$

This shows $u.v^{-1}$ is a root of x^n-1. So $u.v^{-1} \in H$.

Hence (H, o) is group of order n.

We know that H is a finite sub-groups of the multiplicative group of the splitting field K. Hence it is cyclic group of order n.

Theorem 12.4.2: *Let F be a field of characteristic zero. Let $K \supseteq F$ be a splitting field of $x^n - 1$ over F, then the Galois group (A (K/F), o) is isomorphic to a sub-group of the residue class modulo n group (M, t_n) of integers relatively prime to n.*

Proof: We know that $(H,.)$ is the cyclic group of all n roots of $x^n - 1$. If u is the generator of H, the order of u is n. So $u^n = 1$ and u is a primitive nth root of unity. Thus $k = F(u)$. Let σ be any F-automorphism of K. Therefore $H = \{u^0, u^1, u^2,...., u^{n-1}\}$ *i.e.*, every element of H is a power of u. For any nth root λ of unity $\lambda^n = 1$. gives $1 = \sigma(1) = \sigma(\lambda^n) = [\sigma(\lambda)]^n \Rightarrow [\sigma(\lambda)]^n \in H$. So $u^n = 1 \Rightarrow [\sigma(u)]^n = 1 \Rightarrow \sigma(u) \in H$.

Since σ is one-to-one mapping, then the restriction of σ to H *is* an automorphism of the group $H = (u)$ Since $[\sigma(u)]^n = 1$, $\sigma(u)$ is also generator of H. If $\sigma(u) = u^t$, thus $(t, n) = 1$, *i.e.*, t and n are relatively co-prime, let $[t]$ be the residue class modulo n of the integer t then $[t] \in M$, the set of residue classes modulo n. We define a mapping

$f : A(K/F) \to M$ by

$f(\sigma) = [t]$, where $\sigma(u) = u^t$.

f is a well defined mapping. For, let $\sigma(u) = u^r$. *Then* $(r, n) = 1$ and $u^r = u^t \Rightarrow u^{t-r} = 1$. Since u is the primitive nth root of unity, $u^n = 1$, $u^m \neq 1$, $m < n$. So $n \mid t - r$ which means $[t - r] = [0]$ or $[t] = [r]$. For any two $\sigma, \rho \in A(K/F)$ let $f(\sigma) = [t_1]$ and $f(\rho) = [t_2]$. Then $\sigma(u) = u^{t_1}$ and $\rho(u) = u^{t_2}$.

$f(\sigma) = f(\rho) \Rightarrow u^{t_1} = u^{t_2}$

$\Rightarrow u^{t_1 - t_2} = 1$

$\Rightarrow n \mid t_1 - t_2$

$\Rightarrow [t_1 - t_2] = [0] \Rightarrow [t_1] = [t_2]$.

Hence f is one-to-one mapping

$(\sigma \, o \, \rho)(u) = \rho(\rho(u)) = \sigma(u^{t_2}) = (u^{t_2})^{t_1} = (u^{t_2})^{t_1}$ Thus $f(\sigma \, o \, \rho) = [t_1 \, . \, t_2] = [t_1][t_2] = . f(\sigma) . f(\rho)$. This shows f is an isomorphism from the group $(A(K/f), 0)$ to the group $(M, \odot_n)$.

Remark: Let F_1 be a splitting field of $x^n - a$ over F. Let K be an extension field of F such that $F_1 \subset K$ and let K also contain a primitive nth root of *unity, say* u. Then the set H of nth roots of unity

$H = (\{1, u, u^2,..., u^{n-1}\}$. Let λ be any other root of the polynomial $x^n - a$. Then $\lambda^n = a$. For each u^t, $t = 0, 1, 2,..., n - 1$ $(u^t \lambda)^n = (u^t)^n \lambda^n = (u^n)^t \lambda^n = 1$ and $\lambda^n = a = u^t \lambda$, $t = 0, 1, 2,..., n - 1$ is a root of $x^n - a$.

Thus the set of n roots of $x^n - a$ is the set

$\{\lambda, u\lambda, u^2\lambda,..., u^{n-1}\lambda]$ Since F_1 is the splitting field of $x^n - a$ over F contained in K, $\{\lambda, u\lambda, u^2\lambda, ..., u^{n-1}\lambda\} \subset F_1$. Hence $F(u, \lambda) \subseteq F_1$. But $F_1 = F(u, u\lambda, u^2\lambda,..., u^{n-1}\lambda) \subseteq F(u, \lambda)$. Here $F_1 = F(u, \lambda)$.

Lemma 12.4.2: *Suppose that the field F has all nth roots of unity (for some particular n) and suppose that $a \neq 0$ is in F. Let $x^n - a \in F[x]$ and let K be its splitting field over F, then:*

(1) $K = F(\lambda)$ where λ is any root of $x^n - a$.

(2) The Galois group of $x^n - a$ over F is abelian.

Proof: **(1)** Since F contains all nth roots of unity, it contains primitive nth root u of unity. We note that $u^n = 1$ and $u^m \neq 1$, $0 < m < n$. In the remark above we have seen that if λ is the root of the polynomial $x^n - a$ over F, the polynomial $x^n - a$ has its n roots λ, $u\lambda, u^2\lambda,...., u^{n-1}\lambda$.

These all roots are distinct. This follows from that if $0 \leq i, j < n$, $u^i \lambda = u^j \lambda$ $\Rightarrow (u^i - u^j)\lambda = 0$

$\Rightarrow u^i - u^j = 0$ since $\lambda \neq 0$

$\Rightarrow u^i = u^j$

CHAPTER 12

$\Rightarrow u^{i-j} = 1$ which is impossible for $i - j < n$.

Since $u \in F$, all of $u, u\lambda, u^2\lambda, \ldots, u^{n-1}\lambda$ are in $F(\lambda)$ which are the roots of $x^n - a$. Thus $F(\lambda)$ is the splitting field of $x^n - a$ over F. Hence $K = F(\lambda)$.

2. If σ, ρ are any two elements in the Galois group of $x^n - a$, that is, if σ, ρ are two F-automorphism of $K = F(\lambda)$, we note that

$\lambda^n - a = 0$, since λ is the root of $x^n - a$.

So $$\sigma(\lambda^n - a) = \sigma(0) \Rightarrow \sigma(\lambda^n) - \sigma(a) = 0$$
$$\Rightarrow (\sigma(\lambda))^n - a = 0$$

This shows $\sigma(\lambda)$ is the root of $x^n - a$. Similarly $\rho(\lambda)$ is also root of $x^n - a$, and $(\sigma(\lambda) = u^i\lambda$ and $\rho(\lambda) = w^j\lambda$ for some i and j. thus

$$\begin{aligned}(\sigma o \rho)(\lambda) &= \sigma(\rho(\lambda)) = \sigma(w^j\lambda)\\ &= \sigma(w^j)\,\sigma(\lambda)\\ &= u^j\,\sigma(\lambda) \text{ since } u^i \in F\\ &= u^j u^i \lambda\\ &= w^{j+i}\lambda\end{aligned}$$

Similarly $(\rho o \sigma)(\lambda) = u^{i+j}\lambda$

$i + j = j + i$, $\rho o \sigma = \sigma o \rho$. Hence $(A(K/F), 0)$ is an abelian group.

Definition 12.4.2: The field K is a *radical extension of* F if and only if $K = F_n = F_{n-1}[x]/(m_n(x)), \ldots, F_1 = F[x]/(m_1(x))$, where $m_i(x)$ is an irreducible polynomial in $F_{i-1}[x]$ of the form $x^r - c$.

If w is the root of $x^r - c$ which does not lie in F_{i-1}, then $F_i = F_{i-1}(w)$ is the splitting field of $x^r - c$ over F_{i-1} where $w^{ri} \in F_{i-1}$.

Thus we can say that the field K is a radical extension of F if there exists a finite sequence of fields $F_1 = F(w_1), F_2 = F_1(w_2), \ldots, F_n = F_{n-1}(w_n)$ such that $w_1 \in F$, $w_2^{r_1} \in F_1, \ldots, w_n^{rn} \in F_{n-1}$. It is clear that

$F \subset F_1 \subset F_2 \subset F_3 \subset \ldots \subset F_n = K$.

Definition 12.4.3: Let $f(x) \in F[x]$ with degree $f(x) \geq 1$. The equation $f(x) = 0$ is solvable by radicals if a splitting field K of $f(x)$ over F lies in a radical extension of F.

Example 12.4.1: Consider the quadratic monic polynomial $f(x) = x^2 + bx + c$ over the field Q of rational numbers. It has two roots $\dfrac{-b + \sqrt{b^2 - 4c}}{2}$ and $\dfrac{-b - \sqrt{b^2 - 4c}}{2}$ none of the roots lie in Q. So it is irreducible in Q. If we put $w = \sqrt{b^2 - 4c}$, then $w^2 = b^2 - 4c \in Q$. If we adjoin w to the field Q, we get the extension field $F_1 = Q(w)$ which is the splitting field of $= x^2 + bx + c$ over Q and F, is a radical extension of Q. Hence $f(x) = x^2 + bx + c$ is solvable by radicals.

Example 12.4.2: Consider the cubic polynomial $f(x) = a x^3 + 3b x^2 + 3c x + d$ over the field Q of rational numbers.

Putting $z = a x + b$, we get the transformed equation

$g(z) = z^3 + 3 B_1 Z + C_1 = 0$, $B_1, C_1 \in Q$

$B_1 = ac - b^2$ and $C_1 = a^2 d - 3abc + 2b^3$

Suppose $Z = \dfrac{1}{p^3} + \dfrac{1}{q^3}$ etc. then by solving it and comparing with (1) we get

$p + q = -C_1$ and $\dfrac{1}{p^3}+\dfrac{1}{q^3} = -B$, *i.e.*, $pq = -B^3$, p and q are given by

$$P = \frac{-C+\sqrt{C^2+4B^3}}{2},\ Q = \frac{-C-\sqrt{C^2+4B^3}}{2}$$

By cardans formula the roots of $f(x)$ are $\dfrac{1}{p^3}+\dfrac{1}{q^3}, w\,\dfrac{1}{p^3}+w^2q^{1/3}, w^2\dfrac{1}{p^3}+w\dfrac{1}{q^3}$, where w is the cube root of unity.

Now
$$p^{1/3} = 3\sqrt{\frac{-C}{2}+\sqrt{B^3}+\frac{C^2}{4}}$$

and
$$q^{1/3} = 3\sqrt{\frac{-C}{2}-\sqrt{B^3}+\frac{C^2}{4}}$$

we take
$$\alpha_1 = B^3 + \frac{C^2}{4}, \alpha_2 = 3\sqrt{\frac{-C}{2}+\alpha_1}, \alpha_3 = \sqrt{-3} \text{ then}$$

$$F_1 = Q(\alpha_1),\ F_2 = F_1(\alpha_2),\ F_3 = F_2(\alpha_3) \text{ with}$$

$\alpha \in Q$, $\alpha_2^3 \in F_1$, $\alpha_3^2 \in F_2$. It is clear that $Q \subset F_1 \subset F_2 \subset F_3$. Further $g(z)$ contains all the roots of $f(z)$. Here $g(z)$ is solvable by radicals and consequently $f(x)$ is also solvable by radicals.

Every polynomial of degree five and higher cannot be solved by radicals. So here we shall give condition for solvability by radicals. Here we assume that F is a field which contain all nth roots of unity for every integer n.

Here we note a very important thing that if F has all nth roots of unity then adjoining one root of $x^n - a$ to F, where $a \in F$, gives us the whole splitting field of $x^n - a$ Thus by definition 12.1.7 this must be a normal extension of F.

Theorem 12.4.3: *If p (x) is solvable by radicals over F then its Galois group is solvable.*

Proof: *Let* $p(x)$ be solvable by radicals. Then the splitting field K of $p(x)$ will lie in the radical extension of F.

So there exists a finite sequence of fields.

$F \subset F_1 = F(w_1) \subset F_2 = F_1(w_2) \subset F_3 = F_2(w_3) \subset \ldots\ldots \subset F_k = F_{k-1}(w_k)$,

where $w_1^{r_1} \in F_1, w_2^{r_2} \in F_1$, $w_3^{r_3} \in F_2 \ldots w_k^{r_k} \in F_{k-1}$ where $K \subset F_k$. It is clear that F_k is also normal extension of F. Since F_k is a normal extension of F, F_k is also a normal extension of any field F_i, $i = 1, 2, \ldots, k-1$.

By the above lemma each F_i is a normal extension of F_{i-1}, and since F_k is normal over F_{k-1}, by fundamental theorem 12.2.1 of Galois theory, $(A(F_k/F_i), o)$ is a normal sub-group in $(A(F_k/F_{i-1}), o)$. Consider the chain.

$A(F_k/F) \supset A(F_k/F_1) \supset \ldots\ldots \supset A(F_k/F_{k-1}) \supset (e)$.

Since in the chain each sub-group is a normal sub-group in the one preceding it. Since F_i is a normal extension of F_{i-1}, by the fundamental theorem 12.2.1 of Galois theory the

CHAPTER 12

automorphism group of F_i over F_{i-1}, $(A\,(F_i/F_{i-1}), o) \cong \left(\dfrac{A(F_k/F_i-1)}{A(F_k/F)}, o\right)$. However, by the above lemma $(A\,(F_i/F_{i-1}), o)$ is an abelian group. Thus each quotient group $(A\,(F_k/F_{i-1})/A\,(F_k/F_i), o)$ of the chain is abelian.

Thus the group $A\,(F_k/F)$ is solvable. Since $K \subset F_k$ and K is a normal extension of F because K is the splitting field of $p\,(x)$ over F, by theorem 12.2.1 $(A\,(F_k/K), o)$ is a normal sub-group of $(A\,(F_k/F), 0)$ and $(A\,(K/F), 0) \cong (A\,(F_k/F)/A\,(F_k/K, 0)$. Thus $A\,(K/F)$, is a homomorphic image of $A\,(F_k/F)$ which is a solvable group. So $A\,(K/F)$ must be a solvable group. $(A\,(K/F), 0)$ is the Galois Group of $p\,(x)$ over F.

Thus if $p\,(x)$ is solvable by radicals , then its Galois group is solvable.

In general polynomial $f\,(x)$ of degree n is given by $f\,(x) = x^n + a_1 x^{n-1} + ... + a_n$, then we see by the following theorem whether it is solvable by radicals.

Theorem 12.4.4: *The general polynomial of degree $n \geq 5$ is not solvable by radicals.*

Proof: We have seen in the example of Galois group that the Galois group of a polynomial $f\,(x)$ is a permutation group of the roots of $f\,(x)$. If deg $f\,(x) = 5$, then Galois group is $(S_5, 0)$. But the group $(S_5, 0)$ is not solvable and consequently the polynomial $f\,(x)$ with deg $f\,(x) \geq 5$ is not solvable by radicals.

PROBLEMS

1. Prove that if the Galois group $(A\,(K/F), 0)$ of $p\,(x) \in F\,[x]$ is solvable, then $p\,(x)$ is solvable by radicals.
2. If a field K is a finite extension of F by radicals then prove that there exists a finite normal extension K containing K such that K is an extension of F by radicals.
3. Prove that S_4 is a solvable group.
4. If $p\,(x)$ is solvable by radicals over F, prove that we can find a sequence of fields $F \subset F_1 = F(w_1) \subset F_2 = F_1(w_2) \subset .... \subset F_k = F_{k-1}(w_k)$ where $w_1^{r_1} \in F, w_2^{r_2} \in F_1, ..., w_k^{r_k} \in F_{k-1}$ containing all the roots of $p\,(x)$ such that F_k is normal over F.
5. Prove that the alternating group (the group of even permutations in S_n $(A_n, 0)$ has no nontrivial normal sub-group for $n \geq 5$.

Index

D

E

F

G

H

I

J

K

L

M

N

O

P

Q

R

S